Eco-Cities

A Planning Guide

Applied Ecology
and Environmental Management

A SERIES

Series Editor
Sven E. Jørgensen
Copenhagen University, Denmark

ADDITIONAL VOLUMES IN PREPARATION

Eco-Cities
A Planning Guide

Edited by **Zhifeng Yang**

CRC Press
Taylor & Francis Group
Boca Raton   London   New York

CRC Press is an imprint of the
Taylor & Francis Group, an **informa** business

CRC Press
Taylor & Francis Group
6000 Broken Sound Parkway NW, Suite 300
Boca Raton, FL 33487-2742

© 2013 by Taylor & Francis Group, LLC
CRC Press is an imprint of Taylor & Francis Group, an Informa business

No claim to original U.S. Government works

Printed in the United States of America on acid-free paper
Version Date: 20120823

International Standard Book Number: 978-1-4398-8322-8 (Hardback)

Library of Congress Cataloging-in-Publication Data

Eco-cities : a planning guide / editor, Zhifeng Yang.
     p. cm. -- (Applied ecology and environmental management)
    Includes bibliographical references and index.
    ISBN 978-1-4398-8322-8 (hbk. : alk. paper)
    1. Sustainable urban development. 2. Urban ecology (Sociology) 3. City planning--Environmental aspects. I. Yang, Zhifeng.

HT241.E35 2013
307.1'416--dc23   2012028119

Visit the Taylor & Francis Web site at
http://www.taylorandfrancis.com

and the CRC Press Web site at
http://www.crcpress.com

Contents

SECTION I   Theoretical Bases

SECTION II   *Case Studies*

Preface

Cities are undergoing vast changes in the galloping process of industrialization, urbanization, and globalization, which have brought mounting environmental problems, including climate change, acid rain, water shortage, pollution, hazardous waste, smog, ozone depletion, loss of biodiversity, and desertification that pose severe challenges to sustainable development of our human life. Such changes provide environmental considerations that assume greater importance to the urban planning processes of an increasing number of governments around the world. Researchers and urban planners of urban systems are increasingly concerned about whether urban areas are capable of adapting to these drastic biological, geophysical, and social changes. A widespread paradigm shift in response to the changes urban areas face is a move toward sustainability, which can be defined based on two standards: (1) the ability to improve the quality of human life while living within the capacity of ecosystem support; and (2) the ability to meet contemporary needs without compromising the ability of future generations to meet their needs. Both definitions invoke three equal facets: social equity, economic viability, and environmental functionality. Eco-cities planning knowledge is crucial to advancing sustainability, and sustainability places eco-cities planning knowledge in the context of integrated socio-ecological dynamics.

The emerging paradigm of sustainability in eco-cities planning worldwide is signaled by policies enacted by specific cities, counties, regions, and states. In this book, *Eco-Cities: A Planning Guide*, the sustainability paradigm is reflected in sustainability plans aimed at adapting to changing environmental, social, and economic conditions in the cities we study. Such eco-cities plans themselves have become part of the changing local and regional context, and like climate change, economic globalization, regional and international migration, and other large forcing functions, they must be taken into account in understanding eco-city plans.

Eco-Cities: A Planning Guide will provide an overview of urban ecosystem structure, function, change, and how to successfully accomplish eco-city planning in the face of government requirements. It will add a new dimension to the understanding and application of the concept of urban sustainability, based on hypotheses about the social and biogeophysical processes in several cities, such as Guangzhou, Baotou, etc. which can help adapt to the local sustainability policies and effects of climate change. The research in this book employs methods such as experimentation, comparison, long-term measurement, and modeling. Hypothetical models of feedback between social and biogeophysical processes linked through ecosystem services of ecological flow quality and quantity and storage identify variables and spatial patterns to be measured. The feedback models also support eco-cities' development of future scenarios. Three theories new to socioeconomic models—the locational choices made by households and firms, an urban version of the stream continuum concept, and an application of metacommunity theory to the fragmented urban biota—suggest new urban planning questions and stimulate integrated modeling.

The urban planning modeling draws on existing social, vegetation, ecohydrological, and ecosystem service modules but is refined and operated for enhanced cross-disciplinary integration and prediction. *Eco-Cities: A Planning Guide* enhances an understanding of eco-cities and eco-landscape as integrated, spatially extensive, complex adaptive systems and offers a sampling of planning practice common in this field.

Acknowledgments

This work is supported by the National Natural Science Foundation of China (Grant No. 40871056), the National Ministry of Science and Technology (Grant No. 2007BAC28B03), and Fundamental Research Funds for the Central Universities.

Editor

Zhifeng Yang is a professor and the dean of the School of Environment at Beijing Normal University. In 1989, he graduated from the Department of Water Conservancy and Engineering, Tsinghua University. He has long been working on urban planning and environmental impact assessment. He won the State Grade II Prize of Science and Technology (in 2008 and 2012) and the First Prize of Science and Technology Progress (in 2003, 2004, and 2005, respectively) set by the Ministry of Education, China. He is a productive scholar who has authored more than 10 books on water resources management, urban planning, and ecological engineering, and has published over 300 peer-reviewed articles as well.

Dr. Yang is also active in professional activities. He is a branch chairman of the International Environmental Informatics Association, a branch chairman of the Environmental Geography of Chinese Society for Environmental Sciences, the director of the Environmental Consulting and Appraisal Committee, the executive director of the Beijing Environmental Society, a committee member of the Man and the Biosphere (MAB) Programme in China, and a member of the Science and Technology Committee of Ministry of Education, China. He is now an associate editor of the *Journal of Environmental Informatics* and the *Journal of Environmental Sciences* and an editorial member of the *Journal of Hydrodynamics, Communications in Nonlinear Science and Numerical Simulation* and *Frontiers of Environmental Science & Engineering in China*. He has served as the chairman or a member of program committees for a number of international academic conferences in past years.

Three publications closely related to this book are

Z.F. Yang, L.Y. Xu et al. 2008. *Urban Ecological Planning.* Beijing Normal University Publishing Group, Beijing (in Chinese).

Z.F. Yang et al. 2004. *Planning and Sustainable Development in Ecocities.* Science Press, Beijing (in Chinese).

Z.F. Yang et al. 2004. *Environmental Planning Theory and Practice in Ecocity Zone.* Chemical Industry Press, Beijing (in Chinese).

Contributors

Bin Chen
School of Environment
Beijing Normal University
Beijing, People's Republic of China

Sven Erik Jørgensen
Department of Pharmaceutics and
	Analytical Chemistry
University of Copenhagen
Copenhagen, Denmark

Gengyuan Liu
School of Environment
Beijing Normal University
Beijing, People's Republic of China

Jiansu Mao
School of Environment
Beijing Normal University
Beijing, People's Republic of China

Michela Marchi
Department of Chemistry
University of Siena
Siena, Italy

Meirong Su
School of Environment
Beijing Normal University
Beijing, People's Republic of China

Guangjin Tian
School of Environment
Beijing Normal University
Beijing, People's Republic of China

Sergio Ulgiati
Department of Sciences for the
	Environment
Parthenope University of Naples
Naples, Italy

Linyu Xu
School of Environment
Beijing Normal University
Beijing, People's Republic of China

Zhifeng Yang
School of Environment
Beijing Normal University
Beijing, People's Republic of China

Lixiao Zhang
School of Environment
Beijing Normal University
Beijing, People's Republic of China

Yan Zhang
School of Environment
Beijing Normal University
Beijing, People's Republic of China

Yanwei Zhao
School of Environment
Beijing Normal University
Beijing, People's Republic of China

Section I

Theoretical Bases

1 Eco-City Planning Theories and Thoughts

Meirong Su, Linyu Xu, Bin Chen, and Zhifeng Yang

CONTENTS

1.1 INTRODUCTION AND DEFINITION OF ECO-CITY

1.1.1 URBAN DEVELOPMENT STAGES AND CHARACTERISTICS

Urban evolution has its own stages, not only concerning social and economic development levels but also for emerging environmental problems in the socioeconomic background. The World Bank classified urban eco-environmental problems into two types: problems "related to poverty" and those "related to economic growth and richness" (Yang et al. 2004). In contrast, Satterthwaite (1997) classified urban eco-environmental problems into five types: environmental hazards, excessive exploitation of renewable resources, excessive depletion of nonrenewable resources, huge waste, and excessive utilization of environmental capacity. We summarize and classify urban eco-environmental problems into the following three types: problems related to poverty, production, and consumption. Each type of eco-environmental problem is concentrated in a specific stage of urban development, as shown in Figure 1.1.

Generally speaking, cities will seek an ideal developmental mode after the aforementioned three stages, when influenced by both internal conditions and external

3

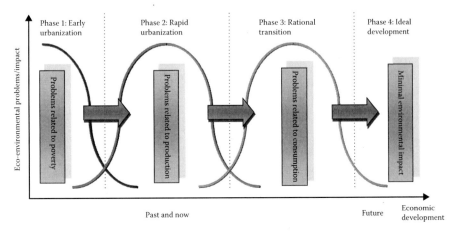

FIGURE 1.1 (**See color insert.**) Stages of urban development.

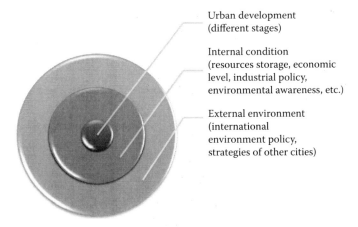

FIGURE 1.2 Impacting factors of urban development.

surroundings (see Figure 1.2). From the perspectives of environmental protection and sustainable development, the final ideal stage of urban evolution is a mature stage named "eco-city," in which economic development, social progress, and environmental protection develop in a harmonious way; there are no problems related to poverty and production; and the impact of problems related to consumption is minimal.

1.1.2   ECO-CITY PERSPECTIVE: DEFINITION AND CHARACTERISTICS

After reflecting on the urban developmental stages and emerging eco-environmental problems since the advent of industrialization, the eco-city concept has been regarded as an urban development paradigm in the global wave of ecological civilization. In an eco-city, it is believed that the environment will be properly protected and maintained while the society and economy develop smoothly, which promotes human development. Seriously considering the relationships between humans and nature led to the final conclusion that humans must develop in harmony with nature to realize their own sustainable development.

There are different understandings of what exactly an eco-city is. Yanitsky (1981) states that an eco-city is an ideal habitat with a benign ecological circulation in which technology and nature fully merge; human creativity and productivity reach a maximum level; the residents' health and environmental quality are well protected; and energy, materials, and information are efficiently used. Register (1987) regards an eco-city as an ecologically healthy city in which the objective of ensuring the health and vigor of man and nature reasonably guides human activities. Influenced by the theory of the social–economic–natural complex ecosystem proposed by Ma and Wang (1984), Chinese scholars have generally considered eco-city as a stable, harmonious, and sustainable complex ecosystem that makes possible "all-win" development among social, economic, and environmental factors; full fusion of technology and nature; maximal motivation of human creativity; increasingly improved urban civilization; and a clean and comfortable urban environment.

In addition, there are also different emphasized points for eco-city planning and construction. One of the report of Man and Biosphere, a program launched by UNESCO, puts forward five key points of eco-city planning: an ecological protection strategy, ecological infrastructure, residents' living standard, protection of history and culture, and merging nature into the city (Yang et al. 2004). Wang (2001) states that eco-city construction includes a high-quality environmental protection system, efficient operation system, high-level management system, good greenbelt system, and high social civilization and eco-environmental consciousness.

Referring to the definition and understanding of an eco-city, we summarize the characteristics of eco-cities, combining our understanding of urban ecosystems and, especially, eco-cities into the following seven points:

1. *Health and harmony*: In an eco-city, the human support system is healthy and sustainable so that it can provide enough and consistent ecosystem services. Further, all economic, social, and natural components are organized in a reasonable way, that is, in a harmonious ecological order in the temporal and spatial dimensions.
2. *High efficiency and vigor*: The "high consumption," "high emission," "high pollution," and "low productivity" developmental modes are altered into more environmentally friendly modes in an eco-city. For instance, energy and materials are used with high efficiency, all industries and departments cooperate within a harmonious relationship, and the productivity of the system is correspondingly high.
3. *Low-carbon orientation*: Faced with the ever-present threat of climate change, low-carbon development should also be emphasized. This can be exemplified by higher resource productivity (i.e., producing more with fewer natural resources and less pollution), as well as by developing leading-edge technologies, by creating new businesses and jobs, and by contributing to higher living standards (Department of Trade and Industry 2003).

4. *Sustaining prosperity*: Regarding sustainable development as a basic guideline, resources will be reasonably located both spatially and temporally. In other words, the development of the current generation cannot jeopardize the development of the next generation. Thus, prosperity will be sustained in an eco-city.

5. *High ecological civilization*: In an eco-city, the concept of ecological civilization is displayed in and permeates all fields, including industrial production, human day-to-day activities, education, community construction, and societal fashion.

6. *Holism*: Eco-cities do not emphasize the improvement of single factors (e.g., economic growth or a good environment) but pursue optimal holistic benefits by integrating social, economic, and environmental factors. Aside from economic development and environmental protection, holism emphasizes the comprehensive improvement of human living standards.

7. *Regionality*: Urban development depends on regional foundations in terms of natural conditions, the supply of resources, and the environmental capacity. Thus, the optimal development mode of each city is different from that of all others due to these different regional characteristics.

1.2 ECO-CITY PLANNING THEORIES

Based on an understanding of the characteristics of an eco-city, several basic theories and principles have been established to guide the overall procedure of eco-city planning.

1.2.1 ECO-PRIORITY THEORY

Because many factors must be considered in eco-city planning at the same time, the eco-priority theory was established to guide eco-city planning when there are conflicts among different factors. The eco-priority theory advocates that eco-environmental construction and reasonable usage of resources have priority among all types of socio-economic developmental activities on the basis of a win-win situation between economic and natural processes (Xu et al. 2004). The main ideas of eco-priority theory are expressed by the cube in Figure 1.3, and its concrete meanings are explained in Table 1.1.

1.2.2 BASIC PRINCIPLES: HEALTH, SECURITY, VIGOR, AND SUSTAINABILITY

The basic principles of eco-city planning were also established to guide overall urban design and ensure that urban construction occurs in a proper manner. These principles were generalized from four aspects: health, security, vigor, and sustainability (Yang et al. 2004) (see Figure 1.4).

1. *Health*: It is required that a healthy urban ecosystem realizes both the renewability and the maintenance of the system (composed of natural and artificial environments), and provides enough ecosystem services to ensure human health and promote human development. Comprehensive assessment of urban ecosystem health can define the limiting factors of urban development, which will help to determine the key factors and fields for urban planning.

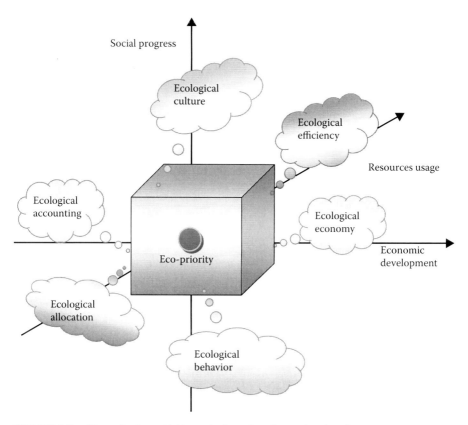

FIGURE 1.3 (**See color insert.**) Eco-priority cube of eco-city planning.

2. *Security*: Urban ecological security is expressed from the aspects of natural, economic, and social systems. Security entails that various abilities and states of the urban ecosystem, such as basic human living demands, population health, social order, and human adaptation to environmental changes, will not be threatened. There are many important thresholds and security layers for urban ecological processes, which induce certain key factors and spatial relationships to form a kind of ecological security pattern. The concept of an ecological security pattern should be extensively considered from the macro to microscale, from portion to holism, and from the present to the future.

3. *Vigor*: A healthy and secure urban ecosystem should also display great vigor. Energy and materials will be efficiently utilized, and economic productivity will be maintained at a high level. Further, the social fashion is active and harmonious and ecological values dominate, both of which are beneficial for human development.

4. *Sustainability*: Urban development should be conducted on the basis of eco-environmental capacity, which is constrained by various resources and environmental factors. Thus, sustainable urban development is achievable between the current generation and the next.

TABLE 1.1
Main Concepts of Eco-Priority Theory

Dimensions		Concrete Concepts of Eco-Priority Theory
Core	Eco-priority principle	Economic growth and environmental improvement should exist in harmony. Eco-environmental construction and reasonable usage of resources have priority among the various socioeconomic activities, and this idea will guide the overall urban ecological planning.
Social dimension	Ecological behavior	Ecological elements are considered in various activities, for example, urban ecological construction should be emphasized, ecological design and planning should be fused into urban planning, and ecological technology should be applied to urban ecological restoration.
	Ecological culture	An ecological perspective should permeate all fields, for example, industrial production, human consumption, education, and community construction. This leads to ecological values and ecological fashion being cultivated in the entire society.
Economic dimension	Ecological economy	An ecological production mode (e.g., circular economy and low-carbon economy) should be established, a green consumption mode should be cultivated, and more ecological investment must be attracted to support the development of ecological agriculture, ecological industry, and ecotourism.
	Ecological accounting	Attention should be paid to ecological values when performing the value estimation. The green gross domestic product (GDP) should also be added into the traditional GDP accounting system.
Resources dimension	Ecological efficiency	Energy and resources are used in a very efficient way. The objective is that minimal consumption of energy and materials satisfy demand to the maximal extent.
	Ecological allocation	The demand of the eco-environment system should be satisfied when basic living demands are assured. The ecosystem services should be improved by protecting the natural ecological process and establishing reasonable ecological relationships.

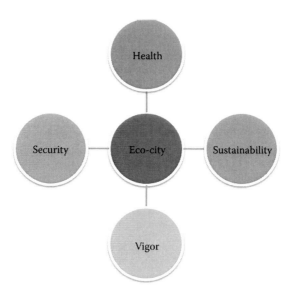

FIGURE 1.4 Basic principles of eco-city planning.

1.3 ECO-CITY PLANNING OBJECTIVES AND INDICATORS

To begin eco-city planning, the planning objectives should first be established. The planning objectives and status quo assessment will affect each other. On one hand, as an expectation of the urban ecosystem, the planning objectives may be used in the status quo assessment as a sort of standard, which can help to define the limiting factors and corresponding key planning fields. On the other hand, as the foundation of the urban ecosystem, the status quo assessment will help to reasonably establish the planning objectives.

1.3.1 HOLISTIC GOALS OF ECO-CITY PLANNING

The limiting factors of urban development are defined according to the status quo assessment of the urban ecosystem. Then, combined with an understanding of the eco-city, the holistic goals of eco-city planning are established. Usually, such goals can be generalized in the following way: guided by the eco-priority theory and basic planning principles of health, security, vigor, and sustainability, comprehensive construction will be performed from multiple aspects during the planning period (e.g., natural resources allocation, economic development, cultivation of the social culture, environmental quality improvement, and ecological restoration). In this way, the eco-city will be formed with well-developed natural, economic, and social subsystems and harmonious relationships between humans and nature.

It should be noted that the planning goals of different cities will vary because each city has its own specific natural condition, economic development level, and social culture characteristics. Taking three typical cities (Baotou, Wanzhou, and Wuyishan) as examples, regulating the industrial structure and establishing a reasonable ecological economy system will be the main goal for Baotou City, restoring the eco-environment and improving human living levels will be the main goal for Wanzhou District in

Chongqing City, and protecting the eco-environment and developing ecological tourism within the environmental capacity limits will be the main goal for Wuyishan City. However, the basic characteristics of reciprocal development among natural, economic, and social factors and harmony between nature and man must be obeyed.

1.3.2 STAGE-BY-STAGE OBJECTIVES OF ECO-CITY PLANNING

According to the holistic goal of eco-city planning, concrete objectives in different periods should be established to realize the ultimate goal of creating an eco-city in a stage-by-stage manner. First, the planning period between the basic year and the objective year is divided into several stages, for which the following three stages are usually adopted: the recent stage, medium-term stage, and long-term stage. Then, the objectives completed during the three stages are confirmed according to the status quo and ultimate goal.

In the recent stage, after determining the key areas and fields of exploitation, rebuilding, restoration, protection, and regulation, the overall eco-city construction begins. Construction in the most important and tough fields must be initiated. Meanwhile, the basic legislation and management systems, as well as the framework of the ecological culture and education, should be established, which can provide proper policy and awareness foundations for the eco-city construction.

During the medium-term stage, construction in all related fields is further emphasized. The key eco-environmental problems will largely be mitigated and the ecological development pattern will basically be formed. Thus, the ecological economy system will be established on the whole. The environmental quality is greatly improved and the ecosystem services are greatly increased. Ecological consciousness is also gradually strengthened.

In the long-term stage, the ultimate goal of eco-city planning is realized. The achievements of the medium-term stage will be further strengthened, and the ecological development pattern will be improved and maintained. The harmonious development among the natural, economic, and social subsystems will be achieved in the urban ecosystem.

1.3.3 PLANNING INDICATORS

To examine the executive effect of eco-city construction and to determine if the staged objective is realized, planning indicators are needed, which are regarded as a valuable quantified representation of planning objectives.

To directly assess the effect of eco-city construction, the planning indicators are usually established in the framework of key objectives and fields. Generally, different urban ecosystems with different characteristics and specific problems have varied planning objectives and key construction fields, which induce different planning indicators for the various urban ecosystems. Although they use different frameworks, it is common that planning indicators for different urban ecosystems consider the basic factors of economy, society, and nature.

Apart from concrete indicators, the planning values of these indicators in different stages must be confirmed on the basis of status quo and planning objectives. The usual indicators are displayed in Table 1.2 as an example.

TABLE 1.2
Planning Indicators of an Eco-City

		Planning Value		
Classification	Indicator	Recent Stage	Medium-Term Stage	Long-Term Stage
Economy	Per capita GDP			
	Annual per capita net income of peasant			
	Annual per capita disposable income of urban residents			
	Proportion of tertiary industry to GDP			
	Per capita GDP energy consumption			
	Per capita GDP water consumption			
	Discharge intensity of SO_2			
	Discharge intensity of chemical oxygen demand			
	Repeated utilization rate of industrial water			
	Comprehensive utilization rate of industrial solid waste			
	Proportion of clean energy			
Society	Popularization rate of junior middle school education			
	Angel's coefficient			
	Registered urban unemployment rate			
	Urbanization rate			
	Per capita house building area of urban residents			
	Per capita road area			
	Popularization rate of gas in built area			
	Popularization rate of biogas digester in rural area			
Eco-environment	Standardized rate of water quality in urban water function zone			
	Treatment rate of urban domestic water			
	Excellent and well-rated air quality			
	Coverage rate of urban noise standardized area			
	Per capita public green areas			
	Decontamination rate of urban house refuse			
	Proportion of investment for environmental protection to the GDP			
	Ratio of protected area to the total area			

1.4 ECO-CITY PLANNING THOUGHTS AND TECHNICAL ROUTE

1.4.1 OVERALL DESIGN FRAMEWORK OF ECO-CITY PLANNING

The overall aspects of eco-city planning can be summarized in the following six points: (1) guideline: ecology theory, and sustainable development theory; (2) basic principle: eco-priority theory; (3) foundation: status quo assessment of the urban ecosystem; (4) main task: construction in defined key fields; (5) implementation: spatial management and optimization; and (6) ultimate objective: health, security, vigor, and sustainability.

During the entire planning course, "Top-down" and "Bottom-up" approaches are combined. For the status quo assessment of the urban ecosystem, the "Top-down" approach is used to first synthesize the holistic situation and analyze the concrete problems and then define the key planning fields. For the concrete implementation, the "Bottom-up" approach is first used for construction in each field and, subsequently, to realize spatial optimization on the whole.

In terms of the key fields, the situations are different for different cities. However, the basic factors of nature, economy, and society must be considered. Choosing Wanzhou District in Chongqing City as an example, the defined key fields are an ecological economy system, an ecological space system, good environmental system, and an ecological human settlements system. With respect to the ecological economy system, the main tasks include regulation of the economic structure, planning of ecological industries, ecological agriculture, green services, and vein industries. With regard to the ecological space system, the main tasks include division of the ecological function zone, formation of an urban landscape pattern, and construction of an urban ecological network. For the environmental system, the main tasks include planning energy security, as well as establishing an ecological land system and water security. Finally, the main tasks of the ecological human settlements system include construction of a greenbelt system, transportation system, infrastructure, and ecological housing. Similarly, for Baotou City, the defined key fields are urban ecological function zoning and landscape pattern construction, energy and resources utilization, environmental quality improvement, and ecological protection and construction. For Wuyishan City, the key eco-city planning fields are an ecological space system, ecological industry system, eco-environment system, and ecological culture system.

1.4.2 TECHNICAL ROUTE OF ECO-CITY PLANNING

Choosing the key construction fields of Wanzhou District in Chongqing City as an example, a schematic of the technical route of eco-city planning, which shows the planning thoughts, basic flow, and supporting theories and technologies, is displayed in Figure 1.5.

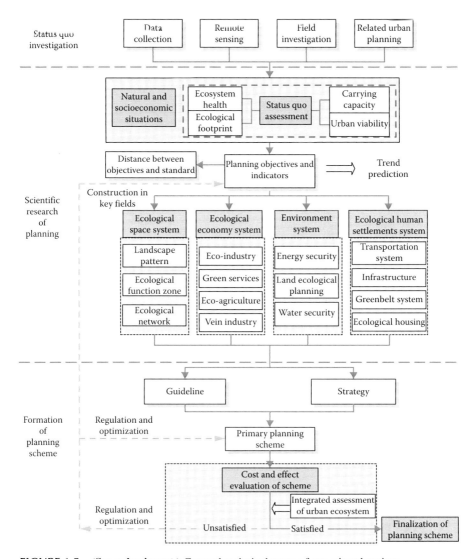

FIGURE 1.5 (**See color insert.**) General technical route of eco-city planning.

REFERENCES

Department of Trade and Industry. *Our Energy Future—Creating a Low Carbon Economy.* Norwich, UK: TSO (The Stationery Office), 2003.

Ma S. J., Wang R. S. The social-economic-natural complex ecosystem. *Acta Ecologica Sinica,* 1984, 4(1): 1–9 (in Chinese).

Register R. *Ecocity Berkeley: Building Cities for a Healthy Future.* Berkeley, CA: North Atlantic Books, 1987.

Satterthwaite D. Sustainable cities or cities that contribute to sustainable development. *Urban Studies*, 1997, 34(10): 1667–1691.

Wang X. R. On the theories, ways and counter measures for the construction of eco-city–A case study of Shanghai, China. *Journal of Fudan University (Natural Science)*, 2001, 40(4): 349–354 (in Chinese).

Xu L. Y., Yang Z. F., Li W. Urban environmental protection plan based on eco-priority rule. *China Population, Resources and Environment*, 2004, 14(3): 57–62 (in Chinese).

Yang Z. F., He M. C., Mao X. Q., Yu J. S., Wu Q. Z. *Programming for Urban Ecological Sustainable Development*. Beijing: Science Press, 2004 (in Chinese).

Yanitsky O. N. Cities and human ecology. In: *Social Problems of Man's Environment: Where We Live and Work*. Moscow: Progress Publishers, 1981.

2 Integrated Urban Ecosystem Assessments

Meirong Su, Zhifeng Yang, Linyu Xu,
Gengyuan Liu, Sergio Ulgiati, Yan
Zhang, and Sven Erik Jørgensen

CONTENTS

2.1 URBAN ECOSYSTEM HEALTH ASSESSMENT*

2.1.1 Review on Urban Ecosystem Health Assessment

2.1.1.1 Concept of Urban Ecosystem Health

The concept of ecosystem health has experienced roughly three development phases, that is, first, focusing on the characteristics of ecosystem itself (Costanza et al. 1998; Karr et al. 1986; Schaeffer and Cox 1992; Ulanowicz 1986; Woodley et al. 1993); second, turning to services for humans (Mageau et al. 1995; National

* This section was contributed by Meirong Su and Zhifeng Yang.

Research Council 1994; Rapport 1989); and third, combined characteristics of eco-system and services for humans (O'Laughlin 1996; Rapport et al. 1999). Building on the previous experience of ecosystem health, the concept of urban ecosystem health combines the ability to satisfy reasonable demand from human society and to maintain its own renewal and self-generative capacity. Therefore, urban ecosystem health is an integrated subject that includes ecological, socioeconomic, and human health perspectives. A few common concepts of urban ecosystem health are listed and analyzed in Table 2.1.

Although there is not any confirmed acknowledged definition for urban ecosystem health, there exist certain basic common characters: (1) ecosystem services maintain a productive capacity, (2) system integrity is the key component of urban ecosystem health, and (3) assessing urban ecosystem health requires a systems perspective. As a complex system composed of natural, societal, and economic components, the urban ecosystem is a network of multiple interactive relationships; thus, its health status should take various factors into account in an integrated way rather than focus only on partial elements such as water, soil, air, or vegetation. Based on the acknowledged need to sustainably integrate reasonable human demands and the ecosystem's ability for renewal, the inclusive factors for a healthy urban ecosystem can be drafted from both the human and the ecological dimensions (Su et al. 2010), as shown in Figure 2.1.

2.1.1.2 Urban Ecosystem Health Standards

The terminology "health" is usually associated with certain physiological standards, such that the system is considered healthy until certain parameters do not conform to the normal range. Similarly, ecosystem health can be measured with respect to standard reference conditions (Campbell et al. 2004). The difficulty is in identify-ing the appropriate state variables to measure and the range of acceptable values for those states (Cabezas and Fath 2002). In one approach, the features of the impacted ecosystem are compared with the one considered undisturbed or pristine (Calow 1993; Rapport 1992, 1993), without any human disturbances (Waltner-Toews 2004). The difficulty is in finding commensurate undisturbed systems.

The problem is even more acute when dealing with urban ecosystems. On natural systems, the human disturbance happens on the original natural background, whereas urban ecosystems are artificially constructed. Therefore, it is much more difficult to assess the intact condition of urban ecosystems. In fact, there does not exist an abso-lute or fixed standard of the urban ecosystem because of the uncertainty caused by the complexity and openness of the urban ecosystem as well as the changing human needs, targets, and expectations of urban ecosystem over time (Odum 1989).

Based on extensive case studies, the International Development Research Centre and the World Health Organization (WHO) have put forward the criteria for what constitutes a healthy urban ecosystem, such as ecological sustainability, social equity, public health, and effective community management (Hancock 2000; Western Pacific Region Office 2000). Their approach is similar to that of finding a pristine natural ecosystem for assessing ecosystem health in that the standard range of the urban eco-system health indicators is based on the conditions of a comparative eco-city, garden city, or those with excellent performance in environmental protection (Guan and Su 2006; Guo et al. 2002; Peng et al. 2007; Sang et al. 2006).

TABLE 2.1
Typical Applications of Urban Ecosystem Health Concept

Person	Conception of Urban Ecosystem Health	Reference	Main Feature
WHO	Defines city health as a system "that is continually creating and improving those physical and social environments and expanding those community resources which enable people to mutually support each other in performing all the functions of life and in developing to their maximum potential."	Hancock and Duhl (1988)	Not only pursues economic development, but urban ecosystem can improve human health. Urban ecosystem integrates healthy lifestyles, environment, and society.
International Development Research Centre (IDRC)	States that urban ecosystem health includes not only the health and integrity of the natural and built environment but also the health of urban resident and the whole society.	Colin (1997)	Urban population under certain socioeconomic, cultural, and political conditions should be emphasized.
Hancock	Based on the relationship among economy, environment, and society, put forward the conceptual framework of healthy cities. And also summarized the six related elements of healthy urban ecosystem, including (1) population health and distribution, (2) societal well-being, (3) government management and social equity, (4) human habitat quality and convenience, (5) natural environment quality, and (6) impact of the urban ecosystem on the larger scale natural ecosystem.	Hancock (2000)	Urban ecosystem health is similar with urban sustainability, which can be represented by the common focus, only the former pay more attention to the human health.
Guo	Understood urban ecosystem health from both ecological and socioeconomic views, in which the former means the complex natural-economic-social urban ecosystem is stable and sustainable and resists external adverse factors, while the latter means the urban ecosystem sustainably provides ecosystem services for urban resident.	Guo (2003)	Healthy urban ecosystems should possess multiple characteristics from various aspects, including vigor from aspect of external representation, diversity and harmony from structural aspect, and regulation and efficiency from function aspect.

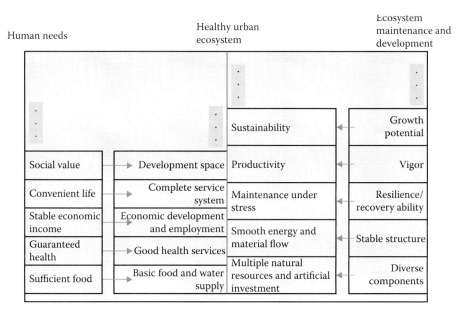

FIGURE 2.1 Basic inclusive factors of urban ecosystem health.

2.1.1.3 Urban Ecosystem Health Indicators

Considering the different views of urban ecosystem health as well as various priorities and objectives, scientists have developed several indicators (Table 2.2), which directly focus on the topic of urban ecosystem health, and others address related researches, for example, Harpham (1996), Takano and Nakamura (1998), and Western Pacific Region Office (2000).

Apart from the main urban ecosystem health indicators mentioned in Table 2.2, certain explanations should be provided for clarity, such as follows. (1) In addition to the WHO (Takano and Nakamura 1998; Western Pacific Region Office 2000), other organizations, such as the United Nations Centre for Human Settlements, the International Institute for Sustainable Development, and the International Joint Commission, have also made efforts to set up indicators of urban sustainable development, which are correlated with the urban ecosystem health indicators (Guo 2003). (2) Besides the conceptual model of PSR (pressure-state-response) (Zeng et al. 2005), others have also been applied to organize urban ecosystem health indicators; for example, DPSEEA (driving force-pressure-state-exposure-effects-action) model, based on Spiegel et al. (2001), which defined the health indicators at the individual, household, and neighborhood levels in the urban ecosystem. (3) Some set up the indicator framework from the features of the urban ecosystem health, such as vigor, function, and structure (Guo et al. 2002; Liu et al. 2009; Su et al. 2009a), while others organized the urban ecosystem health indicators in view of inclusive urban subsystems; for example, natural, economic, and social subsystems (Rong 2009; Wen and Xiong 2008; Zhong and Peng 2003) and ecological, agricultural, production, and living land use subsystems (Zeng et al. 2005). (4) In addition to focusing on the

TABLE 2.2
Main Urban Ecosystem Health Indicators

Indicators	Reference	Pros/Cons
Compared the urban and rural health in Gambia, including economy, environment, public health statistics, health-seeking behavior, health expenditure, and nutrition.	Harpham (1996)	Based on public health statistics, the indicators are available and easily understood and underemphasize economic and cultural factors.
The World Health Organization: (1) Along with the progress of the Healthy Cities Project, proposed 79 healthy urban ecosystem indicators in 1996, from nine aspects including internal character, external performance, influence factor, progress, management and monitoring, proving service, budget and finance, ability development, and community service; (2) Further developed	WPRO (2000)	Human demand and community role are emphasized, through which the urban ecosystem are expected to be more suitable for human development. The indicator system is complicated to encounter data availability and reliability.
459 indicators of a healthy urban ecosystem in 1998, from 12 aspects including human health, urban infrastructure, environmental quality, human housing and living environment, community's role and action, living pattern and prevention performance, health care and environmental sanitation service, education, employment and industry, income and domestic consumption, local economy, and demography statistics.	Takano and Nakamura (1998)	
Taking the classic framework of vigor, organization, resilience, maintenance of ecosystem services, management options, reduced subsides, damage to neighboring system, and human health effects (Mageau et al. 1995; Rapport et al. 1998) used in natural ecosystem health assessment, a similar framework of urban ecosystem health indicators was established using 24 factors.	Guo et al. (2002)	The framework of vigor, organizational structure, resilience, ecosystem service maintenance, and population health has successful tradition in ecosystem health and has been widely used in urban ecosystem health assessment. Whether the indicators are sufficient and typical enough is an open research question.
Organized 30 urban ecosystem health indicators using the framework of natural, economic, and social subsystems in the urban ecosystem. The natural condition is represented by indicators like "atmosphere environmental quality," and the economic development is denoted by indicators like "proportion of the third industry to GDP," while the societal situation is represented by indicators like "house area per capita."	Zhong and Peng (2003)	The traditional method of dividing urban ecosystem into natural, economic, and social subsystems is easily acceptable. But the ad hoc nature of the concrete indicators questions the overall scientific basis.

	Source	
Divided the urban ecosystem into ecological (natural driving forces), agricultural (natural and human driving forces), production and living subsystems (human activities are dominant), and then established a corresponding indicator system within the PSR (pressure-state-response) framework under each subsystem. The ecological land use is expressed by indicators like "forest coverage rate," agricultural land use by indicators like "farmland area per capita," and production and living land use by indicators like "population density."	Zeng et al. (2005)	The viewpoint of different land use subsystems in urban ecosystem is meaningful for practical urban planning and construction. However, the PSR model's ability of reasonably reflecting ecosystem health state is questioned to a certain degree.
Put forward the distance index and the coordination index for the spatial difference of urban ecosystem health into account. The distance index is the gap between the urban developmental status quo of each factor and certain developing objectives, while the coordination index is used to measure the harmonious degree among each factor.	Hu et al. (2005)	The distance index and coordination index are useful to describe the spatial difference with urban ecosystem. But confirming the factors in urban ecosystems and collecting enough data are difficult.
Established a biophysical urban ecosystem health indicator system, by integrating emergy with vigor, structure, resilience, ecosystem service maintenance, and population health, that is, 17 related emergy-based indices are chosen to represent the 5 aspects of the urban ecosystem health status. For example, vigor is represented by indicators like "emergy density," structure is expressed by indicators like "emergy self-sufficiency," resilience is denoted by indicators like "carrying capacity density based on renewable emergy," ecosystem service maintenance is denoted by indicators like "environmental loading ratio," and population health is represented by indicators like "emergy investment ratio."	Su et al. (2009b)	Based on emergy synthesis, urban ecosystem health is assessed on a biophysical foundation. However, it is difficult to put the assessment result into practical management and regulation in an easily understanding way.
Selecting vigor, organizational structure, resilience, and function maintenance as main factors of urban ecosystem health assessment, developed an emergy-based urban ecosystem health indicator, by integrating the indicators "net emergy yield ratio," "environmental loading ratio," "emergy exchange ratio," "emergy density," and "emergy money ratio."	Liu et al. (2009)	By establishing an integrated emergy-based indicator, it seems very clear and simple to assess urban ecosystem health status. However, whether the single indicator can comprehensively represent urban ecosystem health is unknown.

spatial difference within urban ecosystems (Hu et al. 2005; Tian et al. 2009), related indicators emphasizing the temporal dimensional characteristics are also established to denote the urban ecosystem health development over time (Zhang et al. 2006a).

2.1.1.4 Urban Ecosystem Health Assessment Models

Besides the conceptual framework to establish a reasonable indicator system, additional mathematical models are usually needed to treat and process the indicator data to represent the internal characteristics of urban ecosystem health and further satisfy a health assessment.

When considering the current mathematical models of urban ecosystem health assessment, they can be summarized into two categories: one is based on understanding the urban ecosystem health's character while the other faces the problems during the urban ecosystem health assessment. Concretely speaking, modeling urban ecosystem health is difficult due to certain features such as fuzziness, hierarchy, and multiple attributes, and corresponding methods such as fuzzy synthetic assessment model (e.g., Guo et al. 2002; Tao 2008; Zhou and Wang 2005), fuzzy optimal assessment model (Lu et al. 2008; Zeng et al. 2005), fuzzy assessment model combined with analytic hierarchy process (Luo 2006), set pair analysis (SPA) (Su et al. 2009a), relative vector comprehensive assessment model (Sang et al. 2006), attribute theory model (Rong 2009; Wen and Xiong 2008; Yan 2007), and catastrophe progression method (Wei et al. 2008) are applied.

During the course of urban ecosystem health assessment, confirming the weights of various indicators is important, which have a great impact on the final assessment results. The problem of assigning the indicator weights is still an open research question. There are mainly two kinds of methods to define the indicator weight, that is, subjective and objective methods. The widely used subjective method usually defines indicator weights according to human judgments like experts' or professional experiences, for example, the Delphi method and the analytic hierarchy process method (Bi and Guo 2007; Yan 2007). The objective approach is based on the statistical data analysis such as entropy (Shi and Yan 2007; Zhou and Wang 2005), factor analysis (Guan and Su 2006), main component analysis (Lu et al. 2008), and standard deviation analysis methods (Sang et al. 2006). Although the objective method seems and tries to be more scientific, sometimes it does not work well in practice because it ignores the experts' and professional experiences that sometimes are applicable and useful for the actual management of urban ecosystem.

2.1.2 Basic Procedure of Urban Ecosystem Health Assessment

There is a relatively fixed procedure of urban ecosystem health assessment, which can be summarized into the following four steps: (1) confirming the boundary of urban ecosystem, (2) establishing the health indicators, (3) applying suitable models to calculate the health results, and (4) grading the health levels.

2.1.2.1 Boundary Confirmation

The boundary of urban ecosystem needs to be confirmed first, after which the data can be collected correspondingly. According to the concrete situation of study area, the boundary of urban ecosystem should be distinguished in that sometimes it contains only the built-up area while sometimes it contains all the areas in the administrative arrangement.

2.1.2.2   Indicators Establishment

Assessment indicators are treated as well-suited instruments to reflect the urban ecosystem health status according to their characteristics of abstracting information from a complicated system to reduce the complexity and to connect the theoretical ecological background with related political practical requirements (Müller and Lenz 2006; Müller and Wiggering 1999).

The indicator framework of urban ecosystem health can be set up from the features of the urban ecosystem health, such as vigor, function, and structure (Guo et al. 2002; Liu et al. 2009; Su et al. 2009b) while it can also be organized in view of inclusive urban subsystems; for example, natural, economic, and social subsystems (Rong 2009; Wen and Xiong 2008; Zhong and Peng 2003). In addition, certain conceptual models can also be applied to organize urban ecosystem health indicators; for example, PSR (Zeng et al. 2005) and DPSEEA models (Spiegel et al. 2001).

In Section 2.3, certain concrete indicators in specific framework or conceptual model will be introduced in detail.

2.1.2.3   Mathematical Calculation

Since multiple indicators from aspects of social, economic, ecological, and human health are all considered where the ecological meaning of each individual indicator is ambiguous, certain mathematical approaches are required to deal with the indicator information to get a comprehensive and clear assessment of the urban ecosystem health status.

There are many mathematical models that can be applied to conduct the data processing and calculate the final urban ecosystem health results, such as weighted sum model, fuzzy assessment model, SPA, and attribute theory model. Different models have different advantages and application conditions, and the same objective lies in the health status of urban ecosystem, which can be acquired by integrating various indicator information.

In Section 2.4, a few typical mathematical models will be introduced in more detail to show the data processing flow.

2.1.2.4   Gradation of Health Levels

After obtaining the qualitative results of urban ecosystem health status, the gradation of health levels is usually performed by referring to some standard. The health gradation, which can be divided as very healthy, relatively healthy, critically healthy, relatively unhealthy, and ill, will give a clearer contour of urban ecosystem health status than a series of calculated numbers. Moreover, the health gradation is more understandable and acceptable for the government managers and the public.

2.1.3   Urban Ecosystem Health Indicators

Three kinds of urban ecosystem health indicators, called factor-integrated urban ecosystem health index, urban vitality index, and emergy-based urban ecosystem health index

(EUEHI), are introduced in Sections 2.1.3.1 through 2.13.3, which are the representatives of setting up health indicators from the features of the urban ecosystem health, in view of inclusive urban subsystems, and using certain holistic conceptual model.

2.1.3.1 Factor-Integrated Urban Ecosystem Health Index

Taking the classic framework of vigor, organization, resilience, maintenance of ecosystem services, management options, reduced subsides, damage to neighboring system, and human health effects (Mageau et al. 1995; Rapport et al. 1998) used in natural ecosystem health assessment, a similar framework of urban ecosystem health indicators was established from aspects of vigor, organizational structure, resilience, ecosystem services maintenance, and population health (Guo et al. 2002). The concrete indicators from the five factors are listed in Table 2.3.

2.1.3.2 Urban Vitality Index

To describe the vital characteristics of the urban ecosystem, the analogy of urban vital organism was introduced to vividly and systematically assess the urban ecosystem evolution. The urban vitality index, including productivity power, living status, ecological ascendancy, and vital force (see Figure 2.2), which respectively represents the situation of urban economic subsystem, social subsystem, natural subsystem, and ecological regulatory subsystem (Su et al. 2008), is constructed in Table 2.4.

2.1.3.3 Emergy-Based Urban Ecosystem Health Index

Regarding various energy and materials flowing in the urban ecosystem and the merit of emergy as an embodied energetic equivalent for integrated ecological economic evaluation, an EUEHI can be established to reflect the urban ecosystem health status from biophysical foundation. Following the principle of ecosystem health assessment, four major factors, including vigor (V), organizational structure (O), resilience (R), and function maintenance (F), are integrated to construct EUEHI (Liu et al. 2009).

Firstly, factor of vigor can be measured by three indicators, that is, emergy investment, ratio of electricity to total emergy, and ratio of net emergy yield. Ratio of electricity to total emergy is used to describe the industrialization degree of the urban ecosystem, while ratio of net emergy yield can be regarded as an indicator of environmental impacts to estimate the depletion of emergy feedback and system cycle. As the indicators for organizational structure, the nonrenewable emergy ratio is used to describe the resource utilization structure and emergy exchange rate to characterize the coupling input and output structure of commercial economy. Per capita emergy usage to manifest shows the intensity of energy utilization. The input indicators for restoring force embracing environmental load rate and population carrying capacity respectively represent the environmental pressure and the social imported pressure, while the output ones using waste generation rate reflect the ability of the ecosystem to recycle the waste and reduce the usage of nonrenewable resources, thus quantifying the environmental potential and restoring power under certain environmental and population pressure. Finally, the maintenance of urban ecosystem services dominates the possible resource supply of the urban ecosystem for the urban residents, which can be determined by the emergy self-sufficiency, ratio of emergy to money, and the emergy density.

TABLE 2.3
Indicators from Five Factors of Urban Ecosystem Health

Assessing Factors	Meaning	Representative	Measured Indicator
Vigor	Metabolic and primary productivity	Economic productivity	Per capita GDP
		Material consumption efficiency	Per capita GDP material consumption
		Material consumption efficiency	Per capita GDP energy consumption
Structure	Diversity of economic, societal, and natural components	Economic structure	Proportion of R&D to GDP
			Proportion of information industry to GDP
		Social structure	Population density in the built-up area
			Gini coefficient
		Natural structure	Forest coverage rate
			Greening coverage in the built-up area
			Natural reserve coverage
Resilience	Ability of urban ecosystem maintaining its structure and function	Waste treatment index	Treatment rate of urban domestic water
			Standard-reaching rate of vehicle emission
			Comprehensive utilization rate of industrial solid waste
		Materials recycling rate	Repeated utilization rate of industrial water
		Investment of environmental protection	Proportion of investment for environmental protection to GDP
Ecosystem service maintenance	Urban environmental quality and human living convenience	Environmental quality	Comprehensive index of environmental quality
		Life convenience	Per capita public greenbelt in the built-up area
			Per capita house building area of urban residents
			Per capita road area of urban residents
Population health and education	Human physical health and educational and cultural level	Population health	Engel's coefficient
			Mean human lifetime
			Mortality rate of 0–4-year-old children
		Civilization	Average education time

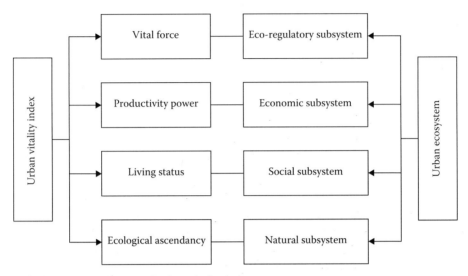

FIGURE 2.2   Framework of urban vitality index.

TABLE 2.4
Index System of Urban Vitality

Criteria	Factor	Index
R₁, productivity power	F₁, economic development level	Per capita GDP
		GDP growth rate
		Annual per capita disposable income of urban residents
		Annual per capita net income of peasant
	F₂, economic structure	Proportion of information industry to GDP
		Growth rate of the secondary industry
	F₃, economic driving force	Proportion of fixed assets investment to GDP
	F₄, economic competitive power	Proportion of foreign investment to GDP
		Proportion of gross export to GDP
R₂, living status	F₅, social justice	Registered urban unemployment rate
		Proportion of receiving the unemployment insurance to the unemployed
		Difference between per capita incomes of rural and urban resident

TABLE 2.4   (Continued)
Index System of Urban Vitality

Criteria	Factor	Index
	F_6, scientific and educational level	Authorized rate of application patent
		Popularization rate of junior middle school education
		Number of college students per 10,000 persons
		Contribution rate of science and technology to economic growth
	F_7, population health	Human mortality
		Number of hospital beds to per 10,000 persons
	F_8, living quality	Per capita house building area of urban residents
		Engel's coefficient
		Automobile per 10,000 persons
		Coverage rate of television
		Telephone popularization rate
R_3, ecological ascendancy	F_9, resources utilization	Per capita water resource quantity
		Forest coverage
		Population density in the built-up area
		Repeated utilization rate of industrial water
	F_{10}, environmental quality	Excellent and good rate of air quality
		Standard-reaching rate of water quality of centralized potable water source
		Treatment rate of urban domestic water
		Comprehensive utilization rate of industrial solid waste
	F_{11}, eco-security	Geologic hazard prevention rate
		Soil erosion treatment rate
R_4, vital force	F_{12}, management and regulatory power	Proportion of investment for environmental protection to GDP
		Popularization rate of environmental protection education
		ISO14000 authorization rate of large-scale enterprises
	F_{13}, system coordination	Coordination coefficient between nature and economy
		Per capita GDP material consumption
		Per capita GDP energy consumption

Subsequently, the new indicator named EUEHI can be defined to estimate the urban ecosystem health as follows:

$$EUEHI = \frac{NEYR \times EER \times ED}{(ELR \times EMR)} \tag{2.1}$$

where NEYR (net emergy yield ratio), ELR (environmental loading ratio), and EER (emergy exchange rate) are on behalf of vigor, organizational structure, and resilience, respectively, and the value of emergy density (ED) divided by emergy money ratio (EMR) can be used to evaluate the function maintenance.

The higher EUEHI is the healthier the urban ecosystem is. Regarding the urban ecosystem health assessment, a comprehensive regulation can be achieved if the use of renewable resources is increased and the economic and social benefits with less environmental pressure are promoted, aiming at the vigor, organizational structure, resilience, and function maintenance of the ecosystem.

2.1.4 Assessment Models of Urban Ecosystem Health

Three typical assessment models, called weighted sum model, fuzzy assessment model, and SPA, are introduced in Sections 2.1.4.1 through 2.1.4.3.

2.1.4.1 Weighted Sum Model

The weighted sum model can be complemented through three calculation steps, that is, data normalization, indicator weight calculation, and weighted sum.

2.1.4.1.1 Data Normalization

Data need to be normalized to unify the units of various indicators and eliminate the effect caused by different orders of magnitude. Concretely speaking, for the positive indicators that denote higher health levels with larger indicator values, the normalization was performed with Equation 2.2:

$$x_i^* = \frac{x_i - x_{i\,min}}{x_{i\,max} - x_{i\,min}} \tag{2.2}$$

where x_i^* is the standardized value of the ith indicator, x_i is the original value of the ith indicator, and $x_{i\,max}$ and $x_{i\,min}$ are the maximum and minimum values of the ith indicator respectively.

In terms of the negative indicators that denote lower health levels with larger indicator values, the normalization was performed using Equation 2.3:

$$x_i^* = \frac{x_{i\,max} - x_i}{x_{i\,max} - x_{i\,min}} \tag{2.3}$$

TABLE 2.5

Judgment Matrix and Weight of the Productivity Power Factors

R_1	F_1	F_2	F_3	F_4	W_A
F_1	1	3	2	2	0.423
F_2	1/3	1	1/2	1/2	0.123
F_3	1/2	2	1	1	0.227
F_4	1/2	2	1	1	0.227

2.1.4.1.2 Indicator Weight Calculation

Many methods (e.g., the analytical hierarchy process, expert consultation, factor analysis, and coefficient of variation) can be applied to acquire indicator weights, among which, each method has its own advantages and shortcomings. The basic method of analytic hierarchy process is introduced here.

According to the basic idea of the analytic hierarchy process, those indicators that are regarded as more important under the background of the assessed problem will have relatively larger weights. After the fixed steps, including establishing a hierarchical structure to represent the characteristics of the assessing system, constructing a judgment matrix, and ordering layers and testing consistency, the weights of different layers (i.e., the criteria, factor, and indicator layers) are calculated. Taking the four factors (F_1, F_2, F_3, F_4) under the criteria of productivity power (R_1) for urban vitality index, the judgment matrix and factor weight are shown in Table 2.5.

2.1.4.1.3 Weighted Sum

Based on the standardized values of the indicators and indicator weights, the comprehensive urban ecosystem health level, marked as H, can be finally obtained by the weighted sum model:

$$H = \sum_{i=1}^{n} w_i \times x_i^*$$ (2.4)

where w_i is the indicator weight of the ith indicator. The urban ecosystem health level is greater with larger values of H.

2.1.4.2 Fuzzy Assessment Model

The urban ecosystem health problem can also be designed as a fuzzy synthetic assessment model: $H = W \times R$, where H is the final urban ecosystem health status matrix; W is the weights matrix for the assessing factors (e.g., vigor, organizational structure, resilience, ecosystem services maintenance, and population health), that is, $W = (w_1, w_2, w_3, w_4, w_5)$; and R is the relative membership degree matrix of each assessing factor to each standard grade (very healthy, relatively healthy, critically healthy, relatively unhealthy, and ill), represented as follows:

$$R = \begin{pmatrix} R_{11} & R_{12} & R_{13} & R_{14} & R_{15} \\ R_{21} & R_{22} & R_{23} & R_{24} & R_{25} \\ R_{31} & R_{32} & R_{33} & R_{34} & R_{35} \\ R_{41} & R_{42} & R_{43} & R_{44} & R_{45} \\ R_{51} & R_{52} & R_{53} & R_{54} & R_{55} \end{pmatrix} \qquad (2.5)$$

in which $R_{ij} = \begin{pmatrix} W_1' & W_2' & \dots & W_k' \end{pmatrix} \times \begin{pmatrix} r_{1j} \\ r_{2j} \\ \dots \\ r_{kj} \end{pmatrix}$, and r_{ij} means the membership degree of

the ith (i = 1, 2, 3, 4, 5) assessing index to the jth standard (j = 1, 2, 3, 4, 5); W_k' means the weights of the kth indicator under corresponding assessing indicators. Based on these values, the relative membership degree matrix in view of each factor and the comprehensive health state, marked as R_{ij} and H respectively, can be calculated to reflect the urban ecosystem health levels. These levels are classified as very healthy, relatively healthy, critically healthy, relatively unhealthy, or ill, according to the largest membership degree value.

2.1.4.3 Set Pair Analysis

SPA, which was proposed by Zhao in 1989 and applied to many fields including system engineering, artificial intelligence, forecasting, and multiattribute assessment (Zhao 1995; Jiang et al. 2004), can be chosen as a possible method to deal with the intrinsic uncertainty of urban ecosystem health.

Grounding on the assessment for urban ecosystem health, the problem space Q based on SPA can be defined as (Su et al. 2009a) follows:

$$Q = \{S, M, H\} \qquad (2.6)$$

$$S = \{s_k\} \quad (k = 1, 2, ..., p) \qquad (2.7)$$

$$M = \{m_r\} \quad (r = 1, 2, ..., n) \qquad (2.8)$$

$$H = (h_{kr})_{p \times n} \qquad (2.9)$$

where S is the assessed interval set composed of several selecting cities, s_k denotes the kth city, M is the indices set, and m_r represents the rth index. The positive index that expresses better situation with larger index value is marked as M_1, while the negative one is M_2. H represents the decision-making matrix about problem Q base on SPA, and h_{kr} is the attribute value of index m_r in the interval s_k.

By collecting the best one of each index, the optimal evaluation set is generated, marked as $U = \{u_1, u_2, ..., u_n\}$, while the worst one is marked as $V = \{v_1, v_2, ..., v_n\}$. u_r and v_r respectively represent the best and the worst values of the index m_r.

For $m_r \in M_1$, the comparative interval is $[v_r, u_r]$. In the domain $X_r = \{h_{kr}, u_r, v_r\}$ ($k = 1, 2, ..., p$), the identity and contrary degree of the set pair $\{h_{kr}, u_r\}$ can be defined as follows:

$$a_{kr} = \frac{h_{kr}}{u_r + v_r} \tag{2.10}$$

$$c_{kr} = \frac{u_r v_r}{(u_r + v_r)h_{kr}} \tag{2.11}$$

where a_{kr} is termed as the identity degree that denotes the approximate degree between h_{kr} and u_r, while c_{kr} is the contrary degree, which means the approximate degree between h_{kr} and v_r.

Similarly, for $m_r \in M_2$, a_{kr} and c_{kr} can also be defined in the comparative interval $[u_r, v_r]$, just by exchanging the Equations of a_{kr} and c_{kr} for $m_r \in M_1$.

Considering the weight of each index, the average identity degree and the contrary degree can be counted by Equations 2.12 and 2.13, in the comparative interval of s_k, that is, $[U, V]$, as follows:

$$a_k = \sum_{r=1}^{n} w_r a_{kr} \tag{2.12}$$

$$c_k = \sum_{r=1}^{n} w_r c_{kr} \tag{2.13}$$

where a_k is the average identity degree expressing the close extent between s_k and U, while c_k describes the average contrary degree representing the close extent between s_k and V. Then, the approximate degree between s_k and U, marked as r_k, can be expressed as given in Equation 2.14:

$$r_k = \frac{a_k}{a_k + c_k} \tag{2.14}$$

With larger value of r_k, the urban ecosystem health situation of the kth city is better.

According to the above procedure, a relative approximate degree of urban ecosystem health to the optimal evaluation set can be derived by combining multiple indices to reflect the relative health levels of various urban ecosystems. Since the optimal assessment set is derived from the concerned urban ecosystems and updated over time, the subjectivity of the health assessment standard can be avoided to some extent.

2.2 ECOLOGICAL CARRYING CAPACITY ASSESSMENT*

2.2.1 REVIEW OF URBAN ECOLOGICAL CARRYING CAPACITY

The concept of "carrying capacity" originated from ecology. It usually refers to the biological carrying capacity of a population level that can be supported for an organism, given the quantity of food, habitat, water, and other life infrastructure present. Along with the phenomena such as land degeneration, environmental contamination, and population expansion, carrying capacity has been gradually cited in urban ecology and became a more complicated and integrated concept. It can also be used to determine urban development density (Kyushik et al. 2005). This concept has been evolved into different terms such as population carrying capacity in natural ecosystem, resource carrying capacity (RCC), and environmental carrying capacity in anthropic ecosystem. The concept, connotation, and meaning of carrying capacity are being developed and perfected along with the development of ecology and society (Carey 1993). Particularly, after the emergence of the complex urban ecosystem theory, the meaning of urban ecosystem carrying capacity has been developed into a more holistic and systematic concept (Carey, 1993).

Presently, some researchers focus only on the capacities of individual components (Xu et al. 2003). However, the urban population and their activities jointly form the core component of the urban system, interlinked with the urban eco-environment. The development of an urban ecosystem is built upon the interactions between environmental carrying capacity (ECC), RCC, and social-economic development capacity (SEDC). It is well established that precisely describing the system characters and variabilities can be prohibitively difficult (Costanza and Cornwell 1992). While significant progress has been made in evaluating carrying capacity, most current methods are nonquantitative and lack analytical rigor (Prato 2001). The modeling of ecological footprint (Wackernagel and Rees 1996) is currently the most representative quantitative method to evaluate carrying capacity, but it is still lack of flexibility and adaptability in forecasting procedures (Zhao 2005). The concept of compound carrying capacity (CCC) is introduced in this chapter and studied as an index of the interactions between ECC, RCC, and SEDC and as a basis for meeting the challenges of urban sustainable development and eco-city building. Considering the diversity and complexity of urban ecosystems, the methodologies for both calculations and adjustment mechanisms of the urban ecosystem compound carrying capacity (UECCC) are outlined, with references to the urban ecosystem health index and the evaluation models of sustainable development. The significance and functioning of the UECCC were interpreted in view of city development perspectives and characteristics of the compound urban ecosystem. In addition, a case study for Guangzhou City, which is located in southern China, is used in this chapter as an example to demonstrate the calculation procedures for UECCC and its interpretation.

---

* This section was contributed by Linyu Xu and Zhifeng Yang.

2.2.2 Theory Model of Urban Ecological Carrying Capacity

2.2.2.1 Defining Urban Ecosystem Compound Carrying Capacity

Ecologists generally consider carrying capacity to be the maximum number of individuals under a certain condition. This environment can support these individuals without damaging its ability to support future generations within the specific area. This study focuses on the ability of urban ecosystem of supporting humans and their activities. Therefore, the UECCC index is defined in this study mainly to address the maintenance of the natural function and the urban ecosystem health. The concept of UECCC in this chapter is different from the traditional definition of carrying capacity.

UECCC is defined as the potential ability to maintain urban ecosystem health, which includes the ability to develop under normal conditions and the ability of resilience under stress conditions. Compared with traditional concepts, the UECCC includes the ability to develop and provides a framework for integrating physical, socioeconomic, and environmental systems into planning for a sustainable environment. Most do agree that changes in technology affect the carrying capacity of a system. Evidently, new technologies affect how resources are consumed, and thus, if carrying capacity depends on the availability of that resource, the value of the carrying capacity would change (Meyer and Ausubel 1999). For example, raising yields has allowed the developed world to support an increasing population while cropping a decreasing amount of land. For these reasons, the limit of UECCC based on fixed resource limits or a single, unchanging carrying capacity is unrealistic.

The urban ecosystem health can be viewed as a state that is in compliance with the suitable target of the city. The essence of the research on urban ecosystem carrying capacity is to evaluate whether the urban environment is able to attain the development target and to describe how the urban ecosystem maintains its health.

2.2.2.2 Biology Immunity Model for Urban Ecosystem

Urban ecosystems resemble organisms in which they have the similar abilities of both self-modulation and self-resilience. The resilience of a system refers to its ability to maintain its structure and pattern of behavior in the presence of stress (Holling 1986). The resilience ability and the resisting ability of urban ecosystem are both important to maintain the health of an urban ecosystem by supplying resources and cleaning contamination generated by large-scale economic activities.

2.2.2.2.1 Comparison with Biological Immunity

The urban ecosystem is constantly exposed to stresses such as resource consumption and pollutant emissions, which have similar side effects as pathogens in the human body. These stresses can be counterbalanced by urban carrying capacity (resource supply and environment self-purification) in a certain time frame at certain level. When the rate and strength of the stresses and the carrying capacity are at balance, the urban ecosystem will maintain normal or healthy state. But once the rate or the strength of pressure is beyond that of the carrying capacity, the detrimental effects of the environment pollution, resource consumption, and ecosystem disturbance will appear. The similarity analysis of urban ecosystem carrying capacity and human immunity is shown in Table 2.6.

TABLE 2.6
Comparative Study of Urban Ecosystems and the Human Immune System

	Human Immunity	Urban Ecosystem Carrying Capacity	Similarities
Definition	Human immunity is the resistance of human body against infection. It can keep the stability of in-environment through recognizing and excluding the foreign matters such as antisubstance (Wei and Zhang 2002)	Urban ecosystem carrying capacity is the potential capacity of an urban ecosystem to maintain its health and to develop stability. It includes the resistance of urban ecosystem against the pressure that would be harmful to its health, the resistant ability when the pressure disappears, and the development ability by which some proper target can be achieved	To sustain main body alive and act normally
Material basis	Human immune system Natural immunity response	Urban ecosystem storage Ability of resource supply and self-purification of natural ecosystem	Interacting with main body as an integral part (a) Congenital (b) Be limited (c) Passive
Functions	Acquired immunity response	Pollution control and artificial eco-environment building	(a) Memory (b) Exceptional function (c) Self-adjustability in normal conditions (d) The supplement of the natural immunity response/self-cleaning (e) Be able to improve immunity by manual work (f) Initiative

When the stresses are beyond the carrying capacity, the resource recovery and the environment self-purification become insufficient. It is then necessary to import resource into the urban system or explore new energy to raise the rate of resource supply and to carry out comprehensive ecological improvements to rebalance the eco-environment. All the above behaviors with obvious urban properties are defined as SEDCs, which include the economic development, the technical innovation, the capital construction, and the eco-planning. Therefore, the UECCC can be viewed as the integration of the sustainable ability of the natural ecological subsystem and the development ability of the socioeconomic subsystem. The sustainability of the natural ecological subsystem that has a similar function as the immune system can be defined as the inner sustainable ability. And it supplies urban ecosystem with nutrition and space. On the contrary, the development ability of the socioeconomic subsystem that has a similar function as the medicament and the clinic operation can be instructed by the consciousness of a human being and is therefore called the outer development ability. It is the most active part of urban ecosystem that affects the inner sustainable ability and promotes urban ecosystem to develop. In summary, both the inner sustainable ability and the outer development ability are keeping the urban ecosystem healthy.

2.2.2.2.2 Biology Immunity Model for Urban Ecosystem

Based on the comparative study of urban ecosystems and the human immune systems, the biological immunity model for urban ecosystem can be constructed. It is assumed that the urban ecosystem is the material input, and its main function is to offer eco-services. The extent of these services is dependent on both the intrinsic carrying capacity and the acquired carrying capacity of the urban ecosystem. Human beings are the receptors of these services. This theoretic model shows the carrier, the carried target, and the carrying mechanism of the UECCC.

Urban ecosystem storage (UES) is an important component of urban ecosystem, which can offer eco-services for urban ecosystem and affect the health of the urban ecosystem. It is also the material basis of UECCC. Depending on the way that the storage offers eco-services, UES can be classified into the source ecosystem storage (SoES), the sink ecosystem storage (SiES), and the channel ecosystem storage (CES). UES has different effects on urban ecosystem during different development stages. At the initial stage of urban development, the UES is the cradle of urban ecosystem, supplying nutrition to the urban system. When the urban system arrives at its period of great prosperity, the UES will advance and support its development; it will also limit the overdevelopment of the city at the upper stage of ecological succession. As a result, UES can also be called "urban ecosystem carrier" with the similar function as the immune system of a human being.

The eco-services of the UES are the end usage of carrying capacity, responding to the demand of urban ecosystem (mainly human being). Because the demand information cannot be transferred immediately or in time from the carried target to the carrier, the eco-services of the UES will not be sufficient for the demand of urban ecosystem in a certain stage. This is the essence of the stress for the urban ecosystem. When the burdened information activates the UECCC, a quick response will be provoked and the potential of the UECCC for recovery and resumption will be exerted.

The UECCC can be categorized into the intrinsic carrying capacity and the acquired carrying capacity.

2.2.2.2.3 Limits of UECCC

Generally, the limits of carrying capacity has always been overemphasized, while its essential aspect that it is a type of ability or potential of ecosystem tends to be ignored. This is because its limits are easier to be expressed and utilized, comparing with its ability. Although the UECCC in this chapter is mainly investigated regarding its ability, which is an abstract concept and difficult to quantify, its limits are discussed herein because they are the main traits of the UECCC.

UECC comprises of not only the natural ecosystem but also the social-economic system, as discussed in Section 2.2.1. The sustainable capacity determines the upper limit of UECCC based on the maximum supply of resource and the maximum sustainable ability of environment. The development ability determines the lower limit of UECCC according to the minimum development ability of the social-economic system or the smallest size of the urban population (this is determined by the population standard of the city in different countries). The ecosystem will function well when the UECCC values is within its upper limit and lower limit.

Moreover, the UECCC is not a constant value, but rather a dynamic value varying across stages along with the development of cities. UECCC is governed by several factors that include the demand of a human being in urban ecosystem for living quality and eco-services, the development target of cities, and the health state of urban ecosystem. This dynamism of UECCC can be illustrated by the demand of a human being for eco-services, that is, a stress on urban ecosystem. Therefore, the intersection angle between the UECCC and the UEPIO (the inner and outer pressure put on urban ecosystem) at a certain stage can be used to illustrate the direction of urban development. There are three development directions and seven statuses of urban ecosystem according to the value of the slope, which can be positive, negative, or zero (see Figure 2.3).

Accordingly, when the slope is positive, the UECCC will take up the advantage niche stepwise and facilitate the urban development quickly; when the slope is zero, the UECCC will reach equilibrium with the UEPIO and the urban system will develop steadily; when the slope is negative, the UEPIO will exhaust the advantage niche and terminate the urban development. It should be noted that the curve of urban development in Figure 2.3 illustrates only the dynamic equilibrium status between the UECCC and the UEPIO, but not the comparison of the absolute value of the UECCC and the UEPIO. In practice, the appropriate level is more important, which should be the exact target of urban development and also the main purpose of this research of UECCC.

2.2.3 Evaluation Methods

The preceding discussion on the concept, connotation, and character of the UECCC revealed that it is a capability or a potential that keeps urban ecosystem healthy and steady. We can determine whether the urban ecosystem is in an equilibrium status or healthy by the comparison between the UECCC and the UEPIO. Therefore, it is the

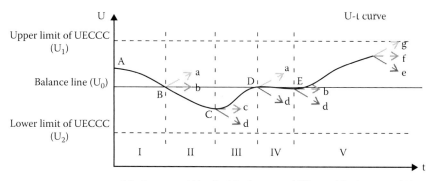

a : Moving toward V   b : Moving toward IV   c : Moving toward VI
d : Moving toward the lower limit of UECCC   e : Moving toward I
f : Moving toward VII   g : Moving toward the upper limit of UECCC

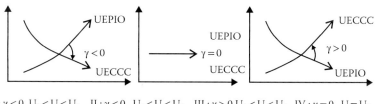

I : $\gamma < 0$ $U_0 < U < U_1$ II : $\gamma < 0$ $U_2 < U < U_0$ III : $\gamma > 0$ $U_2 < U < U_0$ IV : $\gamma = 0$ $U = U_0$
V : $\gamma > 0$ $U_0 < U < U_1$ VI : $\gamma = 0$ $U_2 < U < U_0$ VII : $\gamma = 0$ $U_0 < U < U_1$
A : V -> I   B : I -> II   C : II -> III   D : III -> VI   E : VI -> V

FIGURE 2.3 Illustration of the dynamic equilibrium status of urban development at a certain stage.

immanent function of UECCC to estimate sustainable development. The *sustainable development degree* (Feng and Wang 1997) is an integrated measurement index of sustainable development, which can be referenced to build the measuring model of UECCC.

2.2.3.1 Measuring Model of UECCC

The measuring model of UECCC is established in reference to the biology immunity model for urban ecosystem. The model can be divided into two associated parts, which are measuring models of intrinsic carrying capacity and acquired carrying capacity.

Measure model of intrinsic carrying capacity is given by

$$N = R \cdot \alpha_s^2 \cdot e^{\beta_s} \quad \text{in this,} \quad \begin{cases} R = k_1 \left(\sum_{i=1}^{n} S_i \cdot \log_2 S_i \right) \cdot \sum_{i-1}^{n} S_i \cdot P_i \\ \alpha_s = k_2 \sum_{i=1}^{m} r_i / G \\ \beta_s = \frac{1}{k} \sum_{j=1}^{k} \lambda_j K_j \end{cases} \quad (2.15)$$

where N is intrinsic carrying capacity index; R is resilience index; α_s is resource supplying index; β_s is environment carrying capacity index; r_i is quantity of the resource i; G is GDP, which is calculated as the invariable price in the year 1990; S_i is percent of the ith earth surface cover in the whole urban; P_i is production capacity of the ith earth surface cover; λ_j is proportion of some pollutant; K_j is the jth pollutant letting criterion; k is the category of the pollutant; and k_1 and k_2 are constants, which are used to counteract dimension as relative carrying capacity; their numerical values are not essential.

Measuring model of acquired carrying capacity is given by

$$F = \mu\sigma \cdot \text{Eco} \quad \text{where} \quad \text{Eco} = \frac{\Delta G/G}{\Delta POP/POP} \tag{2.16}$$

where F is postnatal carrying capacity index; μ is technique index; σ is human resource index; Eco is economic capacity index; $\Delta G/G$ is developing rate of GDP; and $\Delta POP/POP$ is population variational rate.

Integrating both intrinsic and acquired carrying capacities

$$UECCC = f(N, F) = r \cdot N \cdot e^F \quad \text{where} \quad \gamma = a \sum_{j=1}^{I} \left(\cos\frac{\pi}{2} \cdot \frac{M_j/POP}{M_{j0}/POP'} \right) + b \tag{2.17}$$

where $UECCC$ is urban ecosystem carrying capacity; r is the character index of those cities with natural resource as main industry, whose value is less than 1; M_j is the exploitation amount of the jth nonrenewable resource; M_{j0} is the consumed amount of the jth nonrenewable resource; and a and b are constants, and $a + b = 1$.

2.2.3.2 Measuring Model of UEPIO

The UECCC exists relatively to UEPIO, and the UEPIO is rooted in the rapid swell of urban population and economic development. The pressure on urban ecosystem should be categorized into inner and outer stress (indirect stress) because of the two-sided trait of the social-economic system. The social-economic system will consume resources and cause environment pollution, which is the so-called inner stress, but it also advances urban development. While the outer stress (or the indirect stress) comes from urban complex ecosystem, it can be illustrated by population, economic action, and living quality demand. To demonstrate the sustainable development of a city, the urban ecosystem pressure index (UEPIO) needs to be measured to obtain the intersection angle (named γ) with UECCC in this chapter. Given that the urban ecosystem is healthy and human beings' living conditions are at a certain standard, (1) when $\gamma > 0$, the increasing rate of UECCC is quicker than that of the UECCC and the urban ecosystem is healthy; (2) when $\gamma = 0$, the urban ecosystem is in the equilibrium status; and (3) when $\gamma < 0$, the UEPIO will be beyond the UECCC, and the ecosystem will break down if this condition holds for a longer time period.

The measuring model is as follows:

$$UEPIO = \alpha_u^2 \cdot e^{\beta_u} \quad \text{where} \quad \begin{cases} \alpha_u = k_3 \sum_{i=1}^{m}(POP \cdot s_i + G \cdot \omega_i)/G \\ \beta_u = \dfrac{1}{k} \sum_{j=1}^{k}\lambda_j(POP \cdot w_j + G \cdot \psi_j) \end{cases} \quad (2.18)$$

where α is resource resuming index; β is environment pollution index; POP is population; s_i is the resuming amount per person of the ith resource; G is GDP, which is calculated as the invariable value of the year 1990; ω_i is the resuming amount per 10,000 yuan RMB; λ_j is the weight of a pollutant; w_j is the exhausting amount per person of the jth pollutant; ψ_j is the exhausting amount per 10,000 yuan RMB of the jth pollutant; k_3 is the constant; and k is the sort of the pollutant.

2.3 EMERGY-BASED URBAN ECOSYSTEM EVALUATION*

2.3.1 INTRODUCTION

2.3.1.1 Coupling Technological Progress, Welfare, and Environmental Care

Human production and consumption activities can amplify the benefits to human society. However, evidence in recent decades of escalating human impacts on ecological system worldwide raises concerns about the spatial and temporal consequences of negative effects to human well-being and ecosystem integrity (Brown and Ulgiati 2005; Sachs 2005). Especially in urban metabolic system, the fastest economic development is planned, coinciding with high rates of environmental change and accelerated species loss. Complex overlap of various factors creates a bumpy road to sustainability. These interactions have caused the trepidation concerning the disruption in the balance of humanity and nature. The Millennium Development Goals point out that sound policy and management interventions can often reverse ecosystem degradation and enhance the contributions of ecosystems to human well-being, but knowing when and how to intervene requires substantial understanding of both the ecological and the social systems involved (Sachs 2005).

To rebalance the social and environmental dimensions of sustainability with the economic one, the socio-environmental damages of the urban system must be quantified. Over the past 20 years, there have been many studies in the analysis focusing on the basic metabolism related to the input side and the environmental impacts (Ayers and Kneese 1969; Daniels and Moore 2002; Fischer-Kowalski 1998; Fischer-Kowalski and Huttler 1998; Haberl 2006; Wolman 1965). However, a large number of studies are focused on urban industrial material metabolism, such as those about Taiwan (Huang 1998), Toronto (Sahely et al. 2003), Nantong (Duan 2004), Sydney (Lenzen et al. 2004), and Paris (Barles 2007). There are

* This section was contributed by Gengyuan Liu and Sergio Ulgiati.

a few studies focusing on household metabolic process (Forkes 2007; Newman et al. 1996) and even less dealing with the associated health burdens to the people and the surrounding ecosystem. Most of the studies, however, use monetary measures to assess natural capital and human capital values and losses. The quantitative measure of urban metabolism must take into proper account both production and consumption processes. As a consequence, there is an urgent need to develop a quantitative methodology that can evaluate the adverse environmental effects of both production and consumption activities, addressing specific damages to human health and ecosystems, and taking into account how they affect the urban system's dynamics and sustainability.

2.3.1.2   Emergy Metrics for Urban Metabolism: The State of the Art

Emergy synthesis is a method of environmental accounting derived from energy system theory that uses the energy (in units of the same kind) required to produce goods or services as a nonmonetary measure of the value or worth of components or processes within ecosystems and the economy (Odum 1996). Until now, a large number of systems have been evaluated by means of the emergy method on regional and national scales (Brown and Odum 1992; Huang and Odum 1991; Lan and Odum 2004; Liu et al. 2010; Odum et al. 1995; Ulgiati et al. 2007; Yan and Odum 1998). Most of these studies, however, did not focus on the impact of emissions on ecosystem and human integrity, although important steps ahead have been made by some authors. Ulgiati et al. (1995) first pointed out that the impact of emissions on natural and human-dominated ecosystems requires additional emergy investment to take care of the damage or altered dynamics and to make a system or process sustainable. Ulgiati and Brown (2002) calculated the additional emergy for the environmental services required to dilute emissions, without considering atmospheric diffusion and chemistry. Hau and Bakshi (2004) first proposed the use of disability-adjusted life years (DALYs) from Eco-Indicator 99 (E.I. 99) impact assessment method to evaluate the emissions' impact on human health of economic sectors by using ecological cumulative exergy consumption (ECEC) analysis. Brown and Ulgiati (2005) used the emergy method to suggest a system view to ecosystem's integrity and also to assess the emergy investment needed to restore ecosystem health. Lei and Wang (2008) tracked the waste treatment processes and calculated the transformity of the fly ash and slag in Macao, as a result of the incineration of municipal solid waste (MSW). Zhang Jiang et al. (2009) integrated dilution and E.I. 99 methods to evaluate the sustainability of Chinese steel production. The research on a single industry was proposed by these authors as an initial case of application on regional scale. Therefore, considering cities as a multi-industry integrated system, emergy-based city studies should investigate the global impacts of emissions and integrate them into the set of existing emergy indicators to provide suitable and scientifically based information for cost-effective abatement strategies and policy decisions. In seeking an effective model in the analysis of emissions, other authors developed hybrid life cycle assessment (LCA)-based methodologies (Udo de Haes and Lindeijer 2001), where emissions are characterized by end point impact factors related to human and ecosystems health.

2.3.2   Methodology

2.3.2.1   Emergy-Based Urban Metabolic Model

A typical diagram describing an urban system is shown in Figure 2.4, where the energy system symbols are used (Odum 1996). At the planetary level of organization, there are no substantial exchanges with the larger system except for solar and gravitational energy entering the system from external sources. Within the large box indicating the spatial boundaries of the urban system, solar and gravitational emergies (R) entering from outside processes results in rain, wind, tides, waves, and so on. Nature also does work, which indirectly supports the activities of the world socioeconomic system (e.g., the photosynthesis of natural ecosystems that fixes carbon and replenishes oxygen in the atmosphere, which is necessary for all lives, the movement of clean air that replaces contaminated air over cities, and water flows that provide the capacity to dilute municipal wastes). Natural products are used by ecosystems, but some of these products, for example, soil, timber, and groundwater, are appropriate for use by the socioeconomic system. The emergy provided by fuels and electricity is modeled on a separate pathway that acts on the material products by arranging and ordering them. Humans do work on the environmental system to extract and process the slowly renewed material products of natural work, that is, fossil fuels and minerals. These inflows are considered to be nonrenewable because they are being used by the socioeconomic system at a rate that is much greater than their natural renewal rate. Human work is also used to carry out economic production using these raw materials and to carry out the other processes and functions of the society. A special category of human work is recognized in this model; that is, the work performed to extract

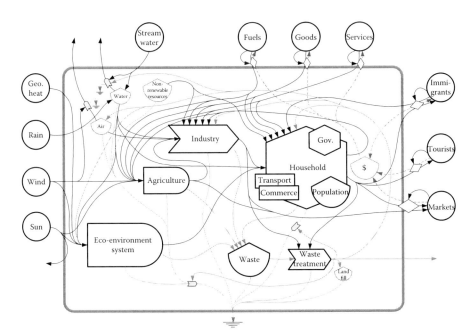

FIGURE 2.4   **(See color insert.)** Emergy diagram of Beijing ecosystem.

renewable energy from nature for direct use in running socioeconomic systems. This work creates, maintains, and operates infrastructures capable of transforming renewable energy into electricity or another high-quality form of energy that can be used to operate the socioeconomic system. The sustainability of our current society depends on the long-term success of human endeavors to magnify the work done on pathway to the point where it can carry out most of the work processes needed to support the socioeconomic system. People build assets and knowledge through carrying out economic processes and using the economic products and services (EPS) produced. In turn, they use these assets and their knowledge to perform the work processes needed to capture more fossil fuel, mineral, and natural energies in the service of society.

In Figure 2.4, money is indicated by dashed lines that flow in the opposite direction to the human work performed. Note also that money flows only track the flows of human work and do not flow counter to the natural work pathways. This diagram shows that economic processes are dependent on the work processes of nature, but that money does not track or account for these natural work processes; therefore, money flows in the market economy are an incomplete measure of the work required to assure the continued and proper functioning of societies and of the value incorporated in EPS.

A crucial feature of this model is that materials in the form of slowly created mineral products of the earth and more rapidly created natural products of the biogeosphere are incorporated into EPS through the expenditure of fossil fuel energy controlled by knowledgeable human actions. In Figure 2.4, minerals, human work, fossil energy, and natural products are all part of the same interaction and thus their flows are not independent, but rather are functions of one another because of their multiplicative interaction. In addition, the flow of money is coupled to this production process through the laws of supply and demand and the price mechanism so that plots of money flow versus energy or emergy flow are plots of a function of x versus x, and thus high positive correlations should be expected.

2.3.2.2 Emergy Evaluation Method

A general introduction to the methods of environmental accounting using emergy can be found in many publications, for example, Odum (1996). Emergy is a quantity based in the second law of thermodynamics because it is a measure that accumulates all the energy used up in the process of creating any item in terms of a common energy unit, that is, an emjoule of solar energy, coal energy, and so on. The transformation to emergy units normalizes all inputs to a production process in terms of an equivalent ability to do work in the system. Emergy is formally defined as all the available energy of one kind previously used up directly and indirectly to make a product or service (Odum 1986, 1988; Scienceman 1987). In the case of environmental accounting, emergy is measured in solar emjoules (seJ). The emergy of any product or service can be quantified by obtaining data on the available energy or mass of the product or service and then multiplying this value by the appropriate emergy per unit value, that is, the transformity (seJ/J) for energy or the specific emergy (seJ/g) for mass. For the purpose of exploring the change of urban energetics during its evolutionary process, emergy and emergy indices are incorporated into the urban metabolic model for simulation, as indicated in Figure 2.5. The equations for simulating emergy flows to study the characteristics of urban energetics are summarized in this figure.

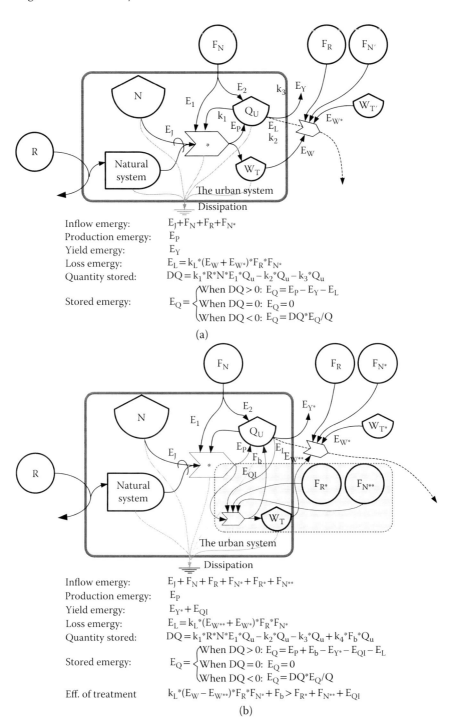

FIGURE 2.5 Urban metabolic system diagram and equations for simulating emergy flow (a) without and (b) with waste treatment system.

Emergy analyses are carried out using transformities, specific emergies, and other factors that are determined relative to a particular planetary baseline (Odum 1996; Ulgiati and Brown 2002), which is determined from the solar equivalences of the three primary energy inputs to the biogeosphere; that is, solar radiation, residual heat, and deep heat of the earth, and the gravitational attraction of the sun and the moon. In this study, transformities were converted from global emergy baseline of 9.44E+24 to 15.83E+24 seJ/year recommended by Ulgiati and Brown (2002).

2.3.2.2.1   Renewable Sources

Renewable resources are replenished on a regular basis as a result of the use of planetary emergy inflows in solar radiation, the deep heat of the earth, and gravitational attraction of the sun and the moon. These primary planetary emergy inflows and the continuously generated coproducts of their interactions in the biogeosphere comprise the renewable resources of the earth. In general, all renewable resources known to be important inputs to a system are evaluated, and the emergy contributed to the system by each is determined. While all renewable energies known to be important are calculated and included in the table, not all of them are included in the emergy base for a system. If all the coproducts of a single interconnected planetary system are counted, some of the emergy inflow will be counted twice; therefore, only the largest of any set of coproducts is counted in the emergy base for a given area of the earth.

Rain carries two kinds of energy, the chemical potential energy that rainwater has by virtue of its purity relative to seawater and the geopotential energy of the rain at the elevation at which it falls. Renewable energy also enters a state or other system through cross-border flows of energy and materials in rivers. Renewable energy inflows to the system can be determined at two points: (1) the point of entry and (2) the point of use. The first of these two flow measurements gives the emergy received by the system and the second gives the emergy absorbed or used in the system. For example, the incident solar radiation is received by the system and the incident solar radiation minus the surface albedo is absorbed. The geopotential energy of rain on land at the elevation it falls is the geopotential energy received by the system, whereas the geopotential energy of the runoff relative to the elevation at which it leaves the state is used on the landscape to create landforms. The chemical potential energy of the rain that falls on the land is received, but the water transpired is actually used by the vegetation to create structures on the landscape. In some cases, almost all the emergy received by the system is absorbed, for example, almost all tidal energy received is dissipated in estuaries and on the continental shelf.

2.3.2.2.2   Evaluating Nonrenewable Resources

Nonrenewable resources are raw materials that have been built over a long time by environmental processes, but that are being used by human activities at a rate much faster than they can be renewed. Coal mining or groundwater withdrawals in excess of the recharge rate are examples of nonrenewable resources. An emergy evaluation does not determine the contribution of a nonrenewable resource by the price paid for the raw material—a ton of coal for instance—because this is not the value of the coal itself. It is the price someone is willing to pay for the labor and machinery required to mine the coal. When evaluating coal as an emergy input, it is important to evaluate or take into account the energy required to make the coal. The solar emergy required to make a joule of coal is its solar transformity in seJ/J. A material flow is multiplied by its specific emergy (seJ/g) or converted to energy and then multiplied by its transformity to obtain an emergy flow. All storages in the system that are being used faster than they are being replaced contribute to the nonrenewable emergy supporting the system. This includes storages that can be used renewably; for example, soil, groundwater, and timber.

2.3.2.2.3   Evaluating Exports and Imports

Emergy is imported and exported in three forms: (1) emergy in services separate from any material flows (consulting, data analysis, financial services, etc.), (2) emergy in materials entering and leaving the state, and (3) emergy in the human service associated with the material inflows and outflows (collecting, refining, manufacturing, distributing, shipping, and handling). The data sources and methods used to evaluate imports and exports will vary depending on the system. The methods used for the calculation of the emergy in the mass of imported materials is as follows:

1. Choose a base year for the tonnage calculation. Prices must be available for detailed categories of imports in the base year. We used data from the 1999 Beijing's commodity flow survey to determine average prices of 12 commodity categories in the base year.
2. Calculate the average tonnage moving per dollar for each commodity category in 1999 and use this as the price ($/g) of that commodity. At this point, we have the average price of goods in each import class in 1999.
3. Identify the price index data that can be applied to the average 1999 prices to estimate commodity prices from 1999 to 2006. Although import price indices were available for some commodities in some years in the Beijing Statistical Yearbook, we did not have a consistent data set for all commodity classes of interest over the entire time period. And we adjusted the 1999 prices using the Producer Price Index.
4. Calculate the tonnage moving in each commodity category.
5. Multiply the mass imported in each category in each year by the specific emergy or convert the mass to energy and multiply by the transformity to determine the emergy gained through the import of material goods.
6. Sum all the categories except the one containing fuels and minerals to get the total emergy of the goods imported.

2.3.2.3   Emergy-Based Environmental Impact Assessment Model

Every economic process generates useful products and undesired impacts at the same time (to the ecosystem, to human health, and to human assets). Ulgiati et al. (1995) focused on the emergy resources required to prevent or fix reversible damages. Moreover, these authors pointed out that (1) additional emergy resources are needed to replace the lost assets or units when irreversible damages occur and that (2) when replacement is not possible, at least a conservative estimate of the natural or human capital loss should be attempted, based on the resources previously invested for its generation to ascertain the true cost of a process product. Following Ulgiati et al. (1995) and Ulgiati and Brown (2002), additional emergy cost terms should be included to account for (a) dilution and abatement of emissions by natural processes; (b) abatement, uptake, and recycle of emissions by means of technological devices; (c) repair of damages to human-made assets by means of maintenance activities; (d) reversible and irreversible damages to natural capital (e.g., loss of biodiversity); and finally, (e) reversible and irreversible damages to human health. As a consequence, the total emergy cost U ($U = used$) can be calculated as follows:

$$U = R + N + F + F_1 + ... + F_n \qquad (2.19)$$

where R and N are respectively the locally renewable and the nonrenewable emergy resources and F is the emergy of imported goods and commodities (including their associated services), where the F_i ($i = 1, 2, ..., n$) terms include the environmental or human-driven emergy investments ($F =$ feedback) needed to prevent or fix the damages occurred and charged to the process:

$F_1 = S_j F_{1,j} =$ the sum of all jth input flows to prevent or fix damage 1;

...

$F_n = S_k F_{n,k} =$ the sum of all kth input flows to prevent or fix damage n.

For the sake of clarity, if combustion emissions damage the facades of urban buildings, such a damage can be assessed in terms of the emergy investment F_i needed to restore it, that is, $F_i = A \times S_k f_{n,k}$, where A is the damaged surface and $f_{n,k}$ is the emergy investment per unit surface (chemicals, paints, and labor) needed to restore the facade.

Disregarding the additional resource investments due to impact prevention or repair would underestimate the real demand for the process to occur and be sustainable.

The main focus of this chapter is to apply such a framework to the sustainable development and management of an urban system, taking the city of Beijing (China) as a case study. Such a goal requires that specific procedures are identified and applied to calculate the additional resources needed for sustainable development of the urban system by removing those factors that affect human and environmental health.

Figure 2.6 shows two patterns for release of emissions: (a) without and (b) with waste treatment systems. The figure represents only a subsystem of the Beijing urban system of Figure 2.4; that is, the waste released and its interaction with the urban system itself. Air and water emissions and solid waste are controlled based on additional input of fuels, goods, and labor force. The terms $F_1, ..., F_n$ in Equation 2.19 are indicated in Figure 2.6 as $L_{w,n}$ ($n = 1, 2, 3$) to specifically point out their nature of emergy losses (L) associated to a process waste (w) generation. Without treatment,

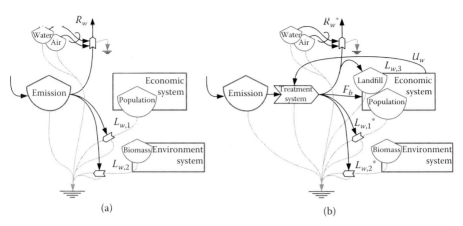

(a)   (b)

FIGURE 2.6 Direct and indirect emergy inflows from environment and economic system (a) without and (b) with waste treatment system. R_w, emergy of ecological services needed to dissipate the emissions; $R_w{}^*$, emergy of ecological services needed to dissipate the emissions after treatment, $L_{w,1}$, emergy of the human health losses caused by the emissions; $L_{w,1}{}^*$, emergy of the human capital losses caused by the emissions after treatment; $L_{w,2}$, emergy of the natural capital losses due to the emissions; $L_{w,2}{}^*$, emergy of the natural capital losses due to the emissions after treatment; $L_{w,3}$, emergy of the human capital losses caused by land occupation; U_w, emergy of waste treatment input; F_b, emergy of feedback of useful products to the upstream process.

the emergy loss associated to damaged human capital is indicated as $L_{w,1}$, which means that some emissions cause pathological impacts on human beings that in turn require additional investment for replacement or fixing; meanwhile, other kinds of emissions, such as acid rain and lake eutrophication, may lead to loss of flora and fauna. The emergy loss associated to the degradation of natural capital is indicated as $L_{w,2}$. Untreated emissions need ecological services to render them harmless, such as dilution and abatement, and these emergies are indicated as R_w. To prevent or minimize further pollution damage, a waste treatment system can be applied as designed in Figure 2.6b. The waste treatment system could effectively reduce waste (not to zero) through additional resources input. The new (lower) human and natural capital emergy losses after waste treatment are denoted as $L_{w,1}{}^*$ and $L_{w,2}{}^*$ (being respectively $L_{w,1}{}^* < L_{w,1}$; $L_{w,2}{}^* < L_{w,2}$). Furthermore, the damage associated to solid waste disposal can be measured by land occupation and degradation, the emergy of which (i.e., the emergy value of land, irreversibly degraded) is denoted as $L_{w,3}$ (Cherubini et al. 2009). The additional emergy investment for treatment is denoted as U_w and should be in principle lower than the damage-related losses $L_{w,n}$, to be feasible and rewarding. The waste treatment system is designed to recycle and reuse part of the emissions (flow F_b) through the use of eco-technologies. Such a recycle flow should allow a proportional decrease of the total emergy cost U, by decreasing the use of local nonrenewable resources N or by decreasing the imports F in Equation 2.19. However, this improvement was not accounted for in the present study because the proposed pattern is not yet fully implemented in Beijing urban waste management policy.

2.3.2.4   Evaluation of the Impacts of Emissions

2.3.2.4.1   Quantifying Ecological Services

Emissions are sometimes rendered harmless due to services provided by the eco-system, which dilute or abate the emissions to an acceptable concentration or state. The emergy value of these ecological services may be calculated from knowledge of the concentration and nature of the emissions and the transformity of the relevant ecological services. For example, the emergy required to dilute nitrogen dioxide in air may be determined with information about the concentration of the emissions, the acceptable or the background dilution concentration, and the transformity of the wind.

Ecological services for diluting airborne and waterborne pollutants can be calculated as follows (Ulgiati and Brown 2002):

$$M_{air/water} = d \times \left(\frac{W^*}{c} \right) \qquad (2.20)$$

where $M_{air/water}$ is the mass of dilution air/water needed, d is the air/water density, W^* is the annual amount of the ith pollutant, and c is the acceptable concentration from agreed regulations or scientific evidence. Equation 2.20 should be applied to each released pollutant flow. Using the "acceptable concentration" assumes that small amount of pollution is acceptable. Instead, if the background concentrations were used for "c," this would have implied a pollution level down to a level that is more or less the level before the industrial era. Much more environmental services would be needed than actually available, thus placing a constraint to the acceptability of emissions: no emissions that cannot be absorbed or abated by the environment. Once the dilution mass of air or water is known, the energy value of needed environmental services referred to in Equation 2.18 is determined by calculating the energy of the dilution air or water. These flows can be of kinetic nature, if only their pollutant transport service is considered, or even of chemical nature, if their ability to drive chemical reactions and abate the pollutants is accounted for. Typical equations can be as follows:

Release of chemicals into the atmosphere:

$$\left[F_{w,air} = R_{w,air} = N_{kinetic} \times tr_{air} = \left(M_{air} \times v^2/2 \right) \times tr_{wind} \right]_i \qquad (2.21)$$

Release and conversion of chemicals into water bodies:

$$\left[F_{w,water} = R_{w,water} = N_{chem} \times tr_{chem,water} = \left(M_{water} \times G \right) \times tr_{chem,water} \right]_i \qquad (2.22)$$

Equations 2.21 and 2.22 are applied to the ith released pollutant; M_{air} is the mass and $N_{kinetic}$ is the kinetic energy of dilution air moved by the wind, tr_{air} is assumed to be the transformity of wind, v is average wind speed, N_{chem} is the chemical available energy of water (equal to its ability to drive a chemical transformation), $tr_{chem,water}$ is the transformity of water chemical potential (Odum et al 2000, folio 1, Table 2), and G is the Gibbs free energy per unit mass of water relative to reference seawater (4.94 J/g).

If the pollutant is waste heat (assumed release to the atmosphere), we must consider the service of cooling in addition to the service of dilution of chemicals. The cooling calculation procedure starts from the total amount of heat released by the system (roughly, the total energy used by the system itself and converted to degraded heat). The heat released to the air increases its temperature from average environmental temperature T_o to a higher new-equilibrium temperature T_e considered acceptable by the present legislation or the scientific community. Assuming that the acceptable T_e is only 1°C higher than the average environmental temperature, the following equation should be used:

$$M_{air/water} = \frac{Q_{released}}{\rho \times (\Delta T)} = \frac{Q_{released}}{\rho \times (1°C)} \qquad (2.23)$$

where M is the heat-dilution mass required to lower the emission temperatures to the accepted temperature and ρ is an average thermal capacity of air gases. Once the heat-dilution mass for cooling service is known, it can be used in Equation 2.23 to calculate the additional cooling emergy required.

Finally, the total environmental support needed to treat the chemical and heat emissions can be calculated as follows:

$$R_w^* = \text{Max}\left(R_{w.air.i}^* \right) + \text{Max}\left(R_{w.water.i}^* \right) \qquad (2.24)$$

It is worth mentioning that this method is proposed without considering—for the sake of simplicity—the diffusion and the chemistry processes in the atmosphere and that it relies on the implicit assumption that the available dilution air/water is always sufficient (which may not be true and would place a limit to the emissions or require technological treatment).

2.3.2.4.2 Quantifying Ecological and Economic Losses

A number of methods have been developed in previous studies for assessing the environmental impact of emissions. It would be a very useful further step to integrate such methods within a procedure capable to describe and quantify the actual damage to populations or assets in emergy terms; that is, in terms of lost biosphere work. Examples of such a natural capital and human capital losses are the decreased biodiversity due to pollution or ecosystem simplification or the economic losses related to damages to human health, land occupation and degradation, and human-made assets, among others.

In this study, a preliminary damage assessment of losses is performed according to the framework of the E.I. 99 assessment method (Goedkoop and Spriensma 2000). Such a method, similar to all end point life cycle impact assessment methods, suffers from very large uncertainties intrinsically embodied in its procedure for assessment of final impacts. Yet, it provides a preliminary—although uncertain—estimate of impacts to be used in the calculation procedure of total emergy investment. Damages to natural capital are expressed as the potentially disappeared fraction (PDF) of species in the affected ecosystem, while damages to human health are expressed as DALY, according to Murray et al. (1994), Goedkoop and Spriensma (2000), and Ukidwe and Bakshi (2007).

Using concepts from E.I. 99 (PDF and DALY) to quantify a process impact on ecosystems and human health has the advantage that the assessment relies on damages that can, in principle, be measured or statistically calculated. Unfortunately, the available data in these ecological models are restricted to Europe (in most cases, to the Netherlands), and their use to assess other countries requires adjustments (Zhang et al. 2010) and calls for urgent database improvement. Moreover, the dose-response relationship considered in the E.I. 99 is linear instead of logistic (Ukidwe and Bakshi 2007). The latter characteristics suggest the method only to apply to slow changes of pollutants' concentration and are not suitable for large emission fluctuations such as environmental accidents.

The impact of emissions on human health can be viewed as an additional indirect demand for resource investment. Human resources (considering all their complexity: life quality, education, know-how, culture, social values and structures, hierarchical roles, etc.) can be considered as a local slowly renewable storage that is irreversibly lost due to the polluting production and use processes. Societies support the wealth and relations of their components to provide shared benefits. When such wealth and relations are lost, the investment is lost and such a loss must be charged to the process calling for changes and innovation. The emergy loss can be calculated as

$$L_{w,1}^* = \sum m_i^* \times \mathrm{DALY}_i \times \tau_H. \tag{2.25}$$

Here, $L_{w,1}^*$ is the emergy loss in support of the human resource affected, i refers to the ith pollutant, m^* is the mass of chemicals released, DALY is its E.I. 99 impact factor, and τ_H is the unit emergy allocated to the human resource per year, calculated as τ_H = total annual emergy/population. The rationale here is that it takes resources to develop a given expertise or work ability and societal organization; when it is lost, new resources must be invested for replacement (not to talk of the value of the individual in itself that is not quantifiable in physical terms).

PDF is the acronym for potentially disappeared fraction of species (E.I. 99, Goedkoop and Spriensma 2000). Such effects can be quantified as the emergy of the loss of local ecological resources, under the same rationale discussed earlier for the human resource:

$$L_{w,2}^* = \sum m_i^* \times \mathrm{PDF}(\%)_i \times E_{Bio} \tag{2.26}$$

Here, $L_{w,2}^*$ is the emergy equivalent of impact of a given emission on urban natural resource and PDF (%) is the fraction potentially affected, measured as PDF $\times$ m^2 $\times$ year $\times$ kg^{-1}. A damage of 1 in E.I. 99 means all species disappear from 1 m^2 during 1 year or 10% of all species disappear from 10 m^2 during 1 year, and so on. E_{Bio} is the unit emergy stored in the biological resource (seJ $\times$ m^{-1} $\times$ year^{-1}), which is presented as the emergy of local wilderness, farming, forestry, animal husbandry, or fishery production.

As previously noted, additional emergy loss $L_{w,j}$ should be included to also account for pollution-induced damage to the city assets (e.g., facades of buildings and corrosion of monuments) according to Ulgiati et al. (1995). This is not, however, included in the present study due to lack of sufficient data.

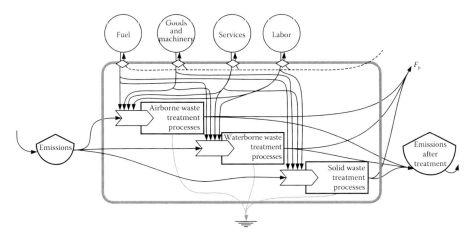

FIGURE 2.7 Aggregated diagram of emergy contribution from different sources to airborne, waterborne, and solid waste treatment system.

2.3.2.4.3 Quantifying Emergy Investment for Treatment

According to Ulgiati et al. (2007) and Cherubini et al. (2009), an additional emergy investment for safe abatement or disposal of waste materials is accounted for comparing advantages from decreased damage-related emergy losses. In this study, all the relevant input flows are contained within the total purchased emergy. Accordingly, in the case of waste treatments, all the emergy required (E_w) is not added to the urban total emergy consumption to avoid double counting. Also, the emergy derived from recycled and reused material (flow F_b) are not accounted into the exports.

The emergy of the city's wastes (W) in our analysis included industrial waste, MSW, sewage, and gaseous emissions that result from the combustion of fossil fuels and the incineration of MSW. To evaluate urban waste emergy, the emergy inputs in the form of labor, fuel, water, electricity, and capital (machines) must be accounted for, in addition to the emergy of all the wastes that represent the inputs and outputs in the treatment processes (Figure 2.7). Due to the uncertainty of available data, only reused materials in solid waste treatment processes (methane and compostable matter) are calculated.

Finally, damage associated to solid waste generation can be measured by land occupation for landfill and disposal. This may be converted to emergy through the emergy/area ratio (upper bound, average emergy density of economic activities) or even through the emergy intensity of soil formation (lower bound, average environmental intensity). Thus, the related emergy loss ($L_{w,3}$) can be obtained using the total occupied land area multiplied by the economic or environmental emergy intensity of such an area (choice depends on the area of the investigated system).

2.3.2.5 Corresponding Emergy-Based Performance Metrics

Based on emergy accounting and quantification of the emissions' impacts, several performance metrics can be evaluated (Brown and Ulgiati 1997; Odum 1996). These performance metrics can be listed as follows.

2.3.2.5.1 Emergy Yield Ratio

$$EYR = \frac{U}{\left(F + G + P_2 I + P_2 I_3\right)} \tag{2.27}$$

Here, U is the total emergy used ($U = R + N + F + G + P_2 I + P_2 I_3$), R is the locally renewable environmental resource, N is nonrenewable resource, F is imported fuel, G is imported good and mineral, $P_2 I$ is purchased service, and $P_2 I_3$ is emergy paid for imported labor.

EYR being the ratio of total emergy input to imported emergy, it indicates the efficacy of the system to make use of economic investment. By comparing EYR values, one can understand the reliance of a process on local resources or its dependence on imports. The higher the value of EYR, the higher its ability to exploit local renewable or nonrenewable resources. Of course, if renewable resources are exploited, the process is sustainable; if nonrenewables are exploited, an excess exploitation rate may make the process not sustainable.

When additional emergy input flows associated to natural capital or human capital losses are accounted for, the ratio becomes as in Equation 2.28, where emergy losses are considered as indirect input flows to be provided again for the replacement of the lost capital and the system to be sustainable.

$$EYR' = \frac{\left(U + E_w^* + L_{w,1}^* + L_{w,2}^* + L_{w,3}\right)}{\left(F + G + P_2 I + P_2 I_3 + E_w^* + L_{w,1}^* + L_{w,2}^* + L_{w,3}\right)} \tag{2.28}$$

2.3.2.5.2 Environmental Loading Ratio

The environmental loading ratio (ELR) is defined as in Equation 2.29. It is the ratio of the sum of local nonrenewable emergy and purchased emergy (including services) to the locally renewable emergy. ELR being the ratio of nonrenewable and imported resources to locally renewable, it indicates the intensity of the indirect environmental resource contribution to a metabolic system. A system with a higher ratio depends more heavily on indirect resources, compared to a fully natural system that only depends on locally renewables R. The higher the ratio, the greater the stress on the local environmental resource.

$$ELR = \frac{\left(N + G + F + P_2 I + P_2 I_3\right)}{R} \tag{2.29}$$

Equation 2.30 expresses a modified ELR accounting for the additional emergy input flows associated to natural capital or human capital losses.

$$ELR' = \frac{\left(N + G + F + P_2 I + P_2 I_3 + E_w^* + L_{w,1}^* + L_{w,2}^* + L_{w,3}\right)}{R} \tag{2.30}$$

2.3.2.5.3 Emergy-Based Sustainability Index

This index is calculated with Equations 2.27 through 2.30.

$$ESI = \frac{EYR}{ELR} \qquad (2.31)$$

$$ESI' = \frac{EYR'}{ELR'} \qquad (2.32)$$

This index is an aggregate measure of the economic benefit (*EYR*) per unit of environmental loading. Equation 2.21 applies when losses of natural and human-made capital are also included.

2.3.3 Calculation Process

Here, we choose Beijing as a case study to show how to calculate.

2.3.3.1 Determination of Pollutants

Our study will deal with the harmful emissions for the human health and ecosystem listed in Table 2.7. Air emissions discharge from both urban production and use include SO_2, dust, NO_x, and CH_4 (respiratory disorders) and CO_2, N_2O, and CH_4 (climate change). The data related to SO_2, dust, and NO_x were collected from governmental publications such as the Beijing Statistical Yearbook and the Chinese Environmental Statistical Yearbook (BSY 2000–2007; CESY 2000–2007). Data about CO_2, N_2O, and CH_4 are calculated as greenhouse gases released at local and

TABLE 2.7

Lists of Emissions and Environmental Impacts

		Source[a]	Damage Category of Human Health	DALY/kg of Emission	Damage Category Ecosystem Quality	PDF × m² × year
Airborne pollution	CO_2	p/c	Climate change	2.10E–07		
	NO_x	p/c	Respiratory disorders	8.87E–05	Acidification	5.71E+00
	SO_2	p/c	Respiratory disorders	5.46E–05	Acidification	1.04E+00
	Dust	p/c	Respiratory disorders	3.75E–04		
	N_2O	p/c	Climate change	6.90E–05		
	CH_4	p/c	Respiratory disorders	1.28E–08		
	CH_4	p/c	Climate change	4.40E–06		

(Continued)

TABLE 2.7 (Continued)
Lists of Emissions and Environmental Impacts

		Source[a]	Damage Category of Human Health	DALY/kg of Emission	Damage Category Ecosystem Quality	PDF × m² × year
Waterborne pollution	Mercury	p			Ecotoxic emissions	1.97E+02
	Cadmium	p	Carcinogenic effects	7.12E–02	Ecotoxic emissions	4.80E+02
	Hexavalent chromium	p	Carcinogenic effects	3.43E–01		
	Lead	p			Ecotoxic emissions	7.39E+00
	Arsenic	p	Carcinogenic effects	6.57E–02	Ecotoxic emissions	1.14E+01
	Volatile phenol	p	Carcinogenic effects	1.05E–05		
	Cyanide	p	Carcinogenic effects	4.16E–05		
	Oil	p	Carcinogenic effects	4.16E–05		
	Chemical Oxygen Demand (COD)	p/c	Eutrophication[b]	n.a.	Eutrophication[b]	n.a.
	NH₄-H	p/c	Eutrophication[b]	n.a.	Eutrophication[b]	n.a.

Notes:
[a] p means pollutions come from urban production, c means pollutions come from urban use process.
[b] The ecological losses caused by COD and NH₄-N were not considered due to the lack in the corresponding data.

global scales, based on direct and indirect energy consumption, that in turn are evaluated according to the Embodied energy analysis method (Herendeen 2004; Slesser 1974). The embodied energy of materials and energy flows is calculated by multiplying local inputs by appropriate oil equivalent factors.

2.3.3.2 Emergy Calculation Process

Tables 2.8a, b and c list the evaluated emergy values of the detailed flows, reflecting the general economic situation of Beijing. The input to the process is divided into five categories: free renewable environmental resources (R), exploited local nonrenewable resources (N), imported fuels and minerals (F), imported goods (G), and purchased services (P_2I). Correspondingly, the operation for all the above-mentioned processes will inevitably produce environmental impacts.

The indirect flow was calculated as given in Tables 2.9 through 2.12.

TABLE 2.8A
Emergy Flows Supporting Urban Metabolic System in 2006

Items	Units	Raw Amount	Transformity[a] (seJ/unit)	Ref. Trans.	Emergy (seJ/year)
Free Renewable Environmental Resources					
1. Sun[b]	J/year	7.02E+19	1	By definition	7.02E+19
2. Kinetic energy of wind[c]	J/year	4.87E+16	2.51E+03	Odum et al. (2000)	1.22E+20
3. Rainfall (geopotential energy)[d]	J/year	1.25E+15	1.74E+04	Odum et al. (2000)	2.19E+19
4. Rainfall (chemical potential)[e]	J/year	1.12E+16	3.05E+04	Odum et al. (2000)	3.43E+20
5. Geothermal heat[f]	J/year	1.79E+16	5.76E+04	Odum et al. (2000)	1.03E+21
Exploited Local Nonrenewable Resources					
6. Top soil loss[g]	J/year	3.17E+14	1.23E+05	Odum et al. (2000)	3.90E+19
7. Coal[h]	J/year	2.04E+17	6.69E+04	Odum et al. (2000)	1.37E+22
8. Minerals[i]					
Limestone	g/year	1.52E+13	1.68E+09	Brandt-Williams (2001)	2.55E+22
Sand and gravel	g/year	1.02E+13	1.68E+09	Brandt-Williams (2001)	1.70E+22
Iron ore	g/year	1.68E+13	1.44E+09	Odum et al. (2000)	2.41E+22

Notes:

[a] References for transformities (conversion from global emergy baseline of 9.44E+24 to 15.83E+24 seJ/year):

Calculations:

[b] Sun: average insolation = 5.36E+09 J/m²/year, total area of Beijing region = 1.64E+04 km², continental albedo = 0.201. Solar energy received = (total area of Beijing region) (average insolation) (1−albedo) = (1.64E+04 km² × 10⁶) (5.36E+09 J/m²/year) (1−0.201) = 7.02E+19 J/year.

(Continued)

TABLE 2.8A (Continued)
Emergy Flows Supporting Urban Metabolic System in 2006

c Kinetic energy of wind: air density = 1.3 kg/m³, wind velocity (annual average) = 2.5 m/s, observed winds are about 0.6 of geostrophic wind, drag coefficient = 1.00E−03, time frame = 365 × 24 × 60 × 60 = 3.15E+07 s/year. Wind energy = (air density) (drag coeff.) (geostrophic wind velocity)³ (total area) (time frame) = (1.3 kg/m³) (1.00E−03) (2.5 m/s/0.6)³ (1.64E+04 km² × 10⁶) (3.15E+07 s/year) = 4.87E+16 J/year.

d Rainfall (geopotential energy): total agricultural area of Beijing = 1.64E+10 m², rain (annual average) = 0.318 m/year, average elevation = 43.5 m, runoff rate = 56.40%. Energy = (total area) (rainfall) (% runoff) (avg elevation) (gravity) = (1.64E+04 km² × 10⁶) (0.318 m/year) (56.40%) (43.5 m) (9.8 kg/m²) = 1.25E+15 J/year.

e Rainfall (chemical potential energy): water density = 1.00E+06 g/m³, mass of rainfall water = (rainfall) (total area) (water density) = (0.318 m/year) (1.64E+04 km² × 10⁶) (1.00E+06 g/m³) = 5.22E+15 g/year, fraction of water that is evapotranspired = 44%, Gibbs free energy of water = 4.94 J/g. Energy = (evapotranspired water) (Gibbs free energy per gram water) = (5.22E+15 g/year) (44%) (4.94 J/g) = 1.12E+16 J/year.

f Geothermal heat: average heat flow per area = 3.50E−02 J/m²/s. Energy = (land area) (heat flow per area) = 1.79E+16 J/year.

g Net loss of organic matter in topsoil: soil erosion rate = 8.15E+02 g/m²/year, average% organic in soil = 0.02, assuming water content in organic matter = 0.7, energy content of dry organic matter = 5.00 kcal/g. Energy = (total agricultural area) (erosion rate) (% organic) (1 − water content in organic matter) (energy content of dry organic matter) (4186 J/kcal) = (1.64E+10 m²) (8.15E+02 g/m²/year) (0.02) (1−0.7) (5.00 kcal/g) (4186 J/kcal) = 3.17E+14 J/year.

h Fuels input from local region: coal = 6.42E+06 t/year, coal energy = (6.42E+06 t/year) (3.18E+10 J/t) = 2.04E+17 J/year; Oil = 0.00E+00 t/year (BSY, 2007), oil energy = (0.00E+00 t/year) (4.30E+10 J/t) = 0.00E+00 J/year; natural gas = 0.00E+00 m³ (BSY, 2007), natural gas energy = (0.00E+00 m³) (0.7174 kg/m³) = 0.00E+00 J/year.

i Constructed local input: cement quantity of production = 1.27E+07 t/year, assuming 1.2 t limestone and 1.6 t sand and gravel are needed to produce 1 t cement and 50% of sand and gravel is from local regain, limestone = 1.52E+13 g/year, sand and gravel = 1.02E+13 g/year, iron ore = 1.68E+07 t = 1.68E+13 g/year.

TABLE 2.8B

Emergy Imports for Urban Metabolic System in 2006

Items		Units	Raw Amount	Transformity (sej/unit)	Ref. Trans.	Emergy (sej/year)
9.	Hydroelectricity[a]	J/year	2.30E+14	3.36E+05	Odum et al. (2000)	7.74E+19
10.	Stream flow[b]	J/year	8.81E+15	3.05E+04	Brandt-Williams (2001)	2.69E+20
11.	Fuels import[c]					
	Coal	J/year	7.04E+17	6.69E+04	Odum et al. (2000)	4.83E+22
	Coke	J/year	4.72E+16	1.10E+05	Bastianoni et al. (2009)	5.18E+21
	Crude oil	J/year	3.45E+17	9.08E+04	Bastianoni et al. (2009)	3.13E+22
	Gasoline	J/year	9.20E+16	1.05E+05	Bastianoni et al. (2009)	9.64E+21
	Kerosene	J/year	1.23E+17	1.10E+05	Bastianoni et al. (2009)	1.36E+22
	Diesel oil	J/year	8.61E+16	1.10E+05	Bastianoni et al. (2009)	9.48E+21
	Fuel oil	J/year	4.42E+15	1.10E+05	Bastianoni et al. (2009)	4.87E+20
	Liquefied petroleum gas (LPG)	J/year	6.66E+15	1.11E+05	Bastianoni et al. (2009)	7.37E+20
	Natural gas	J/year	1.58E+17	9.85E+04	Romitelli (2000)	1.56E+22
12.	Electricity[d]	J/year	1.47E+17	1.74E+05	Odum et al. (2000)	2.57E+22
13.	Imported goods					
13.1.	Imported food, livestock, and products[e]					
	Grain	J/year	1.91E+16	1.14E+05	Yan and Odum (1998)	2.18E+21
	Rapeseed	J/year	8.23E+16	8.88E+04	Odum et al. (1987)	7.31E+21
	Vegetable	J/year	1.42E+14	7.37E+04	Odum et al. (1987)	1.05E+19
	Fruit	J/year	2.30E+13	8.88E+04	Ulgiati et al. (1994)	2.04E+18
	Meat	J/year	2.75E+09	5.31E+06	Yan and Odum (1998)	1.46E+16
	Milk	J/year	2.36E+11	3.35E+06	Yan and Odum (1998)	7.90E+17

(Continued)

TABLE 2.8B *(Continued)*
Emergy Imports for Urban Metabolic System in 2006

Items		Units	Raw Amount	Transformity (sej/unit)	Ref. Trans.	Emergy (sej/year)
13.2.	Imported raw and processed materials[f]					
	Wood	J/year	1.51E+15	5.36E+04	Odum et al. (2000)	8.11E+19
	Iron ores	g/year	4.68E+13	1.44E+09	Odum et al. (2000)	6.72E+22
	Sand and gravel	g/year	1.02E+13	1.68E+09	Brandt-Williams (2001)	1.70E+22
	Paper and paperboard	J/year	1.20E+15	7.37E+04	Lan and Odum (2004)	8.85E+19
	Silk	J/year	6.39E+11	1.12E+07	Odum et al. (2000)	7.18E+18
	Wool, animal hair	J/year	1.32E+14	7.37E+06	Odum et al. (2000)	9.70E+20
13.3.	Imported goods					
	Polythene (PE)	g/year	7.30E+10	4.69E+09	Brown and Ulgiati (2004)	3.43E+20
	Polypropylene (PP)	g/year	1.60E+10	4.69E+09	Brown and Ulgiati (2004)	7.51E+19
	Polystyrene (PS)	g/year	1.10E+10	4.69E+09	Brown and Ulgiati (2004)	5.16E+19
	Other coke chemicals	g/year	2.54E+10	4.89E+09	Brown and Ulgiati (2004)	1.24E+20
	Other petroleum products	g/year	1.16E+12	4.89E+09	Brown and Ulgiati (2004)	5.69E+21
	Iron and steel	g/year	2.70E+13	3.16E+09	Bargigli and Ulgiati (2003)	8.53E+22
	Aluminum and articles	g/year	1.20E+12	7.74E+08	Odum et al. (2000)	9.29E+20
13.4.	Other metals and articles	g/year	2.16E+11	4.74E+09	Odum et al. (2000)	1.02E+21
13.5.	Hi-tech products, machinery and electrical equipment					
	Steel	g/year	3.65E+09	3.16E+09	Bargigli and Ulgiati (2003)	1.15E+19
	Aluminum	g/year	1.65E+09	7.74E+08	Odum et al. (2000)	1.28E+18
	Copper	g/year	1.20E+09	3.36E+09	Brown and Ulgiati (2004)	4.05E+18

Other metals	g/year	4.20E+09	Odum et al. (2000)	1.89E+19
Ceramics/glasses	g/year	1.69E+10	Brown and Ulgiati (2004)	5.37E+19
Plastics	g/year	6.09E+09	Odum et al. (2000)	4.39E+19
13.6. Transport equipment				
Steel	g/year	1.88E+10	Bargigli and Ulgiati (2003)	5.94E+19
Aluminum	g/year	3.21E+09	Odum et al. (2000)	2.48E+18
Rubber and plastic material	g/year	2.29E+08	Odum et al. (2000)	1.65E+18
Copper	g/year	6.87E+08	Brown and Ulgiati (2004)	2.31E+18
13.7. Electronic goods (estimated from component materials)				
Ferrous metal	g/year	1.25E+09	Bargigli and Ulgiati (2003)	3.94E+18
Silica/glass	g/year	1.62E+09	Odum et al. (2000)	5.16E+18
Copper	g/year	4.36E+08	Brown and Ulgiati (2004)	1.47E+18
Plastics	g/year	1.43E+09	Odum et al. (2000)	1.03E+19
Aluminum	g/year	8.72E+08	Odum et al. (2000)	6.75E+17
Other metal	g/year	4.98E+08	Odum et al. (2000)	2.36E+18
14. Imported human labor (commuters)	$/year	7.30E+08	This study, country emergy/$ ratio	3.65E+21
15. Services associated to imports				
From other provinces	$/year	1.80E+10	This study, country emergy/$ ratio	9.02E+22
Import	$/year	1.05E+10	This study, world emergy/$ ratio	1.19E+22
Size of Specific Sectors				
16. Tourism	$/year	2.30E+10	This study, country emergy/$ ratio	1.15E+23

(Continued)

TABLE 2.8B *(Continued)*
Emergy Imports for Urban Metabolic System in 2006

Notes:

Calculations:

a Hydroelectricity: Hydroelectricity = 6.40E+07 kwh/year. Energy = (6.40E+07 kwh/year) (3.60E+06 J/kwh) = 2.30E+14 J/year.

b Stream flow: upstream inflow = 1.78E+09 m³/year, coefficient = 4.94E+06 J/m³. Energy = (upstream inflow) (coefficient) = (1.78E+09 m³/year) (4.94E+06 J/m³) = 8.81E+15 J/year.

c Fuel import: coal = 2.68E+07 t/year, coal energy = (2.68E+07 t/year) (3.18E+10 J/t) = 7.04E+17 J/year; coke = 1.66E+06 t/year, coke energy = (1.66E+06 t/year) (2.85E+10 J/t) = 4.72E+16 J/year; crude oil = 8.09E+06 t/year, oil energy = (8.09E+06 t/year) (4.30E+10 J/t) = 3.45E+17 J/year; gasoline = 1.97E+06 t/year, gasoline energy = (1.97E+06 t/year) (4.67E+10 J/t) = 9.20E+16 J/year; kerosene = 1.23E+17 t/year, kerosene energy = (1.23E+17 t/year) (4.30E+10 J/t) = 1.23E+17 J/year; diesel oil = 2.00E+06 t/year, diesel oil energy = (2.00E+06 t/year) (4.30E+10 J/t) = 8.61E+16 J/year; fuel oil = 1.04E+05 t/year, fuel oil energy = (1.04E+05 t/year) (4.26E+10 J/t) = 4.42E+15 J/t; LPG = 1.56E+05 t/year, LPG energy = (1.56E+05 t/year) (4.26E+10 J/t) = 6.66E+15 J/t; natural gas = 4.06E+09 m³, natural gas energy = (4.06E+09 m³) (3.89E+07 J/m³) = 1.58E+17 J/year.

d Electricity: electricity = 4.10E+10 kwh/year. Energy = (4.10E+10 kwh/year) (3.60E+06 J/kwh) = 1.47E+17 J/year.

e Imported food, livestock, and products: grain = 1.32E+06 t/year, grain energy = (1.32E+06 t/year × 1000) (1.45E+07 J/kg) = 1.91E+16 J/year; rapeseed = 3.29E+06 t/year, rapeseed energy = (3.29E+06 t/year × 1000) (2.50E+07 J/kg) = 8.23E+16 J/year; vegetable = 1.01E+04 t/year, vegetable energy = (1.01E+04 t/year × 1000) (1.41E+07 J/kg) = 1.42E+14 J/year; fruit = 1.00E+04 t/year, fruit energy = (1.00E+04 t/year × 1000) (2.30E+06 J/kg) = 2.30E+13 J/year; meat = 3.94E+02 t/year, meat energy = (3.94E+02 t/year × 1000) (6.99E+06 J/kg) = 2.75E+09 J/year; milk = 8.04E+04 t/year, milk energy = (8.04E+04 t/year × 1000) (2.93E+06 J/kg) = 2.36E+11 J/year.

f Imported raw and processed materials: wood = 1.89E+05 m³/year, wood energy = (1.89E+05 m³/year) (8.00E+09 J/m³) = 1.51E+15 J/year; paper and paperboard = 6.00E+07 t/year, paper and paperboard energy = (6.00E+07 t/year) (2.00E+07 J/t) = 1.20E+15 J/year; silk = 3.40E+07 kg/year, silk energy = (3.40E+07 kg/year/1000) (1.88E+07 J/t) = 6.39E+11 J/year; wool, animal hair = (7.00E+06 t/year) (1.88E+07 J/t) = 1.32E+14 J/year.

TABLE 2.8C

Additional Resources Input for the Waste Treatment Processes in 2006

Items		Units	Raw amount	Transformity (seJ/unit)	Ref. Trans.	Emergy (seJ/year)
Waterborne Waste Treatment Processes						
17.	Electricity[a]	J/year	1.48E+15	1.74E+05	Odum et al. (2000)	2.58E+20
18.	Chemical products[b]					
	Phosphorus removal reagent	kg/year	4.26E+06	4.44E+12	Grönlund et al. (2004)	1.89E+19
	Flocculating reagent	kg/year	3.53E+05	4.44E+12	Grönlund et al. (2004)	1.57E+18
	Hydrochloric acid	kg/year	8.24E+06	4.44E+12	Grönlund et al. (2004)	3.66E+19
	Sodium chlorate	kg/year	2.05E+06	4.44E+12	Grönlund et al. (2004)	9.10E+18
19.	Labor[c]	$/year	1.72E+07	7.47E+12	This study	1.28E+20
20.	Service embodied in fuels and goods	$/year	1.30E+08	7.47E+12	This study	9.74E+20
Airborne Waste Treatment Processes						
21.	Electricity[d]	J/year	1.06E+15	1.74E+05	Odum et al. (2000)	1.84E+20
22.	Chemical products[e]					
	Desulfurizer	kg/year	9.52E+07	4.69E+12	Brown and Ulgiati (2004)	4.47E+20
23.	Labor[f]	$/year	1.29E+07	7.47E+12	This study	9.62E+19
24.	Service embodied in fuels and goods	$/year	5.04E+07	7.47E+12	This study	3.77E+20
Solid Waste Treatment Processes						
25.	Electricity[g]	J/year	6.64E+12	1.74E+05	Odum et al. (2000)	1.16E+18
26.	Garbage truck[h]					
	Garbage truck (steel)	g/year	5.58E+10	3.16E+09	Bargigli and Ulgiati (2003)	1.76E+20

(Continued)

TABLE 2.8C (Continued)
Additional Resources Input for the Waste Treatment Processes in 2006

Items	Units	Raw amount	Transformity (sej/unit)	Ref. Trans.	Emergy (sej/year)
Solid Waste Treatment Processes					
Garbage truck (plastic and tires)	g/year	6.20E+09	7.21E+09	Odum et al. (2000)	4.47E+19
Diesel for truck	J/year	1.10E+13	1.10E+05	Odum et al. (2000) and Bastianoni et al. (2009)	1.21E+18
27. Auxiliary fuel for incineration[i]					
Coal	J/year	1.29E+14	6.69E+04	Odum (1996)	8.61E+18
Oil	J/year	8.35E+12	9.08E+04	Odum et al. (2000) and Bastianoni et al. (2009)	7.58E+17
28. Chemical products for incineration[j]					
Limestone	g/year	2.94E+09	1.68E+09	Brandt-Williams (2001)	4.93E+18
Carbonate	g/year	2.94E+08	1.68E+09	Brandt-Williams (2001)	4.93E+17
29. Labor[k]	$/year	4.87E+07	7.47E+12	This study	3.64E+20
30. Service embodied in fuels and goods	$/year	1.45E+07	7.47E+12	This study	1.08E+20
Recycle and Reuse Part of the Emissions					
31. Methane[l]	kg/year	1.30E+07	5.22E+04	Odum et al. (1996)	6.78E+11
32. Fertilizer[m]	kg/year	4.90E+10	2.68E+09	Odum et al. (2000)	1.31E+20

Notes:

Calculations:

[a] Electricity: Electricity consumption rate for waterborne waste treatment processes = 2.50E–01 kwh/t, treated water = 1.65E+09 t/year. Energy = (2.50E–01 kwh/year) (1.65E+09 t/year) (3.60E+06 J/kwh) = 1.48E+15 J/year).

[b] Chemical products: Rough estimate of the chemical products input is given based on simplistic assumption that all the plants adopt the craft of Orbal oxidation ditch. Thus, phosphorus removal reagent = (2.59E–09 kg/t) (1.65E+09 t/year) = 4.26E+06 kg/year (Zhang et al. 2010), flocculating reagent = (2.15E–10 kg/t) (1.65E+09 t/year) = 3.53E+05 kg/year (Zhang et al. 2010), hydrochloric acid = (5.01E–09) (1.65E+09 t/year) = 8.24E+06 kg/year (Zhang et al. 2010), sodium chlorate = (1.25E–09) (1.65E+09 t/year) = 2.05E+06 kg/year (Zhang et al. 2010).

c Labor: Labor in waterborne waste treatment = 4.50E+03 se, average wage = 2.98E+04 yuan/se/year. Labor = (4.50E+03 se) (2.98E+04 yuan/se/year)/(7.81 yuan/$) = 1.72E+07 $/year.

d Electricity: Coal-fired generators = 2.10E+10 kwh/year, electricity consumption per electricity generation for airborne waste treatment processes = 1.4%. Energy = (2.10E+10 kwh/year) (1.4%) (3.60E+06 J/kwh) = 6.64E+12 J/year).

e Chemical products: Desulfurizer consumption rate = 1.40 t/t SO_2, SO_2 remove = 6.80E+04 t/year. Desulfurizer = (1.40 t/t SO_2) (6.80E+04 t/year) × 1000 = 9.52E+07 kg/year.

f Labor in airborne waste treatment = 3.38E+03 se, labor = (3.38E+03 se) (2.98E+04 yuan/se/year)/(7.81 yuan/$) = 1.29E+07 $/year.

g Electricity: Electricity for compost = 9.42 kwh/t. Energy = (9.42 kwh/t) (1.96E+05 t/year) (3.60E+06 J/kwh) = 6.64E+12 J/year).

h Garbage truck: Garbage truck number = 6197, average mass of trucks = 1.00E+4 kg, fraction of trucks is steel = 90%, fraction of trucks is plastic material and tires = 10%, average lifetime = 10 year, average diesel for truck = 57,000 × 0.725 kg/year. Energy diesel = (57,000 × 0.725 kg/year) × 6197 × (43,000 J/kg) = 1.10E+13 J/year.

i Auxiliary fuel for incineration: Incineration = 9.80E+04 t/year, auxiliary coal per incineration = 5.00E–02 t/t, auxiliary oil per incineration = 2.00E–03 t/t.

j Chemical products for incineration: Limestone per incineration = 3.00E–02 t/t, carbonate per incineration = 3.00E–03 t/t.

k Labor: Labor in solid waste treatment = 3.80E+04 se, average wage of sorting worker = 1.00E+04 yuan/se/year, labor = (3.80E+04 se) (1.00E+04 yuan/se/year)/(7.81 yuan/$) = 4.87E+07 $/year.

l Methane: Landfills (garbage) = 4.68E+06 t/year (BESY 2007), methane yield per landfills = 90 m³/t, density = 0.77 kg/m³, collection rate = 4%. Methane = (4.68E+06 t/year) (90 m³/t) (0.77 kg/m³) (4%) = 1.30E+07 kg/year.

m Fertilizer: Fertilizer yield rate = 25%. Fertilizer = (4.68E+06 t/year) × 25% × 1000 = 4.90E+10 kg/year.

TABLE 2.9
Ecological Services Needed to Dilute Some Airborne and Waterborne Pollutants (seJ/year)

		Ref. Concentration	2006
$R_w{}^*$-c			
1	SO_2	2.00E−02 mg/m³	3.95E+19
2	Dust	8.00E−02 mg/m³	4.21E+18
3	NO_x	5.00E−02 mg/m³	2.70E+19
4	Heat released	Assumption	4.91E+18
5	COD	1.50E+01 mg/L	1.01E+21
6	NH_4-N	1.50E−01 mg/L	1.21E+22
$R_w{}^*$-c-air	Max (1:4)		3.95E+19
$R_w{}^*$-c-water	Max (5:6)		1.21E+22
$R_w{}^*$-p			
7	SO_2	2.00E−02 mg/m³	4.53E+19
8	Dust	8.00E−02 mg/m³	5.43E+18
9	NO_x	5.00E−02 mg/m³	1.81E+19
10	Heat released	Assumption	8.90E+18
11	Cadmium	1.00E−04 mg/L	3.01E+17
12	Chromium	1.00E−02 mg/L	1.37E+18
13	Lead	1.00E−02 mg/L	3.16E+17
14	Arsenic	1.00E−02 mg/L	0.00E+00
15	Volatile phenol	2.00E−03 mg/L	5.85E+19
16	Cyanide	1.00E+00 mg/L	1.51E+16
17	COD	1.50E+01 mg/L	9.30E+19
18	Oil	5.00E−02 mg/L	2.19E+20
19	NH_4-N	1.50E−01 mg/L	6.49E+20
$R_w{}^*$-p-air	Max (7:10)		4.53E+19
$R_w{}^*$-p-water	Max (11:19)		6.49E+20

Notes: $R_{w-i}{}^*$: emergy of environmental services needed to dilute *i* pollutant to an acceptable level; p means pollutions from urban production, and c means pollutions from urban use process.

TABLE 2.10
Additional Emergy Input for Waste Treatment and Loss Reduction (seJ/year)

	2006
E_w	3.24E+21
F_b	1.31E+20
Losses reduction	1.65E+22

TABLE 2.11

Emergy Losses Caused by the Solid Pollutants (seJ/year)

	2006
$L_{w,3\text{-p}}$	1.69E+19
$L_{w,3\text{-c}}$	4.51E+19

Notes: p means pollutions come from urban production, and c means pollutions come from urban use process.

TABLE 2.12

Summary of Flows of the City 1999–2006

Variable	Unit	Item	2006
R	seJ/year	Renewable sources	1.03E+21
N	seJ/year	Nonrenewable resources, $N = N_0 + N_1$	1.37E+22
N_0	seJ/year	Dispersed rural source	3.90E+19
N_1	seJ/year	Concentrated use	1.37E+22
G	seJ/year	Imported goods	1.89E+23
F	seJ/year	Imported fuels	1.60E+23
P_2I	seJ/year	Purchased services	2.02E+23
P_2I_2	seJ/year	Emergy for tourism	1.15E+23
P_2I_3	seJ/year	Emergy paid for imported labor	3.65E+21
$(P_2I + P_2I_3)_R$	seJ/year	Renewable fraction (10%)	2.06E+22
$(P_2I + P_2I_3)_N$	seJ/year	Nonrenewable fraction (90%)	1.85E+23
U	seJ/year	$U = R + N + G + F + P_2I + P_2I_3$	4.69E+23
POP		Population	1.58E+07
GDP	$/year	Gross domestic product	1.01E+11
$R_w^{\;*}$	seJ/year	Emergy of ecological services needed to dissipate the emissions	1.42E+22
$L_{w,1}^{\;*}$	seJ/year	Emergy of the human life losses caused by the emissions	2.58E+21
$L_{w,2}^{\;*}$	seJ/year	Emergy of the ecological losses due to the emissions	3.01E+21
$L_{w,3}$	seJ/year	Emergy of the land occupation caused by the emissions	4.98E+19
U_w	seJ/year	Emergy investment for waste treatment	3.24E+21
F_b	seJ/year	Feedback emergy	1.31E+20
EYR		$U/(G + F + P_2I + P_2I_3)$	1.03E+00

(Continued)

TABLE 2.12   (*Continued*)
Summary of Flows of the City 1999–2006

Variable	Unit	Item	2006
EYR'		$(U + R_w^* + I_{w,1}^* + I_{w,2}^* + I_{w,3} - F_b)/$ $(G + F + P_2I + P_2I_3 + R_w^* + I_{w,1}^* +$ $I_{w,2}^* + I_{w,3} - F_b)$	1.03E+00
ELR		$(N + G + F + (P_2I + P_2I_3)_{RN})/$ $(R + (P_2I + P_2I_3)_R)$	2.54E+01
ELR'		$(N + G + F + (P_2I + P_2I_3)_N + R_w^* +$ $I_{w,1}^* + I_{w,2}^* + I_{w,3} - F_b)/(R + (P_2I +$ $P_2I_3)_R)$	2.63E+01
ESI		EYR/ELR	4.06E–02
ESI'		EYR'/ELR'	3.92E–02

2.4 ECOLOGICAL NETWORK ANALYSIS OF URBAN SYSTEMS*

2.4.1 STRUCTURE AND MECHANISM OF THE URBAN ECOSYSTEM

2.4.1.1 Structure

The urban ecosystem is composed of natural and socioeconomic subsystems. The urban ecosystem, a complex system, is composed of natural, social, and economic components. It has a structure similar to that of natural ecosystems. However, the large impact caused by human activities means that it is inconsequential to copy simply the modalities of a natural ecosystem to describe the urban ecosystem. Instead, developing a structure that matches the urban ecosystem's unique characteristics is indispensable. Here, an adapted structure and mechanism is proposed in Figure 2.8.

The flows of substances, energy, and information in the urban ecosystem are governed by the three main categories of "actors" that shape these flows:

1. Producers create the substance and energy of the urban ecosystem. These include the outputs from farming, forestry, animal husbandry, and fishing in the hinterlands as well as mineral resources (industrial raw materials and fossil energy sources), solar and geothermal energy, and human knowledge and craftsmanship.
2. Consumers consist of the humans and enterprises of the urban ecosystem and both consume the resources produced by the producers.
3. Regenerators are also necessary in the urban ecosystem, where the biological degradability of wastes is low and most wastes must be artificially disposed of or decomposed. Thus, the urban ecosystem must include enterprises involved in waste recovery, disposal, and utilization. In addition, the ecological environment is a regenerator because it serves a vital role in the

---

* This section was contributed by Yan Zhang.

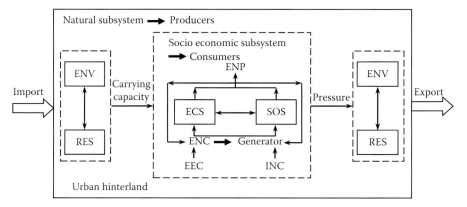

FIGURE 2.8 The structure and mechanisms of the urban ecosystem. ECS, economic subsystem; SOS, social subsystem; ENV, environment; RES, resources; ENP, environmental pollution; ENC, environmental construction; EEC, eco-environmental construction; INC, infrastructure construction.

metabolic processes of the urban ecosystem by providing service functions such as moderating the climate, purifying the atmosphere, protecting water quality, and preventing soil erosion.

2.4.1.2 Mechanism

The urban ecosystem possesses dissipating structures that can absorb substances and energy from the external environment and can export products and wastes to maintain order within the system. At the same time, the urban ecosystem can perform optimization, recycling, and regeneration functions by means of metabolic processes that resemble those of a biological organism (Boyden et al. 1981; Haughton and Hunter 1994; Jordan and Vaas 2000; Wolman 1965; Xiong et al. 2003).

The concept of urban metabolism has been widely accepted and adopted since it was proposed. However, research of urban metabolism's deep meaning and related models are still in the half black box style, which limits the application of the concept of urban metabolism. As there are no in-depth studies of internal metabolic process, in this part, we define the system from the point of bionics.

2.4.2 Urban Metabolic Process

According to the characteristics of material use, internal components were defined as well. We describe the urban metabolic process that consists of three stages and two main lines. The description places important emphasis on the internal metabolic mechanism of urban metabolic system. In this part, we try to break through the black box model to demonstrate the concept and create a new model of urban metabolism. Through the analysis mentioned earlier, we expect to construct a basis for an in-depth study of the application of urban metabolic system's internal structure and functions and to be a reference for the scientific management of cities.

In summary, urban metabolic system was regarded as a special scale of societal metabolism, and it has a complicated system boundary. Based on the analogy with biology, urban metabolism system can be defined as a metabolic system that uses its administrative boundaries as metabolic boundaries and includes the natural environment of the area. As the metabolic actor, societal economic system can be divided according to its material use characteristics to obtain system's metabolic components. These components can be generally divided into artery industry, venous industry, and domestic consumption and then further divided into agriculture, materials and energy transformation industry, mining, recycling, domestic sector, processing and manufacturing, construction industry, and so on. Urban metabolic process is described as the process of providing resources and energy to urban ecosystem, through transmission, transformation, and recycling within cities, finally outputting products and wastes. The process can be parsed into three phrases as anabolism, catabolism, and recycling metabolism, or it is parsed as two main lines—resource metabolism and waste metabolism, in accordance with its metabolic objects.

All these compartmentalizations and analyses improved the past black box model description on urban metabolic system, and they provided a theoretical support for further application research.

2.4.2.1   Definition of the System Boundary

2.4.2.1.1   Urban Metabolic System

In accordance with system's scale, the concept of metabolism in societal economic system can be divided into global societal metabolic system, national societal metabolic system, provincial societal metabolic system, urban metabolic system, industrial park metabolic system (or enterprise metabolic system), and so on. As the research object of urban metabolic system, the city has its own specificity (Figure 2.9).

The internal structure and functions of urban metabolic system are close to each other and they cannot be divided into independent and mutually equal spatial units. That is to say, on discussing internal structure and functions of metabolic system

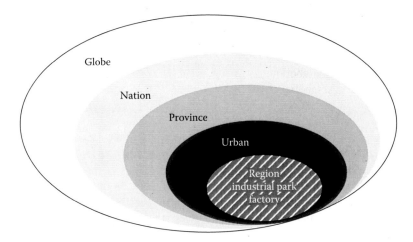

FIGURE 2.9   Multiscale metabolic system.

above urban scale, the system can be divided from the point of metabolic components or it can also be classified according to the city or urban agglomeration, while
for the metabolic system below urban scale, we could only divide it from the perspective of system's internal structure and do research on the structure and functions
between heterogeneous components.

In addition, the city and its surrounded area are heterogeneous, and there is a non-
artificial border between them, causing urban metabolic system much different from
other systems and the boundary's duality of the urban metabolic system.

2.4.2.1.2 Dualism of the System Boundary

On the one hand, cities can be defined as nonagricultural industries and nonagricultural
population gatherers that form the larger settlements in densely populated areas, which
are naturally formulated; on the other hand, we can also define cities as administrative
divisions under the scope of municipal area, and this is man-made. Duality of definition leads to duality of system boundary. There are narrow and broad divisions of both
the urban metabolic system and the system boundary. The urban metabolic system
can be defined as within the region or regions within the administrative boundaries.

From the research point, the former definition of system boundary is stricter, and
urban metabolic system under this definition is more typical and centralized. The
system under this division is large scaled, high intensified, and heavily dependent on
the external environment. The latter definition could only be an approximate system
boundary that is operational, meets statistical standards, and is easily accessed of
data. Meanwhile, most management and decisions are under administrative division,
and so the results of this kind of research have significant meaning on urban management and economy controlling.

Nonurban areas within the administrative boundaries of the cities are the main
difference between the two definitions of system boundary. The point is whether the
producing and consuming activities of rural areas within the city zoning should be
considered as the support of external environment to the urban metabolic system or
one of system's components. Therefore, generally, the higher the city's urbanization
level, the more the similarity between urban metabolic system's strict boundary and
administrative boundary.

2.4.2.1.3 A Bionic Definition of the Boundary

Concept of urban metabolism is arising from the analogy of city and organism.
Therefore, urban metabolic system is a complex ecosystem that includes the natural
environment within the city.

Specifically, the natural environment within the urban metabolic system and
areas without the system are analogous to the organism's internal environment and
external environment. The internal environment, which belongs to the organism, is
also the living space for the metabolic members such as cells and tissues. Metabolic
members of the organism exchange materials with its internal environment, which
forms the internal metabolism of the organism; the external metabolism is formed
when the organism exchanges material with its external environment. These two
kinds of metabolism coexist and both of them should be considered when doing
research on the organism's metabolic process.

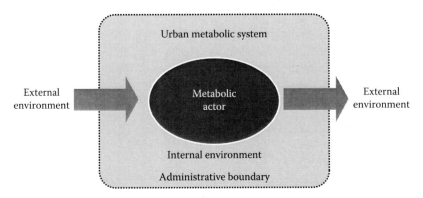

FIGURE 2.10   Boundary of urban metabolic system.

Similarly, as for urban metabolic system, there are societal economic systems, natural environment within cities, and areas outside the cities, which respectively represent metabolic members, internal environment, and external environment (Figure 2.10). The metabolic process is constituted by internal metabolism and external metabolism. Internal metabolism is the material and energy exchange between urban metabolic members and its internal environment, while the material and energy exchange between urban metabolic members and its external environment makes external metabolism. The internal and external metabolic processes both have significant meaning for studying the material and energy exchange between components and environment, and they impact the urban metabolic system differently. Therefore, it is more reasonable to define the urban metabolic system as a complex ecosystem, and it is possible to divide the internal and the external environments so that we can do targeted research on both internal and external metabolism of cities to get more extensive conclusions.

According to this definition, the external environment of urban metabolic system includes natural resources in and out of the administrative districts, ecological environment, and societal economic system. The urban societal economic system, which mainly consists of industry system and consuming system, is the main metabolic member of the system. In accordance with the societal metabolism theory, metabolic members refer to some basic units that could swallow and spit materials independently (Tao 2003). Therefore, the metabolic members could only include human beings in cities, industries, domestic animals, man-made infrastructures, and so on. Whereas the air, water, soil, minerals, and plants constitute the system's internal environment.

2.4.2.2   Components of the Urban Metabolic System

2.4.2.2.1   Compartmentalizing Principle

In this part, units of metabolic actors are classified according to some rules, and these classified units are called metabolic components. Units contained in one kind of component have some common characteristics in the metabolic process. The consuming process cannot be divided further, and so we pay more attention on the industry system to analyze metabolic components.

At present, there are many kinds of industry classification. The United Nations has adopted the 10-category classification, which is used in international standard industrial classification of all economic activities (1971); many countries such as China and Japan adopted the three industries classification, resource-intensive industry classification, and industry status classification. All these methods tend to focus on the economic nature of the industry and ignore the industry's metabolic characteristics. For instance, recycling and disposal of wastes, together with other manufacturing industries, all belong to the manufacturing industry categories, but their material use patterns are apparently different.

Some scholars divided the societal economic system without environment into seven sectors: extraction sector, agriculture sector, conversion sector, industry sector, tertiary industry, transportation sector, and domestic sector. The exergy of these sectors is accounted as well (Chen and Qi 2007). Under this kind of division, in the sector of transportation and tertiary industry, there are no other material use and transformation processes except energy consumption. They look more like the middle part of material delivery. The left part of tertiary industry is the final consumption of materials and energy and there is no producing process in this part, and so it is in the same category with domestic consumption. This kind of division cannot be applied to the research on urban metabolic system related to metabolic process. Some other scholars are also working on urban metabolic system and they divided the metabolic actors into industry, agriculture, and consumer (Zhang et al. 2009a). Although these three components have totally different characteristics, it is a too general dividing way to reflect system's internal structure and functions. These three components need to be divided further to obtain more specific results.

In this part, metabolic components' characteristics stand for their material use characteristics including material input and output; producing and consuming processes; similarity and differences in material use; as well as different transferring ways of products and wastes.

As shown in Figure 2.11, industry A mainly uses natural resources from both internal and external environments, while industry B uses not only natural resources but also secondary resources produced and processed by other industries. Industry C mainly uses secondary resources to do deep processing. Besides, there

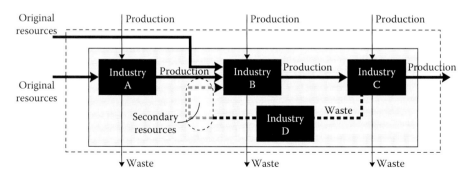

FIGURE 2.11 Several typical material use characteristics.

are also venous industries like industry D that mainly transfer wastes of other industries into secondary resources to supply other industries. Meanwhile, these above-mentioned industries may use products provided by the external environment, and some resulting wastes are discharged into the internal environment or the external environment.

We can see that there are significant differences in the characteristics of material use. Each typical material use pattern contains several industries and the industries can be classified effectively. Therefore, urban metabolic actors can be divided according to the different material use characteristics of metabolic components.

2.4.2.2.2 General Division

As seen from Figure 2.11, industry D is greatly different from other industries; its material use process is a reversed reducing process. Therefore, we can call industries A, B, and C as artery industries. Artery industry transfers resources into products to supply other industries or support domestic consumption. While industries like industry D are called venous industry, the transferred resources could be reused by venous industry.

Urban metabolic actors can be generally divided into three sectors: venous industry, artery industry, and residential consumption. Integrating system's internal and external environments, their relationship is shown in Figure 2.12.

On the one hand, although urban artery industry produces products, it still needs resources just like residential consumption, while venous industries could transfer

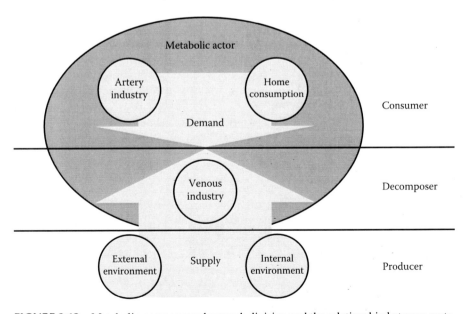

FIGURE 2.12 Metabolic components by rough division and the relationship between metabolic actor and internal and external environments.

wastes into resources, and so they are suppliers of resources, which is similar to internal and external environments. On the other hand, as urban metabolic system itself is a product analogous to natural ecosystem, there are similar eco-relationships in this system. External and internal environments are producers, artery industry and domestic sectors are consumers, and venous industry is the reducer. For those cities that output resources, its external environment may also be a consumer. The more closer the relationship between urban metabolic system and natural ecosystem, the more stable is the urban metabolic system.

2.4.2.2.3 Detailed Division

Urban artery industry stands for those industries that mainly use wastes, and so it can be called *recycling processing industry*, which mainly contains recycling and disposal of waste and sewage treatment, and so on. Urban artery industry can be divided into mining and quarrying and agriculture, both of which bring primary resources into the system; primary processing industries that mainly use primary resources; material and energy transformation industries that mainly use primary resources; manufacturing and processing industries that mainly use secondary resources; special industries that use plenty of renewable resources; and construction industries that transfer resources into storage. Table 2.13 shows typical sectors each component contains.

Mining and quarrying industry directly brings primary resources into urban metabolic system and links up system's metabolic actors and metabolic environment. For those nonresource cities, there is a very low ratio of mining and quarrying

TABLE 2.13
Fine Division and Typical Sectors of Artery Industry

Artery Industry	Features of Material Use	Types of Industry	Sectors
Bring primary resources into the system	Nonrenewable resources	Mining	Mining and washing of coal
			Extraction of petroleum and natural gas
			Mining and processing of metal ores mining and processing of nonmetal ores
			Production and distribution of water
			…
	Renewable resources	Agriculture and sideline	Farming
			Fishery
			Forestry
			Animal husbandry
			…

(Continued)

TABLE 2.13   (*Continued*)
Fine Division and Typical Sectors of Artery Industry

Artery Industry	Features of Material Use	Types of industry	Sectors
Mainly use primary resources	Produce products	Primary processing industry	Processing of food from agricultural products
			Processing of timber, manufacture of wood, bamboo, rattan, palm, and straw products
			Processing of petroleum, coking, processing of nuclear fuel
			Manufacture of raw chemical materials and chemical products
			Manufacture of cement
			Smelting and pressing of metals
			…
	Produce energy	Material and energy transformation industry	Production and distribution of electric power and heat power
Mainly use secondary resources	Mainly use primary products	Manufacturing and processing industry	Manufacture of foods
			Manufacture of textile wearing apparel, footwear, and caps
			Manufacture of furniture
			Printing, reproduction of recording media
			Manufacture of medicines
			Manufacture of plastics
			Manufacture of machinery
			…
	Use plenty of renewable resources	Special processing industry	Manufacture of paper and paper products
			…
	Transfer resources into storage	Construction	Construction of building and civil engineering
			…

industry in the total industries. Resources directly come into metabolic components belonging to processing and manufacturing industries by transportation.

For agricultural and sideline industries, it is part of the environment but not part of the metabolic actors in some researches. Therefore, agricultural products that are used for supplying industry and consumption are playing the role of environment to support the metabolic system, and this could greatly simplify the work of data processing and accounting. However, this method demotes those industries that produce and manufacture food and drink to the same level with mining and quarrying industry. This is not good for the system's overall architecture and hard to understand.

Manufacturing is a complicated system that could be divided into two categories from the point of material use. One is *primary processing industry* that mainly uses primary resources and the other is *processing and manufacturing industry*, which mainly uses secondary resources. A considerable part of output from primary processing industry is inputted to processing and manufacturing industry as raw material. In addition to products from primary processing industry, secondary resources also include wastes deoxidized by venous industry. Processing and manufacturing industry using secondary resources are called *special processing industry*, of which recycled paper producing industry is a typical example.

Production and distribution of electric power and heat power is quite similar to primary processing industry in material input characteristics, but they are totally different in material output characteristics. The common primary processing industry produces materials, while production and distribution of electric power and heat power produces energy. Therefore, it can be called material-energy transformation industry.

The construction industry consists of developing and constructing irrigation facilities and municipal infrastructure such as housing and roads construction. Similar to processing and manufacturing industry, construction inputs materials at the inputting end, while at the outputting end, construction does not output some materials as products to join further circulation or goes out of the system, and its wastes cannot be deoxidized by venous industry either. There is plenty of input but little output; most inputted materials become material storage the of system's components.

2.4.2.3 Description of the Metabolic Process

2.4.2.3.1 Common Aspect

Urban metabolism was firstly described as the process of inputting materials, energy, food, and so on, to urban ecosystem and outputting products and wastes from this system (Wolman 1965). This is a general description of metabolic process. Many scholars after Wolman expanded and deepened the concept from different angles, but they all stuck on its basic characteristics and focused on overall input of resources and output of wastes. Determined by this kind of description, its research content, research models, and research methods were all limited by black box analysis. In this part, *urban metabolism is defined as the process of providing resources and energy to urban ecosystem, then delivering, transforming, and recycling within cities, finally outputting products and wastes*, as shown in Figure 2.13. The description is based on general division of urban metabolic process. It can clearly reflect the most basic and important characteristics and differences of the metabolic processes.

As it is shown in Figure 2.13, the system inputs resources and some products from the environment and then outputs wastes and new products. More importantly, metabolic actors have been through the process of producing products, consuming

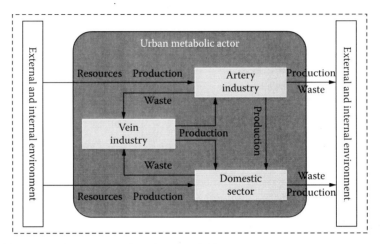

FIGURE 2.13 General process for urban metabolism.

products, and disposing wastes, as well as part of the wastes being reused to make products that can be reused or consumed directly. This is the point of urban metabolic process.

During the metabolic process, there are two patterns that metabolic components acted on metabolic objects: transmitting and transforming. In transmission, the substance is transferred between different metabolic components, but the material itself does not change, while when transformation takes place between components, the modality of the material would change. The transmitting process stands for the input and the output, both within and without the system, and also the process of material transmission between different components of metabolic actors; while transformation means processing in artery industry, domestic consumption, and the deoxidization in venous industry. All these processes aggregate and linkup to form the urban metabolic process.

2.4.2.3.2 The Three Phases

Based on the general description of metabolic process, we have done further analysis on metabolic process. Materials have been through several transferring processes from being inputted to outputted metabolic actors, which could be classified into three phases: anabolism, catabolism, and regulating metabolism, as shown in Figure 2.14. Dotted lines in material transformation indicate material exchange between different components.

As shown in Figure 2.14, anabolism and catabolism, both of which follow the overall material input and output direction, can complete materials' transmission and transformation of the metabolic system, while the regulating metabolism process constitutes a reverse loop to regulate and control the input and the output through recycling materials within the system. The small square in the lower right corner of the figure indicates different functions that venous industry has with other metabolic components. It determines whether materials flow into catabolism, regulating metabolism or anabolism.

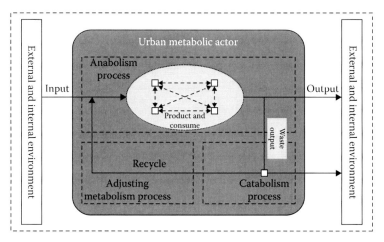

FIGURE 2.14   Urban metabolic process based on three phases.

2.4.2.3.3   The Two Venation

Some scholars analyzed urban metabolic process according to the main line of material flow metabolism; they proposed concepts of product metabolism and waste metabolism, and their main lines are respectively material flow and waste flow (Duan 2004). Generally speaking, *products* and *wastes* are defined from the angle of metabolic components' output; that is, they are named by the roles they play when they leave the last metabolic process. To coordinate with the material use characteristics of components' division, concepts of components input end were adopted in this part: *resources* and *wastes*. These two concepts are defined from the perspective of the roles they are playing when they go into one metabolic process. Resource metabolism and waste metabolism are separately expressed in red and blue lines, shown in Figure 2.15. The dotted lines indicate possible situations existing in a few urban metabolic systems.

From the results shown in Figure 2.15, there are some conclusions as follows:

1. Resource metabolism and waste metabolism are two coexisting processes; they both have relatively independent transmission and transformation chains, but they could still be unified as a whole in the urban metabolic system.
2. Resource metabolism and waste metabolism are linked up by different functions of metabolic components. On the whole, artery industry and domestic consumption have similar function on linking up resource metabolism and waste metabolism, while venous industry plays a more important role. Venous industry is able to change wastes into resources, which greatly influences the whole system's metabolic characteristics. And this is consistent with its function in the decomposing stage.
3. Studying from the perspective of the structure and functions within urban metabolic system, by parsing metabolic processes into resource metabolism and waste metabolism, we can further clarify the metabolic characteristics

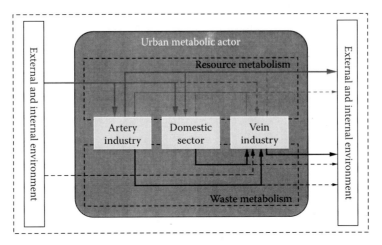

FIGURE 2.15   **(See color insert.)** Urban metabolic process based on two venation.

of different components. As seen in Figure 2.15, resource metabolism connects the utility and effective output of resources, and it tends to reflect resource utilizing efficiency, while waste metabolism connects wastes' generation, reuse, and final emission, which incline to reflect the environmental impact of components.

4. Besides, for resource metabolism, the metabolic object *resource* can be further divided, part of which is the primary resources from nature; the other part is the secondary resources that experience transformation artificially. These secondary resources include not only resources that experienced anabolic "secondary resources" but also renewable resources that experienced regulating metabolism and are reduced by venous industry, some materials even go through several regeneration. (In a sense, primary resources from nature can also be considered as secondary resources and renewable resources compounded or regenerated in nature.) The material type differentiation of components at the inputting end reflects differences of components' material use structure, and it also corresponds with the foundation of metabolic system's detailed division.

2.4.3   ECOLOGICAL NETWORK MODEL OF THE URBAN METABOLIC SYSTEM

2.4.3.1   Ecological Network Model of Urban Whole Metabolism

By analyzing the components of the urban metabolic system, it is possible to determine the direction of the eco-flows through the system and to define its metabolic pathways. The multiple roles played by the components of the system create complicated functions for these components, which mean that the direction of eco-flows among the components is not a chain but rather a network. For this reason, we have constructed a conceptual model of an ecological network for the urban metabolic system.

Ecological networks resemble biological networks and describe the structure of the flows of materials, energy, and currency between different components of the system. The basic units of ecological networks are "compartments" and "pathways." Compartments perform a specified function and thus serve as the functional units of the ecosystem, whereas pathways serve as transmission channels for materials, energy, and currency between compartments.

In this part, we have developed a five-compartment ecological network model for the urban metabolic system. In this model, compartment 1 represents the internal environment of the urban metabolic system, compartment 2 represents its external environment, compartment 3 represents the agricultural sector, compartment 4 represents the industrial sector, and compartment 5 represents the domestic sector (i.e., domestic life of the city's citizens). We have defined 19 metabolic pathways that reflect the exchange of eco-flows among these five compartments (Figure 2.16). Note that in this figure, f_{ij} represents the flow from compartment j to compartment i, and z_i represents the overall input flows through compartment i, for example, $z_1 = f_{12} + f_{13} + f_{14} + f_{15}$. Some case studies about urban whole metabolism are shown in Y. Zhang et al. (2009a, b, c, 2011b, 2012).

2.4.3.2 Ecological Network Model of Urban Energy Metabolism

2.4.3.2.1 Processes Involved in Urban Energy Metabolism

Using models to analyze an urban system is a direct and effective method (Zhang et al. 2006a, b). By abstracting and summarizing the traditional energy utilization systems, this approach can identify the key links in urban energy metabolic processes, permitting the development of a conceptual model of these processes (Figure 2.17).

The conceptual model of urban energy metabolic processes reflects the basic relationships involved in exploitation and utilization of urban energy. In this model, there are links among three key trophic levels that are analogous to those of a natural ecosystem: the energy exploitation sector functions as an energy producer; the energy transformation sector functions as a primary consumer of this energy; and terminal consumption sectors (here, industry and households) function as secondary consumers. In urban energy metabolic processes, the energy produced by the energy exploitation sector is the primary energy source and provides energy for both the transformation and the terminal consumption sectors; it can also produce outputs to areas outside the system. The energy transformation sector, which includes oil refining, power generation, and cogeneration, can utilize the primary energy produced locally or inputted (imported) from regions outside the system to produce the energy that will be used by secondary consumers, and part of this production is output (exported) to regions outside the system. The terminal consumption sectors, which include both industries that utilize energy and households within the city, utilize the primary and the secondary energies from internal and external sources. In addition, urban energy metabolism includes recovery processes related to the recovery of byproduct resources, including the recovery of energy from primary and secondary energy production processes and from industrial and household processes.

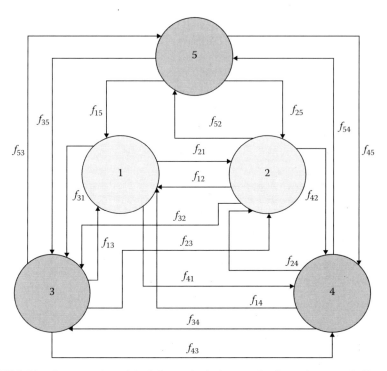

FIGURE 2.16 Conceptual model of the ecological network of an urban metabolic system. Compartment 1, the internal environment; compartment 2, the external environment; compartment 3, the agricultural sector; compartment 4, the industrial sector; compartment 5, the domestic sector; f_{21} represents the transboundary transfer of pollutants; f_{31} and f_{41} represent the resources provided by the internal environment for agriculture and industry respectively; f_{51} represents the service function the environment provides for domestic life, here it need not be considered; f_{12} represents the total resource inputs from the external environment; f_{13}, f_{14}, and f_{15} represent the pollutants discharged by agriculture, industry, and domestic life respectively; f_{32}, f_{42}, and f_{52} represent the resource inputs from the external environment for agriculture, industry, and domestic life respectively; f_{23}, f_{24}, and f_{25} represent the output of agricultural products, industrial products, and labor services respectively; f_{43} and f_{53} represent the agricultural raw materials consumed for industrial production and domestic consumption respectively; f_{34} and f_{35} represent the industrial products and labor services consumed by agricultural production respectively; f_{54} represents the industrial products consumed by domestic consumption; and f_{45} represents the labor services used for industrial production.

2.4.3.2.2 Ecological Network Model of the Urban Energy Metabolism

By analyzing urban energy metabolism processes in the conceptual model, an ecological network model of the system can be developed. The model's components represent the basic links among urban energy exploitation, transformation, consumption, and recovery. The model divides the urban metabolic system into 17 components (sectors): (1) energy exploitation; (2) coal-fired power; (3) heat supply; (4) washed coal; (5) coking; (6) oil refinery; (7) gas generation; (8) coal products; (9) agricultural; (10) industrial; (11) construction; (12) communication, storage, and postal service; (13) wholesale, retail, accommodation, and catering; (14) household; (15) other consumptions; (16) recovery; and

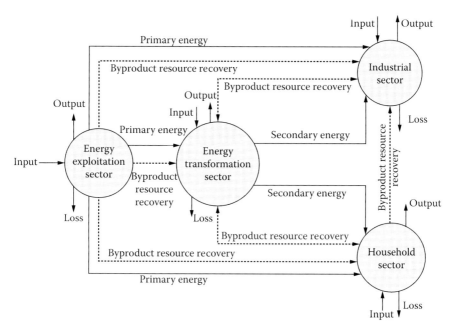

FIGURE 2.17 Conceptual model of urban energy metabolic system's processes.

(17) energy stocks. Based on the structure of natural ecological systems, we used the roles of the components of the urban metabolic system to divide these 17 components into producers, consumers, and decomposers (which recover energy lost from other components of the system). As in a natural ecological system, the decomposers in the urban metabolic system are also considered to be producers because of their role in recovering energy and returning it to the system. The consumers at all levels must obtain energy not only from producers but also from the decomposers and energy stocks. Thus, the recovery and energy stocks are both considered to be producers. The metabolic categories and their corresponding components are listed in Table 2.14.

Based on these compartments and their ecological trophic levels, the energy flows can be described by directional lines that connect the nodes in the network, resulting in an ecological network model for the urban energy metabolism system (Zhang et al. 2009a, b). In the model in Figure 2.18, we have defined 73 metabolic pathways that reflect the exchanges of flows among these 17 compartments. Note that in this figure, f_{ij} represents the flow from compartment j to compartment i, z_i represents the flow into compartment i from outside of the energy metabolic system, and y_i represents the boundary outflows (i.e., flows outside the system) from compartment i. Some case studies about urban energy metabolism are shown in Y. Zhang et al. (2010b, 2011a) and J. Y. Zhang et al. (2011).

2.4.3.3 Ecological Network Model of Urban Water Metabolism

2.4.3.3.1 Processes Involved in the Urban Water Metabolism

Using the trophic levels of natural ecosystems as a reference, we defined the compartments of the urban system as producers, consumers, and reducers and

TABLE 2.14

Metabolic Categories and Their Corresponding Components

Metabolic category	Components
Producers	Energy exploitation sector, recovery sector, and energy stocks sector
Primary consumers	Energy transformation sectors, including coal-fired power, heat supply, washed coal, coking, oil refinery, gas generation, and coal products
Secondary consumers	Industrial and household sectors, including agricultural sector, industrial sector, construction sector; communication, storage, and postal service sector; wholesale, retail, accommodation, and catering sector; living sector; and other consumer sector

determined the water flows among the system's components. Although urban systems are clearly not the same as natural systems, comparing them to natural ecosystems provides a simple metaphor that makes it easier to understand the meaning of the components and the flows among them. Based on this research, we developed a conceptual model of the processes in the urban water metabolism (Figure 2.19). In this model, the producers are the ecological environment and the artificial rainwater collection system; the consumers are the industrial, agricultural, and domestic sectors; and the reducer is the wastewater recycling system. Due to the complex chain of relationships among these components, each component may play different roles at different times; for example, although the ecological environment serves as a producer, it must also consume water resources to sustain its own operation and it must reduce the wastewater that it receives from the urban system. Similarly, the wastewater recycling subsystem (the reducer) both purifies urban wastewater and provides regenerated (recycled) water to support the operation of the urban system (i.e., acts as a producer). These changes in the roles of components result in a reticular system structure rather than a linear structure. Although metabolism is a purely biological concept, it can be applied by way of analogy to cities because the urban water metabolic system is also a mechanism for processing resources and producing wastes. In this sense, cities function as "urban superorganisms" (Park 1936) that exhibit metabolic processes. Using the trophic levels of natural ecosystems as a reference that makes the large flows of matter and energy less abstract, we defined the compartments of the urban water metabolic system as producers, consumers, and reducers and determined the water flows among the system's components.

In our model, there are clear links among the three key trophic levels: the local ecological environment, the terminal consumption sectors, and the wastewater recycling sector. The model follows all flows of water resources among these levels, but mainly reflects the utilization of freshwater, recycled water, and rainwater, as well as the reuse of water and the discharge of wastewater. The local ecological environment

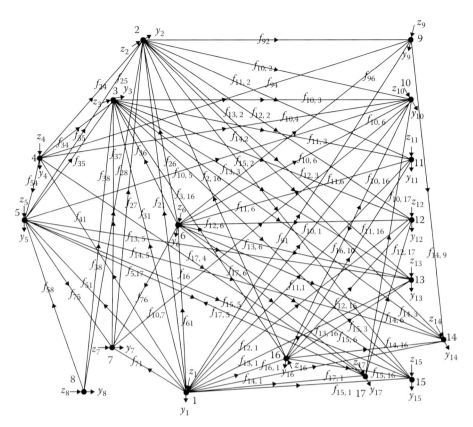

FIGURE 2.18 The ecological network model of the urban energy metabolism system used in the present study. In this network, f_{ab} represents the flows of energy from compartment b into compartment a. We used the following network compartments in this analysis: 1, energy exploitation; 2, coal-fired power; 3, heat supply; 4, washed coal; 5, coking; 6, oil refinery; 7, gas generation; 8, coal products; 9, agricultural; 10, industrial; 11, construction; 12, communication, storage, and postal service; 13, wholesale, retail, accommodation, and catering; 14, household; 15, other consuming; 16, recovery; and 17, energy stocks. The energy types are as follows: 1, raw coal; 2, cleaned coal; 3, other washed coal; 4, mould coal; 5, coke; 6, coke oven gas; 7, other gas; 8, crude oil; 9, gasoline; 10, kerosene; 11, diesel oil; 12, fuel oil; 13, liquid petroleum gas (LPG); 14, refinery gas; 15, natural gas; 16, other petroleum products; 17, other coking products; 18, heat; 19, electricity; and 20, other energy (in tonnes of coal equivalent). z_1 to z_{17} represent input energy from the environment; y_1 to y_{17} represent output energy into the environment.

provides freshwater for the industrial, agricultural, and domestic sectors, but sometimes must also receive water from the external environment. The external environment includes neighboring regions located upstream from the study area within the same basin and other regions that are transferring their water resources to the study area as a result of large-scale hydrological engineering projects. In the case study, we will subsequently discuss for Beijing; the external environment therefore includes upstream regions near the study area, such as the upper basin of the Hai River in

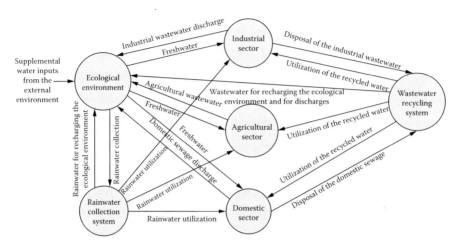

FIGURE 2.19   A conceptual model of the water flows in the urban water metabolism.

areas such as Hebei Province, Shanxi Province, and Inner Mongolia. Water is also being transferred to Beijing through projects such as the Gangnan, Huangbizhuang, Wangkuai, and Xidayang reservoirs in the Hebei Province. Other transfers include the project to export water from the Chetian Reservoir in Shanxi Province, the Yangtze River, and the Huanghe River under the South-to-North Water-transfer Project.

Under the currently limited water supply, the city must consider in depth how best to utilize freshwater and reuse wastewater. The industrial wastewater and domestic sewage are all discharged into the wastewater recycling system. Part of the treated wastewater can be recycled to recharge the ecological environment and for irrigation of municipal green space, washing of streets, agricultural irrigation, and industrial utilization. Rainwater can also be collected to recharge the ecological environment and provide supplemental water for the industrial, agricultural, and domestic sectors. In addition, the industrial sector reuses much of its water to solve the problem of high water consumption. During the utilization of water resources, there is discharge of wastewater produced by the industrial, agricultural, and domestic sectors.

2.4.3.3.2 Ecological Network Model of the Urban Water Metabolism

By analyzing the urban water metabolism processes in the conceptual model, we developed an ecological network model of the system. All water flows among compartments can be represented by directional lines that connect nodes in the network, resulting in an ecological network model for the urban water metabolic system that consists of a series of directional flows along metabolic pathways (Zhang et al. 2009a, b). In the model in Figure 2.20, we have defined 18 metabolic pathways that reflect the flows among these six compartments. Note that in Figure 2.20, f_{ij} represents the flow from compartment j to compartment i, and z_i represents the flow into compartment i from outside the water metabolic system. Some case studies about urban water metabolism are shown in Zhang et al. (2010a) and Li et al. (2009).

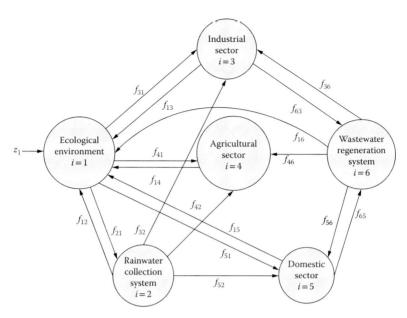

FIGURE 2.20 Ecological network model of the urban water metabolic system. Based on these compartment definitions, we included the following flows in the model: f_{12}, rainwater for recharging the ecological environment; f_{13}, wastewater discharged into the ecological environment by the industrial sector; f_{14}, wastewater discharged into the ecological environment by the agricultural sector; f_{15}, sewage discharged into the ecological environment by the domestic sector; f_{16}, recycled water used to recharge the ecological environment and discharged wastewater; f_{21}, rainwater collection from the ecological environment; f_{31}, freshwater utilized by the industrial sector; f_{32}, rainwater utilized by the industrial sector; f_{36}, recycled water utilized by the industrial sector; f_{41}, freshwater utilized by the agricultural sector; f_{42}, rainwater utilized by the agricultural sector; f_{46}, recycled water utilized by the agricultural sector; f_{51}, freshwater utilized by the domestic sector; f_{52}, rainwater utilized by the domestic sector; f_{56}, recycled water utilized by the domestic sector; f_{63}, wastewater disposal by the industrial sector; and f_{65}, sewage disposal by the domestic sector.

2.4.4 Structure and Relationship Analysis

2.4.4.1 Structure Analysis

As a tool for describing the trophic levels and the interactions among compartments, the ecological network approach allows quantitative analyses of the actors and relationships involved in the components of an ecological network, thereby revealing the integration and complexity of ecosystem behaviors.

Network throughflow analysis is similar to input–output analysis. In the present part, we chose network throughflow analysis to study the flows in the urban energy metabolic system. Nondimensional, input-oriented intercompartmental flows from compartment j to compartment i (g_{ij}) are defined as follows:

$$g_{ij} = \frac{f_{ij}}{T_j} \qquad (2.33)$$

where f_{ij} is the flow from compartment j to compartment i and T_j is the sum of the intercompartmental and boundary outflows from compartment j. In an urban energy metabolism system, the sum of flows into the ith compartment equals the sum of the flows out of the ith compartment, $T_{i(in)} = T_{i(out)}$. From the matrix $G = (g_{ij})$, the dimensionless integral flow matrix $N = (n_{ij})$ can be computed using the following convergent power series:

$$N = (g_{ij}) = G^0 + G^1 + G^2 + G^3 + \ldots + G^k + \ldots = (I - G)^{-1} \tag{2.34}$$

where I is the identity matrix, n_{ij} represents the integral dimensionless value of g_{ij}, which is calculated using a Leontief inverse matrix (Fath 2007), and the matrix N represents the integrated flows of actions between any of the 17 compartments in the network (i.e., the flow g_{ij}). The self-feedback matrix (G^0) reflects flows that originate in and return to a compartment, the matrix G^1 reflects the direct flows between any pair of compartments in the network, G^2 represents the flows that pass through two compartments, k represents the maximum number of steps in the system's pathways, and G^k $(k \geq 2)$ reflects the indirect flows of length k between compartments.

The diagonalized throughflow vector diag($\mathbf{T}$) can be redimensionalized by post-multiplying it by the dimensionless integral utility intensity matrix, such that the dimensional utility flow matrix $\mathbf{Y} = \text{diag}(\mathbf{T})\, \mathbf{U}$. By calculating the sum of each row of matrix $\mathbf{Y}$, the column vector of matrix $\mathbf{Y}$, $\mathbf{y}_i = (\mathbf{y}_{i1}, \mathbf{y}_{i2}, \ldots, \mathbf{y}_{i7})^T$, can be obtained. From matrix $\mathbf{Y}$, a weight can be computed using the following formula:

$$W_i = \frac{\displaystyle\sum_{j=1}^{7} \mathbf{y}_{ij}}{\displaystyle\sum_{i=1}^{7} \mathbf{y}_{ij} \sum_{j=1}^{7} \mathbf{y}_{ij}} \tag{2.35}$$

where $\displaystyle\sum_{j=1}^{7} \mathbf{y}_{ij}$ is the sum of row i of matrix $\mathbf{Y}$ and reflects the demand for inputs from other components. $\displaystyle\sum_{i=1}^{7} \mathbf{y}_{ij} \sum_{j=1}^{7} \mathbf{y}_{ij}$ is the sum of all the rows and the columns of the matrix $\mathbf{Y}$ and reflects the input for the whole system.

W_i reflects the contribution of component i to the system and it does not matter if component i contributes the same material or products to two or more components of the system. In effect, the weight refers to the proportion of the total integrated flows of inputs of the system accounted for by component i of the system. The weight determined by the flows can then be applied to the actual measured values. This lets us derive the contribution of each component and determine the support provided to every component by materials and energy in the societal metabolic system. The overall contribution can therefore fully reflect the position and functions of each component in the system and can be used to characterize its ecological trophic level.

2.4.4.2 Relationship Analysis

Network utility analysis is an ecological network approach that was first introduced by Patten (1991) to express the relative benefit to cost relationships in networks. In this method, a direct utility matrix is constructed and used to analyze the functions within the network (Fath 2007; Fath and Borrett 2006). From the urban network structure that we derived, we analyzed the mutual relationships between elements of the network. In the network utility analysis, d_{ij} represents the utility of an intercompartment flow from compartment j to compartment i and can be expressed as follows:

$$d_{ij} = \frac{(f_{ij} - f_{ji})}{T_i} \tag{2.36}$$

where f_{ij} represents the flow from compartment j to compartment i, f_{ji} represents the flow from compartment i to compartment j, and T_i is the sum of the intercompartmental and boundary inputs into compartment i. From the matrix D, which contains all d_{ij} values, a dimensionless integral utility intensity matrix $U = (u_{ij})$ can be computed from the following convergent power series:

$$U = (u_{ij}) = D^0 + D^1 + D^2 + D^3 + \ldots + D^k + \ldots = (I - D)^{-1} \tag{2.37}$$

where I is the identity matrix, u_{ij} represents the integral dimensionless value of d_{ij}, which is calculated using a Leontief inverse matrix, and the matrix U represents the flows of integrated relations between any pair of compartments in the network (i.e., the flow, d_{ij}). The identity matrix (D^0) reflects the self-feedback of flows through each compartment, the matrix D^1 reflects the direct flow utilities between any two compartments in the network, D^2 represents the indirect flow utilities that pass along two steps, and D^k ($k \geq 2$) reflects the indirect flow utilities along k steps.

The matrix U reflects the intensity and pattern of integrated relations between any pair of compartments in the network (i.e., the utility, u_{ij}). In network utility analysis, the sign of an element in matrix U can be used to determine the pattern of interaction between compartments in the network. In general, the signs in the main diagonal of sgn(U), which represents the sign matrix for matrix U, are positive, which means that each compartment is self-mutual and receives a self-promoting positive benefit from being part of the network (Patten 1991). If we designate the sign of the utility of any element in U as su, the subscripts 12 and 21 represent the flow of utility from compartment 2 to compartment 1 and from compartment 1 to compartment 2 (respectively), and this lets us consider the nature of the relationship between the two compartments. If $(su_{21}, su_{12}) = (+, -)$, compartment 2 exploits compartment 1. By analogy with a natural ecosystem, this means that compartment 2 benefits from the relationship (receives more utility than it transfers to compartment 1), but compartment 1 suffers (receives less utility than it transfers to compartment 2). If $(su_{21}, su_{12}) = (-, +)$, compartment 2 is exploited by compartment 1. By analogy, this means that compartment 1's ability to transfer utility controls the flow of utility from compartment 2. If $(su_{21}, su_{12}) = (-, -)$, then compartment 1 competes with compartment 2, leading to negative impacts for both compartments, whereas if $(su_{21}, su_{12}) = (+, +)$, the relationship between the two compartments represents mutualism, in which both compartments benefit from their interaction (Fath 2007); neither compartment benefits or suffers as a result of the relationship.

In this part, we established a mutualism index (M) for the urban energy metabolic system that reflects the proportions of positive and negative signs in the sign matrices. If we take the integral utility matrix as an example, the mutualism index of the urban energy metabolic system can be expressed as follows:

$$M = J(U) = \frac{S_+}{S_-} \qquad (2.38)$$

Here, $S_+ = \sum_{ij} max(sign(u_{ij}), 0)$ and $S_- = \sum_{ij}(-min(sign(u_{ij}), 0))$ (Fath 2007). If the matrices have more positive signs than negative signs, this means that the urban ecological system exhibits mostly positive relationships between compartments and thus represents network mutualism (Fath 2007). Conversely, if there are more negative signs than positive signs, the system exhibits mostly negative relationships between compartments and many problematic relationships must be solved or mitigated.

During the analysis of urban metabolic relationships based on the sign distribution and the ratio of the numbers of positive and negative signs in the network utility matrix, we can therefore identify three intercompartmental ecological relationships: competition, exploitation, and mutualism. Using the results of this analysis, we can identify potential directions for optimizing a city's energy metabolic system toward greater mutualism.

2.5 APPLICATION OF ECO-EXERGY AND CARBON CYCLING MODELS FOR THE ASSESSMENT OF SUSTAINABILITY*

2.5.1 INTRODUCTION

There is a clear need for indicators to assess the sustainability, to be able to give a quantitative answer to the question: Is the development of, for instance, a city sustainable? This chapter presents two indicators that could and should be used to answer this question quantitatively: eco-exergy (the work energy capacity) and a model of the complete carbon cycling in the city. To answer the raised question completely, it is probably in most cases necessary to use supplementary indicators, but the calculations of eco-exergy and the erection of a carbon cycling model are compulsory under all circumstances to be able to give a complete answer because

1. Energy can be divided into exergy or work energy and anergy, which cannot do work. It is therefore crucial to set up not only an energy balance but also an exergy balance because our focus is of course on the capacity to do work. We use energy in various forms because we want to perform work, for instance, by the use of vehicles for transportation and different

---

* This section was contributed by Sven Erik Jørgensen.

machinery for various tasks in the industries. It is therefore important to know how much of the energy can be used to do work and how much is lost as anergy. By calculations of the work capacity (exergy), it will be possible to know how much work energy is available, which is the core question. Therefore, exergy calculations will give us the efficiency of our energy use.

2. It is agreed that a sustainable development requires that the emission of greenhouse gases (mainly carbon dioxide and methane) is reduced to a minimum because the greenhouse gases accumulate in the atmosphere and thereby change our climate, which is a completely accepted theory today. The main source of greenhouse gases is our increasing use of fossil fuel, but there are also other sources of greenhouse gases (e.g., methane emission from wetlands or carbon dioxide emission from soil), and there are several important sinks of the greenhouse gases (first of the photosynthesis). These processes must of course be taken into account, when the sustainability of the development is evaluated.

These two important indicators are presented in the following to give the readers a clear understanding of the definition of the indicators, of their application in city planning, and how much the indicators tell us about the sustainability of various plans for a further development of a city.

2.5.2 Exergy and Eco-Exergy

Exergy is defined as the work the system can perform when brought into thermodynamic equilibrium with the environment, considering the difference in temperature (heat energy), the difference in pressure (pressure or expansion energy), the difference in altitude (mechanical potential energy), the difference in voltage (electrical energy), and the difference in chemical potential (chemical energy) (see Table 2.15 for the most applied energy forms). The work energy is found as a gradient in an intensive variable times the extensive variable, for instance, expansion work is equal to the difference in pressue times the volume, electrical work is the difference in

TABLE 2.15
Different Forms of Energy and Their Intensive and Extensive Variables

Energy Form	Extensive Variable	Intensive Variable
Heat	Entropy (J/K)	Temperature (K)
Expansion	Volume (m^3)	Pressure ($Pa = kg/s^2\, m$)
Chemical	Moles (M)	Chemical potential (J/mol)
Electrical	Charge (Ampere second)	Voltage (Volt)
Potential	Mass (kg)	(Gravity) (Height) (m^2/s^2)
Kinetic	Mass (kg)	0.5 (Velocity)^2 (m^2/s^2)

Note: Potential and kinetic energy is called as mechanical energy.

voltage times the charge, and potential energy is the difference in altitude times the mass and times the gravity constant.

This form of exergy is denoted as technological exergy. Technological exergy is not practical to use in the ecosystem context because it presumes that the environment is the reference state, which means the next ecosystem for an ecosystem. As the work energy embodied in the organic components and the biological structure and information contributes far most to the exergy content of an ecosystem, there seems furthermore no reason to assume a (minor) temperature and pressure difference between the ecosystem and the reference environment. Eco-exergy (application of exergy or work energy in ecological context) is defined (Figure 2.21) as the work the ecosystem can perform relatively to the same ecosystem at the same temperature and pressure but at thermodynamic equilibrium, where there are no gradients and all components are inorganic at the highest possible oxidation state. Under these circumstances, we can calculate the exergy, which has been denoted as eco-exergy to distinguish it from the technological exergy, as coming entirely from the chemical energy of the many biochemical compounds in the ecosystem. Eco-exergy has been successfully used to develop structurally dynamic models (see Jørgensen 2002; Jørgensen and Fath 2011) as a holistic ecological indicator (see Jørgensen 2006; Jørgensen et al. 2007) and as an important variable to describe ecosystem dynamics (Jørgensen 2012).

Eco-exergy represents the nonflow biochemical exergy. It is determined by the difference in chemical potential ($\mu^c - \mu^{co}$) between the ecosystem and the same

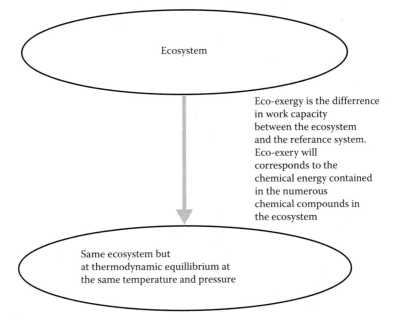

FIGURE 2.21 Definition of eco-exergy is shown. The work capacity in the ecosystem in form of the chemical energy of the many different and complex chemical compounds relative to the reference system is the eco-exergy. The reference system is the same ecosystem but at the thermodynamic equilibrium; that is, a homogeneous system without life. All the chemical compounds are inorganic and there are no gradients.

system at thermodynamic equilibrium. This difference is determined by the activities (approximated by the use of the concentrations) of the considered components in the system and in the reference state (thermodynamic equilibrium) as it is the case for calculations of all chemical processes. We can measure or determine the concentrations in the ecosystem, but the concentrations in the reference state (thermodynamic equilibrium) are more difficult to find; however it is possible to find good estimations, as will be shown later. Eco-exergy is a concept close to Gibb's free energy; eco-exergy has a different reference state from case to case (from ecosystem to ecosystem), and it can furthermore be used far from thermodynamic equilibrium, while Gibb's free energy in accordance to its exact thermodynamic definition is a state function close to thermodynamic equilibrium. In addition, eco-exergy of organisms is mainly embodied in the information content (see also the more detailed discussion in Jørgensen et al. 2010).

As $(\mu^c - \mu^{co})$ can be found from the definition of the chemical potential replacing activities by concentrations, we get the following expressions for the exergy:

$$E_x = RT\sum_{i=0}^{i=n} C_i \ln \frac{C_i}{C_{i,o}} \tag{2.39}$$

where R is the gas constant (8.317 J/K moles = 0.08207 L atm/K moles), T is the temperature of the environment, C_i is the concentration of the ith component expressed in a suitable unit, $C_{i,o}$ is the concentration of the ith component at thermodynamic equilibrium, and n is the number of components. $C_{i,o}$ is of course a very small concentration (except for $i = 0$, which is considered to cover the inorganic compounds), corresponding to a very low probability of forming complex organic compounds spontaneously in an inorganic soup at thermodynamic equilibrium. $C_{i,o}$ is even lower for the various organisms because the probability of forming the organisms is very low with their embodied information that implies that the genetic code should be correct.

By using this particular exergy based on the same system at thermodynamic equilibrium as reference, the eco-exergy becomes dependent only on the chemical potential of the numerous biochemical components.

It is possible to distinguish in Equation 2.39 between the contribution to the eco-exergy from the information and from the biomass. We define p_i as c_i/A, where

$$A = \sum_{i=1}^{n} c_i \tag{2.40}$$

is the total amount of matter density in the system. With the introduction of this new variable, we get:

$$E_x = A \cdot RT\sum_{i=1}^{n} p_i \ln \frac{p_i}{p_{i,o}} + A \ln \frac{A}{A_o} \tag{2.41}$$

As $A \approx A_o$, eco-exergy becomes a product of the total biomass A (multiplied by RT) and Kullback measure:

$$K = \sum_{i=1}^{n} p_i \ln \frac{p_i}{p_{i,o}} \tag{2.42}$$

where p_i and $p_{i,o}$ are the probability distributions, a posteriori and a priori to an observation of the molecular detail of the system. It means that K expresses the amount of information that is gained as a result of the observations. For different organisms that contribute to the eco-exergy of the ecosystem, the eco-exergy density becomes c RT ln $(p_i/p_{i,o})$, where c is the concentration of the considered organism. RT ln $(p_i/p_{i,o})$, denoted as β, is found by calculation of the probability to form the considered organism at thermodynamic equilibrium, which would require that organic matter is formed and that the proteins (enzymes) controlling the life processes in the considered organism have the right amino acid sequence. These calculations can be seen in the works by Jørgensen and Svirezhev (2004) and Jørgensen (2012). In the latter reference, the latest information about the β values for various organisms is presented; see also Table 2.16. For humans, the β value is 2173, when the eco-exergy is expressed in detritus equivalent or 18.7 times as much or 40635 kJ/g if the eco-exergy should be expressed as kiloJoules and the concentration unit g/unit of volume or area. The β value has not surprisingly increased as a result of the evolution. To mention a few β values from Table 2.16: bacteria 8.5, protozoa 39, flatworms 120, ants 167, crustaceans 232, Mollusca 310, fish 499, Reptilia 833, birds 980, and Mammalia 2127. The evolution has, in other words, resulted in a more effective transfer of what we could call the classical work capacity to the work capacity of the information. A β value of 2.0 means that the eco-exergy embodied in the organic matter and the information are equal. As the β values are much larger than 2.0 (except for virus, where the β value is 1.01—slightly more than 1.0), the information eco-exergy is the most significant part of the eco-exergy of organisms.

In accordance to Equations 2.39 and 2.40 and the above-presented interpretation of these equations, it is now possible to find the eco-exergy density for a model as follows:

$$\text{Eco-exergy density} = \sum_{i=1}^{i=n} \beta_i c_i \tag{2.43}$$

The eco-exergy due to the "fuel" value of organic matter (chemical energy) is about 18.7 kJ/g (compare with coal, about 30 kJ/g, and crude oil, 42 kJ/g). Notice that for these calculations, the same value is found for technological exergy and eco-exergy. The chemical energy of detritus, coal, and oil can be transferred to other energy forms, for instance, mechanical work directly, and be measured by bomb calorimetry, which requires destruction of the sample (maybe an organism), however. The information eco-exergy = (β − 1) × biomass or density of information eco-exergy = (β − 1) × concentration. The information eco-exergy is taken care of by

TABLE 2.16
β Values = Exergy Content Relatively to the Exergy of Detritus

Organisms	Plants		Animals
Detritus		1.00	
Viroids		1.0004	
Virus		1.01	
Minimal cell		5.0	
Bacteria		8.5	
Archaea		13.8	
Protists	Algae	20	
Yeast		17.8	
		33	Mesozoa, Placozoa
		39	Protozoa, *Amoebe*
		43	Phasmida (stick insects)
Fungi, molds		61	
		76	Nemertina
		91	Cnidaria (corals, sea anemones, jelly fish)
	Rhodophyta	92	
		97	Gastrotricha
Porifera, sponges		98	
		109	Brachiopoda
		120	Platyhelminthes (flatworms)
		133	Nematoda (round worms)
		133	Annelida (leeches)
		143	Gnathostomulida
	Mustard weed	143	
		165	Kinorhyncha
	Seedless vascular plants	158	
		163	Rotifera (wheel animals)
		164	Entoprocta
	Moss	174	
		167	Insecta (beetles, flies, bees, wasps, bugs, ants)
		191	Coleoidea (sea squirt)
		221	Lepidoptera (butterflies)
		232	Crustaceans
		246	Chordata
	Rice	275	
	Gymnosperms (incl. pinus)	314	
		310	Mollusca, bivalvia, gastropoda
		322	Mosquito
	Flowering plants	393	

(Continued)

TABLE 2.16   (*Continued*)

β Values = Exergy Content Relatively to the Exergy of Detritus

Organisms	Plants	Animals
	499	Fish
	688	Amphibia
	833	Reptilia
	980	Aves (birds)
	2127	Mammalia
	2138	Monkeys
	2145	Anthropoid apes
		Homo sapiens

Note: β values = eco-exergy content relatively to the eco-exergy of detritus (From Jørgensen, S. E. et al., *Ecol. Modell.*, 185, 165, 2005.).

the control and function of the many biochemical processes. The ability of the living system to do work is contingent upon its functioning as a living dissipative system. Without the information, the organic matter could only be used as a fuel similar to fossil fuel. But due to the information eco-exergy, organisms are able to make a network of the sophisticated biochemical processes that characterize life. The eco-exergy (of which the major part is embodied in the information) is a measure of the organization (Jørgensen and Svirezhev 2004). This is the intimate relationship between energy and organization that Schrödinger (1944) was struggling to find.

The eco-exergy is a result of the evolution and of what Elsasser (1981, 1987) calls recreativity to emphasize that the information is copied and copied again and again in a long chain of copies where only minor changes are introduced for each new copy. The energy required for the copying process is very small, but it has of course required a lot of energy to come to the "mother" copy through the evolution, for instance, from prokaryotes to human cells. To cite Margalef (1977) in this context, the evolution provides for cheap—unfortunately often "erroneous," that is, not exact—copies of messages or pieces of information. The information concerns the degree of uniqueness of entities that exhibit one characteristic complexion that may be described.

The application of eco-exergy is based on what could be considered a translation of Darwin's theory to thermodynamics. Biological systems have many possibilities for moving away from thermodynamic equilibrium, and it is important to know along which pathways among the possible ones a system will develop. This leads to the following hypothesis sometimes denoted as the ecological law of thermodynamics (ELT) (Jørgensen 2002, 2006, 2012; Jørgensen et al. 2007): If a system receives an input of exergy (free energy, for instance, from the solar radiation), then it will utilize this exergy to perform work. The work performed is first applied to maintain the system (far) away from the thermodynamic equilibrium whereby exergy is lost as anergy by transformation into heat at the temperature of the environment. If more exergy is available than needed for maintenance, then the system is moved further away from the thermodynamic equilibrium, reflected in growth of gradients. If there is more than one pathway to depart from the equilibrium, then the one yielding the highest eco-exergy storage (denoted as Ex) will tend to be selected. Or expressed differently, among the many ways for ecosystems to move away from the thermodynamic equilibrium, the one maximizing dEx/dt under the prevailing conditions will have the propensity to be selected.

This hypothesis is supported by several ecological observations and case studies (see Jørgensen 2002, 2012; Jørgensen et al. 2000, 2007). Survival implies maintenance of the biomass, and growth means increase of biomass and information. It costs exergy to construct biomass and gain information. Therefore biomass and information possess exergy. Survival and growth can therefore be measured by use of the thermodynamic concept eco-exergy, which may be understood as the work capacity the ecosystem possesses.

By development of an exergy balance, it is possible to use eco-exergy because technological exergy and eco-exergy are the same for fossil fuel, chemical energy of any form, and electrical exergy, except for expansion exergy and exergy of heat due to a temperature difference. For these two energy forms, the technological exergy should be applied because the eco-exergy does not consider differences in temperature and pressure as it is defined.

The work capacity or exergy balance is an important supplement to an *energy balance* because the work capacity balance considers only the energy that can do work (useful energy) and excludes the waste energy or anergy balance—heat energy released to the environment. The balance is used to indicate the activities that are not giving a satisfactory use of the energy to do work and how it would be possible to improve the work capacity balance by increasing the work energy efficiencies of these activities. As the work capacity is a measure for the sustainability (see Jørgensen 2006, 2010), the development of sustainability is determined too. To express it differently, the work capacity balance determines together with the energy balance not only the use of energy but also the efficiency of this use. Moreover, the work capacity considers also the changes of work energy in a city and in nature due to the management of nature and agriculture.

2.5.3 MODELING THE CARBON CYCLING

The carbon compounds causing the greenhouse effect are not only coming from the use of fossil fuel but are adsorbed and emitted from soil, landfills, incineration of waste, wetlands, agricultural land, and the entire nature. The carbon compounds include carbon dioxide and methane and other possible carbon-containing greenhouse compounds. Therefore, control of climate change requires that we consider the entire carbon cycle.

A carbon cycle model is set up for the area, for instance, a city. The model will describe the pools of carbon in the area and how these pools exchange carbon and how much they release or uptake carbon to or from the environment. The carbon pools included in the model—they are the state variables of the model—are the parks; the green areas and other natural vegetation (bushes, grass, etc.); other biomass pools; waste, food, and feed for pets and domestic animals; and the carbon of the soil (should probably be divided into, for instance, five or more pools covering different soil types in different parts of the city, including asphalt, concrete, and different types of soil). The carbon cycle is furthermore influenced by the exchange with the environment: import and consumption of fossil fuel, photosynthesis (nature, parks, and green areas), respiration of humans and animals, carbon dioxide emission by incineration (for instance, biomass used for heating), export and import of carbon (for instance, in vegetables), emission and uptake of carbon by nature, and emission and uptake from soil. The state variables are connected by transfer processes and are connected to the environment by the exchange processes mentioned earlier. Figure 2.22 shows a conceptual diagram of a carbon model erected for the Danish island Samsø. The carbon model that is going to be developed for a specific city will of course be different, and the carbon model will never be the same for two different cities. The model must be changed according to the available information. The state variables are indicated as boxes in the figure, the processes as thick arrows (both the

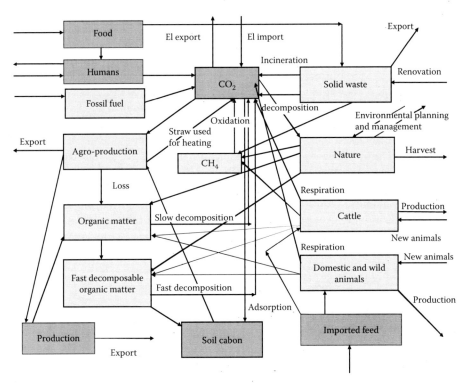

FIGURE 2.22 Conceptual diagram of the proposed carbon cycle model. The boxes indicate the state variables that describe the cycle and the arrows indicate the processes either between two state variables or the influence from the environment. When the model is used as a management tool, it can be used to answer questions such as, for instance, if we change a specific process (green field during the winter, incineration of waste instead of landfill, or more forest, to give three concrete examples), how will it change the carbon dioxide emission? The corresponding changes in the work capacity balance are also easily calculated. The soil carbon pool may be divided into two or more boxes according to the different soil characteristics.

transfer between boxes and the processes exchanging carbon with the environment), and the thin arrows indicate transfer of information within the model. It should be stressed that all these three components in the conceptual model will be different from city to city and of course different from the conceptual diagram showed, which is presented in this chapter to illustrate the idea behind the erection and application of a carbon model. The mathematical formulation of the processes and the corresponding coefficients (parameters) are known from various sources including general ecology. The model should always be calibrated and validated by the use of the collected data, which are considered as very important tests of the model. The model results can be translated to carbon-ecological footprints, expressed in grams of carbon dioxide equivalents per hectare per year. Methane has 23 times higher greenhouse effect than carbon dioxide but is decomposed by a biological half-life time of 7 years in the atmosphere. These properties of methane are taken into account by the translation of methane emission to carbon dioxide equivalents. The carbon cycle model is applied to overview all the carbon sources and sinks. It is thereby possible to assess the processes that should and could be changed to improve the carbon cycle and reduce the net carbon emission. The model can be used to answer questions such as, how much would it influence the net carbon emission if we erect some more 5-ha park, restore a wetland of 3 ha, and use as a recreational area or change the waste treatment system from use of landfills to incineration followed by use of the generated heat? In this context, the release and uptake of carbon by soil and vegetation are important processes. An increased accumulation of carbon in the soil is considered beneficial for the fertility of the soil, for instance, for parks and green areas.

2.5.4 CONCLUSIONS

It is recommended that a sustainability analysis of a city is developed to produce a manual to explain the background knowledge that has been applied to develop the analysis and how the available information has been applied to obtain the exergy balance and the carbon model. If the analysis is supplemented with other indicators that indeed can be recommended, for instance, by the use of biodiversity, the additional indicators must of course also be explained in the manual. Moreover, a manual explaining the application of the developed tools in full detail and also demonstrating how to use the tools in as many details as possible is very beneficial. The manual should of course also demonstrate how the work capacity calculations and the carbon cycle model and the calculated carbon-ecological footprints could be applied as powerful tools to manage development of a sustainability society and to reduce the carbon dioxide emission of the focal area.

It is often in an analysis of the sustainability of a city that it is possible to make a complete flow diagram containing (a) the sources to solid waste, (b) the transfer processes for the collected waste, and (c) the various treatments of the solid waste. It is furthermore recommendable to make a flow diagram for the use of water, covering (a) the sources of water, (b) where the water is used, (c) how the water is treated as wastewater, and (d) where the treated wastewater is discharged. These flow diagrams can be used to assess how it would be possible to reduce the amount of solid waste, to what extent it would be possible to recycle and reuse the waste, and where it may be feasible and beneficial to change the treatment of waste. Similarly, the water flow diagram can be used to see where it would be possible to reduce the water consumption and where the water has a quality that makes it possible to recycle and reuse the water for other uses than tap water. These possibilities of recycling, reuse, and changes of the waste and the water should be elucidated by the application of eco-exergy and the carbon cycle model to give the best basis for an environmentally sound choice.

REFERENCES

Ayers R., Kneese A. V. Production, consumption and externalities. *The American Economic Review*, 1969, 59: 282–297.

Bargigli S., Ulgiati S. Emergy and life-cycle assessment of steel production. Biennial Emergy Evaluation and Research Conference, 2nd, Gainesville, FL. Emergy Synthesis 2: Theory and Applications of the Emergy Methodology, 2003.

Barles S. Feeding the city: Food consumption and flow of nitrogen Paris 1801–1914. *Science of the Total Environment*, 2007, 375(1–3): 48–58.

Bastianoni S., Campbell D. E., Ridolfi R., Pulselli F. M. The solar transformity of petroleum fuels. *Ecological Modelling*, 2009, 220: 40–50.

Bi D. S., Guo X. P. An evaluation on the urban ecosystem health of Changjiang Delta. *Ecological Economy*, 2007, 2: 327–330 (in Chinese).

Boyden S., Millar S., Newcombe K., et al. *Cities as Ecosystems: Opportunities for Local Government*. Toronto: ICLEI, 1981.

Brandt-Williams, S. L. *Handbook of Emergy Evaluation: Folio #4, Center for Environmental Policy. Environmental Engineering Sciences*. Gainesville: University of Florida, 2001.

Brown M. T., Odum H. T. Emergy synthesis perspectives, sustainable development and public policy options for Papua New Guinea. In: *A Research Report to the Cousteau Society*. Gainsville, FL: Center for Wetlands, University of Florida, 1992, pp. 111–124.

Brown M. T., Ulgiati S. Emergy-based indices and ratios to evaluate sustainability: Monitoring economies and technology toward environmentally sound innovation. *Ecological Engineering*, 1997, 9(1–2): 51–69.

Brown M. T., Ulgiati S. Emergy analysis and environmental accounting. *Encyclopedia of Energy*, 2004, 2: 329–354.

Brown M. T., Ulgiati S. Emergy, transformity and ecosystem health. In: Jørgensen S. E., Costanza R., Xu F. L. (Eds.), *Handbook of Ecological Indicators for Assessment of Ecosystem Health*. Boca Raton, FL: CRC Press, 2005, pp. 333–352.

Cabezas H., Fath B. D. Towards a theory of sustainable systems. *Fluid Phase Equilibria*, 2002, 194–197: 3–14.

Calow P. Ecosystems not optimized. *Journal of Aquatic Ecosystem Health*, 1993, 2(1): 55.

Campbell D., Meisch M., Demoss T., Pomponio J., Bradley M. P. Keeping the books for environmental systems: An emergy analysis of West Virginia. *Environmental Monitoring and Assessment*, 2004, 94: 217–230.

Carey D. I. Development based on carrying capacity. *Global Environmental Change*, 1993, 3(2): 140–148.

Chen G. Q., Qi Z. H. Systems account of societal exergy utilization: China 2003. *Ecological Modelling*, 2007, 208(2–4): 102–118.

Cherubini F., Bargigli S., Ulgiati S. Life cycle assessment (LCA) of waste management strategies: Landfilling, sorting plant and incineration. *Energy*, 2009, 34(12): 2116–2123.

Colin M. Indicators of urban ecosystems health. International Development Research Centre, Ottawa, 1997. http://archive.idrc.ca/ecohealth/indicators_e.html. Accessed 18 June 2012.

Costanza R., Cornwell L. The 4P approach to dealing with scientific uncertainty. *Environment*, 1992, 34: 12–20.

Costanza R., Mageau M., Norton B., Patten B. C. Predictors of ecosystem health. In: Rapport D. J., Costanza R., Epstein P. R., Gaudet C., Levins R. (Eds.), *Ecosystem Health*. Malden and Oxford: Blackwell Science, 1998, pp. 240–250.

Daniels P. L., Moore S. Approaches for quantifying the metabolism of physical economies. Part I: Methodological overview. *Journal of Industrial Ecology*, 2002, 5(4): 69–93.

Duan N. Urban material metabolism and its control. *Research of Environmental Sciences*, 2004, 17(5): 75–77 Chinese.

Elsasser W. M. A form of logic suited for biology? In: Rosen R. (Ed.), *Progress in Theoretical Biology*, Vol. 6. New York: Academic Press, 1981, pp. 23–62.

Elsasser W. M. *Reflections on a Theory of Organisms. Holism in Biology*. Baltimore: John Hopkins University Press, 1987, 160 pp.

Fath B. D. Network mutualism: Positive community-level relations in ecosystems. *Ecological Modelling*, 2007, 208(1): 56–67.

Fath B. D., Borrett S. R. A MATLAB function for network environ analysis. *Environmental Modelling & Software*, 2006, 21(3): 375–405.

Feng Y. G., Wang H. D. The quantitative study on the sustainable development of regional population resources environment economy system. *China Environmental Science*, 1997, 17(5): 402–405.

Fischer-Kowalski M. Society's metabolism: The intellectual history of materials flow analysis. Part I. 1860–1970. *Journal of Industrial Ecology*, 1998, 2(1): 61–78.

Fischer-Kowalski M., Huttler W. Society's metabolism: The intellectual history of materials flow analysis. Part II. 1970–1998. *Journal of Industrial Ecology*, 1998, 2(4): 107–136.

Forkes J. Nitrogen balance for the urban food metabolism of Toronto, Canada. *Resources, Conservation & Recycling*, 2007, 52(1): 74–94.

Goedkoop M., Spriensma R. *The Eco-indicator 99: A Damage Oriented Method for Life Cycle Impact Assessment: Methodology Report*. Amersfoort, The Netherlands: Pre Consultants, 2000.

Grönlund E., Klang A., Falk S., Hanæus J. Sustainability of sewage treatment with microalgae in cold climate, evaluated with emergy and socio-ecological principles. *Ecological Engineering*, 2004, 22(3): 155–174.

Guan D. J., Su W. C. Study on evaluation method for urban ecosystem health and its application. *Acta Scientiae Circumstantiae*, 2006, 26(10): 1716–1722 (in Chinese).

Guo X. R. Urban ecosystem health assessment-case study on guangzhou city. Doctoral degree dissertation, Beijing Normal University, Beijing, 2003 (in Chinese).

Guo X. R., Yang J. R., Mao X. Q. Primary studies on urban ecosystem health assessment. *China Environmental Science*, 2002, 22(6): 525–529 (in Chinese).

Haberl H. The global socioeconomic energetic metabolism as a sustainability problem. *Energy*, 2006, 1(31): 87–99.

Hancock T. Urban ecosystem and human health: A paper prepared for the Seminar on CIID-IDRC and urban development in Latin America, Montevideo, Uruguay. April 6–7, 2000. http://www.idrc.ca/lacro/docs/conferencias/hancock.html.

Hancock T., Duhl L. J. Promoting health in the urban context. WHO Healthy Cities Papers No.1, 1988.

Harpham T. Urban health in the Gambia: A review. *Health & Place*, 1996, 2(1): 45–49.

Hau J. L., Bakshi B. R. Promise and problems of emergy analysis. *Ecological Modelling*, 2004, 178: 215–225.

Haughton G., Hunter C. *Sustainable Cities, Regional Policy and Development*. London: Jessica Kingsley, 1994, 357 pp.

Herendeen R. A. Energy analysis and emergy analysis—A comparison. *Ecological Modelling*, 2004, 178: 227–237.

Holling C. S. Resilience of terrestrial ecosystems: Local surprise and global change. In: Clark W. C., Munn R. E. (Eds.), *Sustainable Development of the Biosphere*. Cambridge, UK: Cambridge University Press, 1986.

Hu T. L., Yang Z. F., He M. C., Zhao Y. W. An urban ecosystem health assessment method and its application. *Acta Scientiae Circumstantiae*, 2005, 25(2): 269–274 (in Chinese).

Huang S. L. Urban ecosystems, energetic hierarchies and ecological economics of Taipei metropolis. *Journal of Environmental Management*, 1998, 52(1): 39–51.

Huang S. L., Odum H. T. Ecology and economy: Emergy synthesis and public policy in Taiwan. *Journal of Environmental Management*, 1991, 32: 313–333.

Jiang Y. L., Xu C. F., Yao Y., Zhao K. Q. Systems information of set pair analysis and its applications. *Proceedings of the Third International Conference on Machine Learning and Cybernetics*, Shanghai, China, 2004, pp. 1717–1722.

Jordan S. J., Vaas P. A. An index of ecosystem integrity for Northern Chesapeake Bay. *Environmental Science and Policy*, 2000, 3: 59–88.

Jørgensen S. E. *Integration of Ecosystem Theories: A Pattern*. Dordrecht: Kluwer, 2002, 386 pp.

Jørgensen S. E. *Eco-Exergy as Sustainability*. Southampton, UK: WIT Press, 2006, 220 pp.

Jørgensen S. E. Ecosystem services, sustainability and thermodynamic indicators. *Ecological Complexity*, 2010, 7: 311–313.

Jørgensen S. E. *Fundamentals of Systems Ecology*. Boca Raton, FL: CRC Press, 2012, 320 pp.

Jørgensen S. E., Fath B. *Fundamentals of Ecological Modelling*, 4th edition. Amsterdam: Elsevier, 2011, 400 pp.

Jørgensen S. E., Fath B., Bastiononi S., Marques J. C., Mueller F., Nielsen S. N., Patten B. C., Tiezzi E., Ulanowicz R. *A New Ecology*. Amsterdam, Oxford: Elsevier, 2007, 276 pp.

Jørgensen S. E., Patten B. C., Straškraba M. Ecosystems emerging: 4. Growth. *Ecological Modelling*, 2000, 126: 249–284.

Jørgensen S. E., Svirezhev Y. *Toward a Thermodynamic Theory for Ecological Systems*. Amsterdam: Elsevier, 2004, 366 pp.

Jørgensen S. E., Ladegaard N., Debeljak M., Marques J. C. Calculations of exergy for organisms. *Ecological Modelling*, 2005, 185: 165–176.

Jørgensen S. E., Ludovisi A., Nielsen S. N. The free energy and information embodied in the amino acid chains of organisms. *Ecological Modelling*, 2010, 221: 2388–2392.

Karr J. R., Fausch K. D., Angermeier P. L., Yant P. R., Schlosser I. J. *Assessing Biological Integrity in Running Waters: A Method and Its Rationale*. Champaign: Illinois Natural History Survey, 1986.

Kyushik O., Yeunwoo J., Dongkun L., Wangkey L., Jaeyong C. Determining development density using the Urban Carrying Capacity Assessment System. *Landscape and Urban Planning*, 2005, 73: 1–15.

Lan S. F., Odum H. T. Emergy evaluation of the environment and economy of Hong Kong. *Journal of Environmental Sciences*, 2004, 6(4): 432–439.

Lei K. P., Wang Z. S. Emergy synthesis and simulation for Macao. *Energy*, 2008, 33: 613–625.

Lenzen M., Dey C., Foran B. Energy requirements of Sydney households. *Ecological Economics*, 2004, 49(3): 375–399.

Liu G. Y., Yang Z. F., Chen B., Ulgiati S. Emergy-based urban health evaluation and development pattern analysis. *Ecological Modelling*, 2009, 220(18): 2291–2301.

Liu G. Y., Yang Z. F., Chen B., Zhang L. X., Zhang Y., Zhao Y. W., Jiang M. M. Emergy-based urban ecosystem health assessment: A case study of Baotou, China. *Communications in Nonlinear Science and Numerical Simulation*, 2009, 14(3): 972–981.

Li Y., Chen B., Yang Z. F. Ecological network analysis for water use systems—A case study of the Yellow River Basin. *Ecological Modelling*, 2009, 220(22): 3163–3173.

Lu Y., Zhu X. D., Li Y. F., Sun X. An improved method and its application for urban ecosystem health assessment. *Environmental Protection Science*, 2008, 34(5): 46–48, 59 (in Chinese).

Luo F. Q. The appraisal of urban ecosystem health: A case study of Nanjing city. Master's degree thesis. Hohai University, Nanjing, 2006 (in Chinese).

Mageau M. T., Costanza R., Ulanowicz R. E. The development and initial testing of a quantitative assessment of ecosystem health. *Ecosystem Health*, 1995, 1(4): 201–213.

Margalef R. *Ecologia*. Barcelona: Omega, 1977, 951 pp.

Meyer P. S., Ausubel J. H. Carrying capacity: A model with logistically varying limits. *Technological Forecasting and Social Change*, 1999, 61: 209–214.

Müller F., Wiggering H. Environmental indicators determined to depict ecosystem functionality. In: Pykh Y., Hyatt E., Lenz R. J. M. (Eds.), *Environmental Indices. Proceedings of the International Conference INDEX*. St. Petersburg, 1999, pp. 64–82.

Müller F., Lenz R. Ecological indicator: Theoretical fundamentals of consistent applications in environmental management. *Ecological Indicators*, 2006, 6: 1–5.

Murray C. J. L., Lopez A. D., Jamison D. T. The global burden of disease in 1990: Summary results, sensitivity analysis and future directions. *Bulletin of the World Health Organization*, 1994, 72(3): 495–509.

National Research Council. *Rangeland Health: New Methods to Classify, Inventory and Monitor Rangelands*. Washington, DC: National Academy Press, 1994, 180 pp.

Newman P. W. G., Birrel R., Holmes D. *Human settlements in state of the environment Australia. State of the Environment Advisory Council*. Melbourne: CSIRO Publishing, 1996.

Odum E. P. *Ecology and Our Endangered Life-Support Systems*. Sunderland, MA: Sinauer Associates, 1989.

Odum H. T. *Environmental Accounting: Emergy and Environmental Decision Making*. New York: John Wiley and Sons, 1996.

Odum H. T., Diamond C., Brown M. T. Emergy analysis and public policy in Texas, policy research project report. *Ecological Economics*, 1987, 12: 54–65.

Odum H. T., Brown M. T., Brandt-Williams S. B. (Eds.). *Handbook of Emergy Evaluation: A Compendium of Data for Emergy Computation in a Series of Folios, Folio. #1*. University of Florida, FL: Center for Environmental Policy, 2000.

O'Laughlin J. Forest ecosystem health assessment issues: Definition, measurement, and management implications. *Ecosystem Health*, 1996, 2(1): 19–39.

Park R. E. Human ecology. *American Journal of Sociology*, 1936, 42: 1–15.

Patten B. C. Network ecology: Indirect determination of the life–environment relationship in ecosystems. In: Higashi M., Burns T. P. (Eds.), *Theoretical Studies of Ecosystems: The Network Perspective*. Cambridge: Cambridge University Press, 1991.

Peng J., Wang Y. L., Wu J. S., Zhang Y. Q. Evaluation for regional ecosystem health: Methodology and research progress. *Acta Ecologica Sinica*, 2007, 27(11): 4877–4885.

Prato T. Modeling carrying capacity for national parks. *Ecological Economics*, 2001, 39: 321–331.

Rapport D. J. What constitute ecosystem health. *Perspectives in Biology and Medicine*, 1989, 33(2): 120–132.

Rapport D. J. What is clinical ecology? In: Costanza R., Norton B. G., Haskell B. D. (Eds.), *Ecosystem Health: New Goals for Environmental Management*. Washington DC: Island Press, 1992, pp. 144–156.

Rapport D. J. Ecosystems not optimized: A reply. *Journal of the Aquatic Ecosystem Health*, 1993, 2(1): 57.

Rapport D. J., Gaudet C., Karr J. R., Baron J. S., Bohlen C., Jackson W., Jones B., Naiman R. J., Norton B., Pollock M. M. Evaluating landscape health: Integrating societal goals and biophysical process Rapport. *Journal of Environmental Management*, 1998, 53: 1–15.

Rapport D. J., Böhm G., Buckingham D., Cairns J., Jr., Costanza R., Karr J. R., de Kruijf
H. A. M., Levins R., McMichael A. J., Nielsen N. O., Whitford W. G. Ecosystem health:
The concept, the ISEH, and the important tasks ahead. *Ecosystem Health*, 1999, 5: 82–90.

Romitelli M. S. Emergy analysis of the new Bolivia—Brazil gas pipeline. In: Brown,
M. T. (Ed.), *Emergy Synthesis: Theory and Applications of the Emergy Methodology.*
Gainesville, FL: Center for Environmental Policy, University of Florida, 2000,
pp. 53–69.

Rong S. H. Urban ecology system evaluation using attributive theory. *Journal of North
China Institute of Water Conservancy and Hydroelectric Power*, 2009, 30(3): 92–95
(in Chinese).

Sahely H. R., Dudding S., Kennedy C. A. Estimating the urban metabolism of Canadian cities:
Greater Toronto Area case study. *Canadian Journal of Civil Engineering*, 2003, 30(2):
468–483.

Sang Y. H., Chen X. G., Wu R. H., Peng X. C. Comprehensive assessment of urban ecosystem health. *Chinese Journal of Applied Ecology*, 2006, 17(7): 1280–1285 (in
Chinese).

Schaeffer D. J., Cox D. K. Establishing ecosystem threshold criteria. In: Costanza R.,
Norton B. G., Haskell B. D. (Eds.), *Ecosystem Health: New Goals for Environmental
Management*. Washington, DC: Island Press, 1992, pp. 157–169.

Schrödinger E. *What is Life? The Physical Aspect of the Living Cell*. Cambridge: Cambridge
University Press, 1944, 90 pp.

Shi F., Yan L. J. Application and research of unascertained measure model for urban ecosystem
health assessment. *Bulletin of Science & Technology*, 2007, 23(4): 603–608 (in Chinese).

Slesser M. *Energy Analysis Workshop on Methodology and Conventions*. Stockholm, Sweden:
IFIAS, 1974, p. 89.

Spiegel J. M., Bonet M., Yassi A, Molina E., Concepcion M., Mast P. Developing ecosystem
health indicators in centro Habana: A community-based approach. *Ecosystem Health*,
2001, 7(1): 15–26.

Su M. R., Fath B. D., Yang Z. F. Urban ecosystem health assessment: A review. *Science of the
Total Environment*, 2010, 408(12): 2425–2434.

Su M. R., Yang Z. F., Chen B., Zhao Y. W., Xu L. Y. The vitality index method for urban ecosystem assessment. *Acta Ecologica Sinica*, 2008, 28(10): 5141–5148.

Su M. R., Yang Z. F., Chen B. Set pair analysis for urban ecosystem health assessment.
Communications in Nonlinear Science and Numerical Simulation, 2009a, 14(4):
1773–1780.

Su M. R., Yang Z. F., Chen B., Ulgiati S. Urban ecosystem health assessment based on emergy
and set pair analysis—A comparative study of typical Chinese cities. *Ecological
Modelling*, 2009b, 220(18): 2341–2348.

Su M. R., Fath B. D., Yang Z. F. Urban ecosystem health assessment: A review. *Science of the
Total Environment*, 2010, 408(12): 2425–2434.

Takano T., Nakamura, K. Indicators for Healthy Cities. WHO CC HCUPR Monograph No. 2.
Toyko: World Health Organization Collaborating Center for Healthy Cities and Urban
Policy Research, 1998.

Tao X. Y. Synthetic assessment of ecosystem health in typical resource-exhausted cities in
China. 2008 International Workshop on Education Technology and Training & 2008
International Workshop on Geoscience and Remote Sensing, 2008, pp. 193–196.

Tao Z. P. *Eco-Rucksack and Eco-Footprint*. Beijing: Economic Science Press, 2003.

Tian G. E., Lu Y. H., Gu F., Liu N. N., Chen X. Q. Methodology of urban ecosystem health
assessment. *Journal of Chinese Urban Forestry*, 2009, 7(1): 57–60 (in Chinese).

Udo de Haes H. A., Lindeijer E. The conceptual structure of life cycle impact assessment, final
draft for the second working group on impact assessment of SETAC Europe (WIA-2),
Brussels, Belgium, 2001.

Ukidwe N. U., Bakshi B. R. Industrial and ecological cumulative exergy consumption of the United States via the 1997 input–output benchmark model. *Energy*, 2007, 32(9): 1560–1592.

Ulanowicz R. E. *Growth and Development, Ecosystem Phenomenology*. New York: Springer Verlag, 1986.

Ulgiati S., Odum H. T., Bastianoni S. Emergy use, environmental loading and sustainability. An emergy analysis of Italy. *Ecological Modelling*, 1994, 73: 215–268.

Ulgiati S., Brown M. T., Bastianoni S., Marchettini N. Emergy-based indices and ratios to evaluate the sustainable use of resources. *Ecological Engineering*, 1995, 5: 519–531.

Ulgiati S., Brown M. T. Quantifying the environmental support for dilution and abatement of process emissions: The case of electricity production. *Journal of Cleaner Production*, 2002, 10: 335–348.

Ulgiati S., Bargigli S., Raugei M. An emergy evaluation of complexity, information and technology, towards maximum power and zero emissions. *Journal of Cleaner Production*, 2007, 15(13–14): 1354–1372.

Wackernagel M., Rees W. *Our Ecological Footprint: Reducing Human Impact on the Earth*. Gabriola Island, BC: New Society, 1996.

Waltner-Toews D. *Ecosystem Sustainability and Health: A Practical Approach*. Cambridge: Cambridge University Press, 2004.

Wei T., Zhu X. D., Li Y. F. Ecosystem health assessment of Xiamen City: The catastrophe progression method. *Acta Ecologica Sinica*, 2008, 28(12): 6312–6320 (in Chinese).

Wei W., Zhang G. H. Artificial immune system and its applications in the control system. *Control Theory and Applications*, 2002, 19(2): 157–160, 166.

Wen X. M., Xiong Y. Assessment on urban ecosystem health based on attribute theory. *Systems Engineering*, 2008, 26(11): 42–46 (in Chinese).

WHO Regional Office for Western Pacific Region. *Regional Guidelines for Developing a Healthy Cities Project*. Manila: Western Pacific Region Office, 2000.

Wolman A. The metabolism of the cities. *Scientific American*, 1965, 213: 179–185.

Woodley S., Kay J., Francis G. *Ecological Integrity and the Management of Ecosystems*. Ottawa: St. Lucie Press, 1993.

Xiong D. G., Xian X. F., Jiang Y. D. Discussion on ecological footprint theory applied to regional sustainable development evaluation (in Chinese, with English summary). *Progress in Geography*, 2003, 22 (6): 618–626.

Xu L. Y., Yang Z. F., Li W. Review on urban ecosystem carrying capacity. *Urban Environment & Urban Ecology*, 2003, 16(6): 60–62 (in Chinese).

Yan W. T. Research on urban ecosystem health attribute synthetic assessment model and application. *System Engineering, Theory & Practice*, 2007, 27(8): 137–145 (in Chinese).

Zeng Y., Shen G. X., Guang S. F., Wang M. Assessment of urban ecosystem health in Shanghai. *Resources and Environment in the Yangtze Basin*, 2005, 14(2): 208–212 (in Chinese).

Zhang J. Y., Zhang Y., Yang Z. F. Ecological network analysis of an urban energy metabolic system. *Stochastic Environmental Research and Risk Assessment*, 2011, 25(5): 685–695.

Zhang L. X., Chen B., Yang Z. F., Chen G. Q., Jiang M. M., Liu G. Y. Comparison of typical mega cities in China using emergy synthesis. *Communications in Nonlinear Sciences and Numerical Simulation*, 2009, 14: 2827–2836.

Zhang X. H., Jiang W. J., Deng S. H., Peng K. Emergy evaluation of the sustainability of Chinese steel production during 1998–2004. *Journal of Cleaner Production*, 2009, 17(11): 1030–1038.

Zhang Y., Yang Z. F., Li W. Analyses of urban ecosystem based on information entropy. *Ecological Modelling*, 2006a, 197(1–2): 1–12.

Zhang Y., Yang Z. F., Yu X. Y. Measurement and evaluation of interactions in complex urban ecosystem. *Ecological Modelling*, 2006b, 196(1–2): 77–89.

Zhang Y., Yang Z. F., Yu X. Y. Ecological network and emergy analysis of urban metabolic systems: Model development, and a case study of four Chinese cities. *Ecological Modelling*, 2009a, 220(11): 1431–1442.

Zhang Y., Yang Z. F., Yu X. Y. Evaluation of urban metabolism based on emergy synthesis: A case study for Beijing (China). *Ecological Modelling*, 2009b, 220(13–14): 1690–1696.

Zhang Y., Zhao Y. W., Yang Z. F., Chen B., Chen G. Q. Measurement and evaluation of the metabolic capacity of an urban ecosystem. *Communications in Nonlinear Science and Numerical Simulation*, 2009c, 14(4): 1758–1765.

Zhang Y., Yang Z. F., Fath B. D. Ecological network analysis of an urban water metabolic system: Model development, and a case study for Beijing. *Science of the Total Environment*, 2010a, 408(20): 4702–4711.

Zhang Y., Yang Z. F., Fath B. D., Li S. S. Ecological network analysis of an urban energy metabolic system: Model development, and a case study of four Chinese cities. *Ecological Modelling*, 2010b, 221(16): 1865–1879.

Zhang Y., Li S. S., Fath Brian D., Yang Z. F., Yang N. J. Analysis of an urban energy metabolic system: Comparison of simple and complex model results. *Ecological Modelling*, 2011a, 22(1):14–19.

Zhang Y., Yang Z. F., Liu G. Y., Yu X. Y. Emergy analysis of the urban metabolism of Beijing. *Ecological Modelling*, 2011b, 222(14): 2377–2384.

Zhang Y., Liu H., Li Y. T., Yang Z. F., Li S. S., Yang N. J. Ecological network analysis of China's societal metabolism. *Journal of Environmental Management*, 2012, 93(1): 254–263.

Zhao K. Q. Disposal and description of uncertainties on set pair analysis. *Information & Control*, 1995, 24(3): 162–166 (in Chinese).

Zhao S., Li Z., Li W. A modified method of ecological footprint calculation and its application. *Ecological Modeling*, 2005, 185: 65–75.

Zhong Y. X., Peng W. Assessment of urban ecosystem health. *Jiangxi Science*, 2003, 21(3): 253–256 (in Chinese).

Zhou W. H., Wang R. S. An entropy weight approach on the fuzzy synthetic assessment of Beijing urban ecosystem health, China. *Acta Ecologica Sinica*, 2005, 25(12): 1344–1351 (in Chinese).

3 Planning of Ecological Spatial Systems

Guangjin Tian and Lixiao Zhang

CONTENTS

3.1 ECOLOGICALLY FUNCTIONAL ZONING

By definition, ecological function zoning refers to the process of dividing a certain study area into various ecological function zones according to the similarities and dissimilarities of regional eco-environmental characteristics, eco-environmental sensitivities, and the importance of the eco-services (Cai et al. 2010; Zhang 2009; Zhang et al. 2007). Good ecological function zoning can completely overcome the shortcoming of traditional spatial planning that ignores local carrying capability of resources and environment, making regional spatial planning based on its resource and environment carrying capability and more rational and scientific development. It is another significantly fundamental work associated with ecological environment protection after natural zoning, agricultural zoning, and ecological zoning. The goal of ecological functional zoning is to identify the important zone in terms of ecological security and protection, find the key ecological and environmental problems, and finally provide some effective suggestion for industrial distribution, ecological protection, and construction planning (Fu et al. 1999). It will supply a scientific basis for regional ecological protection, zone-based ecosystem management, and sustainable development. It will maintain regional ecological safety, utilize the resources, optimize the agricultural and industrial allocation, and preserve ecologically sensitive regions.

3.1.1 AIMS AND PRINCIPLES

The objectives of ecological functional zoning are to distinguish ecologically fragile zones, ecological environmental problems, and determine prior protection areas
according to ecological principles and methods (Yan 2007; Yan and Yu 2003).
Specifically, work including the following three aspects should be finished during
the process of ecological functional zoning:

1. A good understanding of local eco-environmental conditions including
   ecosystem types, ecosystem structure, ecological service function, the
   characteristics of their spatial distribution, and so on
2. An identification of key environmental problems of the study area and an
   understanding of their cause and spatial distribution
3. A determination of the functional orientation of each ecologically functional zone providing some advice for regulation

In addition, to accomplish the aims mentioned earlier as well as the requirement
of "the Twelfth Five-Year" planning outline of China, some principles need to be
observed in the process of ecological functional zoning. They are as follows (Fu
et al. 2001; Gao et al. 1998; Liu and Fu 1998):

- *Sustainable development principle:* Ecological functional zoning is
  supposed to promote more reasonable exploitation and utilization of
  resources, avoid blind development, enhance the supporting ability of the
  ecological environment to develop, increase the resilience of social and
  economic development, and finally achieve sustainable development.
- *Genetic principles*: Refers to the identification of the principal factors, ecological sensitivity, ecological service function, and their relationship with
  the structure, process, and pattern of the ecosystems.
- *Principle of regional relativity:* On a spatial scale, any kind of eco-service
  is related to the natural environment and socioeconomic factors of the local
  area, or even a much larger area. Therefore, ecological functional zoning should consider all kinds of physical, social, and economic factors. In
  other words, from the perspective of policy, it is necessary to fully consider
  national policy, upper-level planning, sectoral planning, and so on, and thus
  have a close connection with the master plan of relevant provinces and cities, land-use plans, national economy, and social development planning.
- *Similarity principle:* The formation and differences of ecosystems are usually based on the natural environment, which results in the similarity of the
  ecological environment in a certain area. However, due to the differences of
  natural factors and human behavior, some similarities and dissimilarities often
  happen to the structure, process, and eco-service of the ecosystem. Ecological
  functional zoning is planned on the basis of these similarities and dissimilarities. Hence, it is important to note that the similarities mentioned previously
  are often relative. For different range of areas, the standard of similarity is
  usually different.

- *Regional conjugacy:* Every part of the zoned area should be unique and maintain its spatial integrality. Owing to the rapid development of society and economy, the administrative division has an obvious impact on the naturally geographical division based on the geomorphology and vegetation. Therefore, both should be considered in the process of zoning from the perspective of eco-environmental protection and socioeconomic development. In line with the principle of practicability and manipulability, the ecological functional zoning studied in this chapter attempts to break through the limitations of traditional physical geography division, and, on the premise of keeping administrative divisions, integrates the ecosystem type and administrative division together.

3.1.2 LEVEL AND BASE OF ECOLOGICAL FUNCTIONAL ZONING

Urban systems are characterized by a complicated structure in terms of space. An urban system can be divided into many different levels, from macroscale to microscale. In consideration of the importance, similarities, and dissimilarities of eco-service as well as the characteristic of local eco-environment and socioeconomy, two levels are created with regard to ecological functional zoning.

Level 1 is obtained according to "the Eleventh Five-Year" planning outline of China, which divided the land into four categories, that is, exploitation-prohibited zone, exploitation-limited zone, exploitation-optimized zone, and key exploitation zone. According to the different functional orientations of different areas, different regional policy and performance assessment will be applied to norm the order of spatial development and finally form a rational structure of spatial development.

Exploitation-optimized zone refers to those areas where environmental carrying capability becomes weaker due to a higher degree of development. For these areas, efforts should focus on transforming the traditional developing model, which comes at the cost of large amounts of land occupation, resource consumption, and pollution emission. During the process of development, emphasis should be put on quality and benefit improvement and level advancement in the participation of the world's competent and labor division. Finally, let these areas continue to be dominant in driving the national development of the socioeconomy and participating in economic globalization.

Key exploitation zone refers to those areas with a higher eco-environment carrying capability and under better conditions in terms of economy and population. For these areas, efforts should focus on improvement in infrastructure, investment environment, industrial cluster development, economic scale, and acceleration of industrialization and urbanization. Moreover, they are expected to undertake the industry transfer from the exploitation-optimized zone and population shift from the exploitation-limited zone and exploitation-prohibited zone, finally becoming the important carrier to support national economic development and population aggregation.

Exploitation-limited zone refers to those areas with a weaker eco-environment carrying capability and a high concentration of population and economy, which have a significant influence on ecological security in a large area or even nation-wide. For these areas, developing principles are identified as first taking eco-environmental protection and then developing it appropriately, focusing on special industry suitable for local conditions within the eco-environment carrying capability. At the same time, efforts should be put on strengthening environmental protection and ecological restoration, guiding overloading population migration step by step, and gradually making them the important national or regional ecologically functional zones.

Exploitation-prohibited zone refers to natural reserves by law. For these areas, exploration that is not in accordance with its ecologically functional orientation is forbidden and the disturbance of human behavior should be controlled.

The aforementioned four categories are included in level 1 of an ecologically functional zone, within which a detailed division in level 2 will also be done for specific regulation according to local ecosystem type, structure, and eco-service. The details on the dividing method will be introduced in Sections 3.1.3 and 3.1.4.

3.1.3 PROCEDURE

To achieve ecologically functional zoning, the specific procedure is as follows:

- First, a survey is conducted to understand the current situation of the local socioeconomy and eco-environment, such as the topography, geomorphology, land use, vegetation, and so on. It is fundamental work for the zoning process. Its aim is to identify the key ecological problems and get the relevant digital data used for sensitivity evaluation.
- Second, data processing and a detailed analysis are made, including sensitivity analysis of a single factor and the overall factors. Specifically, the first step is to identify the key ecological problems and set up a relevant evaluation index system. Then each indicator will be divided into four grades; that is, the most sensitive, more sensitive, sensitive, and less sensitive. Finally, what needs to be done is to import all these data into ArcGIS software and calculate them from the perspective of a single evaluation of each ecological problem and a comprehensive evaluation, respectively.
- Finally, ecological functional zoning according to the method mentioned in "the Eleventh Five-Year" planning outline of China was accomplished based on the result of sensitivity analysis, ecosystem carrying capability analysis, and eco-service assessment.

After obtaining the four major ecologically functional zones, a detailed division in level 2 will also be made in consideration of ecosystem type, eco-service, and ecosystem sensitivity. Based on the functional orientation of different areas, some regulation and advice associated with environmental protection will be proposed (Figure 3.1).

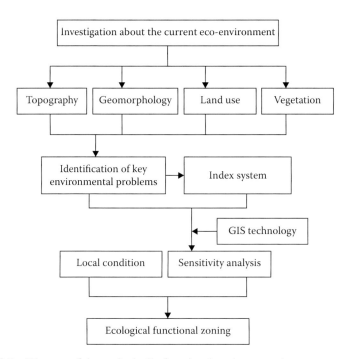

FIGURE 3.1 Diagram of the ecologically functional zoning procedure.

3.1.4 METHOD

The aim of ecologically functional zoning is to protect the environment diversely based on the spatial differences of eco-environmental characteristics, eco-environmental sensitivities, and the importance of the eco-services of the area. So, the first step of ecological function zoning is the assessment of eco-environmental sensitivity and the importance of eco-services spatially (Ouyang and Wang 2005; Yang et al. 2004; Yang and Xu 2007).

3.1.4.1 Eco-Environmental Sensitivity Analysis

Eco-environmental sensitivity refers to the changing probability and degrees of the eco-environment when affected by human activity or the adaptability of eco-environmental factors confronted with pressure or disturbances from outside. Ecological sensitivity analysis is to study the regional dynamic pattern of ecological problems based on the potential possibility impacted by human interactions. It will identify the major ecological and environmental problems and their formation mechanism. It will appraise ecological problems.

Eco-environmental sensitivity analysis is a comprehensive assessment in terms of many kinds of eco-environmental problems that can occur after a series of works, including finding their forming mechanism, analyzing their sensitivity and distribution characteristics, and determining which certain sensitive problems to evaluate. As a result, eco-environmental sensitivity depends on the responsiveness of regional natural conditions, ecosystem type, and ecosystem structure to particular eco-environmental problems. Therefore, the sensitivity analysis of a local ecosystem from the perspective of integrated evaluation will not only provide some references for ecological functional zoning, making it meet the requirements of environmental protection and ecological conservation, but also propose some effective measures on how to protect the environment according to ecological spatial differences.

Based on the above-mentioned procedure, the first step for ecological sensitivity analysis is to identify the key ecological problems severely influencing the local condition. Second, an index system capable of measuring these identified problems of the ecological environment is determined. Finally, the grade diagram of ecological sensitivity is determined by adding the value of the indicator by weight on a spatial scale (Zheng and Tang 2007).

1. *Identification of key environmental problems:* Identification of key environmental problems is the fundamental work of sensitivity analysis. Generally, environmental problems occurring in China include the following aspects: soil erosion, desertification, stone desertification, salinization of soil, biodiversity, cultural heritages, and so on. However, during the process of ecologically functional zoning, not all the relevant problems are considered but only the important ones. In addition, because of the regional differences of the environmental problems, particular identification should be conducted confronting the different areas. This work is done according to the current condition of these problems and the related influence factors.

2. *Index system:* The ability of a local ecosystem to respond to human behavior is usually influenced by the regional conditions of the natural environment. The better its natural condition, the stronger its adjustment ability, and correspondingly the weaker its ecological environmental sensitivity, which occurs if a region has a better natural condition, its ecosystem will be more complicated and its ability of self-control will be better. On the other hand, the worse the regional condition of the natural environment, the worse its ability to adapt to the change outside and the stronger its environmental sensitivity. However, an important problem to solve is how to evaluate the condition of the regional environment or sensitivity. For that, some indicators have to be chosen:

   • *Soil erosion:* Sensitivity evaluation of soil erosion aims to identify the area that is easy to change and analyze the sensitivity degree of the soil to human behavior. The influence factors mainly include surface runoff, slope, soil type, vegetation, and so on.

   • *Desertification:* Desertification is the process by which natural or human causes reduce the biological productivity of drylands (arid and semiarid lands). Decline in productivity may be the result of climate change, deforestation, overgrazing, poverty, political instability, unsustainable irrigation practices, or combinations of these factors. The concept does not refer to the physical expansion of existing deserts but rather to the various processes that threaten all dryland ecosystems, including deserts, as well as grasslands and scrublands. Its indicators include rainfall, average wind velocity, soil, and land-use type.

- *Stone desertification:* Stone desertification refers to a land degradation process with serious vegetation degradation and soil erosion, large-area exposure of the bedrock, and cumulative gravels, owing to human disturbance. It happens mostly in tropic and subtropic areas with humid climates, and Karst developed areas. Its influence factors include rainfall, slope, soil thickness, surface vegetation, and so on.
- *Salinization of soil:* Soil salinization is one of the main forms of land degradation in arid and semiarid regions. Soil salinity is a worldwide agricultural problem. It severely influences plant growth and development, and then crop productivity. Its influence factors include topography, tillage methods, and so on.
- *Biological diversity:* Biodiversity includes ecosystem diversity, species diversity, gene diversity, and landscape diversity. Biospecies diversity is the foundation of existence and development for human society. It is needed to protect the key national parks of China, national forest parks, and so on. These areas are regarded as the most sensitive zone in the sensitivity analysis.
- *Cultural heritages:* Similar to biological diversity, areas regarded as cultural heritages by law will be regarded as the most sensitive zones in the sensitivity analysis.

3. *Sensitivity analysis:* Geographic Information System (GIS) technique-based ecological sensitivity analysis includes the support of ArcGIS or Acrview software, from which an integrated sensitivity diagram of spatial distribution was obtained by building a model and adding the value of the influence factor by weight on the space. The whole process is mainly divided into three parts: (a) Data import and rasterization. The first step of this is to import all the relevant data into the ArcGIS system. Due to the requirement of raster data for calculation, the second step is to transform this original data into grid data using the software of ArcGIS. For instance, the slope data is obtained from the elevation data using the transformative function of this software. (b) Weight import and evaluation of sensitive influence factors. (c) Calculation and data export. Through the above-mentioned process, a sensitivity analysis diagram will be obtained and the difference of regional sensitivity is expressed instinctively.

3.1.4.2 Ecologically Functional Zoning

Ecologically functional zoning is supposed to start with a current situation with regard to spatial pattern structure and characteristics of the regional environment, and is based on the requirements of "the Eleventh Five-Year" planning outline as well as the local overall plan to form a kind of spatial and environmental protection structure characterized by distinctive features and reasonable functional pattern. Specifically, its accomplishment is according to the spatial differences of regional development levels and productivity distributions, in line with principles such as considering the local environment systematically as well as spatial differences specifically, the heterogeneity of the landscape, the within environmental carrying capability, and so on. On the other hand, rational ecologically functional zones should be based on an integrated consideration of the urban overall plan, administrative division, and land use plan.

With the help of current eco-environmental factors such as road, river, administrative division, and so on, the region can be divided into four ecologically functional zones; that is, zone in the red line or exploitation-prohibited zone, zone in the yellow line or exploitation-limited zone, zone in the green line or exploitation-optimized zone, and zone in the blue line or key exploitation zone.

Of these four parts, zones in the red line or exploitation-prohibited zones mainly include national nature reserves, world cultural and nature heritage sites, key national parks of China, national forest parks, and so on. These areas refer to the areas that lie in the most sensitive grade in the sensitivity analysis. They are significant for the whole condition of eco-environmental security, but vulnerable. So, they should be protected by priority, and activity associated with exploration and construction should be forbidden. Zones in the yellow line or exploitation-limited zone are usually the areas that lie in the more sensitive grade in the sensitivity analysis. These areas are relatively more sensitive to human activity and often have a great impact on the regional eco-environment of a larger range. Therefore, they should be paid much attention regarding eco-environmental construction and protection and development within the carrying capability of eco-environment. Zones in the green line or exploitation-optimized zones are the areas belonging to the sensitive grade in the sensitivity analysis. These areas are usually overdeveloped and optimized with regard to industrial level, population, industrial structure, and so on, and should be focused upon. Zones in the blue line or key exploitation zones are the areas belonging to the less sensitive grade in the sensitivity analysis. These areas are usually characterized by abundant resources or a better eco-environmental condition but are less developed. Therefore, these areas should be given priority to develop and also should accept industry or population transferred from other areas.

After the primary zoning of ecological functions, a detailed division in level 2 will be made for regulation according to ecosystem structure, function, and sensitivity of each zone.

3.2   LANDSCAPE ECOLOGICAL PLANNING

Since the 1970s, land use planning, ecological planning, and landscape ecology have developed dramatically. Their combination has resulted in landscape ecological planning. The theory of landscape ecology has been applied to regional planning and regional development. Regional land use and human ecological systems are designed by ecological theory. Remote sensing (RS), geographical information system (GIS), and computer-aided design have been effective tools to implement landscape ecological planning. At present, landscape ecological planning has been widely employed in land use planning and management, nature protection regions and national forest garden planning, urban landscape planning, rural landscape planning, biodiversity protection planning, and so on.

Landscape ecological planning is used to put forward the programs and suggestions for the optimization of landscape ecological planning, using the principles and relative theoretical methods of landscape ecology, on the basis of ecological landscape characteristics and their relationship with human activities as well as ecological analysis, synthesis and evaluation of the landscape, studying landscape patterns and ecological processes, and interactions between human activities and landscape, aiming to coordinate harmonious and sustainable development between landscape ecological functions and humanity. Landscape ecological planning focuses on landscape resources and environmental characteristics and emphasizes the role of human disturbance on the landscape and the fact that humanity is a part of the landscape (Fu et al. 2011; Xiao et al. 2003). Meanwhile, landscape ecological planning also stresses the design of spatial concepts and adoption of a variety of planning strategies (protective, defensive, offensive, or opportunistic) (Ning 2008).

Landscape ecological planning, landscape planning, and ecological planning are closely linked. There are both common ground and differences, but they focus on different things. Landscape planning focuses on small-scale spatial and architectural configuration of the monomer (planning or design). Ecological planning emphasizes the importance of medium- and large-scale analysis and evaluation of ecological factors. Landscape ecological planning regards the use and configuration of large- and medium-scale landscape units as the main objective based on concern about regional ecological characteristics. Generally speaking, landscape ecological planning emphasizes the interactions between spatial patterns and ecological processes.

3.2.1 Major Principles of Landscape Ecological Planning

- *Natural priority principles:* Protection of natural landscape resources and maintenance of natural landscape ecological processes and functions are the premise of biodiversity conservation and rational development and utilization of resources as well as the basis of landscape sustainability. When protecting natural landscape resources, we should pay attention to the protection of environmentally sensitive areas. Environmentally sensitive areas are the areas which are of particular value to humans or possess potential natural disasters. These vulnerable areas are frequently subject to negative environmental effects due to improper human development activities, which generally can be divided into ecologically sensitive areas, culturally sensitive areas, resource production sensitive areas, and natural disaster sensitive areas. Because of the relation between fragility and loss of irreversible change and stability, people should pay attention to protecting and planning environmentally sensitive areas in landscape ecological planning (Yu et al. 2007).
- *Sustainable principles:* Landscape ecological planning should be based on the current state and take the long term into account. Landscape sustainability can be viewed as the spatial expansion of coordination between human and landscape, and the coordination should be established to meet basic human needs and to maintain the ecological integrity of the landscape. Considering

the landscape as a whole, landscape ecological planning performs comprehensive analysis and multilevel design of the entire landscape to adapt the structure, pattern, and proportion of landscape use types in planning regions to the regional natural features and economic development, seeking harmonization and simultaneous development of ecological, social, and economic benefits so as to achieve overall optimized use of the landscape.

- *Targeted principles:* Landscapes in different regions have different structures, patterns, and ecological processes as well as planning purposes, which are the objective requirements for regional differentiation rules. Therefore, we should select different analytical indicators and establish different methods for evaluation planning according to the planning purposes specific to particular landscape planning.
- *Diversity principles:* Diversity refers to the variability and complexity of environmental resources in a characterized system. Landscape diversity means landscape unit diversity in structural and functional aspects. It reflects the complexity of the landscape, including patch diversity, type diversity, and landscape diversity. Diversity is both the criteria for landscape ecological planning and the result of landscape management.
- *Heterogeneous principles:* Heterogeneity is the variation in landscape components' type, composition, and properties, which is the most significant feature by which landscape differs from other life forms. Heterogeneity is not only the source of landscape stability but also a crucial way to improve landscape aesthetics. The maintenance and development of landscape spatial heterogeneity is an important principle for landscape ecological planning and design.
- *Integrated and wholly optimized principles:* Landscape ecological planning is comprehensive research work. Its integration includes two aspects: one is that landscape ecological planning is based on understanding the origins of the landscape, the existing form, and changes, which requires the collaboration of multidisciplinary professional teams (including landscape planners, land and water resources planners, landscape architects, ecologists, soil scientists, forest scientists, geographers, and other professionals) (Ning 2008); the other is to carry out intervention in the landscape purposefully according to internal landscape structures, landscape processes, socioeconomic conditions, and needs for human values. This requires considering socioeconomic conditions such as local economic development strategies and population issues and carrying out EIA after planning implementation on the basis of comprehensive and integrated analysis of natural landscape conditions. Only in this way can we implement landscape planning objectively and strengthen the scientificity and practicability of planning achievements.

3.2.2 Aims and Tasks of Landscape Ecological Planning

By analyzing and forecasting the landscape and its elements' structures as well as the current status, dynamic changes, and trends of spatial patterns, landscape ecological planning aims to determine the targets for management, maintenance, rehabilitation, and construction in landscape structures and spatial patterns and to draft the planning for landscape management and construction, which regards the maintenance and improvement of landscape multiple values as well as maintenance of landscape stability, continuity in ecological processes, and landscape security as its core. Meanwhile, landscape ecological planning aims to realize sustainable use of the landscape through the implementation of guided planning (Guo and Zhou 2007). Tasks for landscape ecological planning can be summarized as follows:

- Analyze landscape composition structures as spatial pattern status.
- Discover the main factors that constrain landscape stability, productivity, and sustainability.
- Determine the optimum composition of landscape structures.
- Determine landscape spatial structures and ideal landscape patterns.
- Find the technical measures for the adjustment, restoration, construction, and management of landscape structures and spatial patterns.
- Put forward proposals for funds, policies, and other external environmental assurance to achieve landscape management and construction targets.

3.2.3 Procedure and Method of Landscape Ecological Planning

To improve the work level of landscape ecological planning and guarantee quality, setting up a set of rules, procedures, and work system is important. The landscape ecological planning process should emphasize full analysis of natural environment characteristics, landscape ecological process, and its relationship with human activities; should pay attention to local landscape resources and social-economic potential and advantages; and should coordinate with the adjacent area, thus improving the ability of landscape sustainable development. Landscape ecological planning is a comprehensive system of methodology and its content almost always involves regional landscape ecological investigation, landscape ecological analysis, and synthesis and evaluation of all aspects (Fu et al. 2011). The general process of landscape ecological planning is as follows (Figure 3.2):

- *Identify the planning objectives and scope:* It must be clear about the area and the problem in a region before planning. Determine the overall goal of the planning and design and gradually decompose the overall goal; the relationship between the goals must be clear. Finally, according to the objectives of the planning, identify the character and type of the planning, for example, biological diversity protection of landscape ecological planning, landscape resources reasonable development planning, landscape restoration and construction planning, urban and suburban landscape pattern, or landscape structure adjustment planning (Guo and Zhou 2007).
- *Landscape ecological investigation:* The objective of landscape ecological investigation is to collect the material and data of the planning region, which aims to understand the landscape structure and the natural process of the planning region, ecological potential, and social status to gain an overall understanding of the regional landscape ecological system and to lay the

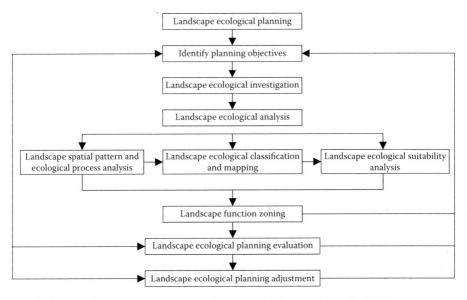

FIGURE 3.2 The general process of landscape ecological planning (Fu, 2011).

foundation for landscape ecological classification and ecological suitability analysis. Basic material usually can be divided into historical material, field investigation, social investigation, and RS and computer database information. Such information includes the names and evaluations of biological and nonbiological components, the landscape ecological process and associated ecological phenomena, landscape impact results and the degree to which it is affected by humans, and so on (Fu et al. 2011). These include (1) geology, hydrology, climate, biological, and other natural geographical factors, (2) the land structure, natural features and cultural characteristics, and topography factors, (3) social influence, political and legal constraints, and economic factors, such as cultural factors.

• *Landscape spatial pattern and ecological process analysis:* Through the combination or the introduction of new landscape elements, landscape ecological planning can adjust or build new landscape structures, which could increase landscape heterogeneity and stability. Landscape patterns and process analysis of landscape ecological planning have an important significance (Guo and Zhou 2007). Dynamic analysis of the landscape pattern is the foundation of studying the interrelation between landscape pattern and ecological process. Landscape pattern is formed by the complex processes of all kinds of environmental conditions and social factors (Fu et al. 2011). The cause of formation and interaction mechanisms is comprehended by studying the characters, which provides a basis for the human impact on the ecological environment. It will develop to benign direction and provide service for rational use of resource and environmental management. Changes in resource use, environmental, and social-economic development can be identified by the dynamic analysis of the landscape pattern. The spatial distribution of resources, environmental, and social economic features can be studied through the dynamic analysis of the landscape pattern. Hence, dynamic analysis of the landscape pattern is one of the research centers of the landscape pattern and ecological process analysis.

- *Landscape ecological classification and mapping:* This is the foundation of landscape ecological planning and management. Because the landscape ecosystem is a complex geographical combination composed of interrelated elements with an orderly internal structure, the function of different landscape ecosystems with different internal structures naturally is different.

  - *Landscape ecological classification:* Focusing on the function and starting from the structure, landscape ecological classification emphasizes the structural integrity and function of unity and determines the classification of the landscape unit. Through this classification, it fully reflects the spatial differentiation and internal relationships and reveals the spatial structure and the ecological functional characteristics. Landscape ecological classification includes three steps. First, according to the interpretation of the RS image (aerial photographs, satellite images) and topographic maps and other graphic text data, coupled with field survey results, select and determine the dominant landscape eco-classification elements and indicators to initially identify the scope and type of individual units. Second, analyze in detail various qualitative and quantitative indicators and list a variety of features, through clustering or other statistical methods, to determine the classification results. Finally, determine the ownership of the different functional units through discriminant analysis based on the type of unit index, which makes as the functional classification (Fu et al. 2011).

  - *Landscape ecological mapping:* According to the results of landscape ecological classification, objectively and generally reflect the spatial distribution and area percentage of landscape ecological classification, which is called landscape ecological mapping. Landscape ecological mapping can be divided into a number of specific spatial units, and each unit has unique biological and nonbiological factors, including the influence of human activities, the unique energy flow, the logistics rule, and the unique structure and function. Draw up specific measure systems for each one of the spatial units to gain economic efficiency and the social efficiency unification in the premise of achieving environmental and ecological benefits (Guo and Zhou 2007).

- *Landscape ecological suitability analysis and landscape evaluation:* Landscape ecological suitability analysis is the core of landscape ecological planning; its goal is to classify the suitability level of the landscape. According to the regional landscape resources and environmental characteristics and development needs and resources use requirements, choose representative ecological characteristics (such as precipitation, soil fertility, tourism value, etc.), and analyze the inner quality of a certain type of resource quality and the relationship with the adjacent landscape (repulsion or compatibility) to determine the type of landscape suitability and restrictions due to the following aspects: the uniqueness of the landscape (rarity and the possible destruction of recovery time scale), diversity (patch diversity, species diversity, and pattern diversity), efficacy (biological productive capacity, economic density, etc.), and agreeableness or aesthetic value (Fu et al. 2011). The commonly used methods include factor overlay method and mathematical combination.

The process of landscape ecological suitability analysis is as follows: first, the planning aims to select each factor (or factor classification), such as land use types, which can be divided into forest, grassland, water, and so on; slope is divided into >15°, 5° −15°, <5°, and so on, thereby giving different weights to various kinds of selected factors on the suitability of various human activities; second, overlay the single-factor layers by the overlay technique to gain all levels of mapping composites; finally, calculate the suitability of various factors to determine the optimal level of landscape.

- *Landscape function zoning:* The division of functional areas comes from landscape space structure and is to meet landscape ecological system and environmental services, biological production, culture, and aesthetics of the four basic functions. It forms a reasonable landscape pattern with the surrounding area spatial pattern linked to achieve improvement of ecological conditions, socioeconomic development, and an increasing ability of sustainable development (Fu et al. 2011).

- *Landscape ecological planning program and evaluation:* According to landscape suitability analysis and the landscape evaluation result, based on the principles of landscape ecological planning, determine the landscape management, recovery, use, and construction of policy and objective, and determine the best composition structure, space landscape structure, and landscape ideal pattern (Guo and Zhou 2007). Landscape ecological planning is mainly the scheme and measurements that are determined based on landscape ecological suitability analysis coupled with the natural characteristics of the landscape. This does not mean that there is no social economy development; it promotes social and economic development and meanwhile seeks the most appropriate landscape pattern. Therefore, it needs cost–benefit analysis and the regional sustainable development ability of the analysis.

- *Landscape ecological planning program implementation and adjustment:* To ensure the smooth implementation of the scheme, it needs to put forward the landscape structure and spatial pattern adjustment, recovery, construction, and management of specific technical measures and proposed protections of landscape management capital, policies, and other external environmental factors. With the passage of time and the change of objective conditions, it needs to constantly correct the original planning to meet the changing circumstances and to achieve the optimal management of the landscape resources of sustainable utilization (Guo and Zhou 2007).

3.2.4 STRATEGIC POINTS IN LANDSCAPE ECOLOGICAL PLANNING

The scope of landscape ecological planning is enormous, and different types of landscape ecological planning have different emphases; spatial network analysis, scaling analysis, and ecological feedback are the three important elements in different types of landscape ecological planning.

- *Spatial network analysis:* Landscape components of the same character and function compose the network ecological system (Liu 2011). It is beneficial to the maintain biodiversity, improve the ecologic environment, and adjust the regional climate. Spatial network analysis includes ecological corridor analysis, connectivity, and node analysis (Wu 2007). During spatial network analysis, we should fully understand the patch, corridor, and matrix as the three basic types of landscape to identify environment resource landscape types, disturbance landscape types, remnant landscape types, and introduced landscape types, which can only make landscape ecological planning better with the source-sink principle. The length and width of the ecological corridors should be increased or reduced based on the actual situation. The property of greenbelts should be increased. The network connectivity of ecological corridor should be enhanced. The disconnection of different corridors should be reduced. The distribution of all the ecological corridors should be more even.

- *Scaling analysis:* The task of landscape ecological planning is to reveal the formation, structure, and function of the ecological system. However, there are some differences for different scales of the description of the ecological system. Hence, landscape ecological planning can be divided into four different levels in the space of scale: patch, corridor, area, and region (Zhang et al. 2007). Only in this way can the similarities and differences of the ecological area be fully realized, which makes landscape ecological planning more complete (Liu and Fu 1998). Scaling analysis in landscape ecological planning includes upscaling and downscaling. Upscaling combines similar zones successively to advanced units according to similarity principles and one unity principles. Downscaling is based on spatial heterogeneity and locates the dominant feature and thus divides the upper zone unit. It gradually downscales the ecological function zone according to minimized differences within the region and maximized differences outside the region.

- *Ecological feedback:* Ecological feedback is a pre-evaluation of ecological effects implemented through the views of direct and indirect, positive and negative, long term and short term to guide ecological spatial planning in advance (Yang 2005). Hence, ecological feedback can eliminate the potential undesirable effects and propose ecological protection programs. Ecological feedback includes ecological risk assessment and ecological benefit assessment. The former focuses on the potential undesirable effects of landscape ecological planning, while the latter focuses on the ecological benefits of landscape ecological planning.

REFERENCES

Cai J. L., Yin H., Huang Y. Ecological function regionalization: A review. *Acta Ecologica Sinica*, 2010, 30(11): 3018–3027.

Fu B. J., Chen L. D., Liu G. H. The objectives, tasks and characteristics of China ecological regionalization. *Acta Ecologica Sinica*, 1999, 19(5): 591–595.

Fu B. J., Chen L. D., Ma K. M., et al. *Landscape Ecology Principles and its Application.* Beijing: Science Press, 2001.

Fu B. J., Chen L. D., Ma K. M., et al. *Landscape Ecology Principles and its Application.* Beijing, China: Science Press, 2011.

Gao J. X., Zhang L. B., Pan Y. Z., et al. *Study on Chinese Ecological Stratagem at 21st Century.* Guiyang, Guizhou: Guizhou People's Press, 1998.

Guo J. P., Zhou Z. X. *Landscape Ecology.* Beijing, China: China Forestry Press, 2007.

Liu G. H., Fu B. J. The principle and characteristics of ecological regionalization. *Acta Ecologica Sinica*, 1998, 6(6): 67–72.

Liu G. H., Fu B. J. The principle and characteristics of ecological regionalization. *Acta Ecologica Sinica*, 1998, 6(6): 67–72.

Liu K. *Ecological Planning: Theory, Method, and Application.* Beijing, China: Chemical Industry Press, 2011.

Ning Z. R. *Landscape Ecology.* Beijing, China: Chemical Industry Press, 2008.

Ouyang Z. Y., Wang R. S. *Regional Ecological Planning: Theory and Method.* Beijing: Chemical Industry Press, 2005.

Wu J. G. *Landscape Ecology: Pattern, Process, Scale and Hierarchy.* Beijing: Higher Education Press, 2000.

Wu J. G. *Landscape Ecology: Pattern, Process, Scale and Hierarchy.* Beijing, China: Higher Education Press, 2007

Xiao D. N., Li X. Z., Gao J., et al. *Landscape Ecology.* Beijing, China: Science Press, 2003.

Yan N. L. *Ecosystem Delineation on Priority Ecosystem Services and Ecosystem Management: Theory Framework and Demonstration Research.* Shanghai: Shanghai Academy of Social Sciences Press, 2007.

Yan N. L., Yu X. G. Goals, principles, and systems of eco-functional regionalization in China. *Resources and Environment in the Yangtze Basin*, 2003, 12(6): 579–585.

Yang P. F. *The Research on the Theory and Approach of Urban and Rural Spatial Eco-planning.* Beijing, China: Science Press, 2005.

Yang Z. F., Li W., Xu L. Y., et al. *Environment Planning Theory and Practice of Ecological Urban.* Beijing: Chemical Industry Press, 2004.

Yang Z. F., Xu L. Y. *Urban Ecological Planning.* Beijing: Beijing Normal University Press, 2007.

Yu X. X., Niu J. Z., Guan W. B., et al. *Landscape Ecology.* Beijing, China: Higher Education Press, 2007

Zhang H. Y. Landscape planning: Concept, origin and development. *Chinese Journal of Applied Ecology*, 1999, 10(3): 373–378.

Zhang H. J., Liu Z. E., Cao F. C. *Ecological Planning: Scaling, Spatial Pattern, and Sustainable Development.* Beijing, China: Chemical Industry Press, 2007.

Zhang J. E. *Ecological Planning.* Beijing: Chemical Industry Press, 2009.

Zheng D. X., Tang X. H. *Ecological Function Regionalization in Fujian Province.* Beijing: China Environmental Science Press, 2007.

4 Planning of Industry System

Jiansu Mao

CONTENTS

4.1   INTRODUCTION

4.1.1   INTRODUCTION TO INDUSTRY SYSTEM

4.1.1.1   Definition of Industry System

In general, industry refers to the production of an economic good or service within an economy. There are four key industrial economic sectors: the primary sector, largely raw material extraction industries such as mining and farming; the secondary sector, involving refining, construction, and manufacturing; the tertiary sector, which deals with services (such as law and medicine) and distribution of manufactured goods; and the quaternary sector, a relatively new type of knowledge industry focusing on technological research, design, and development, such as computer programming and biochemistry. A fifth, quinary, sector has been proposed encompassing nonprofit activities. Industries are also any business or manufacturing activities.

Industries can be classified on the basis of raw materials, size, and ownership. They may be agriculture based, marine based, mineral based, and forest based on the basis of raw materials; they can be classified as small, medium, and large on the basis of size; private sector, state owned or public sector, joint sector, and cooperative sector on the basis of ownership.

4.1.1.2   Industry Classification System

As different countries may have different industry structures for their specific development state, the industry classification systems vary with country and governmental department. For instance, many developing and semideveloped countries depend significantly on industry, and their economies are always interlinked in a complex web of interdependence. Industries are divided into four sectors. (1) Primary: Involve the extraction of resources directly from the earth, which includes farming, mining, and logging. They do not process the products at all. They send it off to factories to make a profit. (2) Secondary: Involve in the processing of products from primary industries. They include all factories—those that refine metals, produce furniture, or pack farm products such as meat. (3) Tertiary: Involve the provision of services. They include teachers, managers, and other service providers. (4) Quaternary: Involve the research of science and technology. They include scientists. Therefore, different countries may follow different industry classification standards.

TABLE 4.1

Summary of Industry Classification System

Section	ISIC Rev.4.0	GB/T 4754-2002
A	Agriculture, forestry, and fishing	Agriculture, forestry, animal husbandry, fishery
B	Mining and quarrying	Mining
C	Manufacturing	Manufacturing
D	Electricity, gas, steam, and air conditioning supply	Electric power, gas, and water production and supply
E	Water supply; sewerage, waste management, and remediation activities	Construction
F	Construction	Transport, storage, and post
G	Wholesale and retail trade; repair of motor vehicles and motorcycles	Information transfer, computer services, and software
H	Transportation and storage	Wholesale and retail trades
I	Accommodation and food service activities	Hotels and catering services
J	Information and communication	Financial intermediation
K	Financial and insurance activities	Real estate
L	Real estate activities	Leasing and business services
M	Professional, scientific, and technical activities	Scientific research, technical service, and geologic prospecting
N	Administrative and support service activities	Management of water conservancy, environment, and public facilities
O	Public administration and defense; compulsory social security	Resident services and other services
P	Education	Education
Q	Human health and social work activities	Sanitation, social security, and social welfare
R	Arts, entertainment, and recreation	Culture, sports, and entertainment
S	Other service activities	Public management and social organization
T	Activities of households as employers; undifferentiated goods- and services-producing activities of households for own use	International organization
U	Activities of extraterritorial organizations and bodies	

Both the international and China's national industry classification systems are shown in Table 4.1, the former is based on ISIC Rev.4.0 and the latter on the national standard GB/T 4754-2002 of China.

4.1.1.3 Constitutes of Industry Division

Within an industry system, several sections related to the exploitation of raw material and its further fabrication and manufacture are termed as sections B, C, D in GB/T 4754-2002, which are the main part of secondary industry. Further divisions of the industry in ISIC Rev.4.0 and GB/T 4754-2002 are described in Table 4.2.

TABLE 4.2
Description of Industry Divisions in ISIC and GB/T 4754-2002

GB/T 4754-2002		ISIC Rev.4.0	
Division	Description	Division	Description
06	Mining and washing of coal	05	Mining of coal and lignite
07	Extraction of petroleum and natural gas	06	Extraction of crude petroleum and natural gas
08	Mining and processing of ferrous metal ores	07	Mining of metal ores
09	Mining and processing of nonferrous metal ores	08	Other mining and quarrying
10	Mining and processing of nonmetal ores	09	Mining support service activities
11	Mining of other ores	10	Manufacture of food products
13	Processing of food from agricultural products	11	Manufacture of beverages
14	Manufacture of foods	12	Manufacture of tobacco products
15	Manufacture of beverages	13	Manufacture of textiles
16	Manufacture of tobacco	14	Manufacture of wearing apparel
17	Manufacture of textiles	15	Manufacture of leather and related products
18	Manufacture of textile wearing apparel, footwear, and caps	16	Manufacture of wood and products of wood and cork, except furniture; manufacture of articles of straw and plaiting materials
19	Manufacture of leather, fur, feather, and related products	17	Manufacture of paper and paper products
20	Processing of timber, manufacture of wood, bamboo, rattan, palm, and straw products	18	Printing and reproduction of recorded media
21	Manufacture of furniture	19	Manufacture of coke and refined petroleum products
22	Manufacture of paper and paper products	20	Manufacture of chemicals and chemical products
23	Printing, reproduction of recording media	21	Manufacture of pharmaceuticals, medicinal chemical, and botanical products
24	Manufacture of articles for culture, education, and sport activities	22	Manufacture of rubber and plastics products
25	Processing of petroleum, coking, processing of nuclear fuel	23	Manufacture of other nonmetallic mineral products
26	Manufacture of raw chemical materials and chemical products	24	Manufacture of basic metals
27	Manufacture of medicines	25	Manufacture of fabricated metal products, except machinery, and equipment

TABLE 4.2 (*Continued*)
Description of Industry Divisions in ISIC and GB/T 4754-2002

GB/T 4754-2002		ISIC Rev.4.0	
Division	Description	Division	Description
28	Manufacture of chemical fibers	26	Manufacture of computer, electronic, and optical products
29	Manufacture of rubber	27	Manufacture of electrical equipment
30	Manufacture of plastics	28	Manufacture of machinery and equipment, and so on
31	Manufacture of nonmetallic mineral products	29	Manufacture of motor vehicles
32	Smelting and pressing of ferrous metals	30	Manufacture of other transport equipment
33	Smelting and pressing of nonferrous metals	31	Manufacture of furniture
34	Manufacture of metal products	32	Other manufacturing
35	Manufacture of general purpose machinery	33	Repair and installation of machinery and equipment
36	Manufacture of special purpose machinery	35	Electricity, gas, steam, and air conditioning supply
37	Manufacture of transport equipment	36	Water collection, treatment, and supply
39	Manufacture of electrical machinery and equipment	37	Sewerage
40	Manufacture of communication equipment, computers, and other electronic equipment	38	Waste collection, treatment, and disposal activities; materials recovery
41	Manufacture of measuring instruments and machinery for cultural activity and office work	39	Remediation activities and other waste management services
42	Manufacture of artwork and other manufacturing		
43	Recycling and disposal of waste		
44	Production and supply of electric power and heat power		
45	Production and supply of gas		
46	Production and supply of water		

Industry always plays a vital role in national economy and environmental issues and thus should be emphasized in environmental management.

4.1.2 FRAMEWORK OF INDUSTRY SYSTEM AND ITS ENVIRONMENT

An industry system is a complex system with a specific structure and components. However, it runs as a whole with a performance of transforming raw materials into

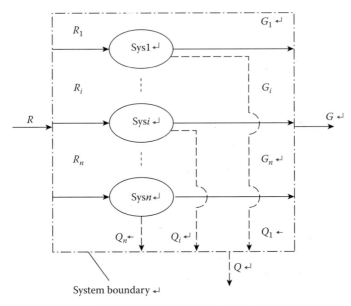

FIGURE 4.1   Conceptual framework showing the relationship between national industrial system and its environment.

final products, meanwhile, it emits residues and wastes into the environment due to industrial metabolic activities. The main relationship between an industry system and its environment can be shown in Figure 4.1.

In Figure 4.1, G refers to industry product; that is, the output of an industry system, which can be represented as economic production (e.g., gross domestic product [GDP]) or material product (amount of specific products) in a certain period of time. R refers to resource consumption; that is, the input of an industry system, including the material and energy resource put into an industry system. Q refers to environmental emissions; that is, the unexpected but unavoidable industry output (industry waste water, industry solid waste, for instance) that will be emitted into the environment. Both R and Q can be expressed as the quantities of certain resource consumption or certain environmental emissions in a certain period.

In national statistics, the total industry output is the sum of its subindustries output of all the subsystems that constitute the industry system, so do for industry resource consumption and environmental emissions, the relationships that exist among the parameters in Figure 4.1 can be expressed as follows:

$$R = \sum_{i=1}^{n} R_i \qquad (4.1)$$

$$Q = \sum_{i=1}^{n} Q_i \qquad (4.2)$$

$$G = \sum_{i=1}^{n} G_i \qquad (4.3)$$

where i is the subsystem index and n is the number of subsystems under consideration.

Because different industry sectors may have different environmental performance and thus lead to different environmental problems, the performance of an industry system as a whole might be improved through the optimization of industry constitutes.

4.1.3 MAIN ENVIRONMENTAL IMPACTS OF INDUSTRY SYSTEM

In industrial activity, the industrial production department takes material resource as production material to form industrial products with a specific function that is able to meet certain human demand by processing and manufacturing. In this process, because of industry's metabolic activity, it is impossible for every material element input to production to be transferred into designed product, and those material elements not entering the prepared product are inevitably discharged into the environment in the way of industrial waste, thus disturbing the environment. This disturbance of the industrial system for the environment is described as environmental impact in series standards of ISO14040, and this impact is divided into three categories related to natural resource consumption, human health, and ecosystem destroy (see LCA content in Section 4.3 for details). In regular practice of environmental management work, however, it is divided into two categories related to natural resource consumption (including consumption of material resource and energy) and industry pollutant discharge, while the latter is further divided into industry waste gas, liquid waste, solid waste, noise, and other pollution subdivisions according to pollutant pattern. Different categories of environmental impact will result in various environmental problems. Major environmental impacts and relevant environmental problems in the industry are described in the following sections.

4.1.3.1 Resource Consumption and Natural Resource Shortage

Material resource is both the production object and fundamental condition of industrial activity. The material resource in industrial activity is ultimately from natural resource; for example, the raw material for textile could be directly from natural resource; like cotton produced from plantation and also from chemical fiber-product of oil processing industry, in which case, its indirect resource is the natural resource of oil. The enormous industrial activities consume plenty of natural resources. Consider water use in China for an example, according to the data provided in *China Statistical Yearbook*, Chinese industrial activities consume around 510 billion tons of fresh water each year, which contributes nearly 86% of the total national water use. The details are shown in Table 4.3.

Among various resources, what deserves concern is nonrenewable resource, especially the metal mineral resources, which play an important role in the development of the modern national economy. Due to its nonreproducibility, people's use of those resources in their production activities will decrease their reserve year after year. If there is no reasonable development and protection, along with human's long-term use, it certainly will result in resource exhaust. Table 4.4 shows the world reserve and service years for several metal mineral resources. The table also shows the service years of many metal materials are only several decades, equal to one or two generations.

TABLE 4.3
Industrial Water Use of China in 2009

Items	Value
Total water use (100 million cubic meter)	5965.2
In agriculture	3723.1
In industry	1390.9
Sum of agriculture and industry, as percentage of total water use (%)	85.7

Source: National Bureau of Statistics of China (NBSC), 2010. China Statistics Press, Beijing.

TABLE 4.4
Main Metal Resource Reserves and Service Life in 2008

Metals	Metal Production (Kt of Metal Content)	Reserves (Kt of Metal Content)	Service Life (year)
Zn	10,100	220,000	22
Cu	14,900	470,000	31
Ni	1,500	62,000	40
Pb	3,280	67,000	21
Ag	20.3	270	13

4.1.3.2   Energy Consumption and Greenhouse Gas Discharge

Energy is the motive power resource for industrial activities and other important fundamental conditions of industrial activities.

The energy consumed in industrial activity mainly includes crude oil, coal, natural gas, water energy, wind energy, and so on. Among these, part is the primary energy, such as coal and water used in power generation; part is secondary energy, which is formed, for convenience of human's use, through human's production and processing of the primary energy, such as electricity, gas, and so on. The secondary energy is ultimately from natural resource; for example, gas is from petroleum's splitting and electricity could be from hydropower or thermal power with fire coal as the principal. Therefore, industrial production activity consumes a good deal of natural primary energy resource.

Among many energy resources consumed in industrial activity, what specifically deserves concern is the mineral fuel, such as petroleum, coal, and so on. On one hand, such a resource has high energy value and big storage density convenient for development and use, which is the widely used energy type in modern national economy; on the other hand, due to its nonreproducibility, people's use of those resources in their production activities will decrease their reserve year after year. If there is no reasonable development and protection, along with human's long-term use, it certainly results in resource exhaust. Consider China as an example, according to the data provided in *China Statistical Yearbook*, 3.06 billion tons of standard coal is consumed yearly and 92.2% of that belongs to nonrenewable mineral fuels. The details are shown in Table 4.5.

TABLE 4.5

Main Energy Consumption of China in 2009

Items	Value
Total energy consumption (10,000 tons of SCE)	306,647
Percentage of total energy consumption (%)	
In coal	70.4
In crude oil	17.9
In natural gas	3.9
Sum of coal, crude oil, and natural gas	92.2

Source: National Bureau of Statistics of China (NBSC), 2010. China Statistics Press, Beijing.

The other problem in energy use, which deserves special concern, is that the energy production process may discharge many kinds of environmental pollutants, especially greenhouse gases (GHGs) discharge. This discharge is closely related to global warming, the current key international environmental problem. Take stationary combustion as an example; its GHGs discharge could be simply estimated according to the following equation:

$$Q_i = \sum_{j=1}^{m}\left(F_{i,j}^{CO_2} \cdot \varepsilon_{i,j}^{CO_2}\right) + 25\sum_{j=1}^{m}\left(F_{i,j}^{CH_4} \cdot \varepsilon_{i,j}^{CH_4}\right) + 298\sum_{j=1}^{m}\left(F_{i,j}^{N_2O} \cdot \varepsilon_{i,j}^{N_2O}\right) \qquad (4.4)$$

where F is the actual amount of final fuel consumption and takes the unit of joule based on IPCC method. ε represents the emission factor of fuels. Different fuel may have different emission factor for different GHGS, the emission factors for various fuels are listed in Table 4.6. The subscript j is the fuel index, which may take value from 1 to 9 to represent coal, coke, crude oil, gasoline, kerosene, diesel oil, fuel oil, natural gas, and electricity, respectively, according to the energy statistics of China (DITS and NBS 2008). m is the number of fuel variety and takes a value of 9. In Equation 4.10, 1, 25, and 298 is the GWPS per kilogram of CO_2, CH_4, and N_2O emissions, respectively; they all take units of kilogram CO_2 equivalent.

It means that 2.07 mg CO_2 eq. of global warming potential (GWP) will be formed during the combustion of 1 ton of coal.

As far as China is concerned, research indicates that energy consumption and its related GHGs discharge occurred in industrial activities accounting for 79.6% (household consumes the other 21% of the total energy) and 92.9% of the total, respectively. Among various industrial activities, the industry division is the most important division in energy consumption and GHGs emission and accounts for 79.3% and 91.4% of the total in China. Details are listed in Table 4.7.

4.1.3.3 Industrial Wastes and Environmental Pollution

As mentioned previously, it is impossible for every material element input to production to be transferred into the designed product due to industrial metabolism,

TABLE 4.6

Emission Factors for Given GHG in Fuel Combustion

Energy	CO_2 (Mg CO_2/TJ)	CH_4 (kg CH_4/TJ)	N_2O (kg N_2O/TJ)
Coal	98.3	1	1.5
Coke	107	1	1.5
Crude oil	73.3	3	0.6
Gasoline	70.0	3	0.6
Kerosene	71.9	3	0.6
Diesel oil	74.1	3	0.6
Fuel oil	77.4	3	0.6
Natural gas	56.1	1	0.1
Electricity	0	0	0

Source: IPCC. IGES. Japan, 2006, http://www.ipcc-nggip.iges.or.jp/public/2006gl/vol2.html.

TABLE 4.7

Contribution by Industrial Subsystems to the Total at Economic Production, Energy Consumption, and GWP of China in 2007(%)

Items	Primary	Industry	Construction	Tertiary
Percentage in economic production	10.84	45.16	5.41	38.60
Percentage in energy consumption	3.47	79.30	1.69	15.20
Percentage in GWP	1.48	91.41	0.41	6.70

Note: The economic production was calculated based on the value-added present price in statistics.

and those material elements not entering the prepared product are inevitably discharged into the environment as industrial wastes.

The industrial wastes can be categorized into the following main types according to their physical states: waste gases, waste water, solid waste, noise, and so on. The industrial solid wastes can be further classified into poisonous wastes, heavy metals, and so on, according to their possible environmental risks to human health. The main industrial waste gases are SO_x, CO_2, NO_x, and so on. For instance, in power station, some quantity of CO_2 will be formed and emitted into the atmosphere in burning of coal while much of SO_x will be emitted in the case of high-sulfur coal burning. These waste gases may come from different industrial sectors and closely relates to different environmental problems. For instance, the emission of NO_x may mainly happen in burning of oil fuels in vehicles and cars, which may cause photochemical smog in certain climatic conditions; the emission of CO_x and SO_x addresses global warming and acid deposition, respectively. The "eight public nuisance" events that happened in the mid-twentieth century are typical industrial pollution cases, which are briefly described in Table 4.8.

TABLE 4.8
"Eight Public Nuisance" Events in the Mid-Twentieth Century

Name of Event	Major Pollutants	Place and Time	Poison Situation	Causes
Mass valley smoke events	Smoke and SO_2	Belgium, December 1930	Attacked thousands of people, 60 people were dead	Intensive factories, large dust discharge;
Donora smog	Smoke and SO_2	Donora Town, the United States, October 1948	Attacked 43% (about 6000 people) and 17 people were dead in 4 days	abnormal weather with temperature invasion, and dense fog
London smog	Smoke and SO_2	London, England, December 1952	4000 people were dead in 5 days	
Los Angeles photochemical smog episode	Photochemical smog	Los Angeles, the United States, May to November each year		Meteorological conditions including automobile exhaust gas, sun and calm wind, and so on
Minamata disease event	Methylmercury	Minamata Town, Kumanoto Prefecture, Japan, discovered in 1953	Discovered people dead in 1953; 180 people were sick and 50 people were dead in 1972	Welder's waste water containing mercury catalyst was discharged to ocean to pollute fish and shellfish
Toyama event (itai-itai disease)	Cadmium	Jinzu valley, Toyama, Japan, discovered in 1931	Over 280 patients from 1931 to 1972 with 34 people dead	The zinc plant's waste water containing cadmium polluted drinking water and farmland
Yokkaichi event	SO_2 and heavy metal dust	Yokkaich, Japan, in the year of 1970	500 people fell ill (asthma, etc.) with 10 people dead	Large amount of SO_2 and heavy metal dust discharged from factory
"The Rice Bran Oil" event	PCB	Aichi Prefecture, Kyushu, Japan, in the year of 1968	Over 5000 patients, 16 people were dead, and 10,000 people were affected	PCB was mixed into rice bran due to improper management

Environmental pollution events continued in this century; for example, the Bohai bay oil spill event in June 2011. The oil field is jointly developed by ConocoPhillips Company and CNOOC. Among that, Penglai 19-3 oil field is the biggest offshore oil and gas field in China now. The operation started in 2002, and it is predicted to reach output peak this year with 60,000 barrels of crude oil each day. In June of 2011, Platform B and Platform C in 19-3 oil field operation area in Bohai Bay had two oil spilling incidents successively. State Oceanic Administration issued the order to stop production. The initial estimated loss is about 0.3 billion yuan of RMB. Furthermore, environmental pollution also occurred in the food industry; for example the milk powder pollution event in China in 2008. It was discovered that many infants suffered from kidney stones after the consumption Sanlu milk powder. Later, the chemical raw material of melamine was found in the milk powder, resulting in thousands of infants falling ill, even several deaths.

Furthermore, the huge anthropogenic material flow has been disturbing the related biogeochemical cycle. A comparison of anthropogenic metal cycle with its biogeochemical cycle is shown in Table 4.9, which indicates that the anthropogenic cycles for some metals have several times overload their natural cycles.

Industrial environmental impacts may cause various environmental problems and may closely address to global sustainable development. The grand objectives and related industrial environmental concerns are presented in Table 4.10, which can be used as the basic objective for information in eco-planning of industrial systems.

TABLE 4.9
Comparison of Anthropogenic Metal Cycle with Its Biogeochemical Cycle

Metals	Concentration in Soil (mg/kg)	Anthropogenic Flow Rate/ Biogeochemical Flow Rate
Al	72,000	0.048
Fe	26,000	1.4
Mn	9,000	0.028
Ti	2,900	0.096
Zn	60	8.3
Cr	54	4.6
Cu	25	24
Ni	19	4.8
Pb	19	12
Mo	0.97	8.5
Cd	0.35	3.9
Hg	0.09	11

Sources: Azar C.,Holmberg J., Lindgren K. *Ecological Economics*, 1996, 18(2): 89–112.
Karlsson S. *Journal of Industrial Ecology*, 1999, 3(1): 23–40.

TABLE 4.10
Grand Objectives and Related Main Industrial Environmental Concerns

Grand Objective	Environmental Concern
Human species existence	Global climate change
	Human organism damage
	Water availability and quality
	Resource depletion: fossil fuels
	Radionuclides
Sustainable development	Water availability and quality
	Resource depletion: fossil fuels
	Resource depletion: nonfossil fuels
	Landfill exhaustion
Biodiversity	Water availability and quality
	Loss of biodiversity
	Stratospheric ozone depletion
	Acid deposition
	Thermal pollution
	Land use patterns
Aesthetic richness	Smog
	Esthetic degradation
	Oil spills
	Odor

4.2 MANAGEMENT OF INDUSTRY SYSTEM

4.2.1 IPAT EQUATION

4.2.1.1 Original IPAT Equation

The quantitative relationship between human development and its impacts has been studied extensively. One important achievement can be traced back to the IPAT equation, which states that environmental impact (I) is a function of population (P), affluence (A), and technology (T) and expressed as Equation 4.5:

$$I = PAT \tag{4.5}$$

where I represents the human environmental impact, which is usually equivalent to the annual consumption of natural resources or emissions into the environment; P represents the population; A represents a measure of social affluence; T represents the technology employed to obtain the social affluence and dispose of consumed products. It is found that the concepts of I, A, and T in Equation 4.5 seem ambiguous and need to be defined specifically when used in application.

4.2.1.2   Several Transformed IPAT Equations

It is easy to understand that the value of social affluence times social population can be treated as total social service, which is signed as S. To make environmental impact equal to environmental impact in Equation 4.5, the technology T should be defined as environmental impact per unit of service; that is, Equation 4.6:

$$I = S \cdot T \qquad (4.6)$$

When economic product is used to represent the social service of a country or a certain region, Equation 4.6 will be transformed as Equation 4.7:

$$I = G \cdot T \qquad (4.7)$$

where G refers to the economic product of concerned country or region and can be represented as GDP or industrial added value, and so on.

If we introduce the concept of eco-efficiency (represent by letter e) into the IPAT equation, which is defined as the societal service provided per unit of environmental impact by OCED in 1998 and can be expressed as Equation 4.8:

$$e = \frac{S}{I} \qquad (4.8)$$

then we may find that the value of eco-efficiency is always the reverse of T and thus can be employed to represent the item technology. In this situation, the IPAT equation can be transformed as Equation 4.9:

$$I = \frac{S}{e} \qquad (4.9)$$

We may find from Equation 4.9 that to reduce total environmental impacts the growth of eco-efficiency must be quicker than that of service, otherwise, it is impossible to realize a better environmental quality.

In practice, the IPAT equation still can be transformed into a multitude of other models to suit different applications, if only its essence—that impact = impact on both sides of the equation—remains unchanged. In words, the IPAT equation quantifies the essential overall relationship between the economy and its environment, although an environmental Kuznets curve (EKC) is still being used to examine this relationship in some current studies.

4.2.2   Possibility of a Win-Win Situation

For cities, or regions and countries with worse environmental quality, an eco-city planner should manage to realize a win-win situation in economic development and environmental protection. The possibility will be analyzed in this section.

4.2.2.1 Relationship between Environmental Impact and Eco-Efficiency

In the previous section, the relationship between environmental impact and eco-efficiency has been shown using Equation 4.9; that is,

$$I = \frac{S}{e} \qquad (4.9a)$$

Equation 4.9 is widely used in eco-industrial planning, we name it "the ISE equation" to differentiate it from other forms of IPAT equation.

Both IPAT and ISE equations are system-based and purpose-oriented equations. They might be transformed diversely according to the system selected and the environmental issues concerned in a specific study.

Considering that GDP (or gross national product [GNP]) is usually used to represent the service of a national economic system, we may substitute S in ISE equation with G, which represents the GDP of a country, Equation 4.9a becomes as follows:

$$I = \frac{G}{e} \qquad (4.10)$$

In this case, e means the GDP per unit of environmental impact. Equation 4.10 is termed IGE equation. An obvious limitation in Equation 4.10 is that only the economic system and its economic product can be analyzed.

Transforming Equation 4.10 into dimensionless format, we obtain the following equation:

$$\bar{I} = \frac{\bar{G}}{\bar{e}} \qquad (4.11)$$

where $\bar{I}$, $\bar{G}$, and $\bar{e}$ are the dimensionless environmental impact, GDP, and eco-efficiency, respectively. Equation 4.11 reflects the relationship between the changing values of the three parameters within a certain period.

Furthermore, Equations 4.8, 4.10, and 4.11 are system-based equations and can be used to analyze the quantitative relationship between any economic system and its environment at different system levels. In that situation, we should keep in mind that the meaning of the variables must change in accordance with the specific system and the questions being addressed. For instance, if we chose the industrial system of China as the system in a case study, the economic product and eco-efficiency of that system would be defined as the gross industrial product and the industrial environmental efficiency, respectively.

4.2.2.2 Formulating the Environmental Impact in Economy Growth

If we assume that GDP and eco-efficiency are both growing exponentially at annual average growth rates of ρ_G and ρ_e, respectively, the changing values of GDP and eco-efficiency in year t will be expressed as follows:

$$\bar{G} = e^{\rho_G \cdot t} \qquad (4.12)$$

$$\bar{e} = e^{\rho_e \cdot t} \qquad (4.13)$$

where t is the number of years since the reference year (year 0) and always takes an integer value greater than 1.

From Equation 4.12, we can derive the following equation:

$$t = \frac{\ln \overline{G}}{\rho_G} \tag{4.14}$$

Substituting Equation 4.14 into Equation 4.13 produces the following equation:

$$\overline{e} = \overline{G}^{\frac{\rho_e}{\rho_G}} \tag{4.15}$$

If we define the ratio of ρ_e to ρ_G as the growth rate ratio of eco-efficiency to GDP and express this ratio as the parameter k, we obtain the following equation:

$$k = \frac{\rho_e}{\rho_G} \tag{4.16}$$

Substituting Equations 4.15 and 4.16 into Equation 4.11 transforms Equation 4.11, as follows:

$$\overline{I} = \overline{G}^{1-k} \tag{4.17}$$

Equation 4.17 reflects the relationship between changes in environmental impact and changes in GDP during exponential growth of both eco-efficiency and GDP.

4.2.2.3 Analysis of a Win-Win Possibility

In order to show the relationship between environmental impacts and economic growth more clearly, we have illustrated Equation 4.17 in Figure 4.2 for various values of k. If we concentrate on the region of the graph for which $\overline{G} > 1$, which represents a situation in which economic growth is occurring, then the following details of Figure 4.2 become apparent.

When $k = 1$, the $\overline{I}-\overline{G}$ curve is a horizontal line at $\overline{I} = 1$ (curve 1 in the graph), which means that the environmental impact remains constant and will be independent of economic growth.

When $k < 1$, the relationship between $\overline{I}$ and $\overline{G}$ can be displayed as a series of ascending lines, represented by curves 2 and 3 in the graph. These curves indicate that the environmental impact will increase with increasing GDP, and thus, the environmental quality will deteriorate further. Moreover, from the relative vertical positions of curves 2 and 3 (with k values of 0.5 and 0.2, respectively), we can deduce that the lower the value of k, the faster environmental impacts will increase. Therefore, when the rate of growth in eco-efficiency is less than that of the economy, increasing GDP will cause a deterioration of environmental quality; the bigger the gap between the two rates is, the worse the environmental quality will become and the faster the problem will occur.

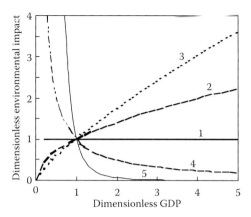

FIGURE 4.2 Curves for changes in environmental impact as a function of GDP. For curve 1, $k = 1$; curve 2, $k = 0.5$; curve 3, $k = 0.2$; curve 4, $k = 2$; curve 5, $k = 5$.

When $k > 1$, the relationship between $\bar{I}$ and $\bar{G}$ can be displayed as series of descending curves, represented by curves 4 and 5 in the graph. These curves indicate that the environmental impact will decrease as GDP increases and that the environmental impact will thus be potentially decreased to a level equal to or below the environmental carrying capacity. In this situation, environmental impacts that exceed the carrying capacity will eventually be mitigated in the future, and deteriorated environments will begin to recover. From this observation, it can be concluded that the quality of the environment will improve only if the rate of growth in eco-efficiency is greater than that of the economy. Moreover, from the relative vertical positions of curves 4 and 5 (with k values of 2 and 5, respectively), we can deduce that the higher the value of k, the faster environmental impacts will decrease; the bigger the gap between the two rates, the greater the possibility for environmental improvement.

It can be proved that the same qualitative outcomes would be obtained under the assumption of linear growth in GDP and eco-efficiency. The only difference will be that the rate of change in the environmental quality would be slower than under the assumption of exponential change.

In practice, changes in the trends for the environmental impacts of an economic system can be determined by comparing the relative growth rates of the economy and of eco-efficiency. The value of k can also be treated as the criterion for determining the relationship between an economic system and its environment. The higher the value of k, the more rapidly the environmental impact will decrease during a period of economic growth and the more the economic system will be in harmony with its environment.

4.2.3 CHARACTERISTICS OF INDUSTRY MANAGEMENT

4.2.3.1 Characteristic Curve for Environmental Impacts

To reduce environmental impacts to the desired level during a period of economic growth, we must first separate the increasing environmental impacts from economic growth by adopting appropriate environmental management and technology. Our goal is to reduce the rate of growth of environmental impacts to a level lower than the rate of economic growth.

Assume that the social GDP grows linearly with time at annual average growth rate, ρ, and that the environmental impact grows at a lower growth rate, $(1-\phi t)\rho$, where ϕ is the coefficient of environmental impact growth and t represents the elapsed time since the reference year. In this case, the changing values of social GDP and environmental impact in the year t will be expressed as follows:

$$\overline{G} = 1 + \rho t \tag{4.18}$$

$$\overline{I} = 1 + \rho t - \phi \rho t^2 \tag{4.19}$$

If we depict Equation 4.19 graphically, a parabola that resembles the reverse U-shaped curve will be obtained (Figure 4.3). In which, we assumed values of $\rho = 0.1$ and $\phi = 0.02$ in our calculation. This curve is called "the characteristic curve for environmental impacts."

The curve in Figure 4.3 shows that the curve for environmental impacts will pass through the following four special positions:

Point A, the "separation" point, represents the starting point at which the environmental impact separates from the economic growth curve. (The value of the curve at this point is set to 1.0 and represents an index against which subsequent values are compared.) At this point, the environmental impact begins to increase at a slower rate than the economic growth. The separation point is always treated as occurring during the starting year (i.e., the reference year) in which environmental management begins to reduce impacts.

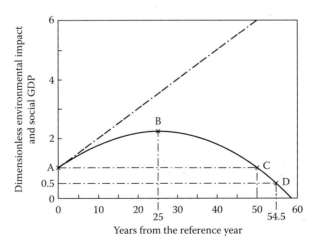

FIGURE 4.3 Curves for environmental impact (solid line) as a function of time and social GDP (dashed line). A, separation point; B, culmination point; C, descent point; D, recovery point.

Point B, the "culmination" point, occurs where $\partial \bar{I}/\partial t = 0$. This point represents the maximum (peak) point on the environmental impact curve. Before this point, the environmental impact increases with time; afterward, the total environmental impact decreases with time. The culmination point thus represents the turning point for environmental impacts.

Point C, the "descent" point, is the critical point where, after the initial process of increase and decrease, the value of the environmental impact equals that in the reference year (i.e., $\bar{I}_C$ equals $\bar{I}_A$), and both equal 1. After this point, the value of the environmental impact will become lower than that in the reference year.

Point D, the "recovery" point, is the critical point at which the environmental impact has been reduced to a level equal to the environmental carrying capacity after a long-term decrease. If we assume that k_1 expresses the ratio of the environmental impact in the reference year to the environmental carrying capacity, $\bar{I}_D$ would equal to $1/k_1$. After this point, the extent to which the environmental impact exceeds the carrying capacity will begin to decrease gradually and environmental quality will improve continuously.

The values of the environmental impacts at points A, B, C, and D are the vital data required for successful environmental management and are significant for environmental quality. In an application of the theory, the four points are termed the "characteristic points for environmental management."

4.2.3.2 Characteristic Curve for Eco-Efficiency

As mentioned previously, environmental improvements cannot be achieved by means of natural (unassisted) transformations of the economic system. Instead, they require an increase in eco-efficiency, as can be seen in Equations 4.9 and 4.10. Substituting Equations 4.18 and 4.19 into Equation 4.11, we obtain the following equation:

$$\bar{e} = \frac{1+\rho t}{1+\rho t - \varphi\rho t^2} \tag{4.20}$$

Equation 4.20 reflects the changing value of dimensionless eco-efficiency as a function of time. If we assign certain values to φ and ρ (e.g., $\rho = 0.1$ and $\varphi = 0.02$), we can draw the curve in Figure 4.4; this curve shows that eco-efficiency increases exponentially with time. From this curve, it is clear that eco-efficiency must improve continuously to realize the targeted improvement resulting from environmental management.

4.2.3.3 Another Expression of Characteristic Curves

In application, the changes in environmental parameters over time will sometimes be expressed as curves for the environmental parameters that change as a function of economic growth (Figure 4.5), which was calculated using the same conditions as Figures 4.3 and 4.4. In this graph, the meaning of each characteristic point is the same as we have previously described.

There are four kinds of curves; generally, all are called as the characteristic curves for environmental management. Two of them are environmental impact and eco-efficiency change as a function of time, just as the two in Figures 4.3 and 4.4; the other

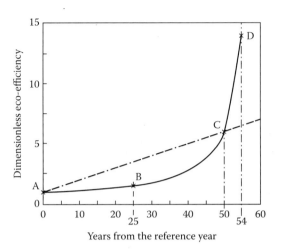

FIGURE 4.4   Curves for eco-efficiency (solid line) and social GDP as a function of time. A, separation point; B, culmination point; C, descent point; D, recovery point.

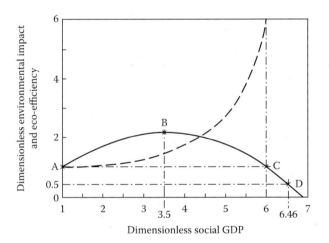

FIGURE 4.5   Curves for dimensionless environmental impact (solid line) and eco-efficiency (dashed line) as a function of social GDP. A, separation point; B, culmination point; C, descent point; D, recovery point.

two are environmental impact and eco-efficiency change as a function of economic growth, as shown in Figure 4.5.

4.2.3.4   Eigenvalues of the Environmental Parameters

At the characteristic points of the curve, the values of four indices (the year of occurrence, the dimensionless social GDP, the dimensionless environmental impact, and the dimensionless eco-efficiency) are the foundation of environmental management and planning and are accordingly called the eigenvalues of the environmental parameters.

TABLE 4.11

Eigenvalues for the Environmental Parameters

Items	Separation Point A	Culmination Point B	Descent Point C	Recovery Point D
t	0	$\dfrac{1}{2\varphi}$	$\dfrac{1}{\varphi}$	$\dfrac{\rho \pm \sqrt{\rho^2 + 4\varphi\rho\left(1-\dfrac{1}{k_I}\right)}}{2\varphi\rho}$
$\bar{G}$	1	$1+\dfrac{\rho}{2\varphi}$	$1+\dfrac{\rho}{\varphi}$	$1+\dfrac{\rho \pm \sqrt{\rho^2 + 4\varphi\rho\left(1-\dfrac{1}{k_I}\right)}}{2\varphi}$
$\bar{I}$	1	$1+\dfrac{\rho}{4\varphi}$	1	$\dfrac{1}{k_I}$
$\bar{e}$	1	$1+\dfrac{\rho}{4\varphi+\rho}$	$1+\dfrac{\rho}{\varphi}$	$k_I\left[1+\dfrac{\rho \pm \sqrt{\rho^2 + 4\varphi\rho\left(1-\dfrac{1}{k_I}\right)}}{2\varphi}\right]$

Based on the features of the environmental impact at these characteristic points, we can calculate the years of occurrence for every characteristic point using Equation 4.14. Substituting the years of occurrence into Equations 4.18, 4.19, and 4.20, in that order, we can calculate the dimensionless values for social GDP, environmental impact, and eco-efficiency at the characteristic points. The results are summarized in Table 4.11 and can be used to calculate the values of the environmental parameters at each characteristic point.

4.2.4 Eco-Efficiency of an Industry System

In this section, the framework shown in Figure 4.1 is employed.

4.2.4.1 Theoretical Analysis

As stated in Section 4.2.1, the eco-efficiency of a system is the output generated per unit of environmental impact. In the present section, we focus on the economic output of the industrial system and primarily consider environmental impacts related to resource consumption and environmental emissions. In other words, we study the resource efficiency (RE) and emission-related environmental efficiency (EE) of the industrial system. Both RE and EE are represented by e, however, e_R and e_E for RE and EE, respectively.

In Figure 4.1, the eco-efficiency of a given subsystem i, can be expressed as

$$e_i = \frac{G_i}{I_i} \tag{4.21}$$

where I refers to R and Q for RE and EE, respectively.

For the national industrial economic system, eco-efficiency is expressed as

$$e = \frac{G}{I} \tag{4.22}$$

where I refers to R and Q for RE and EE, respectively.

Equations 4.21 and 4.22 indicate that higher eco-efficiency means less resource consumption at the same economic output or higher economic output at the same resource consumption.

The relationship between the whole system and its subsystems with respect to resource consumption, environmental emissions, and economic output, expressed in Equations 4.2 and 4.3, and the relationship among the three related parameters, expressed in Equation 4.21, can be substituted into Equation 4.22 to obtain the following:

$$e = \left[\sum_{i=1}^{n} (f_i \cdot e_i^{-1}) \right]^{-1} \tag{4.23}$$

where f_i represents the contribution level of an industry subsystem to the whole and is calculated as a fraction of the subsystem's economic production to GDP, which can be expressed as $f_i = G_i/G$, and $\Sigma f_i = 1$. Equation 4.23 reflects the relationship between the eco-efficiency of the national industrial system and that of its subsystems; the eco-efficiency of a national industrial system is closed related to the structure of the system and the eco-efficiency of its subsystem. Thus, to improve the eco-efficiency of a national industrial system, we should not only improve the eco-efficiency of its subsystems but also adjust its structure by increasing the contributions from those subsystems (industrial sectors) with high eco-efficiencies.

4.2.4.2   Main Industrial Eco-Efficiencies of China

4.2.4.2.1   Energy Efficiency

When we focus on energy consumption in the framework shown in Figure 4.1, the eco-efficiency becomes energy efficiency (ENE). According to the data reported in *China Statistical Yearbook*, we estimated the eco-efficiency of industrial subsystems, which is shown in Figure 4.6.

Figure 4.6 shows that, among the main industrial subsystems, the ENE of construction is the highest with a value of $155.98 million/PJ, meanwhile that of industry is the lowest with a value of $28 million/PJ, only about one-fifth of the ENE of construction and around a half of the ENE of the whole national industrial system. So we may conclude that the low ENE of industry is the key cause for the low ENE of China's industrial system.

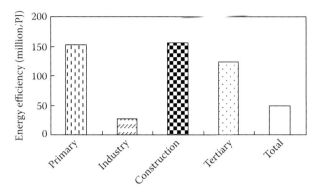

FIGURE 4.6 Energy efficiencies of main industrial subsystems in 2007.

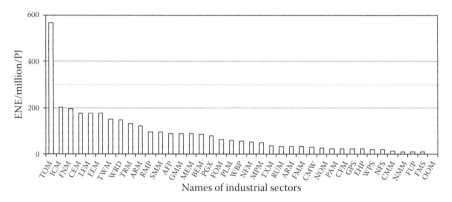

FIGURE 4.7 Energy efficiencies from highest to lowest for China's industry sectors in 2007.

The industries of China is further sorted into 39 industrial sectors according to the national standard GB/T 4754-2002 (see Table 4.2). The ENE of the industrial sectors is estimated based on the reported data in statistics. The results are presented from the highest to the lowest in Figure 4.7.

Figure 4.7 shows that the ENE of TOM is the highest among industrial sectors with a value of $570 million/PJ, followed by ICM and FNM with values of around $200 million/PJ. The ENE of other industrial sectors is less than $200 million/PJ. On the contrary, OOM, FMS, and FUP are the three industrial sectors with the lowest value of ENEs with only around or less than $10 million/PJ of energy efficiencies.

4.2.4.2.2 *GHGs Emission-Related Environmental Efficiency (GEE)*

When we turn the concerned environmental impact to GHGs emission, the eco-efficiency becomes GHGs emission-related environmental efficiency. We estimated these eco-efficiencies of industrial subsystems based on China's statistics in 2007 and show the results as Figure 4.8.

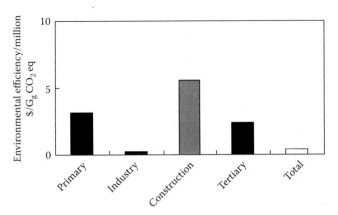

FIGURE 4.8 GHGs emission-related environmental efficiencies of main industrial subsystems in 2007.

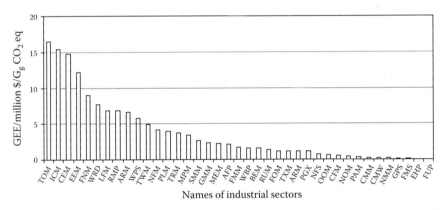

FIGURE 4.9 GHG emission-related environmental efficiencies from highest to lowest for China's industry sectors in 2007.

Figure 4.8 indicates that the GEE of industry is the lowest with a value of $0.21 million/$G_g$ CO_2 eq. among the main industrial subsystems, only 1/250 of the GEE of construction, and less than half of the GEE of the national industrial system. Therefore, industry is the major subsystem of China in GHGs emission.

So as do for ENE, the GEE of the industrial sectors is estimated based on the reported data in statistics and presented from the highest to the lowest in Figure 4.9.

Figure 4.9 shows that the GEE of TOM is the highest among industrial sectors, followed by ICM and CEM with values of around $15 million/$G_g$ CO_2 eq., respectively. On the contrary, FUP, EHP, and FMS are the three industrial sectors with the lowest value of GEES; their GEEs are all less than $0.1 million/$G_g$ CO_2 eq.

The codes used in this section for the various industrial sectors are listed in Table 4.12.

TABLE 4.12
Industry Sectors and Their Codes

Code	Name of Sectors	Code	Name of Sectors
CMW	Mining and washing of coal	MEM	Manufacture of medicines
PGX	Extraction of petroleum and natural gas	CFM	Manufacture of chemical fibers
FMM	Mining and processing of ferrous metal ores	RUM	Manufacture of rubber
NFM	Mining and processing of nonferrous metal ores	PLM	Manufacture of plastics
NOM	Mining and processing of nonmetal ores	NMM	Manufacture of nonmetallic mineral products
OOM	Mining of other ores	FMS	Smelting and pressing of ferrous metals
AFP	Processing of food from agricultural products	NFS	Smelting and pressing of nonferrous metals
FOM	Manufacture of foods	MPM	Manufacture of metal products
BEM	Manufacture of beverages	GMM	Manufacture of general-purpose machinery
TOM	Manufacture of tobacco	SMM	Manufacture of special purpose machinery
TXM	Manufacture of textiles	TRM	Manufacture of transport equipment
TWM	Manufacture of textile wearing apparel, footwear, and caps	EEM	Manufacture of electrical machinery and equipment
LFM	Manufacture of leather, fur, feather, and related products	CEM	Manufacture of communication equipment, computers, and other electronic equipment
WBP	Processing of timber, manufacture of wood, bamboo, rattan, palm, and straw products	ICM	Manufacture of measuring instruments and machinery for cultural activity and office work
FNM	Manufacture of furniture	ARM	Manufacture of artwork and other manufacturing
PAM	Manufacture of paper and paper products	WRD	Recycling and disposal of waste
RMP	Printing, reproduction of recording media	EHP	Production and supply of electric power and heat power
ARM	Manufacture of articles for culture, education, and sport activities	GPS	Production and supply of gas
FUP	Processing of petroleum, coking, processing of nuclear fuel	WPS	Production and supply of water
CMM	Manufacture of raw chemical materials and chemical products		

4.3 PROCEDURE FOR ECO-PLANNING OF INDUSTRY SYSTEM

4.3.1 ECO-INDUSTRIAL PARK CONSTRUCTION

4.3.1.1 Concept of EIP

Eco-industrial parks (EIPs) are a new type of industry organization pattern designed according to circular economy theory, industry ecological principle, and cleaning production requirement and are the aggregation place of ecological industry. EIPs connect different types of plants or enterprises in the form of material circulation flow or energy transmission to resource-shared, by-product, or waste interchangeable industry interdependent combination. It is the "food chain" and "food network" relationship built among "producer–consumer–decomposer" in industrial eco-system upon imitating natural eco-system, making waste or by-product of park-dominate enterprise to become raw material or energy for another enterprise to seek a closed-loop recycle use of material resource, energy gradient utilization, and minimum environmental waste.

The biggest difference between industry intergrowth in EIPs and species intergrowth in a natural eco-system is different causes: the species intergrowth in a natural eco-system is the result of species, natural evolution while the industry intergrowth is mostly produced in the effect of market mechanism or formed through planning. Generally, industry intergrowth system through planning is more beneficial to environment.

4.3.1.2 Pattern for EIPs Construction

Seen from domestic and foreign practice, EIPs are generally constructed in the following patterns:

1. *Enterprise-dominant type:* EIPs which take certain original enterprise or several enterprises as the core to attract relevant enterprises in the industry ecological chain to the park for construction, such as Kalundborg EIPs in Denmark; or EIPs which take enterprise groups as principal and build internal enterprises within the group according to ecological industry and circular economy principle, such as EIPs built by Lubei Petrochemical Enterprise Group in China to form three pieces of industry ecological chains with high degree of correlation, including sulfuric acid and cements production from phosphogypsum, the by-product of phosphoric acid, much use of "seawater" and salinity–alkalinity production.

2. *Enterprise-relevance type:* Parks that connect relevant industries with high industry relevance in ecological concept to fully play complementary effect, such as agriculture ecological park to enhance industry relevance between agriculture and industry to promote sustainable industrial and agricultural development. The industry ecological chains of "sugarcane–sugaring– blackstrap making alcohol–alcohol effluent making compound fertilizer" and "sugarcane–sugaring–bagasse making paper-recovery of black liquor in pulping" in Guigang Guangxi with cane sugar manufacture as the core are EIPs built in this pattern.

3. *Reform and reinforcement type:* To reform and reinforce based on the original industrial parks and hi-tech park to create upgraded EIPs with ecological enterprises concentration.

EIPs have many classification methods. Marian Chertow from Yale University divides EIPs into five categories according to regional size for interchange of material.

1. *Waste-free exchange:* The recyclable materials are provided or sold to other enterprises freely, such as to deliver the waste materials to requesters through a waste materials exchange center. Since such an exchange is spontaneously formed, generally its resource exchange is not sufficient.
2. *Material exchange within enterprise or production departments:* The recyclable materials or industrial by-products are used in a single enterprise or production department rather than in an other enterprise or production department; for example, in a large-scale petrochemical enterprise, the by-product of certain technology could be the raw material of another technology.
3. *Material exchange within enterprises located in the same industrial park:* To form a resource-shared system in close enterprises and carry out energy, water, and material exchange; for example, the industry intergrowth system of "waste materials from brewery–mushroom plantation-raise pig–aquaculture–vegetable planting" has been formed in Monfort Boys Town in Fiji Surva.
4. *Material exchange within close enterprises not located in the same place:* Material exchange or energy gradient utilization is developed within a certain large area, such as in the city. Kalundborg EIPs in Denmark are the typical of such EIPs. Taking heat-engine plant, refinery, and plaster board plant as the core, it is to form the sufficient use of coal, coal ash, sulfur, and other resources and gradient utilization of steam, waste heat, and other energy in Kalundborg.
5. *Material exchange within enterprises in a large area:* This category of EIPs should be a combination of the above various EIPs. At present, there is no successful case reported. The ecological province and stream may be typical of such EIPs.

4.3.2 Eco-Industry System Construction

4.3.2.1 Concept of Eco-Industry System Construction

Construction of eco-industry system in technology means to construct an industry network system according to material circulation principle in ecology, combined with material flow status in industry system in particular area, taking advantage of industry or leading industry as the core and the existing upstream and downstream as branch to promote the evolution of regional industry toward closed circulation ideal status in material flow aspect.

4.3.2.2 Steps in Eco-Industry System Construction

Eco-industry system construction first needs to make deep investigation of industry system in study region, know local advance industry or leading industry as well as main material resource, material flow process and industry technology level of those industries, and know the resource support capability and environmental carrying capability of the used resources in the local place; second to analyze for flow relationship of materials in different life cycle stages in the whole industry system and subsystems of different industries, including in-flow direction, quantity allocation, pattern transition, and space transfer, and so on, and find an unharmonious link (such as the breakpoint in industry in the aspect of material flow) and element (such as mismatching in quantity). Then, it is to make up network defect of ecological industry by introducing a new type of industry to improve relevance among regional industries; properly adjust production scale of industries to promote reasonable allocation of resource utilization; promote reuse of product and extend service life of product upon maintenance and repair means to reduce consumption strength of industry system for raw materials; reduce consumption quantity of raw material by discarding product disassembly and part reuse; and to reduce waste discharge quantity and natural resource consumption quantity through waste recycling. As a whole, it is to achieve "less input, high output, and low discharge" to coordinate and blend with the external resource and environmental system. Finally, in the constructed ecological industry system framework, it is to conduct system management for major material and product of advantage industry, including "design for environment (DFE)," "product development and function substitution and green manufacture," "extend producer's responsibility (EPR)," "green package and transport, recovery and recycle of product after discard," and so on. Since different life cycle stages of products belong to different management departments, it needs to reintegrate the management system according to the constructed ecological industry system to promote a comprehensive management level.

4.3.2.3 Example of Eco-Industry System Construction

As an example, in the work of the Xiamen ecological city planning program in 2005, it is discovered in the earlier phase that the Xiamen industry system takes electro-communication, petrochemical industry, and traffic machinery as pillar industry. In Xiamen ecological city planning, it is to propose the concept to promote the upgrade of the existing industry by constructing electronic and information service ecological industry system, petrochemical industry ecological industry system, and traffic and transportation machinery ecological industry system and form new Xiamen industry by creating ecological ocean industry. Among them, in constructing electronic and information service ecological industry system, taking the original electronic industry of Dell (China) Limited Company as the core component, it is to unite information service industry while extending electronic industry's maintenance service during product use and supplement recycle and after-use disposal industry of electronic products to jointly constitute "semi-life cycle" of electronic products, as shown in Figure 4.10.

Figure 4.10 indicates that the perfect electron and information service ecological industry system should mainly include life service stages, including electronic element supply, product production, product use, discard, and recycle. In the figure,

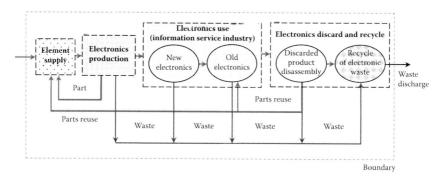

FIGURE 4.10 Electronics and information service ecological industry system concept module.

the dark degree indicates perfection of industry; the darker the column is, the more perfect is the development of relevant industry department.

In the industry system in Figure 4.10, it is to extend the life service of electronics and reduce the consumption strength of the industry system for raw material through the reuse of old products; to reduce the consumption quantity of the industry system for raw material by discarding certain parts in products and reduce waste's discharge quantity by recycling electronics waste while improving the consumption strength of the whole economic activity for natural resource. The whole performance is "less input, high output, and low discharge" and coordination and involvement with external resources and the environmental system.

If further detailing according to the electronics category, it is divided into computer and network product chain, information product chain of telecommunication, digital technology product and "multi-type" household intelligent central product chain, and so on.

Through the on-site research for the Xiamen electronics and information service industry and comparison with the ecological industry module, it is discovered that there are mainly the following differences: the first is the electron elements, main raw materials of Xiamen electronics are mainly from other places; the second is the disposal of waste and old electronics has been in an early form, but still far from mature, especially in blank space in recycling aspect of electronics waste; the third is the reuse of old electronics and old electronics elements is insufficiently regular; and the fourth is electronics types are not sufficient and it should form electronics cluster, making good industry foundation for changing Xiamen as important domestic information product production base and export base. The darkness and the filled extent in Figure 4.10 indicate the difference between industry system status and target status.

4.3.3 ENERGY GRADIENT UTILIZATION

4.3.3.1 Introduction to Energy Gradient Utilization

Energy gradient utilization is a way to utilize the energy properly, which takes the advantage of the energy by different methods according to the grade of the energy, for example, in the cogeneration of heat and power system, high and moderate temperature

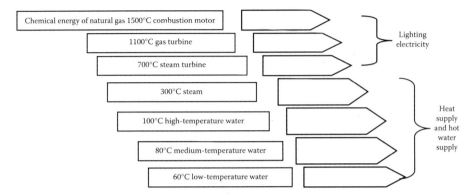

FIGURE 4.11   Example of energy gradient utilization mode.

steam is first used to generate the power or used for the production requiring high-temperature steam while the low-temperature afterheat is utilized for supplying heat to residence. Scale of energy grade refers to level of the energy transforming to mechanical work. High-grade energy mainly consists of electrical power, gas, liquid fuel, and so on. Low-grade energy mainly refers to heat energy, bioenergy, and so on. More often than not, high-grade energy can be converted to low-grade energy and has a higher energy conversion ratio while it is difficult for low-grade energy to transform into high-grade energy, which demands specific technology and considerable energy will be lost during the procedure. Energy gradient utilization can improve the energy utilization efficiency of the whole system and is the key measure in energy conservation as well as the main content aiming at developing circular economy by utilizing energy.

Similar to circular utilization of materials, energy gradient utilization can also be applied in different levels among enterprises and society. The enterprise, based on production energy consumption, usually plans and designs the energy gradient utilization process based on the requirements of each process in the production of different product to make full use of energy. As for the social aspect, the society designs and plans energy gradient utilization mode, technical process, and energy transmit of the area based on requirements on types, grade, amount, and location of energy consumption required by energy users of a specific area such as production and living area to make full use of energy in a specific area. Figure 4.11 is a typical example of energy gradient utilization mode.

4.3.3.2   Main Patterns of Energy Gradient Utilization

Detailed application of energy gradient utilization is present in various forms; however, the most basic form features cogeneration of heat and power and combined cycle power generation.

1. *Cogeneration of heat and power:* Cogeneration of heat and power refers to a combination of heat supply and power generation in the same power plant; it is called CHP for short. The term is putting forward as for the traditional thermal power generation, which only produces one product as the electricity, efficiency being around 35%. This means every 1 MJ electricity

produced will waste 2 MJ heat The wasted energy could be used to heat water and well satisfy needs of heating and bathwater in the surrounding area of the power plant, other factories, and residences. Cogeneration of heat and power usually adopts the steam turbine to generate electricity and uses the waste steam to serve as supplementary heating for existing boiler, whose total yield rate could be 80%. It is clear that cogeneration of heat and power utilizes the heat to be wasted by common power plants to provide inexpensive heating for both industry and domestic purpose, thus greatly improving heating efficiency. Cogeneration of heat and power produces both electricity and heat, possessing numerous advantages compared to separate generation of heat and power, such as reducing energy consumption, decreasing discharge of pollutants, saving energy consumption by utilizing the residual heat, saving land usage, improving heat-supplying quality, making comprehensive usage of resources easier, improving urban image, and reducing security accidents. As cogeneration of heat and power has many advantages, countries around the world have been promoting its development. Thus, the development of cogeneration of heat and power is beneficial.

2. *Combined cycle power generation, CCPG:* Combined cycle power generation refers to generator sets used in power generation and the generating system combining circular system of gas turbine and heat recovery steam generator and circular system of steam turbine. Currently, net efficiency of combined cycle power generation is over 50%, with the highest reaching 56%–57%, which is far beyond the heat efficiency in normal power plants. In the combined cycle power generation system, capacity matching ratio of gas turbine and steam turbine is 2:1; for example, one set of 250 MW gas turbine can be provided with one set of 125 MW gas turbine to work as a 375 MW combined cycle power generation unit or that two sets of 250 MW gas turbine can be provided with one set of 250 MW gas turbine to make a 750 MW combined cycle power generation units.

   In that the generator set of gas turbine combined cycle power generation is a circular system consisting of gas turbine, generator, heat recovery steam generator, gas turbine (condensing power type) or heat-supplying gas turbine (gas pump type or back pressure type), it transforms the high-temperature exhaust gas discharged after the working of gas turbine into steam through the heat recovery steam generator and further sends the gas into the steam turbine to generate electricity or using the dead steam produced after the working of electricity generation for heating. Usual forms are the single-axis combined cycle of gas turbine and gas turbine working together to promote a set of generator and also the multiaxis combined with gas turbine, steam turbine involved separately with the generator. It is mainly used for electricity generation and cogeneration of heat and power. Combined cycle units of gas turbine has the following advantages: (1) High efficiency of electricity generation: electricity generation rate being 57%–58%, it is far higher than common coal-fired power plant (electricity generation rate of 0.75–1000 MW unit is only 20%–48%); (2) Less pollutants discharge: dust, sulfur dioxide, nitric dioxide discharged by coal-fired power plant (concentration reaches 200 PPM) are numerous, which require end-of-pipe equipment for desulfuration, denitration, and electric precipitation while waste gas discharged by heat recovery steam generator of combined cycle power generation plant has almost no dust and an extremely small amount of sulfur dioxide and the nintric dioxide being only 10–25 PPM. (3) Flexible operation mode: starting and shutting time of coal-fired power plant is long, which is suitable for operation of basic load, featuring poor pitch peaking performance. As for gas turbine power plant, not only can it be used for basic load but also can be used for variable load plant; combined cycle generation uses numerous fuels (oil, natural gas, coal, etc.) as the energy for a easier adjustment of variable load according to different energy consumption. (4) Less water consumption: among gas–steam combined cycle power plants, electricity generated by steam turbine only takes up to one-third of the total capacity and water consumption is one-third of that used by coal-fired power plant. In addition, because the burning of hydrogen in CH_4 and oxygen in air can produce carbon dioxide and water, 1.53 kg water can be recollected by burning 1 m^3 natural gas theoretically, which can satisfy water required by the power plant. (5) Less floor area: no piling of coal and dust and usage of air-cooling system makes the floor area only 10%–30% of the coal-fired power plant, which greatly saves land resources. (6) Short construction period: different power plants differ greatly in the construction period; generally, gas turbine takes 8–10 months, combined cycle system takes 16–20 months, while coal-fired power plant takes 24–36 months. Additionally, combined cycle can also be applied for electricity generation using solar energy and combined cycle power generation in the nuclear power station, and so on.

3. *An example of energy gradient utilization in the paper industry:* Consider the paper industry as an example. There are two major forms of energy consumption: one is power consumption that is used in liquid transport, beating, repulping, squeezing, dehydration, and shaping; the second is the steam consumption, which is mainly used in squeezing dehydration and drying of paper. Squeezing through temperature rise by hot cylinder can improve efficiency of squeezing dehydration, and dryness of paper delivered can be 40%–50%. Drying of paper is done by removing residual water contained through the dryer and steaming process to make dryness of paper delivered reaching 92%–95%. This process consumes most of energy in the paper industry, taking up 50% of energy consumption of the whole process. Through analysis of energy consumption in papermaking, two key points can be determined in applying circular economy as for energy consumption in papermaking, one of which is to change the traditional practice of the paper mill to use self-prepared boiler or power plant to supply steam and power into adopting cogeneration to reduce energy consumption and improve the whole utilization efficiency of energy. The second approach is aimed at the paper drying process, which consumes most of the energy to recollect energy by using a heat exchanger to preheat fresh air through the residual air or to heat up water through the heat exchanger for domestic supply.

In summary, energy gradient utilization may greatly improve utilization efficiency of energy, namely saving the energy, preserving natural resources effectively, reducing discharge of pollutants, and bettering environmental quality efficiently, which has made it the key content to promote circular economy and a sustainable development.

4.3.4 CIRCULAR ECONOMY DEMONSTRATION AREA

4.3.4.1 Introduction to Circular Economy

Circular economy demonstration area (hereinafter refer to demonstration area) is a demonstration area taking pollution prevention as a starting point, material circular flow as a feature and sustainable development of society, economy and environment as a final target. It uses ecological rules to organize social and economic activities in the area into several feedback processes of "resource-product-renewable resources," control production of waste in the origin of production and consumption, recycle available product and waste and reasonably dispose for finally unavailable products to achieve material production and consumption of "low exploitation, high utilization and low discharge" and maximize efficient use of resource and energy, reduce pollutant discharge, and promote the harmonious development of the environment and economy.

Comparing with ecological industrial parks, the circular economy parks covers more abundant content and has a wider influence scope. With its essence to be a kind of ecological economy area, in the regional layer, it not only builds industry network among enterprises but also combines material circulation and energy among the primary, secondary, and tertiary industries to achieve the whole circulation of industry in regional economy. The macro economy policy adjustment and legal system reconstruction will a provide strong guarantee for circular economy demonstration area.

The central thought of circular economy demonstration area is to combine circular economy development and regional comparative advantage; pay attention to ecological protection and environmental improvement and reform the present unreasonable regional industry by introducing hi-tech technology and build circular economy production and consumption pattern fit for particular regional advantage.

The construction of circular economy demonstration area should follow "3R" principles (reduce, reuse and recycle). "3R" principles are the core of circular economy and must not be violated in any way of circular economy development, and there is no exception in the construction of circular economy demonstration area. It is indispensable for input end reduction, reuse in the process, and recycle of the whole resource and waste.

4.3.4.2 Main Steps of Circular Economy Construction

China completes the construction of circular economy demonstration area in four steps: demonstration area planning, hardware construction, software support, and index system.

In the aspect of demonstration area planning, the first is to organize planning team with leadership organization and technology organization, then to conduct status investigation to mainly analyze local natural condition, social background, ecological environment and the present economic operation pattern; then to confirm the construction target and specific plan, including the whole framework design, industry development planning, ecological landmark planning, key program selection, law and regulation formulation, and so on. Finally, it is to conduct investment and benefit analysis, mainly meaning investment budget, social, economic and environmental benefit analysis of the construction, and so on.

In the aspect of hardware construction, it is mainly centered in three layers of enterprise, region and society. The enterprise layer needs to achieve cleaning production and minimum pollution discharge, improve technology for energy-saving and emission reduction and develop and utilize waste resources produced by enterprises. The regional layer is to build an ecological industrial park, use ecology and circular economy theory to conduct classified guidance for the present industrial park; promote the level and competitiveness of the present economic and technical development zone; guide reformation of the old industry area, especially to speed up economic transformation in the resource exhaustion area. In the layer of society, it is to build resource recycling society, build classification, recovery and recycling system of urban household rubbish and other waste and old materials, recycled water reuse system in the city and region, ecological industry system and information system, and so on. Considering the primary, secondary, and tertiary industries as a whole, it needs to uniformly plan material circulation and internal energy flow among primary, secondary, and tertiary industries.

The software construction could start from two aspects of law and regulation supporting system design and technical support system design. Law and regulation provide guarantee in policy for demonstration area construction, including law and regulation system construction for circular economy development, preferential policy formulation to promote circular economy development and encouragement for green consumption and procurement. The technical support system is mainly including environmental engineering technology, waste recycling technology and cleaning production technology, and so on. The construction of circular economy pays more attention to key connection technology of development and application of ecological industry to connect various units in the whole demonstration area to really apply hi-tech technology into actual construction.

In the aspect of index system, the construction principle and operation situation of the demonstration area need specific indexes for their evaluation. The index system of the circular economy demonstration area is including four categories: economic development index, circular economy feature index, ecological environmental protection index, and green management index. The economic development index takes the local GDP as the main basis to evaluate local economic increase situation. The circular economy feature index is including resource production rate, recycle use rate, and final disposal quantity, respectively indicating input quantity, production process, and discharge quantity in the demonstration area. The ecological environmental protection index refers to environmental situations of environmental quality, environmental performance, ecological construction, ecological construction potential improvement, and so on. The green management index refers to policy and rule management index and enterprise's management and public consciousness index to promote circular economy.

The construction of the circular economy demonstration area will play a huge role in sustainable industry development of our country in the future. From the angle of resource and energy, the shortage status increasingly appears but must meet people's basic demand. So, recycling of resources and energy is the only road. From the angle of whole industry planning, the circular economy demonstration area effectively combines the primary, secondary, and tertiary industries to promote application of hi-tech technology. From the aspect of the public living level, the construction of the circular economy demonstration area will play a promotional role for protection of local ecological environment, and the beautiful local ecological environment improves people's living standard from the other aspect. Its social benefit is obvious.

4.3.4.3 Implementation of a Circular Economy

Running a circular economy may help to save material and energy uses and to reduce environmental waste and its emission, and thus can be widely used in eco-planning of industrial systems. In general, a circular economy can be developed in the following three main ways in eco-planning of industrial systems: reasonable resource flow organization, advanced resource use technology, and effective management technology.

1. *Resource flow organization:* In organizing resource flow, the circular economy could be developed from the following three layers: (1) minor cycle within economic entity, including enterprise, production base, and so on; (2) middle cycle within enterprises and industries in the industry-centralized area; and (3) the big cycle including production, life, and the whole society. Among them, in the layer of enterprise, the material cycle within enterprise should be considered as the base to constitute the small cycle within enterprises, production base, and other economic entities. In an industry-centralized area, the material cycle should be considered as carrier to constitute the middle cycle among enterprises, industries, and production areas. Considering promotion and application of an ecological park within a certain territorial scope as the main forms, it is to build energy and logistics integration and resource recycle use vertically and horizontally in industry through industry's reasonable organization with the emphasis of waste exchange and integrated resource use so as to achieve low discharge, even "zero discharge," of pollutants produced in the park and form recycle industry cluster or recycle economic zone; achieve sufficient use of resources in different enterprises and industries and build recycle economic industry system with reuse and recycle of secondary resources as the important composition part. In the social level, considering the material circulation in the whole society as the focus, it is to constitute the big recycle including production, life, and the whole society. To balance urban and rural development, production, and life, through building circular economic circle between urban and rural, human's society, and natural environment, it is to construct big material energy recycle between production and consumption, including production, consumption, and recycle;

build social system fit for circular economy; and build resource-saving and environment-friendly society to achieve maximum economic, social, and ecological benefits.

2. *Technology for resource utilization:* In view of technology in resource utilization, the development of circular economy is mainly realized through three paths: (1) efficient utilization of resources; (2) cyclic utilization of resources; (3) harmless emission of industrial wastes.

Efficient utilization of resources: In this aspect, utilization level of resource and output rate of unit element should be improved depending on advance of science and technology and system innovation. For example, exploration of efficient production mode can be adopted, utilize land intensively, make use of water resources and energy, and so on, economically in the field of agricultural production. Promotion of interplant and other efficient cultivation technology and polyculture, the efficient cultivation technology, introduction and cultivation of efficient and high-quality seed, germchit, and cultivation variety, implementation of installation farming, scale, and standardization of agricultural production can all improve the output level of unit land and water surface. Realize water conservation of cultivation through optimization of multiple utilization plans of water resource, perfection of ditch, and other water supply system, improvement of irrigation mode and excavation of agricultural water conservation and other measures. Realize water conservation through development of intensified water conservation in cultivation industry. Second, the quality of land and water body and other resources shall be improved while sustaining power, and bearing capacity of agricultural resource shall be enhanced. Improve soil organic matter, as well as nitrogen, phosphorus, potassium element, and other conditions required by efficient growth of crops and ameliorate soil fertility through returning straw to field, scientific fertilization by the measurement of soil formula and other advanced practical methods. Reform coastal saline–alkali land through acid–base neutralization principle and advanced technology or process long-term soil melioration by planting crop with special efficiency, so as to improve the plantability of saline and alkaline land. Control dosage of pesticide, prohibit high poison pesticides strictly, use fertilizer and agricultural film reasonably, popularize degradable agricultural film, so as to reduce erosion to soil. Ecologization treatment shall be adopted for excrement from livestock and poultry cultivation to reduce pollution to water body. Adjust stocking density and variety in due time, process bait casting and fertilizing reasonably, so as to prevent water and coating quality deterioration of cultivation water area and intertidal zone. Reduce the utilization of antibiotics and other drugs to ensure that crop and animal products can meet health standards.

In the field of industrial production, the improvement of resource utilization efficiency is mainly reflected in energy, water, material, and land conservation, as well as comprehensive utilization of resource and other, which is realized through a series of replacement and substitution of "high" and "low," "new," and "old," while it centers on the level improvement of industrial technology and boosts utilization efficiency of resource through low efficient management and production technology being substituted by the high one, inferior energy being substituted by the high-quality one, low-performance equipment being substituted by the high one, low-functioning material being substituted by the high one, lower layers industrial building being substituted by the higher ones. On the other hand, it centers on reasonable utilization of resource with protogenesis resource being substituted by repair and reproduction of parts and equipments, as well as waste metal, plastics, paper, rubber and other renewable resources, protogenesis material being substituted by recycled material and other resource utilization and other reasonable substitution, "high" substituted by "low," "old" substituted by "new" through surplus heat utilization and recycled water utilization in some production links, so as to realize improvement of resource utilization efficiency. Advocate life style of resource conservation in the field of personal consumption; popularize energy and water conservation appliances. Life style of resource conservation is not to cut down necessary personal consumption but to overcome delinquenent conduct of resource squander, so as to reduce unnecessary resource consumption.

Cyclic utilization of resource: Establish cyclic utilization channel of renewable resource, achieve effective utilization of resource, and reduce demand to natural resource through industrial chain construction for resource cyclic utilization in production and life in cyclic utilization of resource, so as to improve the development of economy and society in natural and harmonious circulation. Implantation of crop as well as cultivation of livestock and poultry shall comply with natural ecology discipline themselves in agricultural production and realize organic coupling agricultural cyclic industrial chain through advanced technology. Establish longitudinal and horizontal industrial link, improve cyclic and recycle utilization of resource taking concentration area of production as key area, regarding industrial by-product, waste material, surplus heat and energy, waste water, and other resources as carrier in industrial production. For example, carry out combined heat and power generation, regional central heating project, develop utilization of surplus heat and energy, as well as energy recycle of organic waste, and form various types of industrial chains of energy step utilization by centering on energy. Construct production of regenerated water and network project of water supply, organize concatenate utilization of waste water reasonably and form industrial chain for reutilization of water resource by centering on waste water. Establish extended industrial chain, recycle processing chain of renewable resources, comprehensive utilization chain of wastes, as well as repair and refurbish process chain of equipments and parts, construct comprehensive utilization chain of renewable and available resources by centering on waste and used materials and by-products. The key point is to construct a recycling network for living waste materials, give full play to circulating function of business service industry, collect and retrieve used product, waste, and used materials or salvaged material in living and service industry, so as to increase probability that the resources return to production link and improve recycle or reclamation of resource.

Harmless emission of industrial wastes: Reduce impact to ecological environment incurred by the activity of production and living in harmless discharge of salvaged material. Carry out cleaning cultivation mainly through popularizing ecological cultivation in agricultural production. Dispose excrement from livestock and poultry by exerting methane ferment technology, so as to turn harm into good and produce methane and organic farming fertilizer. Control aquaculture pesticide, popularize scientific baiting, and reduce water body pollution incurred by aquaculture. Explore ecology complementary type aquatic product cultivation, strengthen harmless treatment of fodder for livestock and poultry, as well as inspection, prevention, and cure of epidemic. Implement agricultural cleaner production, adopt comprehensive control to organism and physics and other plant diseases and insect pests, reduce the usage amount of pesticide, cut down the pesticide residue of crops and accumulation of pesticides poison in soil. Adopt degradable agricultural film and implement recycling of agricultural film and reduce residue in land. Popularize reduction of waste discharge and cleaner production technology, apply dedusting, desulfuration, and denitration technology in industrial production, as well as decomposition, biochemical treatment, incineration treatment and other biosafety disposal of industrial waste oil and water and organic solid for coal-fired boiler, reduce generation of waste gas, liquid, and solid vigorously in the process of industrial production. Enlarge the application ratio of clean energy, cut down discharge of energy production, and used harmful substances. Advocate to reduce consumption pattern of disposable products in personal consumption and cultivate living habit of garbage classification.

3. *Effective management technology:* Effective technical methods to promote circular economy are cleaner production, construction of ecological industrial park (EIP), and circular economy legislation.

   Cleaner production refers to practical production methods and measures that can fulfill people's requirement, use natural resources and energy reasonably, and protect the environment. The essence is a kind of planning and management for human production activity with least consumption of resource and energy. It provides minimization, recycling, and harmless to waste or consume waste during production. Cleaner production is a precondition to carry out a circular economy, which generally refers to single enterprise. Its main purpose is to realize the minimization of pollution. Its main measure is to adopt cleaner production technology in enterprise, so as to achieve advanced level in the consumption of water, electricity, and raw material, as well as pollutant discharge for the product of enterprise; enhance repeat utilization rate of water, realize minimum discharge of waste water; introduce high relevancy technology in large-scale enterprise, form industrial chain and establish circular economical enterprise through step utilization of product, waste, and water.

Ecology industrial park is the second layer of circular economy. Ecology industrial park is a kind of new industrial organization form established according to the idea of circular economy and the principle of industrial ecology. Its objective is to reduce the generation of waste to a great extent and to turn one generated waste or by-product into raw material of another industry, thus, realizing low or zero discharge of waste in one industrial park. Circular economy mode "recovery–recycling–design–production" is followed in ecological park, which realizes cyclic utilization of substance and optimization allocation of resource through planning and designing a reasonable industrial chain. Denmark Kalundborg industrial park is the most typical representative in the world industrial ecological system currently. Power plant, refinery, pharmaceutical factory, and gypsum board manufacturing plant in this park are its core, which reduce the generation amount and disposal cost of waste and achieve favorable economic benefit through effective utilization of waste and by-product generated from other enterprise production.

Circular economy is a kind of economic development mode with win-win of economic interest and environmental benefit for the long run, however, it fails to be realized spontaneously through market owing to "commonality" property of environment. Therefore, every country formulates relevant laws, regulations, and economic policies through lawful guarantee, economic incentive, and other various modes since the advancement of circular economy, so as to change "commonality" property of environment and turn circular economy into conscientious selection of market subject from initial compulsive implementation. Chinese circular economic legislation is mainly reflected in two basic laws presently: One of them was passed through Standing Committee of National People's Congress in June 2002 and Cleaner Production Promotion Law was implemented as of January 1, 2003. The other was passed through Standing Committee of National People's Congress in August 2008 and Circular Economy Promotion Law was implemented as of January 1, 2009.

5 Planning of Sustainable Energy and Air Pollution Prevention

Gengyuan Liu and Linyu Xu

CONTENTS

5.1   SUSTAINABLE URBAN ENERGY PLANNING

5.1.1   ENERGY SECURITY

In the face of increasing energy market volatility affecting both accessibility and affordability, and environmental concerns over continued demand for fossil fuels, the need to improve energy security will be one of the more significant energy challenges facing most jurisdictions this century. Despite its importance and the growing number of energy-related problems confronting many jurisdictions, many people, including politicians, policy makers, and members of the general public, have difficulty in understanding energy-related issues in general and energy security in particular. The framework (Figure 5.1) is listed in the following sections.

5.1.2   PLANNING OBJECTIVES AND INDEX SYSTEM

5.1.2.1   Energy Security Goals

The energy sector used a collaborative process to develop its vision statement and security goals. The energy sector envisions a robust, resilient energy infrastructure in which continuity of business and services is maintained through secure and reliable information sharing, effective risk management programs, coordinated response capabilities, and trusted relationships between public and private security partners at all levels of industry and government (EPAct 2005).

5.1.2.2   Sector Profile

The energy sector includes assets related to three key energy resources: electric power, petroleum, and natural gas. Each of these resources requires a unique set of supporting activities and assets, as shown in Table 5.1. Petroleum and natural gas share similarities in methods of extraction, fuel cycles, and transport, but the facilities and commodities are separately regulated and have multiple stakeholders and trade associations. Energy assets and critical infrastructure components

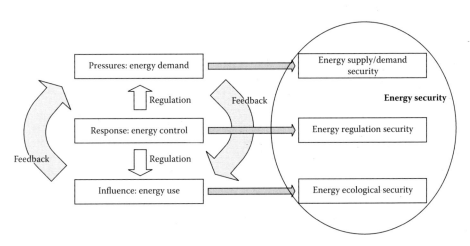

FIGURE 5.1   Energy security framework.

TABLE 5.1

Segments of the Energy Sector

Electricity	Petroleum	Natural Gas
• Generation	• Crude oil	• Production
• Fossil fuel power plants	• Onshore fields	• Onshore fields
• Coal	• Offshore fields	• Offshore fields
• Gas	• Terminals	• Processing
• Oil	• Transport (pipelines)	• Transport (pipelines)
• Nuclear power plants	• Storage	• Distribution (pipelines)
• Hydroelectric dams	• Petroleum processing	• Storage
• Renewable energy	facilities	• Liquefied natural gas
• Transmission	• Refineries	facilities
• Substations	• Terminals	• Control system
• Lines	• Transport (pipelines)	• Gas markets
• Control centers	• Storage	
• Distribution	• Control systems	
• Substations	• Petroleum markets	
• Lines		
• Control centers		
• Control system		
• Electricity markets		

are owned by private, federal, state, and local entities, as well as by some types of energy consumers, such as large industries and financial institutions (often for backup power purposes).

5.1.2.3 Indicator System for Energy Security Planning

Most of the energy security analyses seek to simplify this complex concept by using quantitative indicators. Indicators are proxy signals that—if rightly selected—may reflect a complex system's characteristics and dynamics in simple numbers. The problem is that indicators tend to proliferate: faced with simplifying the complex concept of energy security, analysts tend to throw dozens of indicators in a single energy security analysis. This defeats the main purpose of simplification: confronted by six indicators, a policy maker or an analyst may not feel much wiser than before the quantification process started (see Table 5.2).

5.1.3 PRESSURE OF ENERGY DEMANDS

5.1.3.1 Analysis of Currency Demands

The energy consumption of the delivery district of a power plant depends on many different influence factors. Generally, the energy demand is influenced by climate parameters, seasonal data, and economical boundary conditions. The heat demand of a district heating system depends strongly not only on the outside temperature but also on the additional climatic factors like wind speed, global radiation, and humidity.

TABLE 5.2
Indicator System for Energy Security Planning

Items	Unit	Reference Value
The energy consumption per GDP	tec/10^4 RMB GDP	≤1.2
Clean energy utilization ratio	%	≥50[a]
Rural biogas utilization rate	%	≥28.4[b]
Urban air quality days	day/year	≥333
Carbon emission intensity	kg/10^4 RMB GDP	<5.0
Ecological footprint per capita energy consumption	hm²/capita	0.2–0.5[c]

Notes:
[a] Reference on "Notice on Adjusting Indicators for Quantitative Examination of Comprehensive Control of Urban Environment during the Tenth Five-Year Period."
[b] Reference on the planning objectives of national rural biogas in 2010.
[c] Reference on the average of ecological footprint per capita energy consumption in southern provinces in 2003.

On the other side, seasonal factors influence the energy consumption. Usually, the power and heat demand is higher in summer and winter than spring and autumn. Furthermore, vacation and holidays have a significant impact on the energy consumption. Last but not least, the heat and power demand in the delivery district is influenced by the operational parameters of enterprises with large energy demand and by the consumer's behavior. Additionally, the power and heat demand follows a daily cycle with low periods during the night hours and with peaks at different hours of the day.

5.1.3.2 Energy Forecast Analysis

The quality of the energy demand forecast depends significantly on the availability of historical consumption data and on the knowledge about the main influence parameters on the energy demand.

The analysis of the relationships between energy consumption and climate factors includes the following activities:

- Energy balancing (distribution of the demand)
- Analysis of the main influence factors
- Design of the mathematical model
- Analysis and modeling of typical demand profiles

5.1.3.2.1 Industrial Energy Demand

The industrial sector includes agriculture, mining, construction, and manufacturing activities. The sector consumes energy as an input to processes that produce the goods that are familiar to consumers, such as cars and computers. The industrial sector also produces a wide range of basic materials, such as cement and steel, which are used to produce goods for final consumption. Energy is an especially important input to the production processes of industries that produce basic materials. Typically, the industries

TABLE 5.3
Elasticities of Energy in Wanzhou

Item	Unit	Energy Elasticity Coefficient
Coal	t	−0.01
Washed coal	t	—
Coke	t	−2.96
Natural gas	×10⁴ m³	−0.33
Gasoline	t	1.19
Kerosene	t	−7.30
Diesel	t	4.38
Fuel oil	t	—
Heating power	×10⁸ kJ	—
Electricity	×10⁴ kWh	−0.31

that are energy intensive are also capital intensive. Industries within the sector compete among themselves and with foreign producers for sales to consumers. Consequently, variations in input prices can have significant competitive impacts. The most significant determinant of industrial energy consumption is demand for final output.

The manufacturing industries are modeled through the use of a detailed process-flow or end-use accounting procedure. The dominant process technologies are characterized by a combination of unit energy consumption estimates and "technology possibility curves." The technology possibility curve is an exponential growth trend corresponding to a given average annual growth rate, technology possibility coefficient, or elasticity of energy. The elasticity defines the assumed average annual growth rate of the energy intensity of a process step or an energy end use. The formula is

$$C_E = \frac{\alpha(\text{annual average of energy consumption})}{\beta(\text{annual average of economic output})} \quad (5.1)$$

in which,

$$\alpha = \left(\frac{E}{E_0}\right)^{\frac{1}{t_i-t_0}} - 1 \quad (5.2)$$

$$\beta = \left(\frac{M}{M_0}\right)^{\frac{1}{t_i-t_0}} - 1 \quad (5.3)$$

The elasticities of energy in Wanzhou, as a case study, are listed in Table 5.3.

5.1.3.2.2 Traffic Energy Demand

In terms of primary energy use in 1996, transportation sector carbon emissions, which almost equaled industrial carbon emission levels, were the second highest among the end-use demand sectors. Nearly, 33% of all carbon emissions and 78%

TABLE 5.4

Fuel Efficiency Index and Driving Parameters of Motor Vehicles

Items		Passenger Car	Freight Car	Motorcycle
Fuel economy	2005	9.5	13.7	3.50
standards	2010	8.60	12.3	3.50
(L/100 km)	2015	8.13	11.3	3.50
	2020	7.74	10.2	3.50
Average speed of vehicles (km/h)		30	20	35
Annual travel time (hour)		600	600	600

of carbon emissions from petroleum consumption originate from the transportation sector. In the reference case, carbon emissions from transportation are projected to grow at an average annual rate of 1.9% to 2010, compared with 1.4% for the commercial sector and 1.2% for both the residential and industrial sectors. In addition, transportation is the only sector with increasing carbon emissions projected for the period from 2010 to 2020 in the carbon reduction cases. Therefore, if there are no specific initiatives to reduce carbon emissions in the transportation sector, especially beyond 2010, increasing pressure may have to be exerted in the other sectors in order to reach and then maintain 2010 carbon emissions targets beyond 2010.

Consumers select light-duty vehicles (cars, vans, pickup trucks, and sport utility vehicles) based on a number of attributes: size, horsepower, price, and cost of driving, and weigh these attributes based on their personal preferences (Atkins 1999). This analysis uses past experience to determine the weight that each of these attributes have in terms of consumer preferences for conventional vehicles. Technologies are represented by component (e.g., front wheel drive, electronic transmission type) with each technology component defined by a date of introduction, a cost, and a weight that indicates its impact on efficiency and horsepower. The fuel efficiency indices and driving parameters of motor vehicle are listed in Table 5.4.

5.1.3.2.3 Domestic Energy Demand

The number of occupied households is the most important factor in determining the amount of energy consumed in the residential sector. All else being equal, more households mean more total use of energy-related services.

The end-use services for which equipment stocks are modeled include space conditioning (heating and cooling), water heating, refrigeration, freezers, dishwashers, clothes washers, lighting, furnace fans, color televisions, personal computers, cooking, clothes drying, ceiling fans, coffee makers, spas, home security systems, microwave ovens, set-top boxes, home audio equipment, rechargeable electronics, and VCR/DVDs. In addition to the major equipment-driven end uses, the average energy consumption per household is projected for other electric and nonelectric appliances. The module's output includes number of households, equipment stock, average equipment efficiencies, and energy consumed by service, fuel, and geographic location. The fuels represented are distillate fuel oil, liquefied petroleum gas, natural gas, kerosene, electricity, wood, geothermal, coal, and solar energy.

5.1.4 ENVIRONMENTAL EFFECTS OF ENERGY USE

5.1.4.1 Atmospheric Environment Quality Influence of Energy Use

5.1.4.1.1 Atmospheric Baseline Evaluations

As part of the environmental assessment process, there is a requirement to evaluate the existing baseline in an area proposed for development. This is to develop a general background level so that the cumulative impacts from the development can be assessed in the environmental assessment and/or to determine what the impacts would be with no expansion in place.

5.1.4.1.2 Atmospheric Environmental Impact Forecast Analysis

1. *Atmospheric environmental impact forecast methods:* According to the energy demand and pollution emission forecasting results in base scenario, the atmospheric environmental impact can be forecasted quantificationally.

   a. *Combustion emissions:* Air pollutions are most closely related to energy and fuel consumption. However, some tools may be common to both greenhouse gas (GHG) and air quality analysis. The Air Resources Board's (ARB) EMission FACtors (EMFAC) model, for instance, produces estimates of CO_2 emissions and fuel usage as well as the more traditional mobile source emissions. Project-level CO_2 emissions from highway operation (not construction) can also be estimated using the CT-EMFAC tool. This tool was developed as an interpretation of the California Air Resource Board's EMFAC 2007 model that simplifies the process of developing composite emission factors for highway project air quality analysis. The equation of atmospheric emission after energy consumption is

$$\begin{aligned} \text{Emissions of major pollutants} &= \text{Emission factor} \times \text{Energy consumption} \\ &= \text{Pollutant produce coefficient} \times (1 - \text{reducing} \\ &\quad \text{rate after control}) \times \text{Energy consumption} \end{aligned}$$

(5.4)

The emission factors of fuel combustion are listed in Table 5.5.

TABLE 5.5
Emission Factors of Fuels

Type	Emission Factor (kg/t)		
	SO$_2$	NO$_2$	Dust
Bituminous coal	16 S*	9.08	6
Fuel oil	18.68 S*	8.57	1.8
Diesel	11.97 S*	3.02	2.08

Note: The emission factor reference to national average level. S* means the sulfur in fuels (%).

b. *Vehicle emissions:* Mobile sources account for a large fraction of fossil fuel combustion in most countries. Of this, the largest source is road transport. In 1996, road transport accounted for 24% of CO_2 emissions from fuel use in the United States, while in Europe, the figure was 22%.

Road transport emits mainly CO_2, NO_x, CO, and nonmethane volatile organic compounds (NMVOCs); however, it is also a small source of N_2O, CH_4, and NH_3. Therefore, the only major direct GHG emission is CO_2. Emissions of CO_2 are directly related to the amount of fuel used. Emissions of the remaining gases depend on the amount of fuel used but are also affected by the way the vehicle is driven (e.g., the speed, acceleration, and load on the vehicle), the vehicle type, the fuel used, and the technology used to control emissions (e.g., catalysts). Thus, the simplest way to estimate the emissions of other gases is to use fuel-based emission factors; this is only appropriate where there is insufficient data to use the more complete methods available.

Methodology used depends on national legislation and availability of statistical data. The Intergovernmental Panel on Climate Change (IPCC) guidebook is based on USA (American Automobile Manufacturers Association, 1998) and European* experience.

The equation of vehicle emissions forecasting is

$$Q_{\text{motor vehicle}} = \sum_{i=1}^{n} P_i \times L_i \times K_i \times 10^{-6} \qquad (5.5)$$

where $Q_{\text{motor vehicle}}$ is the total emission of automotive vehicle, t; P_i is the population of motor vehicles i; L_i is trip mileage of motor vehicle i; n is the type of vehicles; K_i is the emission factor of motor vehicle i, g/km.

The emission factors of common motor vehicles are listed in Table 5.6.

TABLE 5.6

Emission Factors of Common Motor Vehicles (Unit: g/km)

Type	Emission Factor				
	SO_2	NO_2	Dust	CO	HC
Large cars	1.47	5.36	1.40	17.39	2.21
Medium cars[a]	0.79	4.6	0.96	51.7	8.1
Small cars	0.11	1.74	0.53	18.54	2.80
Mini cars	0.05	1.50	0.24	33.50	3.34
Motorcycles	0.08	0.17	0.17	14.40	2.0

Note:
[a] The SO_2 and dust emission factors reference to the average values of China.

---

* CO_2 emissions can be estimated from the mileage; however, it is usually best to estimate the total emission from the fuel consumption (as this is the more reliable data) and allocate this emission to the vehicle types by vehicle mileage data and relative fuel efficiencies.

TABLE 5.7

Emission Factors of Unit Fuel in IPCC (Unit: g/MJ)

	CH_4	N_2O	NO_x	CO
Medium Fuel Oil (MFO)	0.007	0.002	1.8	0.18

c. *Power vessel emission:* Power vessel emits mainly NO_x, SO_x, CO_2, CH, and PM_{10}. The simplest way to estimate the emissions is similar to road emissions-based emission factors. The equation of power vessel emissions forecasting is

$$Q_{power\ vessel} = \sum Q_{fuel} \times K \times \alpha \times 10^{-6} \quad (5.6)$$

where $Q_{power\ vessel}$ is the total emission of power vessel, t; Q_{fuel} is the diesel consumption of power vessel i, t; α is the thermal conversion coefficient of diesel, kJ/kg; K is the emission factor of power vessel i, g/hph. IPCC is used to calculate the GHG emissions (see Table 5.7).

2. *Air environmental capacity analysis:* The Division of Air Pollution Control is directed to maintain the purity of the air resources consistent with the protection of normal health, general welfare, and physical property of the people while preserving maximum employment and enhancing the industrial development.

5.1.4.2 Ecological Footprint: A Tool for Assessing Sustainable Energy Supplies

National ecological footprint accounts consist of two measurements. The footprint aggregates the total area that a given population, organization, or process requires to produce food, fiber, and timber; provide space for infrastructure; and sustain energy consumption. The biocapacity aggregates the total area available to supply these demands.

The ecological footprint focuses on six potentially renewable demands: cropland, pasture, forests, built-up land, fisheries, and energy. Activities in these components are deemed sustainable if rates of resource use and waste generation do not exceed a defined limit that the biosphere can support without degrading the resource stock. By including a core set of potentially sustainable activities, the ecological footprint defines minimum conditions for sustainability.

To provide a quantitative answer to the research question of how much regenerative capacity is required to maintain a given resource flow, ecological footprint accounts use a methodology grounded on six assumptions:

1. It is possible to use annual national statistics to track resource consumption and waste generation for most countries.

2. Resource flows can be measured in terms of the bioproductive area necessary for their regeneration and the assimilation of their waste. (Resource and waste flows that cannot be measured are excluded from the assessment.)
3. Bioproductive areas of different productivity can be expressed in a common unit of standardized usable biological productivity. Usable refers to the portion of biomass used by humans, reflecting the anthropocentric assumptions of the footprint measurement.
4. The sum of mutually exclusive areas needed to maintain resource flows expressed in a common unit represents aggregate demand; the sum of mutually exclusive bioproductive areas expressed in a common unit represents aggregate supply.
5. Human demand (footprint) and nature's supply (biocapacity) are directly comparable.
6. Area demand can exceed area supply, meaning that activities can stress natural capital beyond its regenerative capacity. For example, the products from a forest harvested at twice its regeneration rate have a footprint twice the size of the forest. A footprint greater than biocapacity indicates ecological deficit. Ecological deficits are compensated in two ways: either the deficit is balanced through imports (ecological trade deficit) or the deficit is met through overuse of domestic resources, leading to natural capital depletion (ecological overshoot).

Cropland, forests, pastures, and fisheries vary in biological productivity or their capacity to provide ecological goods and services through photosynthesis. One hectare of arid rangeland, for example, has less capacity to recycle nutrients, produce food, and support diverse land-use patterns than one hectare of temperate forest. The ecological footprint, therefore, normalizes each bioproductive area—cropland, pasture, forests, built-up land, and fisheries—into common units of "global hectares" (gha).

Current accounts weight productivity according to agricultural suitability—a function of numerous factors, including temperature, precipitation, soils, and slope. The Food and Agriculture Organization and International Institute for Applied Systems Analysis have created a spatially explicit distribution of these variables in the "suitability index" of Global Agro-Ecological Zones 2000. Recent ecological footprint accounts use this index to translate unweighted hectares into global hectares (see Table 5.8). Other possibilities for weighting productivity include potential primary production and the rate of economically useful biomass production.

5.1.5 RESPONSE: ENERGY CONTROL

Scenario development as an aid to planning is focused on developing alternative visions of the future. Visioning exercises typically look farther into the future (i.e., 10 years or more) than other futures methods. Scenario planning (or scenario learning) has proven to be a disciplined method for imagining possible futures in which decisions may be played out, and a powerful tool for asking "what if" questions to explore the consequences of uncertainty. By working with scenarios of quite

TABLE 5.8
Global Biocapacity

Area	Equivalence Factor (gha/ha)	Global Area (Billion ha)	Biocapacity (Billion gha)
Cropland	2.1	1.5	3.2
Pasture	0.5	3.5	1.6
Forest	1.3	3.8	5.2
Built-up land	2.2	0.3	0.6
Fisheries	0.4	2.3	0.8
Total	1.0	11.4	11.4

Note: The relative productivities of major global bioproductive areas, expressed in actual area (hectares) and productivity weighted area (global hectares). Equivalence factors are based on the suitability index of Global Agro-Ecological Zones 2000 (FAO/IIASA). Data are rounded.

different futures, the analytical focus is shifted away from trying to estimate what is most likely to occur toward questions of what are the consequences and most appropriate responses under different circumstances.

There are various approaches for developing scenarios:

1. Define the topic/problem and focus of the scenario analysis.
2. Identify and review the key factors/environmental influences on the topic.
3. Identify the critical uncertainties.
4. Define scenario logics (often using scenario matrices).
5. Create/flesh out the scenarios.
6. Assess implications for business, government, and the community.
7. Propose actions and policy directions.

5.1.6 CONTROL SCENARIO OPTIMIZATION (EPIC FRAME)

We propose the energy pressure-energy impact-energy control (EPIC) frame (see Figure 5.2).

The overall objective is

$$\min B = \sum_{t=1}^{m} \sum_{i=1}^{n} E_i^t b_i \qquad (5.7)$$

where B is the total cost of energy use in planning year; Et_i is the demand of the energy i in planning year t; b_i is the cost of the energy i.

The first constraint condition is

$$\sum_{i=1}^{n} E_i p_i(h,s,j) \leq Q \qquad (5.8)$$

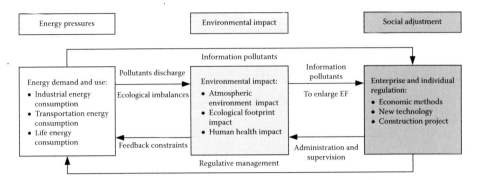

FIGURE 5.2   Energy pressure-energy impact-energy control (EPIC) frame.

where p_i is the emission factor of the emission k of the energy i; h, s, and j are constant coefficients of emission factors; Q is the environmental capacity of the emission k.

The second constraint condition is

$$\frac{\sum_{i=1}^{n} (E_{i1} - E_{i0})}{\sum_{i=1}^{n} E_{i0}} \leq r \frac{G_{t1} - G_{t0}}{G_{t0}} \tag{5.9}$$

where G_{t1} is the gross domestic product (GDP) in target; G_{t0} is the GDP in base year; r is the growth coefficients of energy consumption and GDP, the desired value is 0.5.

The third constraint condition is

$$EC - EF = ES \geq 0 \tag{5.10}$$

where ES is ecological remainder, hm^2; EC is forest area, hm^2; EF is the ecological footprint of waste gas, hm^2.

5.2   PLANNING OF AIR ENVIRONMENTAL QUALITY IMPROVEMENT

5.2.1   Atmospheric Quality Assessment

5.2.1.1   Evaluation Method and Comprehensive Index

Four pollutants including NO_2, SO_2, CO, and total suspended particle (TSP) are chosen as the evaluation factors for evaluating the air quality (Morris and Therivel 1995). The comprehensive index for evaluating the air quality could be calculated as follows:

$$P_i = \frac{C_i}{S_i} \tag{5.11}$$

$$P_{average} = \sum_{i=1}^{n} P_i / n \tag{5.12}$$

TABLE 5.9

Ambient Air Quality Standard in China (GB3095-1996)

	Concentration Limits (mg·m⁻³)	
Type	Time	Grade II
SO_2	Average daily	0.15
NO_2	Average daily	0.08
TSP	Average daily	0.30
CO	Average daily	4.00

where P_i is the air pollution index of the pollution i; $P_{average}$ is the average type comprehensive index of air quality; C_i is the average value of the pollution i per year; S_i is the air quality standard of the pollution i.

5.2.1.2 Evaluation Criterion

The concentrations of NO_2, SO_2, CO, and TSP adopt the Grade II of Ambient Air Quality Standard in China (GB3095-1996) (see Table 5.9).

5.2.2 ATMOSPHERIC ENVIRONMENT FUNCTION DIVISION

The function of the atmospheric environment division is mainly to make clear the process of the transportation of energy and chemical substances particularly in the lower atmosphere. Aims are to reduce uncertainities and uncompleteness of the method to predict future atmospheric environment and its application. The objectives of National Ambient Air Quality Standard are shown in Table 5.10.

5.2.3 ASSESSING AIR QUALITY IMPROVEMENTS

5.2.3.1 Estimating the Improvement Required

Before identifying the options it has available for improving air quality, the local authority will need to determine the overall level of improvement required. This can be calculated simply, in g·m⁻³, as the difference between the total predicted concentration (from the Stage 3 Review and Assessment) and the relevant air quality objective. This can be expressed in terms of concentration units or as a percentage.

It is important that the point of maximum concentration, where exposure is likely, has been identified, and the required improvement is calculated using this information. Having said this, consideration also should be given to the need to allow for some headroom for future development or uncertainty in the overall assessment process. It may be appropriate, therefore, to seek a greater percentage improvement than would otherwise be required just to meet the objective. However, any additional requirement of this type will need to be properly justified as it will almost inevitably have implications for the costs of compliance.

TABLE 5.10
National Ambient Air Quality Objective

Type	Time	Grade I	Grade II	Grade III	Unit
			Concentration Limits		
SO_2	Annual average	0.02	0.06	0.1	mg·m^{-3} (normal state)
	Daily average	0.05	0.15	0.25	
	1 hour average	0.15	0.5	0.7	
TSP	Annual average	0.08	0.2	0.3	
	Daily average	0.12	0.3	0.5	
PM_{10}	Annual average	0.04	0.1	0.15	
	Daily average	0.05	0.1.5	0.25	
NO_x	Annual average	0.05	0.05	0.1	
	Daily average	0.1	0.1	0.15	
	1 hour average	0.15	0.15	0.3	
NO_2	Annual average	0.04	0.04	0.08	
	Daily average	0.08	0.08	0.12	
	1 hour average	0.12	0.12	0.24	
CO	Daily average	4	4	6	
	1 hour average	10	10	20	
O_3	1 hour average	0.12	0.16	0.2	
Pb	Season average	1.5			
	Annual average	1			
B[a]P	Daily average	0.01			
Fluoride	Daily average	7			
	1 hour average	20			
	Month average	1.8	3.0		mg·m^{-3}
	The growing season average	1.2	2.0		

This general approach to assessing the required overall level of reduction works well for conservative pollutants, which do not undergo any significant change as a result of atmospheric chemistry. However, the assessment of the impact of the releases of nitrogen oxides represents a particular problem as the emission is usually mainly NO, which is converted to NO_2 in the atmosphere. A number of empirical relationships have been derived to convert NO_x to NO_2, for example, Derwent et al., 1996 (other examples of NO_x:NO_2 conversion are discussed in (LAQM. G2(00), 1999)). The appropriate factor will vary with source type, and where the ambient concentration is made up of contributions from different sectors it is difficult to combine these in a manner that enables the percentage improvement, expressed as NO_x, to be calculated robustly. Taking into account these technical difficulties, it may be better to present the required reductions of nitrogen oxide compounds in terms of NO_2. For consistency, local authorities should use the same NO_2:NO_x relationship in preparing action plans as was used in their Review and Assessment.

An example of how relative reductions from different sources can be calculated is as follows: Assuming that a 4 μg·m^{-3} reduction in annual average NO_2 (from 44 μg·m^{-3} NO_2 annual average) is required, and that this represents a reduction of 9%, the reduction in NO_x can be calculated using a relevant NO_x:NO_2 conversion relationship. In this example, the assumption is made that

- 44 μg·m^{-3} NO_2 equates to 80 μg·m^{-3} NO_x.
- 40 μg·m^{-3} NO_2 equates to 69 μg·m^{-3} NO_x.
- 4 μg·m^{-3} NO_2 reduction equates to 11 μg·m^{-3} NO_x reduction, which represents 14% reduction of NO_x.

The local authority has identified the relative source contributions of NO_x in this area as 50% road traffic, 20% industry, and 30% background (unaccounted sources). As road traffic appears to be the primary source of NO_x, the local authority has decided to calculate the percentage improvement in road traffic needed to effect a 14% improvement in NO_x (or 11 μg·m^{-3} reduction):

Traffic contributes 50% of the total 80 μg·m^{-3} NO_2; this 50% contribution is equivalent to a contribution of approximately 40 μg·m^{-3} NO_2 from road traffic:

$$(50 \times 80)/100 = \text{Value of } NO_x \text{ contribution from road traffic} = 40 \mu g \cdot m^{-3}$$

As the local authority wishes to achieve an 11 μg·m^{-3} reduction in NO_x, it is the objective to reduce the road traffic contribution by 11 μg·m^{-3} to 29 μg·m^{-3}.

Therefore, if all the emission reduction was expected to come from road traffic sources, the percentage improvement needed for road traffic can be calculated using the equation:

$$\text{Improvement from road traffic} = ((\text{Predicted value} - \text{Required value})/\text{Predicted value}) \times 100$$
$$= ((40 - 29)/40) \times 100 = 28\%$$

Due to uncertainties and allowing a degree of "head space" for future development, this figure is rounded up to 30%. A 30% reduction in NO_x emissions from road traffic is, therefore, required to meet the air quality objectives, assuming that road traffic is the only source from which reductions are required.

5.2.3.2 Selection of Options to Improve Air Quality

The next stage in the selection and development of options is to identify the sources where controls might be effective in reducing concentrations and which make a significant contribution to the exceedance of a particular objective. This would, for example, exclude most background sources. Control options can then be identified for the "relevant" sources. The wall chart provided with this guidance identifies a large number of control options and while the list is by no means exhaustive, it will provide a good starting point.

Where further controls within an action plan are required from industrial processes, the appropriate regulatory authority (either the Environment Agency or Local Authority) may need to seek further information from the operators concerned as to the additional control measures that might be applied. The appropriate level of reduction will then be determined in a manner consistent with the appropriate legislative requirements. It will, therefore, not be possible in most circumstances to identify in advance the level of reduction that can be achieved from industrial processes. To deal with this issue, it is suggested that in discussion with the regulator, a range of possible scenarios are identified representing, for example, high, medium, or low percentage reductions (NSCA 1999).

In a large number of cases, it will not be possible for individual options to deliver the entire reduction required. Indeed, it may be more effective to combine a number of options to deliver the required improvement. It is, therefore, a key task to identify the optimum mix of options taking into account the considerations discussed in this guidance.

5.2.3.3   Further Modeling of Options

Initially, local authorities will wish to assess the potential for a wide range of options to reduce air pollution. However, it will probably not be practicable to use complex modeling software to assess all these and, therefore, a relatively simple screening approach could be used to assemble option packages. These can then be subjected to the other considerations discussed in this guidance, such as cost-effectiveness, non-air-quality impacts, and practicability. Once a "shortlist" of possible emission reduction scenarios have been identified, further detailed dispersion modeling will need to be undertaken to properly assess the improvement in the area of exceedance or Air Quality Management Area. Where a number of control options have been identified, they can be combined into different scenarios and their ability to deliver the required level of improvement is considered.

REFERENCES

American Automobile Manufacturers Association. (1998). *Economic Indicators: The Motor Vehicle's Role in the U.S. Economy.* Detroit, Mich.

Atkins W. S. (1999). *An Evaluation of Transport Measures to Meet NAQS Objectives: Stage 2 Final Report.* WS Atkins Planning Consultants.

Derwent R. G., Jenkin M. E., & Saunders S. M. (1996). Photochemical ozone creation potentials for a large number of reactive hydrocarbons under European conditions. *Atmospheric Environment.* 30, 181–199.

Energy Policy Act of 2005 (EPAct 2005). http://frwebgate.access.gpo.gov/cgi-bin/getdoc .cgi?dbname=109_cong_bills&docid=f:h6enr.txt.pdf.

Local Air Quality Management Guidance (LAQM.G2(00)). (1999). *Developing Action Plans and Strategies: The Principal Considerations.* DETR.

Morris P. & Therivel R. (Eds) (1995). *Methods of Environmental Impact Assessment.* UCL Press, London. ISBN 1-85728-215-9.

National Society of Clean Air and Environmental Protection (NSCA) (1999). The How To Guide: Consultation for Local Air Quality Management. NSCA, Brighton, UK.

6 Urban Water Environment Quality Improvement Plan

Yanwei Zhao and Zhifeng Yang

CONTENTS

6.1   HEALTHY ASSESSMENT ON URBAN WATER ECOLOGY

6.1.1   Definition of Urban Aquatic Health

As an analogy concept of human health, the definition of a water body's health is unspecific yet (Norris and Thomas 1999). The main divergence of the academic opinions between scholars is the involvement of human value. Karr (1999) treated the original ecological integrity of a water body as the health, regarding that the biocenosis is nearly undisturbed and is able to construct the diversity and the functional framework. Norris and Thomas (1999) believed that a water body's health relies on the judgment of the social system, so the requirement of human's welfare should be considered. Meyer (1997) gave a comprehensive illustration, regarding that a healthy water body should be able to sustain the framework and functions of an ecological system, including human and social values. This understanding stated by Meyer is well accepted by most scholars.

City is a highly artificial and human-centered compound ecological system. The operation of the system significantly relies on the continuance of the eco-service function of the city water body. The health of the city water body not only means the reasonable framework, sustainable process, performance, and integrity in ecology but also the services for water supply, flood control, conservation of water and soil, and entertainment. Therefore, the health of the city water body is an integral concept of human development and ecological protection, a statement of menace and despondence between human and water body. Assessing the health of the city water body and setting an administrative goal have to be based on the public and social anticipation and the judgment of human value.

6.1.2   Assessment Method of Urban Aquatic Health

The integrative fuzzy hierarchical assessment model concludes five steps: to aggregate the assessment indices; to build an aggregation of assessment standards; to calculate index weighs by analytical hierarchical process (AHP); to build single-factor judgment matrix; and to carry out fuzzy synthesis and integrative assessment. These five steps are illustrated in details as follows.

6.1.2.1   Aggregation of Assessment Indices

The aquatic health can be described by five elements such as water quantity, water quality, riparian zone, aquatic life, and physical form (Ladson et al. 1999), which are interdependent and interactive and can cover different ecological processes, perform different functions, and form the whole ecosystem.

Water quantity and water quality are the two important attributes of the water source. Water quantity is an important carrier to express how flow regimes are varying, which reflects synthetically the climate attributes of the drainage basin, the coverage attributes of the earth's surface, the landform and topography of the aquatic ecosystem, and how much aquatic ecosystem is disturbed by human activities. Water quality is the fundamental guarantee of social productivity and health of biology and human beings. The organic combination of these two characteristics is essentially required by the existence of aquatic organism and the accomplishment of physical process and biochemical reaction of the aquatic ecosystem. At the same time, their combination also guarantees the development of social economy. Changes of the flow velocity and water level caused by water exploitation and exploitation and utilization rate are the two indices used to describe how much water quantity is disturbed by the socioeconomic activities of human beings. Fluid quality can represent water environmental quality with the water quality index (WQI), while sediment pollution can indicate potential water environmental pollution pressure with the pollution index of the sediment.

Riparian area lies in the ecologically fragile land on the boundary of land–water ecotones and is one of the most heterogeneous and complicated ecosystems (Jungwirth et al. 2002). In addition, it plays an important role in maintaining regional biodiversity, accelerating the exchange of material and energy, resisting flow erosion and infiltration, filtrating and absorbing nutriments, and so on (Mckone 2000), embodying three aspects of ecological function: corridor, buffer, and retaining wall (Zhang 2001). The disturbance suffered by an urban riparian area contains mostly unsound invasion of land use by human beings, changes of the hydraulic disturbance mechanism, landscape gradient destruction, and the corridor disjunction caused by infrastructure construction. The functions of a riparian area are embodied in soil and water erosion control, landscape effects, and flood control. A riparian area is extremely important to the landscape functions, disaster resistance, and biological conservation of the urban aquatic ecosystem and is scaled in the width of a riparian area, vegetation coverage area, effect and reachability of landscape, and standard of flood control.

The condition of aquatic life form is a relatively aggregative expression of the health of an aquatic ecosystem, which reflects intimidation caused by the activities of human beings and accumulative effect caused by natural ecological succession of the aquatic ecosystem. It is indicated by the index of biological integrity (IBI) and the surviving condition of endangered and rare species, and the latter is imported to reflect the protecting requirements of special species.

The change of physical structure is caused directly by the physical rebuilding activities of human beings. The changes of physical structure are caused directly by the rebuilding activities, which can be manifested in four aspects including the exchange ability between water body and river band and river wall, the environment of habitat and migration, physical fitness and connectivity, and be represented by solidified condition of river bank, solidified condition of river wall, riverbed, riverbank stability, connectivity with water body around (lake, marsh, etc.) and ecological patches (greenbelt, park, etc.), connectivity of river corridor, habitat integrality, fishway setting, and hydrological facilities that impede fish migration.

According to the above analysis, the aquatic health assessment indicator system, containing 17 assessment indicators, is established (Table 6.1).

TABLE 6.1

Aquatic Health Assessment Indicator System

Elements	Items	Indicators
Hydrology	Hydrology	Changes of flow velocity and water level caused by water exploitation
	Water quantity	Exploitation and utilization rate
Water quality	Fluid	WQI
	Sediment quality	Sediment pollution index
Aquatic life	Biotic integrity	IBI of fish
	Rare and endangered species	Surviving conditions of rare and endangered species
Riparian zone	Soil and water erosion control	Width of riparian area
		Vegetation coverage of riparian area
	Landscape construction	Area, effect, and reachability of recreation facilities
	Flood control	Flood control guarantee rate
Physical structure	Exchange ability	Solidified condition of bank and course of the water body
	Physical fitness	Bed stability of the water body
		Bank stability of the water body
	Connectivity	Connectivity with natural ecologic patches
		Connectivity of the corridor of the water body
	Inhabitant and migration	Habitat status
		Fishway setting

6.1.2.2 Building the Aggregation of Assessment Standards

Determination of assessment standards is the focus and difficulty for the ecosystem health assessment (Ma et al. 2001). Especially for the aquatic ecosystem, there is not yet an agreed understanding and a uniform standard at present (Karr 1999). Assessment standards vary with the location, size, type, and phase of ecological succession of the aquatic ecosystem and social expectation of different stakeholders. Currently, common methods to determine the aquatic health assessment standard are (1) referring to historical materials, (2) on-the-spot investigation, (3) reference contrast method (4) using the national standard and correlative research results for reference, (5) public participation, and (6) experts judging (Zhao and Yang 2005), of which each method has its own merits and demerits and is applicable to different types of indicators.

In this research, in terms of the nature of illegibility and relativity of the aquatic ecosystem health, the health assessment standard includes five ranks as very healthy, healthy, subhealthy, unhealthy, and sick. Quantificational indices, such as exploitation and utilization rate of water source, WQI, and so on, are commonly rated by using the methods of historical monitoring material for reference, referring to the national standard and correlative research results; nonquantificational indices are marked and carved into five thresholds as <1, 1–2, 2–3, 3–4, and >4. The marks are rated by the experts' judgment, in the base of public participation.

6.1.2.3 Deciding Index Weights

AHP proposed by Zhao et al. (1986) is used usually to decide the weight of each element and the total hierarchical weight aggregation of each indicator. First, the hierarchical structure model is built by regarding the aquatic ecosystem as the total object layer. Second, hydrology, water quality, physical form, riparian zone, and aquatic life are established as the first-level subobject, while hydrology, water quality, fluid sediment quality, and landscape construction as the second-level subobject, and each specific index as the third level. Then, experts in ecology, hydrology, and environmental science and managers in water conservancy, environmental protection, fishery, and municipality departments are invited to mark according to the standard scale table so as to construct the judgment matrix. Finally, the weights of every element and the total hierarchical order of each indicator are rated by the order of importance and coherence verification.

6.1.2.4 Building Single-Factor Judgment Matrix R

In terms of the responding relationship of indicator values to the aquatic ecosystem health conditions, indicators can be classified as two types such as a positive one (the larger the index value is, the healthier the river is) and a negative one (the smaller the index value is, the healthier the river is). The steps to calculate the membership functions of these indicators are as follows (s_{ij} is the j-level health standard of indicator i, and r_{ij} is the relative membership grade of indicator i vs. standard j):

1. Positive index, such as the standard of flood control and width of riparian area.
   a. When x_i is less than its corresponding first-level standard (sick), $r_{i1} = 1$; $r_{i2} = r_{i3} = r_{i4} = r_{i5} = 0$.
   b. When x_i is between its corresponding j-level standard s_{ij} and $j+1$-level standard $s_{i,j+1}$, $r_{ij} = \dfrac{s_{i,j+1} - x_i}{s_{i,j+1} - s_{i,j}}$, $r_{i,j+1} = \dfrac{x_i - s_{i,j}}{s_{i,j+1} - s_{i,j}}$, and the membership grades to other health level are zero.
   c. When x_i is bigger than its corresponding fifth-level standard (very healthy), $r_{i1} = r_{i2} = r_{i3} = r_{i4} = 0$; $r_{i5} = 1$.
2. Negative index, such as WQI and exploitation and utilization rate of water resources.
   a. When x_i is bigger than its corresponding first-level standard (sick), $r_{i1} = 1$; $r_{i2} = r_{i3} = r_{i4} = r_{i5} = 0$.
   b. When x_i is between its corresponding j-level standard $s_{i,j}$ and $j+1$-level standard $s_{i,j+1}$, $r_{i,j} = \dfrac{x_i - s_{i,j+1}}{s_{i,j} - s_{i,j+1}}$, $r_{i,j+1} = \dfrac{s_{i,j} - x_i}{s_{i,j} - s_{i,j+1}}$, and the membership grades to other health level are zero.

c. When x_i is less than its corresponding fifth-level standard (very healthy), $r_{i1} = r_{i2} = r_{i3} = r_{i4} = 0$; $r_{i5} = 1$.

According to the classifications and calculating methods mentioned previously, we need to decide the membership function of each specific indicator and then calculate the judgment matrix of each indicator of five types of assessment elements.

$$R = \begin{bmatrix} R_1 \\ R_2 \\ \cdots \\ R_n \end{bmatrix} = \begin{bmatrix} r_{11} & r_{12} & \cdots & r_{15} \\ r_{21} & r_{22} & \cdots & r_{23} \\ \cdots & \cdots & \cdots & \cdots \\ r_{n1} & r_{n2} & \cdots & r_{n5} \end{bmatrix}$$

6.1.2.5 Carrying out Integrative Fuzzy Hierarchical Assessment

Fuzzy synthesis is accomplished by fuzzy weighting linear transform, namely $B = W \cdot R = (B_1, B_2, B_3, B_4, B_5)$. Then the membership matrixes of an aquatic ecosystem to five health levels, such as very healthy, healthy, subhealthy, unhealthy, and sick, are obtained. The results to conduct the integrative river health assessment are unified, and the membership matrixes of the five elements to the health level can also be elicited by similar methods.

6.2 SECURITY ASSESSMENT OF URBAN WATER ECOLOGY

6.2.1 CONCEPT DEFINITION OF WATER SECURITY

Water resource security, water environment security, and water ecological security are considered the three different aspects of water security, all of which interrelatedly constitute the research contents on water security.

Water resource security refers to the security of the water supply for the social and economic development, including meeting the required water demand of the social and economic development and also ensuring the water use efficiency (quality and quantity) of the social and economic development to reach the safety standard. Water environment security means the effective control of water pollution, as well as the facts that the quantity of pollutant discharged does not exceed the carrying capacity of the water environment, and the water safety of all sectors of the society can be guaranteed. Water ecological security is defined as the conditions that the minimum ecological water demand of the river ecosystem is satisfied, the ecosystem is not collapsed because of human's superfluous ecological water use, the impact of the natural disasters to the aquatic ecosystems is reduced, and the healthy development of the ecosystem is maintained.

According to the previously mentioned definition and connotation of water security, the conceptual framework of water safety in Wanzhou District is proposed (Figure 6.1).

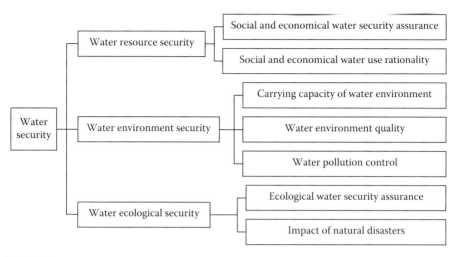

FIGURE 6.1   The conceptual framework of water safety in Wanzhou District.

6.2.2   EVALUATING INDICATOR SYSTEM OF WATER SECURITY

The indicators of water resource security, water environment security, and water ecological security are shown in Table 6.2.

6.2.3   EVALUATION METHOD AND MODEL

The linear weighted sum method is determined as the evaluation method of water security. Its basic formula is as follows:

$$WSI = \sum_{i=1}^{n} w_i x_i$$

In the formula, WSI is water security indicator, w_i is the weight of the i indicator; and x_i is the standard value of the i indicator. The method of calculation is as follows:

$$x_i = \frac{c_i}{c_0}$$

(when the value of the indicator has a positive correlation with the water security level)

$$x_i = \frac{c_0}{c_i}$$

(when the value of the indicator has a negative correlation with the water security level)

$$x_i = 1$$

(when the value of the indicator is up to the standard)

In the formula, c_i and c_0 refer to the status value and standard value of the indicator.

TABLE 6.2

Evaluating Indicator System of Water Security in Wanzhou District

Destination Layer	Criterion Layer	Index Layer	Standard Value
Water resource security	Social and economical water security assurance	Water supply–demand equilibrium index	>0
	Social and economical water use rationality	Consumptive use per ten thousand yuan	≤150 m³/10,000 ¥
		Repeated utilization factor of industrial water	≥50%
		Water use coefficient of field irrigation	≥0.50
Water environment security	Carrying capacity of water environment (0.5)	Water environment capacity utilization	≤90%
	Water environment quality (0.25)	Qualification rate of water functional area	100%, and no water body exceeds IV in the city
		COD emission intensity	<5 kg/10,000 ¥
	Water pollution control (0.25)	Concentrated treatment rate of urban sewage	≥70%
		Qualification rate of the industrial waste water discharge	≥95%
		Harmless treatment rate of urban domestic garbage	100%
Water ecological security	Ecological water security assurance (0.5)	Water resources development ratio	≤30%
	Impact of natural disasters (0.5)	Control rate of water loss and soil erosion	100%

Note: Water supply–demand equilibrium index = (the quantity of available water resources – social and economical water demand)/the quantity of available water resources; the best security value of utilization coefficient of irrigation water refers to the requirements of the water utilization coefficient of the large irrigation field in "Technical Specifications of Water-saving Irrigation (SL207-98)"; the standard value of qualification rate of the industrial waste water discharge is based on "Assessment Index of Environmental Protection Model Cities"; in consultation with the experts, the standard values of the other eight indicators refer to the standard value determination in "Development Indicators of Ecological Provinces, Cities, and Counties."

6.2.4 EVALUATION CRITERIA

The WSI obtained from the linear weighted sum method falls into the interval [0,1]. When the indicator is 1, the water security level is the best; When the indicator is 0, the water security level is the lowest. On the basis of people's cognitive habits to the relation between the grade level and the quality, the nonspacing method is employed to divide the comprehensive index of water security to five levels, namely worst, worse, average, good, and ideal (Table 6.3).

TABLE 6.3
Grade Standard of Water Security Level

WSI	0–0.4	0.4–0.6	0.6–0.8	0.8–0.9	0.9–1
Level	I	II	III	IV	V
Security level	Worst	Worse	Average	Good	Ideal

6.3 SUPPLY–DEMAND ANALYSIS OF URBAN WATER RESOURCES

6.3.1 PRINCIPLES OF SUPPLY–DEMAND ANALYSIS

6.3.1.1 Comprehensiveness

In the supply–demand analysis, the development need of each subsystem in the urban complex ecosystem, including the ecological environment and the water demand systems in life and production, should all be taken into consideration. Meanwhile, the scope of water supply systems should be expanded, such as the reusing of industrial water, the recycling of domestic sewage, and the desalination of sea water, so as to reflect the dynamic role of social and economic systems.

6.3.1.2 Unity

Water resource is the unity of the quality and quantity of water, and only the organic integration of the quality and quantity of water can ensure the corresponding ecosystem services. In order to enhance the virtuous circle and development of the eco-city, the number of water resources and the variation trend of water quality should be considered in the supply–demand analysis. Only the organic unity of the water supply and water quality improvement can ensure the safe and sustainable water supply.

6.3.1.3 Initiative

With the protection, restoration, and improvement of the function of the water body as the operating points, the active response to the potential contradiction between supply and demand is needed. The emphasis on the cyclic utilization of water resource and sewage recycling emphasizing water recycling and water resource, combined with structural measures such as water conservation, source opening, and water transfer, will help to unify the development and protection of relatively independent water resources into supply–demand analysis, providing a fundamental solution to water supply–demand contradiction and promoting urban ecological construction.

6.3.1.4 Ecological Priority

Considering the demand of the development of natural ecosystems for water resource, on the premise that the basic domestic water demand is satisfied, the eco-environmental water supply including urban landscaping and water purification should be put into prior allocation so as to ensure the exertion of the restoring, buffering, and self-regulatory function of the eco-city and to promote the harmony under the protection of the natural ecological support system.

6.3.2 Conceptual Framework of Supply–Demand Analysis

In accordance with the previously mentioned principles, the conceptual framework of supply–demand analysis of water resources was established (Figure 6.2). The water demand system is composed of the life, environment, and production, while the water supply system consists of local surface water and groundwater, transit foreign water, water transfer, and recycling water.

6.3.2.1 Water Supply System

The local surface water and groundwater are the basic supply source of urban development, while with the dual pressures of upstream exploitation and pollution, the supply of the transit foreign water proves to be unstable because of the uncertainty of its water quality and quantity. In order to ease the regional ecological environment problem, the water transfer for environmental purposes has received great attention (Fadali and Shaw 1998); however, the inter-regional water transfer, restricted to costs, political coordination, water rights, inter-regional management, and other factors, easily leads to ecological problems. With technology advances and lower costs of sewage treatment, sewage recycling has gained a rapid development. In parts of the Nordic countries, the household wastewater treatment rate has increased from 70% to 90%, which has become an important water source for urban landscape with low-quality requirements. Compared with the water supply with hydraulic measures, the recycling of water is not affected from the spatial and temporal variability of rainfall and has good stability, increasing the supply and reducing the water pollution load.

The water sources can be categorized as controllable and uncontrollable. The controllable water includes stored water with hydraulic facilities (including

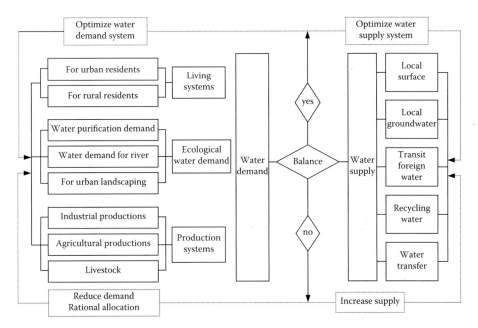

FIGURE 6.2   Conceptual framework of water resources supply–demand analysis.

water transfer) and recycled water, mainly used to meet the life, production, and landscaping water demand. The rest are uncontrollable water, used to provide hydrological conditions for habitat of the aquatic life, to maintain the balance of aquatic ecosystems and water purification, and to ensure the ecological functions of the water body. The utilization of the controllable water resources is completed by the socioeconomic system; therefore, it has high stability. The supply of the uncontrollable water resources is fulfilled through natural processes of water bodies, such as the dilution and diffusion led by the dynamics and the active utilization of the coastal vegetation; thus, it changes with the natural law of the river and has obvious seasonal, regional, and variability characteristics.

6.3.2.2 Water Demand System

1. *Ecological water demand:* Urban ecological water demand refers to the water resources required for maintaining the stability of the river ecosystem structure, the normal functioning of ecosystems, and the water environmental function, as well as ensuring the urban ecological construction. It includes the water demands for the river ecosystem and urban landscaping.

   a. *Water demand for the river ecosystem:* The basic flow of the river is the quantity foundation for the structure stability of the river ecosystem and its effective functioning. Recently, three research methods have been widely applied at home and abroad: standard flow method, hydraulic method based on hydraulics, and the habitat method on the biological basis (Yang et al. 2003). The Tennant method in the flow algorithm (Tennant 1976) (Table 6.4) is applicable to the rivers with seasonal changes, especially large-scale multichannel calculation of the basic flow.

   Water purification demand is considered the quality assurance for the river ecosystem structure stability and effective functioning of the water environment, and the functional requirement for fulfilling the survival of the aquatic life and completing the physical processes

TABLE 6.4
Base Flow Standard in "Tennant" Method

Criterion	October–March	April–September
Best	60–100	60–100
Better	40	60
Good	30	50
Medium	20	40
Bad	10	30
Worse	10	10

Note: The data are the percentage of the average flow.

and biochemical reactions of water bodies. Its quantity characterization reveals the pressure on aquatic ecosystems from pollution load in human's nonrational social and economic activities. Compared with the water demand in life and production, its quantity is generally in a large range, and it changes with the water quality objectives. As a result, it is characterized by the threshold, dynamic, and subjective features, as well as sensitivity to changes of the pollution load, playing a decisive impact on the supply-demand balance. The introduction of this concept is meant to express the threatened status of water quality with the water quantity under a quality standard so that the water quality requirement can be involved in the comparison of supply and demand, making the water quality and quantity in the unity. The formula is as follows:

$$Q = \frac{\left(Q_e - Q_p\right)}{\left(C_o - C_b\right)},$$

where Q refers to the water demand for water purification, Q_e to the total pollution emission of one pollutant, Q_p to the self-purification amount of pollutants in water, C_o to the water quality objectives, and C_b to the background value for the water pollutants.

   The basic flow of river and the water demand for purification are two components of the water demand of the river ecosystem, which have different functions. From the perspective of volume, the two are a natural combination, mutually overlapping each other and assuring each other. The one with more value is taken as the water demand of the river ecosystem.

   b. *Water demand of urban landscaping:* Water demand of urban landscaping points to the water needed for urban purification and vegetation systems such as grass, shelter, artificial gardens, and so on. It is mainly provided by artificial investment in social and economic systems. The water demand is calculated by the quota method, with the pavement area of urban paved roads, public green space, urban garden area, the tree numbers along the roads, and the corresponding water demand.

2. *Water demand of living systems:* It means the water demand of urban and rural residents. The water demand level is involved with the size of cities and towns, living conditions, living habits, urban climate, and other factors. The calculation method can be classified as the quantitative and qualitative predictions. The former includes the time regression method, dynamic correlation method, and combinatorial optimization method, and the latter refers to the comprehensive and classified water indicator method.

3. *Water demand of production systems:* The water demand of production systems includes that of the industrial and agricultural productions. The level of water demand is related to agricultural and industrial structure, technological processes, production management, water conservation level, and others. The calculation methods consist of the trend method, factor

analysis, and water coefficient method, which is the most commonly used. The main indicators contain the total value of the industry output and its corresponding coefficient of water deprivation, agricultural irrigation, livestock numbers, and their water demand quota.

6.3.2.3 Supply–Demand Analysis

The state of equilibrium between supply and demand is determined by the quantity comparison of water supply and water demand, which is, in this study, represented by the equilibrium index, namely the ratio expression of the water demand and supply under a guaranteed rate. If the equilibrium index is less than or equal to 1, the water supply and demand are in balance; if it is more than 1, the water supply and demand are unbalanced. The surplus index, expressed by the ratio of the surplus water and the water demand, is used to describe the surplus degree of water resources. The ratio of water deficiency is the rate of the water shortage and the water demand, showing the severity of the imbalance. In the state of equilibrium, the water supply–demand system needs to be optimized with technical and economic analysis so as to seek a cost-effective water supply–demand model. In the imbalanced condition, the intensity and methods of social and economic activities should be adjusted based on the reasons for imbalance, and many comprehensive measures, including control of pollution load, waste water recycling, water conservation, and water conservancy construction, should be adopted, with the purpose of reducing demand, increasing supply, and the rationally allocating water resources.

6.4 MEASURES OF IMPROVING URBAN WATER QUALITY

6.4.1 Reasonable Configuration of Urban Water Resources

The configuration of urban water resources means that by adhering to the principles of impartiality, high efficiency, and sustainability, taking advantage of various engineering and nonengineering measures, and considering market laws and law of resource configuration, usable water resources are distributed among regions and water-using departments via measures such as reasonably restraining demand, ensuring effective supply, and improving ecological environmental quality. The configuration aims to make out a relatively fair and acceptable water resources distribution plan, on the basis of reducing the effect on the city to a minimum, to maximize the benefits between development, use, protection, and management of water resources (Eker et al. 2003; Sun and Wang 2011).

The reasonable configuration of water resources should be the proper balance of equality and efficiency (Xu and Gao 2008). We should consider the following principles during implementation: (1) ensuring the sustainable use of water resources; (2) placing equal emphasis on broadening water sources and saving water; (3) taking actions that suit local circumstances and giving water use priority to citizens, and (4) maximizing the comprehensive benefits.

The natural water has certain capability of self-purification; thus, the discharge of domestic and product waters should be reduced to a minimum, namely maintaining the discharge within the bearing capability of ecological environment so as to ensure water purification and water basin functions. The optimized configuration of urban water resources aims at meeting temporal, spatial, quantitative, and qualitative demands of coordinated developments between society, resources, environment, and economy to make the limited water resources obtain maximum benefits. The systematic configuration should be the integration of quality, quantity, space, and time of water resources (Feng 2000).

The configuration of urban water resources mainly takes the following measures (Geerse and Lobbrecht 2002; Zhang et al. 2008; Lim et al. 2010):

1. *Multiple use of water:* Multiple use of water mainly refers to regenerating and recycling of urban water resources. The essence of water resource recycling is to absorb waste water caused by human activities within a social production system, rather than exerting pressure to the environment outside it, with wastewater resources regeneration as the most commonly seen form. It can be divided into industrial production recycling (including water recycling within production circle and among production department) and regional water resources recycling. The former can be realized by strengthening environmental management and auditing of environment and production of enterprises, while the latter can be attained by market and industrial policy mechanisms and coordinating water resources between production and domestic departments.

2. *High-quality water to be used in important places:* "High-quality water to be used in important places" has become a tendency in modern urban water resource configuration with the core intention of providing high-quality water to important places and reasonably developing and distributing water resources with minimum investment to make the limited urban water resources produce maximum value.

3. *Water supply by different quality:* On the one side, the potable water system is taken as the main water supply system in a city and on the other side, another pipe-network system is laid for low-quality, recycled, or sea waters for forestry greening, vehicle cleaning, toilet cleaning, road cleaning, and/or industrial cooling. The latter system is called the nonpotable water system, which is usually partially in completeness and is often taken as the supplement of the main water supply system.

4. *Rainwater resource use:* Rainwater resource can be taken for direct or ecological use. By direct use, we mean that after collecting the rainwater from water collecting facilities on architectures or porous pavement, the water collected can be used for cooling or municipal construction after simple treatment or used for drinking after disinfection and other treatments. Ecological use means that the rainwater is stored in a man-made lake or pool as landscape water resources to increase water area and improve local climate.

5. *Seawater resource:* Seawater desalination, due to its decreasing costs, is a new way to solve a freshwater resources crisis. Desalinated seawater can be used for drinking. After sand setting, algae removal, and biocide

treatment, the seawater can be directly used as industrial circulated cooling water. Seawater can also be used for toilet flushing water in island and coastal cities.

6.4.2 Construction of Urban Sewage Treatment System

Urban sewage treatment system includes central treatment in urban sewage processing plant, land ecological engineering treatment, water ecological engineering treatment, and wetland engineering treatment.

Central treatment in urban sewage processing plant: Large-scale urban sewage processing plant usually follows steps of pretreatment, primary treatment, secondary treatment, in-depth treatment, and sludge treatment, in which the secondary treatment is the core step. From the perspective of treatment mechanism, urban sewage treatment technology can be divided into physical, chemical, and biological treatment technologies.

Land ecological engineering treatment, which is also referred as land treatment, uses the self-adjusting mechanism of a soil-plantation system to conduct sewage treatment. It usually includes soil percolation and sewage irrigation technologies.

Water ecological engineering mainly refers to oxidation pond or stabilization pond. Water ecological engineering treatment utilizes the purification capability of nature water to have polluted water slowly flowing or staying for a long time and decreases organic pollutants via metabolic activities. Water ecological engineering can be divided into microorganism stabilization pond and aquatic organism pond; the former includes aerobic ponds, anaerobic pond, facultative pond, and aerated pond; and the latter has aquatic plant pond and aquaculture pond.

Wetland engineering treatment means that the polluted water is intentionally put into wetlands, where they are full of plantations like reeds and bulrush, to make it maintain a saturation state. During its flow in a certain direction, the polluted water is purified after going through hydrophilous plants, soil, and so on. The mechanism recovers the polluted water via physical, chemical, and biological effects of the stroma–plant–microorganism compound ecosystem. Wetland, according to its characteristics, can be divided into natural wetland and constructed wetland.

Compared with the large-scale urban sewage processing plant, the land ecological engineering and water ecological engineering have features of low investment and operation cost, are easy to manage and maintain, and have obvious treatment effect over nutrients and can produce other benefits indirectly. Their defects are that they have large land coverage and two or three plant growth seasons to form a stable plant and microorganism system for treatment; thus, they are not preferred to be used in areas where land resources are limited.

6.4.3 Urban Water System Ecological Management

Urban water ecological management is the concrete application of theory of eco-hydraulics in the urban water system. The management requires the full consideration of relations between creatures and river mechanical property, the understanding of structures and functions of the river eco-system, and to follow physical,

chemical, and biological processes to ensure the hydraulic requirement for protecting river eco-functions and carry out river flow management and river physical construction (Li and Lei 2006).

Urban water ecological management is achieved via the reasonable construction
of an urban water network and scientific distribution of urban water eco-flow. By
building an urban water network, proper distribution of urban water eco-flow and
recovery of urban inner river connections, and building reasonable hydrophytic habitats, the interaction between water creatures and nutrients in urban inner rivers can
be promoted, the complexity and stability of the urban water system structure can be
ensured, and the habitats and ecodiversity can be enhanced.

6.4.4 IMPROVEMENT OF URBAN WATER SELF-PURIFICATION

The improvement is mainly attained by aeration, adding reagents, biomanipulation,
recovering aquatic vegetation, and bioaugmentation.

By conducting re-aeration, the purification capability to pollutants can be
enhanced.

Reagents including chemical and biological reagents are added for improving the
self-purification capacity of water body.

Biomanipulation, by taking advantage of nutrition relations within the lake ecosystem, reduces the quantity of algae and improves water quality via a series of
alterations to coenosis and environment. It operates under the principles of adjusting
sheltering mechanism/fish structure, protecting and developing large animal plank-
ton, using herbivorous zooplankton, and constraining the excessive growth of algae.

Aquatic vegetation recovery includes nature recovery and artificial rebuilding of
aquatic vegetation, the former promotes natural recovery of lake aquatic vegetation
by adjusting water environment with slow recovery process and the latter rebuilds
aquatic vegetation via biological engineering.

Bioaugmentation applies the principle of purifying pollutants by the biofilm
process and puts a proper biocarrier after the bioprocessing to the affected water
for improving the self-purification capability of the water. Main biocarriers include
pebbles and synthetic materials such as elastic fiber and polyethylene plastic ball,
and main methods include biological membrane, artificial contact oxidation, mobile
biological purification, biological grid and biological floating island.

6.4.5 CONTROL OF URBAN WATER INNER POLLUTING SOURCE

The two methods to be used are sediment environmental dredging and sediment
contamination control.

Sediment environmental dredging uses a special dredging equipment to remove
and clean sediments in water so as to create favorable conditions for recovering
urban water ecology. The removed sediments can be sent to forestry lands, used as
fertilizer for green land and for agricultural production (Zhu et al. 2002), which can
also be used as filling materials of river banks after flowing treatment and adding
some consolidation materials.

Sediment contamination control fixes pollutants in sediments by adding chemicals.

6.4.6 Construction of Urban Water Habitats and Fishways

The creation of habitats can mainly be made by alerting the river flow and the physical characteristics of the river bed to make a natural-like flowing way and have the water flow to comprise many speed zones. It protects the natural state of water to have the river has interphase structure of shoals and deep pools so as to build the diversity of water flowing for protecting biodiversity.

Recovering or building fishways aims at avoiding the separation to the river biochannel due to road or urban engineering to create hydrographic and sediment conditions for moving and migrating of fishes and other aquatic organisms. Three commonly seen fishways are gradeless fishways, hydraulic barrier fishways, and nature simulation fishways.

6.4.7 Physical Structural Recovery of Urban Water

Physical structural recovery of urban water mainly includes the building of ecological revetment, straightening of river channels, and so on.

The ecological revetment, by "protecting and creating sound creature living environment and landscape" as basis, changes made-made concrete architectures to the revetment in which water and earth, water and creatures are mutually supported, and fit for creatures to live. The commonly seen ecological revetment materials include plants, ecological concrete, concrete ecological planting medium, geotextile materials greening grid, and soil curing agent (Zhang et al. 2000; Ji et al. 2001).

Straightening of river channels will incur the change of water flow characteristics and thus also makes the changes of sediment and water erosion. The appearance of marginal banks, shoals, or channel bars provide favorable hydrological and sediment environments for water creatures. The biological communities in curving river sections are remarkably more complicated than those of linear river sections; the curving river sections have more habitat types and stronger self-purification. Urban nearshore area has high land development cost and can hardly fulfill complete recovery; we can properly enhance river curvature and expand water area, considering landscape viewing and cultural resource protection.

REFERENCES

Eker I., Grimble M. J., and Kara T. Operation and simulation of city of Gaziantep water supply system in Turkey. *Renewable Energy*, 2003, 28: 901–916.

Fadali E. and Shaw W. D. Can recreation values for a lake constitute a market for banked agricultural water? *Contemporary Economic Policy*, 1998, 16(4): 433–441.

Feng S. Y. *Introductory theory of sustainable use and management of water resources*. Beijing: Science Press, 2000.

Geerse J. M. U. and Lobbrecht A. H. Assessing the performance of urban drainage systems: 'General approach' applied to the city of Rotterdam. *Urban Water*, 2002, 4: 199–209.

Ji Y. X., Lu Y. J., and Mo A. Q. Beach and bank protection works at outer bar and inner stream reach in the Yangtze River estuary. *Bulletin of Soil and Water Conservation*, 2001, 21(1): 15–17.

Jungwirth M., Muhar S., and Schmutz S. Re-establishing and assessing ecological integrity in riverine landscapes. *Freshwater Biology*, 2002, 47: 867–887.

Karr J. R. Defining and measuring river health. *Freshwater Biology*, 1999, 41: 221–234.

Ladson A. R., White L. J., Doolan J. A. et al. Development and testing of an index of steam condition for waterway management in Australia. *Freshwater Biology*, 1999, 41: 453–468.

Li D. W. and Lei X. Q. Principle and method of ecological hydraulic scheduling in the construction of urban water networks. *Yangtze River*, 2006, 37(11): 63–65.

Lim S. R., Suh S. W., Kim J. H., and Park H. S. Urban water infrastructure optimization to reduce environmental impacts and costs. *Journal of Environmental Management*, 2010, 91: 630–637.

Ma K. M., Kong H. M., Guan W. B. et al. Ecosystem health assessment: Methods and directions. *Acta Ecologica Sinica*, 2001, 21(12): 2106–2116. (in Chinese).

Mckone P. D. Streams and their corridors-functions and values. *Journal of Management in Engineering*, 2000, 5: 28–29.

Meyer J. L. Stream health: Incorporating the human dimension to advance stream ecology. *Journal of the North American Benthological Society*, 1997, 16: 439–447.

Norris R. H. and Thoms M. C. What is river health? *Freshwater Biology*. 1999, 41: 197–209.

Sun L. H. and Wang J. An overview on optimizing urban water resources allocation. *Ground Water*, 2011, 33(3): 156–157.

Tennant D. L. Instream flow regimes for fish, wildlife, recreation and related environmental resources. *Fisheries*, 1976, 1(4): 6–10.

Yang Z. F., Cui B. S., Liu J. L. et al. *The theory, method and practice of eco-environmental water requirement.* Beijing: Science Press, 2003.

Zhang J. C. Riparian functions and its management. *Journal of Soil and Water Conservation*, 2001, 15(6): 143–146 (in Chinese).

Zhang J. Y., Zhou D. P., and Li S. C. Brief introduction of study on slope eco engineering for rock slope protection. *Bulletin of Soil and Water Conservation*, 2000, 20(4): 36–38.

Zhang X. H., Zhang H. W., Chen B., Chen G. Q., and Zhao X. H. Water resources planning based on complex system dynamics: A case study of Tianjin city. *Communications in Nonlinear Science and Numerical Simulation*, 2008, 13: 2328–2336.

Zhao H. C., Xu S. B., and He J. S. *Analytic hierarchy—A simple new decision-making methods.* Beijing: Science Publishers, 1986. (in Chinese).

Zhao Y. W. and Yang Z. F. Preliminary study on assessment of urban river ecosystem health. *Advances of Water Sciences*, 2005, 16(3): 349–355. (in Chinese).

Zhu G. W., Chen Y. X., and Tian G. M. Reviews on development of pollution control techniques of sediment. *Journal of Agro-Environment Science*, 2002, 21(4): 378–380.

7 Eco-Habitat and Eco-Cultural System Planning

Yan Zhang and Meirong Su

CONTENTS

7.1 PLANNING AN ECOLOGICAL HABITAT SYSTEM

Human settlements are the basic environments that urban residents live in and are also the basis of urban ecological construction. In this section, we review the connotation and characteristics of urban ecological human settlements and then discuss construction, including residential areas with reasonable layouts, perfectly functioning infrastructure, and ecological residential areas that are environmentally safe.

7.1.1 CONNOTATION AND CHARACTERISTICS OF AN ECO-HABITAT

A human settlement is a place where humans live together; it is the land surface closely related to human survival activities (Wu 2001). Generally speaking, human settlements contain cities, towns, villages, and other organic coalitions of material and nonmaterial structures, which are based on the concepts of human dwellings and environmental science.

In an eco-habitat system, ecology can be defined as a highly harmonious ecological relationship between human beings and the environment. It is a kind of survival and developing mechanism that includes competition, symbiosis, and autogeny as well as a type of systematic function pursued for persistence and harmony through time, space, quantity, structure, and order. It is a specific action moving toward sustainable development. Concerning the connotation of an ecological human settlement, there is no agreed demarcation in the academic community. Some specialists believe that the standards of an ecological human settlement are high efficiency, energy conservation, comfort, and aesthetics (Zhang et al. 2004). However, others believe that an ecological human settlement is a sustainably developing human environment (Ravetz 2000). Together, the connotation of an ecological human settlement can be generalized as one, which is focused on urban areas and settlements and based on the living environment of the local elements of the structure, their function, and the dynamic system used to reduce the pressure on the natural environment to maximize the use of natural materials. In an ecological human settlement, natural state and natural energy sources are used to achieve harmony and unity between human and nature. Such a settlement is also characterized by low consumption and low emissions while meeting the requirements of sustainable development.

Features of an eco-habitat include (1) perfectly functioning settlements with a live-equipped infrastructure, living comfort, and convenience; (2) good environmental quality (air, water, and acoustic environments) of residential quarters reaching the national standard; (3) a high degree of greening in which the green coverage rate is high and three dimensional; (4) an ecological way of life focused on saving energy,

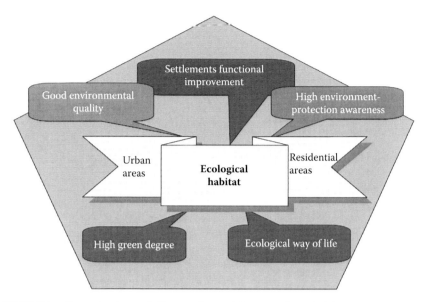

FIGURE 7.1 **(See color insert.)** Connotation and features of eco-habitat.

lowering the levels of resource and energy consumption from water facilities by the living system, and lowering the levels of waste and pollutant emissions in the environment; and (5) high environmental awareness; that is, a green philosophy rooted in the residents' minds (Figure 7.1).

7.1.2 PLANNING IDEAS FOR URBAN ECOLOGICAL HABITAT CONSTRUCTION

Urban eco-habitat planning is the application of eco-living principles and a systems engineering approach for the optimization of planning and organization of cities, urban areas, and settlements, as well as the construction and management of living systems to reduce pressure on the environment while providing a clean, beautiful, and comfortable living space. In short, the perfect addition to an urban ecological living system should include a clean living space, beautiful green ecosystems, and traditional elements. In addition, a full complement of infrastructure, safe and convenient transportation network, and efficient conservation energy systems, as well as advanced management and a reasonable healthy and harmonious ecological culture, should be included.

Urban ecology of human settlements uses a people-first philosophy from residents and needs-based urban development—through macroadjustment layout of urban functions—to build an urban landscape system. In light of this improved urban infrastructure, microconstruction of the urban ecological living area and other optimization of the urban landscape system improve the urban environment and health, enhance the urban transport environment for a full range of services, and promote ecological consumption patterns and cultural ideas, building a sound ecological living system (Figure 7.2).

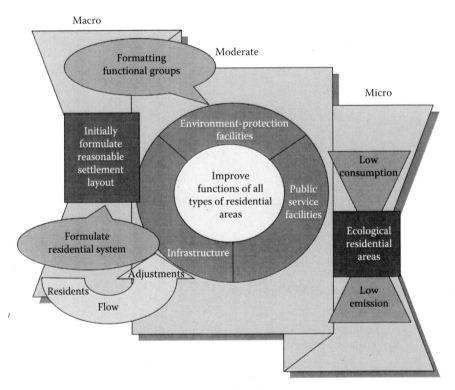

FIGURE 7.2 **(See color insert.)** Planning of ideas for urban ecological habitat construction.

7.1.3 EVALUATION OF URBAN ECO-HABITATS

7.1.3.1 Evaluation Framework

An eco-habitat environmental assessment framework for the construction of an eco-habitat city can be established according to the planning ideas: (1) infrastructure—an eco-habitat's support systems, including transport and communications, waste water and solid waste disposal, and public service facilities; (2) living systems—the city's building capacity, the living conditions, and the level of eco-construction settlements; (3) ecological environment—the habitat's air, water, and sound quality, as well as the efficiency to maintain the vitality of the urban eco-system; and (4) social humanity—the degree of social harmony, including a stable employment environment, good education, and a more satisfactory level of income and consumption.

7.1.3.2 Evaluation Indicator System

According to the evaluation framework established above, we choose indicators corresponding to content for the various elements based on data accessibility. We also established an indicator system including the criteria and feature layers and layer indicators (see Table 7.1) (Han et al. 2009).

TABLE 7.1

Evaluation Indicator System for an Urban Ecological Habitat Environment

Criterion Level	Feature Layer	Index Layer	Standard Value
Infrastructure	Transportation and communications	Paved road area per capita in urban areas	$\geq 10 \ m^2$
		Telephone penetration rate	100%
		Internet coverage rate	>40%
	Sanitation conditions	Urban garbage treatment rate	100%
		Urban sewage centralized treatment rate	$\geq 70\%$
		Urban gas penetration rate	$\geq 92\%$
	Service facilities	The public benefit network/10,000 people	250 units
		Community service facilities/10,000 people	15 units
		Beds in hospital/1000 people	6.7 beds
Living system	Living conditions	Urban housing space per capita	$\geq 30 \ m^2/people$
		Population density	$\leq 100 \ people/km^2$
	Construction capacity	Urbanization level	$\geq 55\%$
		Technology's contribution to economic growth	$\geq 60\%$
	Eco-residential areas	Utilization rate of recycled water	30%
		Utilization rate of energy-saving materials and equipment	30%
Ecological environment	Environmental quality	Urban air quality	330 days
		Urban average environmental noise	$\leq 60 \ dB$
		Centralized drinking water source quality compliance rate	100%
	Ecological maintenance	Soil erosion rate	$\geq 90\%$
		Urban public green area per capita	$7.5 \ m^2$
		Forest coverage rate	$\geq 40\%$
Society and humanity	Stability	Urban registered unemployment rate	<2%
		Poverty rate	<3%
	Educational level	Junior high school dropout rate	0
		College students/10,000 people	≥ 450 people
	Living quality	Engel coefficient	$\leq 40\%$
		Urban residents disposal income per capita	$\geq 14,000$ yuan

Generally, these data are available in the *City Statistical Yearbook*, the *Yearbook of Chinese Cities*, the State of the Environment, Environmental Protection Plan, and elsewhere. The Environmental Protection Agency standard values are based on the ecological county, eco-city, and eco-province construction index (Trial), the National

Sanitary City standards, the Ministry of Construction, the size of settlements, the urban development technology evaluation system (Trial), green residential construction points and technical guidelines, and other leading cities in reference to the domestic selection.

7.1.3.3   Evaluation Model

The formula for the model using the weighted summation index is

$$Q = \sum_{j=1}^{m} N_j W_j \ (j = 1,2,\ldots,m) \tag{7.1}$$

where Q is the evaluation index, N_j is the index for the jth value of the index, W_j is the jth index weight $(0 \leq W_j \leq 1,$ and $\Sigma W_j = 1)$, and m is the number of evaluations.

1. *Index value:* The index value of the indicators is standardized by the raw data obtained. For positive indicators, larger is better, such that $N_j = X_j/Y_j \ (j = 1,2,\ldots,m)$. For negative indicators, smaller is better, such that $N_j = Y_j/X_j \ (j = 1,2,\ldots,m)$ (Huo et al. 2006). X_j is the status of the jth index value, and Y_j is the standard of the jth index value. The indicator is no longer treated with standardized dimensions.

2. *Index weight:* Integration of the analysis hierarchy process (AHP) and entropy method allows the determination of the index weight. Proposed in the 1970s, the AHP is a subjective weighting method in which the essence of the indicators is defined by the evaluators according to the importance of the human endowed right way (Saaty 1977). The advantage of this method is fully reflected in the experience of experts and the lack of sample data required, but the AHP is subjective and arbitrary. Indeed, even if the correlation between the indicators is considered, it is difficult to accurately reflect the relationship between the index systems of the internal structure; for example, AHP index value does not reflect the degree of change over time but rather the entropy of the assignment mechanism emphasizes this information. The statistical physics and thermodynamic term "entropy" was originally conceived in 1948, but Shannon introduced the concept of information entropy into information theory (Yan et al. 2006). Entropy, in the objective weighting method, is a direct indicator of the raw data based on statistical analysis or other mathematical approaches to the weight obtained after the method. It is, to some extent, a subjective weighting method used to avoid the shortcomings of mathematics with a strong theoretical basis, but occasionally when entropy is determined by weights and objective facts, there is a gap. In the assignment mechanism based on entropy weight, the target weight is unchanged over the years. This result may be inconsistent with the actual empowerment, not the actual importance of the response indicators and levels of analysis, but will not occur.

In view of this, a combination of subjective and objective indicators can be used to determine the weights of the variables to correct two kinds of mutually empowering possible bias: AHP to determine the criteria corresponding to the relative importance of each of the elements (such as B1 corresponding to C1, C2, and C3), based on field research to determine when the local government should obtain information and views of local residents, which determine the matrix structure; and then calculate the empowerment weight, in accordance with the formula for calculating entropy corresponding to each element, of the various indicators (such as C1 corresponding to the D1, D2, and D3). The habitat system, infrastructure, housing system, environment, social, and human aspects are equally important, and therefore, the four criteria layer indices (B1, B2, B3, and B4) are given weights of 0.25.

By evaluating the current status, we can identify the infrastructure, housing system, environment, social, and human aspects of urban living environment issues to develop a targeted plan.

7.1.4 PLANNING GOALS OF URBAN ECO-HABITAT CONSTRUCTION

7.1.4.1 Overall Goals

Based on the concept of sustainable development and the suitability, according to the city's natural, social, and economic reality, through the relevant infrastructure and ecological landscape construction, we could create beautiful living environment system of functions appropriate to each group, rational layout, and complete functions. By popularizing ecological living patterns and methods, we can achieve harmony between humans and nature, as well as create a safe, comfortable, clean, and civilized living environment.

7.1.4.2 Stage Goals

7.1.4.2.1 The First Stage

Coordinating land use planning of the overall urban layout (e.g., building orientations) with the development of the urban structure and the function of the characteristics of each group, we can form an initial, reasonable layout of the residential areas. The land use of older residential areas must be adjusted, and new supporting infrastructure must be built. Planning should occur with environmental protection facilities to construct an environmental protection system. This will ensure a solid base for building a clean and beautiful living environment with a reasonable layout.

7.1.4.2.2 The Second Stage

After consolidating the results of the first stage for all types of residential locations and the characteristics of residents, the residential functions are further improved. The infrastructure of the residential areas with high- and mid-range positioning should also be improved, as well as strengthening the environmental pollution control system. A full complement of features for the residents includes public service facilities, initially formed with the surrounding environment, resulting in a comfortable, convenient, clean, and civilized living environment.

7.1.4.2.3   The Third Stage

Various plans result in accomplishing the indicator standards, and through further adjustment of public service and environmental protection facilities, we can build a human settlement system where the central group functions with a high degree of coordination and the environment remains pristine. Eco-housing construction patterns and lifestyles are promoted, focusing on the development of urban energy and water systems and the construction of ecological residential areas. Fully completing the construction of an ecological city district that was built in harmony with nature requires low consumption and low emissions to achieve sustainable development of the system.

7.1.5   CONTENTS OF URBAN ECO-HABITATS CONSTRUCTION

7.1.5.1   Optimize Service Functions and Landscape Patterns of an Urban Area

7.1.5.1.1   Construct a Reasonable Residential District System

In a reasonable living environment for the construction of settlements, the quality of strength is the key. The combination of residential land suitability analysis, optimization of the urban layout of residential land, and suitability for building strength control of urban living determine the living area layout and reasonableness of the system. The layout of residential areas should consider the surrounding environment, network structures, public buildings and residential layout, group composition, green space, and environmental systems and internal relations. The layout should also be relatively independent of the organic whole. Reasonable set of public service facilities will avoid smoke, gas (odor), dust, and noise pollution and, thus, will not disturb the residents. To focus on the integrity of the landscape and space, municipal and other public sites should be combined with arrangements for homes or public buildings. Power supplies, telecommunications, lights, and other pipelines should be buried underground. To environmentally design public spaces, one should coordinate buildings, roads, squares, courtyards, green spaces, and architectural sketches with human activities and their interrelationship(s) with the Ministry of Construction. Gradual control of urban land use with high-density, high-strength construction increases open space, while decreasing road congestion and reducing the impact of construction on urban space and landscapes. Replacing inefficient land use in urban areas improves environmental quality while improving the leisure and ecological environment.

7.1.5.1.2   Optimize Urban Transport Network Construction

Transportation infrastructure is the major infrastructure of the city, an important condition to ensure and promote production activities, in which passenger transport ensures the necessary conditions for normal residential life. An inefficient traffic environment outside of the city will lead to inefficiencies in the city's economy, and inefficient traffic conditions within the city will lead to an inefficient operation of the city, reduce the level of safety, and cause adverse visual effects. These phenomena have an extremely negative impact on the city's living environment. Based on the concept of eco-living environment, the road transport system should meet safe, convenient, and environment friendly requirements. A road network, public transport, and supporting facilities perform three aspects of urban transport networks.

1. *Road networks:* By three-dimensionally promoting the smooth flow of traffic, we can construct a multilayered transport network. Opening up the air connection channel among large-scale facilities reduces unnecessary up-and-down traffic, lowering the ground floor of the traffic density. A flow density air walk system can be established in the area to improve traffic flow efficiency and walking comfort. One should also consider the use of underground space, building underground network connections. With subway stations and parking lots (the hearts of large-scale facilities), urban centers in iconic locations have direct and explicit connections to large urban centers in the city transport network; that is, direct channel flow. An urban road network is mainly used to meet the residential needs of fast and multiple-route trip convenience. Three general traffic networks should be built: (1) rapid transit—communication between the main transport links and capable of a large volume of urban rapid transit; (2) the backbone—the main communication backbone of urban transport links between the various groups and mass transit within the groups (existing roads to the city can be expanded, extended, and used to enable transportation for the blind); and (3) second dryer—the main transport links within the group communicate to connect various residential areas.

2. *Public transport:* Implementation of appropriate public transport incentives should make public transport in the city, not private transportation, a priority to increase its appeal. One must develop an active public information campaign to promote the advantages of public transport, such as how public transport is more conducive to environmental protection and economical. The legal construction of the bus system should also be strengthened.

   The rational layout of the transportation route network and the operating efficiency of the public transportation system and service levels should be enhanced to entice more people to use public transportation as a way of behavioral targeting. The transportation line network design must be updated with particular attention to improve the interface with other passenger transport systems, as well as improving the living area travel conditions to gradually solve the traffic problems of local residents. Based on the services provided by traveling distance and traveling speed, the bus lanes can be divided into three levels: the fast lanes, general lanes, and regional lanes. The bus station service radii are controlled within 500 m, combined with the construction of an integrated transport hub to closely link aviation, ports, railway stations, road passenger transport stations, and other facilities outside of city public transportation. In establishing the status of public transport, on the basis of the development of large, medium volume passenger traffic type gradually forms a functional structured, diversified transport system, the overall public transport.

The development of bus priority channel can be accelerated by improving the serving quality of public transportation to reduce private car service quality trends. Bus lines can be dense on some main and secondary roads and then gradually expanded. While widening roads, the transformation of the original nonmotorized vehicle lanes should be changed to priority bus lanes.

Updates to vehicles to meet the ecological requirements of the living system should also be fast-tracked. Specifically concerning the use of clean energy, vehicle noise standards and the three aspects of accessibility in vehicle factories should be tightly controlled. All buses must conform to the standard implemented by the European II, and road vehicles with excessive emissions are strictly prohibited. Car noise limits should also gradually be increased to create a quiet residential, recreational environment. When updating vehicles, one should choose low-floor, low-entry-type models to help disabled people to participate in social life, as well as to more easily facilitate services for the elderly and children to promote the harmonious development of human settlements. Finally, to realize an intelligent transportation system, the use of GPS technology and electronic bus schedules must be promoted.

3. *Supporting facilities:* Reasonable space for walking should be provided to ensure the safety and comfort of pedestrian traffic. Similarly, sidewalk standards must be improved, sidewalks should be appropriately broadened, and a green recreation belt can be constructed on both sides of the road. In large sections of traffic, construction, bike lanes, implement local, nondiversion to ensure traffic safety and speed. Set sound barrier on both sides according to main trunk.

An eco-city should also gradually expand the scope and number of pedestrian lights, improve intersection designs, and create refuges. Width of the road more than two-way four-lane intersection, or near the commercial center of the road intersection, must be conducted with an additional overpass or tunnel. Clear direction signs should be placed at the road node, focusing on service facilities and government agencies to facilitate residents' travel, as well as promote the smooth flow of traffic.

A professional and standardized parking and public transport management hub should be constructed, setting aside additional bicycle parking facilities in the central areas of commercial streets and in the city center to aid in traffic control. Before entering roads in the central area, free (or low-cost), large parking lots must be built to encourage rural residents to use public transportation into the city, to improve travel efficiency, and to reduce vehicle pollution. In addition, cement-paved parking areas should be avoided in favor of permeable surfaces such as raster bricks to increase green areas.

7.1.5.1.3   Strengthening the Habitat Green Land Network Construction
Through the building of green communities and green units, the development of face-like green space, linear green space, and point-like green spaces as an overall strategy and the completion of the city point, line, surface combine to create the green network.

1. *Maintain facet green land:* Facet green refers to a large private green space in the city. Such spaces are one of the elements of an urban green space network that provide fresh air to improve the urban atmospheric environment. Facet green land plays an important role in the formation of urban ecosystems. One should focus on maintaining and strictly protecting core city parks from encroachment and destruction. Squares and parks to enhance the transformation gradually increase the greening rate to the grass instead of concrete floor, to the trees instead of grass; while promoting a rainwater collection system, green area up to 2 ha of green space should consider the design of rainwater harvesting facilities.

2. *Extend linear green land:* Is a linear green space along the city transportation lines and the old city walls, rivers, and lakes formed by a narrow coastal strip green? Adequate ventilation must be constructed for the exhaust of the construction project to adjust the intensity of the urban heat island effect and provide streetscape, shade shelters, and other services (i.e., leisure and cultural activities) to meet the needs of residents. For green space, urban roads should be constructed to allow trees to be planted on both sides to isolate the residences from traffic. Because this forest is not primarily used for air filtration, the available forests should be increased to 20–30 m in width. Such roads also serve as the intake channel, and low vegetation should be sparsely planted behind the green jungle to facilitate air circulation. Roads should not be built above the tree canopy covering the opening to facilitate the easy discharge of polluted air. Residential green space is composed of shrubs and green corridors, combining various flowers, shrubs, and trees to form a multilayered composite structure that can be used for shade, beautification, and moderate noise and dust reduction (Figure 7.3).

3. *Develop punctiform green grounds:* Point-like garden green building primarily refers to residential areas and green space, a green space system, and residents are the most direct and frequent contact with this type, so it is increasingly important for people living in the urban environment. Green residential areas should include public green space, green space beside houses, and greenery owned by public buildings. Residential areas should be based on the layout of different ways to set the appropriate center for green space, and the elderly, children, and other massive venues, such as public green ribbon and green residential area to carry out network construction, are advised to retain and use planning or modified within the scope of existing trees and green space (Figures 7.4 and 7.5). A green living area shall meet the following requirements: all the land that can be green is green and should develop vertically, especially green roofs; the greenery

FIGURE 7.3 (See color insert.) Schematic diagram of road green.

FIGURE 7.4 (See color insert.) Schematic diagram of vertical green wall.

between houses should be carefully planned and designed; new construction should not be >30% green rate; and the old district reform should not be <25% (Ministry of Construction 2006).

7.1.5.1.4 Improve Infrastructure of Urban Environmental Sanitation

1. *Increase garbage collection and refuse transfer stations; improve the collection system and transportation facilities:* According to the number and characteristics of residential areas to set the number of refuse collection points, and in accordance with road traffic urban functions and features, lots laid waste collection station location. The service radius of refuse collection points should be ≥ 0.8 km in a facility ≥ 80 m^2. Station area collection points should be arranged to meet garbage collection, garbage truck movement, and convenient/safe operation requirements. The architectural design and exterior decorations should be in accordance with the

FIGURE 7.5 **(See color insert.)** Green roof diagram.

surrounding residential/public buildings and the environment. Collection points should establish a certain width of the green belt. Collection points should be allocated to the drainage facilities. RTS should set the convenient transportation, municipal residents with better conditions and less affected areas. Small station (waste <150 t/day): for each 2–3 km², set a land area of ≥800 m². A waste transport distance of >20 km should be set for large (transfer >450 t/day) and medium (transport capacity of 150–450 t/day) transfer stations. RTS is beautiful and should be in harmony with the surrounding environment. The facility should be operated to achieve closure, volume reduction, and compression with advanced equipment. Airborne dust, noise, odor, drainage, and other indicators should be consistent with relevant environmental standards. The green transfer station rate should not exceed 30% (Ministry of Construction 2005).

2. *Classify and obturate garbage discharge and transport:* Waste collection containers should be clearly identified by the type of garbage collection; separated wastes should be separately transported. Garbage collection should be put in the position of a fixed point and should be convenient for residents while not affecting visual amenities. Such efforts will help separate garbage collection operations and meet the mechanization industry requirements. Refuse collection point service radii should not exceed 70 m. Hazardous waste must be collected and transported alone; the waste container should be closed and should have an easily recognizable logo. The capacity of various types of storage containers should be counted according to population, daily discharge volume of wastes, types of waste, and collection frequency. Refuse storage containers must meet usage needs. The waste shall not overflow and affect the environment (Ministry of Construction 2005).

3. *Improve recycling:* Establish a new system of waste recycling facilities to enhance the garbage resource recycling industry. Set up various forms of recycling companies and economic entities for recycling wastes; for example, increase scrap yards and acquisition and multichannel recycling of waste material. Establishing a waste material exchange center is necessary, as well as waste material information networks and databases (Li 2000).

7.1.5.2 Construct Safe and Environmental Eco-Habitats

7.1.5.2.1 *Optimize the Design of Settlement Landscape*

The residential area in the landscape design should be integrated with its environmental and ecological context using rational planning to appropriately simulate the natural landscape and water features, so that residents who have been living in downtown areas can return to nature. We recommend that parking and storage room be constructed underground. Use greenbelts on the surface as much as possible to create a more user-friendly space. Paved with tiles embedded in the grass to make the green area can be fully carried out and can be effectively protected.

1. *Landscape of the communities:* (a) The construction of fitness playgrounds of a certain size and number in the district is required, and urban roads should be constructed between 5 and 10 m of the buffer zone, scattered in the area so that residents living nearby are not affected. In the lounge area, we shall plant shade trees and build a certain amount of seats. (b) Construct a relatively concentrated leisure plaza. The square arrangement of the people settlements and distribution center, according to settlement size of the area, should ensure that there is adequate sunshine and ventilation, is near planted shade trees and flowers court, and so on, and arranged rest seat, permeable paving materials, which mainly have color pattern and some antislip materials. (c) Construct small children's play lots. They should be located in the greenery and be away from main roads and the windows of residents by ≥ 10 m to ensure children's safety and reduce the impact on residents. Not more than planting trees around the person's attention and maintain good through visibility (Figure 7.6).

2. *Community water features' landscape:* Landscape design of the main body of water refers to the water's state and form. Concerning the water's state, the district's water features often become the focus of people's attention. Swimming pools are mostly filled with still water, so accompaniment by water, artificial rock waterfalls, Dushui walls, fountains, and other dynamic water features enrich the ornamental value and create a diverse landscape.

   We can also consider the design of natural water features and garden water features: (a) Natural water features—the design must obey the natural landscape, integrating internal and external landscape elements to create a new living form. Revetment design should be close to the surface, and the height to the water, its type, and timber must be in harmony with the surrounding environment. (b) Garden water features—for artificial water features for more reserved spaces in natural water bodies, we need a comprehensive design to make natural and artificial water features blend with water to create a dynamic effect of vibrant living environment. (c) Decorative water features—through controlling water flow to achieve artistic effect, with the light of changes in the music and the visual impact, further demonstrate the vitality and dynamic body of water to meet the requirements of people hydrophilic.

FIGURE 7.6 **(See color insert.)** Landscape design diagram of settlement water feature.

FIGURE 7.7 **(See color insert.)** Landscape design diagram of settlement simulation.

3. *Imitated landscape:* (a) Pile up rocks in large communities to integrate the urban ecological environment. Using natural stone, the rock pile should not be too large, its composition should be scattered, and the piles should generally be located in the green residential areas (at the entrance) and the center green area to adapt to the configurations of the trees, flowers, and water. (b) Pave with fine sand, like flowing water. White is better. (c) Artificial flooring: use tiles and pebbles to level pavement into the scales regularly watermarks, used for garden path; the use of colored tile, gradual retreat by the light to dark halo in order to create the beach effect (Figure 7.7).

7.1.5.2.2   Construct an Ecological Residence (A House Estate)

As a whole, ecological settlements should have a reasonable layout, suitable scale, complete service facilities, good publicity of environmental protection, and other basic characteristics. The construction of an ecological residential location should be chosen based on the prevailing wind patterns in the city or to avoid the leading direction of the dispersion of pollutants from large industrial areas. One should choose areas with moderate levels of economic development and population density or areas away from the main road but close to the road network. Selected areas are ideally near the water level or better overall green areas. New individual settlements should meet the needs of 3000 families to ensure a per capita living area of ≥20 m². The housing industry center of the Ministry of Construction has compiled technical guidelines for the green construction of small residential areas as a standard for eco-settlement planning (Table 7.2).

TABLE 7.2
Construction Requirements of Ecological Residential Areas

Indicator Categories	Specific Criteria
Energy system	Avoid multiple pipelines into the households; design energy-saving residential envelope and hot and cold systems; encourage the use of solar, wind, and geothermal energy; maintain appropriate indoor temperatures (winter is 20–24°C, and summer is 22–27°C)
Water environment system	Establish a waste-water treatment and rainwater collecting system outdoors; use water-saving appliances and energy-efficient landscape water systems both indoors and outdoors
Air environment system	Outdoor air quality is required to Grade II; ensure natural ventilation indoors and ventilation facilities in bathrooms; kitchen should have gas concentration emission systems and meet indoor air quality standards
Sound environment system	Architectural design requires equipping with noise reduction facilities (outdoor sound in the daytime is <50 dB, sound at night is <40 dB; indoor sound in the daytime is <35 dB, sound at night is <30 dB)
Light environment system	Interior design requires using the daylight hours to best utilize natural light; avoid light pollution indoors and outdoors (e.g., light advertising and glass curtain walls); use energy-efficient lighting indoors and outdoors; promote the use of solar
Green system	Greenery with green renewable mechanisms provided by photosynthesis cleans the air and releases oxygen, regulating temperature and humidity and maintaining biological diversity
Waste disposal system	Garbage should be bagged for sealed storage; implement separate collection; the collection rate is 100%; the classification rate is ≥50%

TABLE 7.2   (*Continued*)
Construction Requirements of Ecological Residential Areas

Indicator Categories	Specific Criteria
Eco-building system	Use 3R materials (reusable, recyclable, and renewable); choose nontoxic, harmless, pollution-free, nonradioactive, nonvolatile, healthy materials; use construction materials approved by the National Environmental Accreditation Board and awarded the environmental label

Source: http://www.docin.com/p-17654846.html.

7.1.5.2.3   *Construct an Eco-Settlement Service System*

1. *Service infrastructure:* The establishment of service centers in settlements is needed; centers should be equipped with a property management department, supermarkets or convenience stores, fitness clubs (or simply sports equipment), playgrounds, small meeting squares, parks, pharmacies or health stations, and fast food restaurants. These are a series of services and facilities needed to meet daily needs to improve the degree of living ease of the residents and to reduce unnecessary travel distance. The service center is mainly provided for residents of the settlements although it takes into account a range of other residents living in the region that have not yet joined the settlements.

   When configuring service facilities, the main considerations are the multilevel needs of the material and cultural lives of the residents in the area, as well as their own requirements on public service facilities management projects, corresponds to the construction project and its service area population size facilitate residents to use and play the greatest economic efficiency project. For the operational step, one should first locate major residential areas in the small service outlets and then gradually increase the number of outlets until there is one service center in each settlement with 5000 people.

   Generally speaking, to meet the daily needs of 3–5 million residents, the residential area should be equipped with police, a street agency, comprehensive supermarkets, barber shops, a comprehensive maintenance department, cultural centers, medical clinics, and schools. To meet the basic needs of 0.7–1.5 million residents, the settlements should be equipped with a nursery, primary and secondary schools, markets, and a medium-sized supermarket. To meet the daily needs of 3000–7000 residents, the settlements should be equipped with a neighborhood committee, parking lots, integrated service stations, convenience stores, breakfast supply points, and health posts.

   According to the usage frequency of the service, clients, traffic conditions, and population density, one can use the service radius as a reference for distribution. An arbitrary public service facility is used as the center, with its best services to the farthest one, the service radius is the circle radius from it. For low-frequency use facilities, one should attempt to attract more users by expanding the service radius to support its operation. Table 7.3 presents the various service facilities and the corresponding radius of the configuration of the reference relationship.

The planning of the service center must be consistent with the appropriate sports facility or premises. Small settlements of <5,000 people should configure small parks and simple fitness equipment for free, settlements of >5,000 people should be configured with fitness clubs, and settlements of >10,000 people should construct a swimming pool, tennis courts, and other facilities. Meanwhile, all types of outdoor venues within the settlements should be combined with walking and green space systems located for easy access.

Generally, for every 1 million people or 0.5 km radius of the settlements (residential area), a supermarket must be built. For every 3,000 people or 0.3 km radius of the settlement (residential area), a convenience store should be constructed.

Use semiunderground space rationally to construct parking lots. Both sides of local roads and buildings' surroundings should maintain a certain percentage of green space. The appropriate gap between the buildings can be used to construct a small landscape and create a beautiful living environment for the region.

2. *Considerate service system:* The property management agencies of settlements should be based on the needs of residents to increase service center services and give priority to laid-off workers. The service centers should be open 24 h/day for telephone request for emergency, repairs to the residents on call; public power, water, sanitation should be regularly maintained; implementing closed-end management of residential area, residential 24-hour security personnel on duty at the entrance of residential area; parking, set special personal management at the park.

TABLE 7.3
Configuration Relationship of Services and Public Facilities

Scope (Radius R, Interval H)	Population Size	Public Facilities Allocation
R: 1–2 km, H: on foot, 15–30 min	1,000 people	Nursery and elderly welfare facilities
R: 4–6 km, H: bicycle 30 min	5,000 people	Clinics, markets, and schools
R: 6–10 km, H: car 30 min	10,000 people	Commercial street, specialized hospitals, and institutions of higher learning
R: 20–30 km, H: car 60 min	20,000 people	General hospital and commercial centers

Source: Chen, C. X. 2001. Wuhan University. Master Thesis. (in Chinese).

For a long-term plan, it is necessary to gradually take advantage of supermarkets and convenience stores in the network for information and goods distribution, transitioning them to carriers of e-commerce to realize the organic integration of virtual networks and physical stores. This will also speed up the use of e-commerce in the settlements (residential zone) and expand commodity sales.

3. *Settlement construction management:* By setting the classification bins and promoting activities to encourage residents to classify wastes before discarding them in order to improve the efficiency of focus and for the implementation of garbage collection, as well as the implementation of renewable waste regularly, sharing recovery; strict sanitation management practices, including the prohibition of arbitrary built structures and in the streets and public channels to put their things and every 2 years on the public channel interior wall structures, unified paint; it is not allowed to open restaurants outside the service centers. It is banned to set up stalls which sell groceries, vegetables, or fast food. Open-air barbecue is strictly prohibited. Smoking is banned in public places.

   Motor vehicle horns are forbidden for use in the district. District vehicle entrances and exits are separate, and one-way cycle lanes are established. Speed brakes are used to control speed. Rational planning arrangements for bicycle and car parking spaces help to avoid crowding the sidewalks and sport/leisure venues. In building entrance and sidewalk barrier-free, we set access and blind and avoid these special channels being occupied by bicycles or debris. A reasonable set of indicative signs is needed to guide correct parking positions and routes. Prohibit high pollution and high-noise motor vehicles; for example, tractors or trucks, in the area. In settlements, one should also build car washes to encourage access to clean vehicles and deny entrance to vehicles that are too dirty.

4. *Building a harmonious community culture:* We should vigorously strengthen the culture of the community's hardware construction, the construction of more community libraries, museums, sports arenas, comprehensive cultural squares, matched with the surrounding cultural facilities to make community cultural activities more hardware for support. Building libraries in the settlements, trading books on a regular basis, and holding art exhibitions in the service center are all aimed at promoting and enhancing the culture of the settlement. Organize folk art society, inherit and carry forward the traditional culture, carry out the nature of recreation and leisure activities in residential areas, enhance ecological thinking are to cultivate the spirit of settlements to encourage residents to actively participate in eco-residential construction, as well as to improve people's awareness of ecological protection and active living and to improve people's cultural taste.

   A multilevel, multispecies, all-around community cultural activity is the soul of the development of a harmonious community. As long as there are activities, people can be united. Therefore, different types of leisure and cultural activities should be carried out, including seminars, trainings, rehearsals, performances, and so on; on the other hand, during festivals, large-scale special activities should be held in cities and districts. Improve the dynamic culture and static culture. A variety of cultural activities should be used to facilitate participation by the elderly, children, the disabled, and other disadvantaged groups, as well as to create more convenience to further encourage the participation of the residents.

Schools within or near the settlements should be consistent with the standards of green schools. From kindergarten onward, the young children's ecological awareness and good living and spending habits should be developed. In the window or Loudao blackboard publicity of small settlements, education for ecological civilization should be developed and regularly updated; from encouraging care for environmental health and then extending to the ecological environment of the city's concern and promoting moderate consumption; organizing for residential landscaping; improved grassroots management organization give full play to the neighborhood or property management departments, who play an exemplary role in ecological civilization construction in the settlements, as well as orgnization and publicity.

7.2 ECOLOGICAL CULTURAL SYSTEM PLANNING

The construction of an "ecological city" is rooted in the city's ecological culture. This section describes the city's ecological and cultural meaning, construction content, and implementation approach, as well as further explores how landscape ecology supports the construction of an ecological culture. The establishment of sound ecological spiritual culture, ecological material culture, and ecological institutional culture can promote the sustainable development of eco-city.

7.2.1 Ecological Cultural Connotation

7.2.1.1 Ecological Cultural Definition

The eco-city, as an ideal development model of an ecological civilization, requires economic development, social progress, and ecological protection to maintain that a high degree of harmony, material, energy, and information is used with high efficiency, promoting the mutual symbiosis between human and nature (Yang et al. 2004). Obviously, the ecological culture is the soul and power source of an eco-city, and thus, eco-city construction is rooted in the city's ecoculture and needs the support of the ecological culture. When building a harmonious society and creating eco-cities, improving the eco-culture increasingly shows its importance.

Eco-culture is ecologically a sound culture that is the direct expression of the ecological relationship between nature and society in a material civilization and spiritual civilization, with the fundamental purpose of achieving harmonious development between human and nature. It requires ecological ethics, ecological value, and aesthetics, and such ecological concepts are embodied in various cultural forms, emphasizing the way of thinking, the mode of producing, life methods, and the way of decision-making in developing and promoting the harmonious coexistence of human and nature and the establishment of appropriate mechanisms of social protection (Ouyang et al. 2002). Corresponding to the level of the cultural elements, the ecological culture should pursue good and sustainable social developmental trends to establish morality, ethics, and values in harmony with nature. Converted to a micro–macro perspective, ecological culture should gradually impact decision-making behavior, management systems, and social morals in the macroscopic view, as well as gradually lead to the transformation of people's values, ways of production, and consumption behavior (Sun et al. 2003).

7.2.1.2 Urban Ecological Cultural Connotation

Urban ecological culture is the special expression of ecological culture on the city scale. Its meaning is same as the connotation of ecological culture but the perpetrators of the implementation of the city. This called for harmonious values, spirit, production methods, behavior regulations in the gradual process of urban development constitute the urban production and life-oriented system, all members of the city gathering place, incentives, constraints, and other dynamic effects (Song and Ruan 2004) provide the impetus for urban construction.

Ecological culture seeks to establish the core values, morality, and the ethics of harmony with nature and thus leads to the sustainable development of a society. Specifically, ecological and cultural construction is a macropolicy to gradually influence and induce behavior management systems in society, gradually changing the micro-people's values, ways of production, and consumption behavior. This will shape a new class of decision-making culture, business culture, consumer culture, community culture, science and technology culture, and other cultural institutions (Sun et al. 2003).

The research of the relationship between culture and ecology and the discussion ecological environment need people to conduct deep culture and value shift and gradually form a new form of cultural—ecological culture. Ecological culture as a culture of sustainable development, reflecting the way, process, and results in human pursuit of harmony between human and nature, harmony between human and society. Broadly speaking, ecological culture refers to the values, activities, mental state, and way of thinking that formed from the contact between people and nature. Therein, the natural environment is the basis for human subjects, the object of the world's people as the main form (Chen 2002; Bai 2003). Ecological culture is a culture displaying the harmonious development of human and nature; it is the mainstream culture of an eco-city (Yi 2001). An ecological management effort can increase the eco-culture to promote eco-city development.

7.2.1.3 Guidelines for Urban Ecological Cultural Construction

From the ecological cultural connotations, ecological culture is extremely broad, involving people's consciousness, ideas, and beliefs, as well as their behavioral, organizational, institutional, regulatory, and other cultural patterns. Thus, it is necessary to draft uniform guidelines for ecological cultural construction; that is, "a core and four principles" (Su et al. 2007)

1. *The Core:* The core guiding principle of ecological culture construction is the ecology principle; that is, being ecologically sound is an important basis for determining building activities, which will foster ecological behavior patterns and values as the ultimate goal of reunification.
2. *The Four Principles:*

   Love: According to eco-center egalitarian ideology of the deep ecology, in the biosphere family all living entities are an integral part of the overall, and they have equal intrinsic value and the same power of the survival and development (Chen and Wang 2003). Thus, human beings as just one in many forms of life should be open to the feelings of love, absorb ecological humanism in ancient China, "Heaven-de," "Virtuous," and other ideas (Zeng 2006), respect for other life existence, development power, and the laws of nature. Specifically, biological diversity should be conserved and historical and cultural sites should be preserved. Natural laws of development, according to local conditions, should be followed to perform construction activities. Respect for the social development process, combined with the level of economic development, population quality, the development of the overall urban planning building measures.

   Efficient: Full and rational use of limited natural resources and environmental capacity resources for social progress and to create the greatest value. The traditional "high consumption," "nonloop" operation mechanisms should be altered. Resource utilization efficiency should be improved to maximize the utility of the expertise and resources. Clean production should be used to establish a recycling economy, respecting ecological consumption and waste recycling.

   Sustainability: With the sustainable development of human society as the goal, social and economic development should be within environmental carrying capacity limits. The controlled use of natural resources will achieve intergenerational equity. Respect for human needs, to achieve the comprehensive and free development, pay attention to balanced development of urban and rural areas, reflect the development of intragenerational equity.

   Features: Full use of the existing natural landscape and cultural landscape, showing the city's personality and charm, to enhance the taste of the city. One should focus on the urban landscape culture, historical culture, folk culture, and other cultural characteristics. Concerning the natural landscape, one must protect the rare animals and plants of the city, protect the

TABLE 7.4

Content of Ecological Cultural Construction by Social Factors

Influencing Factors	Content of Construction
Politics	Establish ecological law and administrative system; perfect public participation mechanism
Economy	Develop ecological agriculture and industry; promote circular economy
Culture	Protect famous historical cities and landscape; develop ecological tourism; cultivate ecological consumption concept and ethics
Education	Propaganda and education for environmental protection

existing scenery, and fully display the city's unique landscape. Concerning the cultural landscape, historical relics should be conserved and restored and unique folk customs promoted (as well as folk art and other cultural forms) to tap the city historical, cultural, architectural, religious, and other cultural connotations.

7.2.1.4 General Content of Ecological Culture Construction

Ecological and cultural connotations broadly determine their content related to human consciousness, idea, belief, behavioral, organizational, institutional, regulatory, and other forms of cultural patterns. According to different basis, the general construction of ecological and cultural content of the different categories can be classified as the following:

1. Social factors, which include political, economic, cultural, educational, and other aspects (see Table 7.4).
2. Industries, which include agriculture, industry, tourism, construction, and other fields, as shown in Table 7.5.
3. Implementing actors, which include governments, enterprises and public institutions, communities, schools, and families, as shown in Table 7.6.

7.2.2 EVALUATION OF URBAN ECOLOGICAL CULTURE CONSTRUCTION

To fundamentally curb further deterioration of ecological crisis, embarked on sustainable development, there must be new cultural choices, the existing culture from the "spirit level," "system level," and "material level" to an update into "ecological level" culture; the result is a culture of human form updated—ecological and cultural (Sha 1997). In an eco-cultural system, material culture is the basis for ecology, the ecosystem culture ensures the ecology and the cultural spirit is the core. To track and monitor the progress of the implementation of an ecological culture, an evaluation system must be established, as shown in Table 7.7.

TABLE 7.5
Content of Ecological Cultural Construction by Industries

Industries	Content of Construction
Agriculture	Develop ecological agriculture; expand the ecological agricultural technology and mode; produce nonpolluted, green, and organic agricultural products; establish a production–circulation–sales chain or cycle between agriculture and neighboring regions and other industries
Industry	Perform cleaner production; develop a circular economy Build an eco-industrial chain between urban and rural areas Disclose enterprise environmental information; grade environmental behaviors; design an enterprise ecological image
Tourism	Make use of unique cultural resources; explore the connotation and form of local special culture and historical culture Combine unique landscape resources; develop ecological tourism; strengthen tourism management; permeate ecological cultural thoughts
Construction	Maintain unique houses and historical buildings; combine traditional style and modern style Select building materials with environmental protection features; emphasize the coordination of buildings and the surrounding environment Pay attention to the relevance between adjacent buildings and the meaning of the groups of buildings

TABLE 7.6
Content of Ecological Cultural Construction by Implementing Actors

Actors	Content of Construction
Governments	Lead ecological cultural training; perfect scientific decision-making mechanism and integrated management mechanism
Enterprises	Implement cleaner production and develop circular economy; ISO14000 management system certificate; adopt staff safety and health standard; disclose enterprise environmental information
Public institutions	Cultivate ecological consciousness in daily management and operation
Schools	Popularize environmental protection knowledge in school education and establish green schools
Communities	Cultivate ecological consumption concepts, water saving, energy saving, and garbage classification in communities
Families	Perform ecological consumption in daily life; use environmentally friendly products such as Freon-free refrigerators and nonphosphate detergents

Select the appropriate indicators, from all levels of ecological and cultural construction to evaluation of its implementation, before the implementation of an eco-culture, the status of various indicators of the value compared with the standard, to identify the need to focus on strengthening the construction aspects. During the construction process, with reference to the stage of the target standards, we could guide and supervise the construction of an ecological culture.

TABLE 7.7

Evaluation Index for Urban Ecological Culture Construction

Level of Construction		Name of Index	Standard
Ecological spirit culture	Psychological culture	Public satisfaction with the environment	>95%
		Environment protection volunteer ratio	>40%
	Cognitive culture	Popularity of environmental protection propaganda and education	>90%
		The proportion of green schools in middle and primary schools	>80%
Ecological material culture	Enterprise production culture	The proportion of enterprises that disclose environment information	>95%
		ISO14000 certificate rate of large enterprises	>90%
		Rate of output value by green and nonpolluted products	>85%
	Community consumption culture	The proportion of garbage classification communities	100%
		The proportion of reclaimed water reuse communities	>60%
		The proportion of families who use environmentally friendly products	>50%
Ecological system culture	Decision-making mechanism	The execution rate of the planned environmental impact assessment	>95%
		The proportion of environmental argument and witness hearings in major decision making	100%
	Management mechanism	The proportion of green office products used by governments	>60%
		The proportion of civil servants who participate in ecological cultural training	>90%

7.2.3 PLANNING GOAL

Based on the understanding of the connotation of ecological culture, the goal and evaluation of eco-cultural development can be established.

7.2.3.1 Overall Goal

One should complete urban ecological and cultural construction in phases, material culture from the ecological, spiritual, and cultural ecology and ecological levels of the institutional culture to start with the formation of ecological production concept, the implementation of ecological consumption habits, cultivate ecological ethics, the establishment of ecological rule of law, improve the mechanism for public participation, and ultimately the integration of urban and rural ecological culture.

7.2.3.2 Stage Goals

7.2.3.2.1 The First Step

The government's eco-cultural training should be strengthened, ecologically sound scientific decision-making mechanisms should initially be formed, and an integrated management system should be established. Environmental protection and education can be popularized to strengthen primary and secondary schools to go green. Environmental education and eco-town public education can be popularized on the basis of the construction and development of ecological agriculture. Various forms of eco-tourism, especially rural eco-tourism, can also be developed.

7.2.3.2.2 The Second Step

Public environmental awareness, environmental protection publicity, and education should continually be enhanced to improve coverage. From the government, enterprises, units, communities, schools, and families at all levels create a full range of ecological and cultural atmosphere. Continue eco-tourism and eco-agriculture to strengthen eco-town culture.

7.2.3.2.3 The Third Step

Integration of urban and rural eco-cultural, ecological concept caught on ecological science policy-making mechanism, and integrated management system fully formed.

7.2.4 Urban Ecological Culture Approach

To improve urban ecological and cultural construction with unified guidance and operations, covering a wide range of ecological and cultural construction, we propose the construction of an eco-cultural framework. This framework can be summed up as a main line, three levels, three approaches, and four brands, as shown in Figure 7.8.

7.2.4.1 Main Line

The harmony between humans and nature, as well as natural and socioeconomic development, creates harmonious, sustainable cities. This is the main line of an urban ecological culture.

7.2.4.2 Three Levels

Ecological culture is established at three levels, that is, spiritual ecological culture, material ecological culture, and institutional ecological culture.

7.2.4.2.1 Ecological Spirit Culture—Ideology

Ecological spirit culture is to change people's traditional concept by means of mentality and cognitive influence, and to make people form ethical values of harmonious relationship between nature and human. This is the ultimate eco-culture, the most fundamental goal.

1. *Mentality culture:* The induction and promotion of a spiritual civilization to align recycling, green practices, and ecological and cultural values with the concept of an eco-deep heart. Promote the traditional concept of change

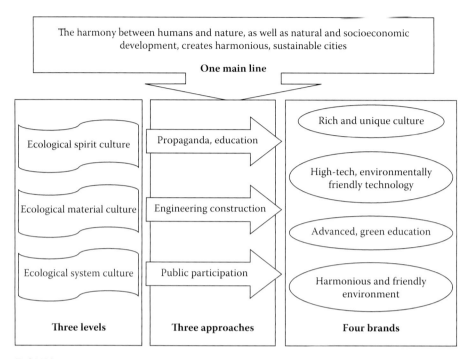

FIGURE 7.8 Construction framework of ecological culture.

and advocate healthy and civilized modes of production and consumption behavior, as well as social habits of ecological change.

2. *Cognitive culture:* By strengthening the ecological philosophy, ecological science, and the aesthetics of ecological education and awareness, a universal ecological knowledge is created that fosters ecological concepts and strengthens ecological awareness. Promote a higher level of civilization by training a generation of educated, idealistic, high-quality ecological community builders.

7.2.4.2.2 Ecological Material Culture—Physical Form

An ecological material culture uses the material form as a carrier to express ecological ideas, including changing the flow of production materials and consumption. This is in accordance with the sustainable development and requires the construction of corporate and community cultures and the establishment of a recycling-based society. This is the fastest that an eco-culture can be reflected and the most important goal.

1. *Corporate culture and the mode of production:* Transform the mode of production to clean production and create a circular economy by organizing material production.

2. *Community culture and consumption:* Changes in people's consumption concepts and the formation of simple, efficient, green consumer habits form the material consumption-induced ecological aspects of production.

7.2.4.2.3 Ecological System Culture—System Form

Speed up decision-making mechanisms and management system reform. Improve urban development regulations and policies to increase the functions of governmental departments and public participation in decision making. Foster sustainable operating mechanisms to achieve systematic management of urban decision making. This is a smooth implementation of eco-cultural development of the necessary protection.

7.2.4.3 Three Approaches

7.2.4.3.1 Propaganda and Education

Using extensive publicity and education for environmental protection, one can improve the ecological awareness of educated citizens (including policy makers, managers, and the general public) to create a good social climate.

7.2.4.3.2 Public Participation

Ecological and cultural construction should cover a broad spectrum and essentially require a new social value and ecological moral system, which requires the participation of the masses. It should fully understand the public demand to expand channels for public participation. One should try to protect the rights of public participation, mobilizing the masses to participate in the eco-cultural construction.

7.2.4.3.3 Engineering Construction

An ecological culture should ultimately be implemented on a range of projects, from the government, enterprises, institutions, communities, schools, and families at all levels to carry out a comprehensive green cells project, the implementation of eco-culture. Create a green government, community, schools, and hotels, units (including corporate/business units), sites, homes (city), and 10-star civilized households. Strengthen the eco-cultural missionary by building an ecological science education base and by the development of eco-tourism to create green bus and green media. Improve the ecological legal system, promote administrative reform, transform government functions, and streamline government agencies. Establish mechanisms for public participation, implement government administrative affairs and corporate environmental information disclosure, and promote eco-environmental protection and the hearing system. Strengthen collaboration and establish dedicated eco-cultural construction of functional departments responsible for the construction of ecological culture, planning, decision making, organization, coordination, and supervision.

7.2.3.4 Four Brands

The results of ecological and cultural construction output are to create the city's cultural, scientific, educational, environmental, and other brands. Cities according to their characteristics have to play on the basis of

1. *Rich and unique culture:* Mining the city for cultural resources, combined with the city's history, culture, music, architecture, landscape, and characteristics will help to integrate various cultures, ideas.

2. *High-tech, environmentally friendly technology:* In the process of building a material and cultural ecology, resource efficient recycling is the principle of high-tech, environmental technology and industry.
3. *Advanced, green education:* Public education ensures the sustainable development of urban ecological and cultural construction. Increased investment in education and construction, with an emphasis on creating green primary and secondary school environments.
4. *Harmonious and friendly environment:* Ecological material culture through the visual manifestation of the silent appeal of spiritual and cultural ecology and ecological protection within the institutional culture jointly create a warm, friendly environment and build a socialist material civilization, institutional civilization, and spiritual civilization and the organic unity of a harmonious society, and thus attract talent, capital, and surrounding partners. Both ensure that people live comfortably, are free to develop, and influence foreign guests consciously to protect the ecological environment, and jointly safeguard the harmonious and friendly urban environment.

7.2.5 Ecological Culture Construction under the Guidance of Landscape Ecology

The cumbersome construction of the ecological and cultural needs of a guiding theoretical framework while the overall integrity of landscape ecology, macroregional guidance for cities, and other features provide the possibility of ecological culture. Based on the substrate-patch-corridor model from landscape ecology, urban ecological culture framework can be constructed. Specific measures are proposed in terms of point, line, and surface to develop urban ecological culture and create a city card (Su et al. 2008).

Competition among cities has emerged as an ecologization trend, meaning that the competition in the urban context of the era of ecological civilization to create a unique image among global urban ecological and cultural cities has become an important strategy (Song and Ruan 2004). In this situation, knowing how to build a distinctive urban ecological culture has become a concern.

7.2.5.1 Feasibility of Landscape Ecology to Guide Urban Ecological Culture Construction

Landscape ecology is a comprehensive subject that arose in recent decades and integrates geography, ecology, systems theory, control theory, and other disciplines (Li 2005; Mao and Wang (2006) construction of the West). Landscape ecology studies the structure of the macroscale landscape (spatial pattern), its functions (ecological processes), and evolution (spatial dynamics) with holistic and integrated macroregional characteristics (Liu 2006). The vitality of landscape ecology as a whole is its integrated ideas and directly involved in human subjects (Yang et al. 2006). Landscape ecology, as an integrated solution to the environmental and resource problems of new scientific theories and methods, has been highly valued in recent years for the rational development and utilization of resources and resource protection, urban development planning, land management and scientific management, environmental protection, and land improvement (Li 2005).

Ecological culture development is a holistic development within a specific region, and closely related to human activities and consciousness. Landscape ecology also study the overall characteristics of a region and pay great attention to human issues. Therefore, Ecological culture development and landscape ecology are similar in some ways. We can apply the basic theory of landscape ecology to guide a city's ecological and cultural construction to provide a framework for the construction of the system.

7.2.5.2 Urban Ecology and Cultural Development Framework under the Guidance of Landscape Ecology

A matrix–patch–corridor is the basic pattern in landscape ecology (Zhang et al. 2005). In understanding the model based on the clear matrix, patches, and corridor landscape ecological significance, we give them the urban ecological meaning of culture (Table 7.8).

A matrix, in an ecological and cultural sense, plays the same role as the macro-context. Specifically, in space, it represents, in addition to patches, corridors outside the geographical area of great scope. Ideologically, means to enhance regional culture based on people's attitude toward culture, ideas, and ideals.

In an ecological culture, plaque refers to space that differs from the surrounding environment. Spatially, a plaque is the gathering of cultures of various characteristics,

---

TABLE 7.8
Ecological Cultural Meaning of Basic Patterns in Landscape Ecology

Items	Meanings in Landscape Ecology	Meanings in Ecological Culture
Matrix	The macroscopic background structure that is most widely distributed, with the largest continuity; background ecosystem or land use pattern, generally present as an area; very important to determine the characteristics of the entire ecosystem	Soil on which various cultures grow; cultural background and basic ideas; very important to determine cultural keynotes and fundamental development direction
Plaque	Space segment that is different from the surrounding environment in appearance and nature but has a certain internal homogeneity; generally present as a patch or block in space; with the property of activate space structure	Reflects cultural points of various cultures; aggregation regions of various unique cultures; the core in ecological cultural construction; the key that determines whether ecological cultural construction succeeds or fails
Corridor	Line or belt structure that is different from the surrounding two sides; very important in promoting material, energy, and information exchange and maintaining ecological processes	Connections between different plaques and the network formed by these connections; very important in the spread and communication of various cultures

spread in various parts of the city. Culturally, it includes the city's history, culture, architectural culture, living culture, production culture, and other various cultural forms.

A corridor is a connectivity structure. Spatially, it is the thread or belt area that establishes contact among plaques, and these links are intertwined to form a network channel. On the content, it includes the green channel, road, and landscape connectivity, which play a role in a variety of hardware and software.

REFERENCES

Bai, G. R. 2003. On ecological culture and ecological civilization. *Human Geography* 18 (2): 75–8 (in Chinese).

Chen, W. H. and Wang, K. 2003. Deep ecology and its aesthetical insight. *Journal of Social Science of Hunan Normal University* 32 (2): 23–6 (in Chinese).

Chen, X. G. 2002. On ecological culture. *Social Science Front* 6: 250–1 (in Chinese).

Han, X., Xu, L. Y., Chen, B., and Yang, Z. F. 2009. Evaluation and optimization of the ecological human settlement in an immigration city. *China Population, Resources and Environment* 19 (5): 22–7 (in Chinese).

Huo, X. J., Pan, Y. Z., Zhang, L. W., and Qu, S.H. 2006. The city's ecological appraisement of sustainable development: A case study in Baotou Inner Mongolia. *Journal of Arid Land Resources and Environment* 20 (1): 140–5.

Li, M. H. 2005. A rapid developing subject-landscape ecology. *Journal of Biology* 22 (5): 8–10, 45 (in Chinese).

Li, W. X. 2000. The measures and strategy for resources recovery and recycling of municipal refuse in China. *Resources Science* 22 (3): 17–9 (in Chinese).

Liu, Q. 2006. Application of landscape ecology into tourism programming. *Resources Environment and Development* 3: 19, 26–28 (in Chinese).

Mao, J. X. and Wang, Z. X. 2006. Discussion on urban landscape design based on acoustic ecology. *Environmental Science & Technology* 29 (1): 94–8 (in Chinese).

Ministry of Construction P. R. China. 2005. Standard for setting of town environmental sanitation facilities (CJJ 27-2005) (in Chinese).

Ministry of Construction P. R. China. 2006. Code of urban residential areas planning and design (GB 50180-2002) (in Chinese).

Ouyang, Z. Y., Wang, R. S., and Zhen, H. 2002. Ecological culture in Hainan ecological construction. *China Population, Resources and Environment* 12 (4): 70–2 (in Chinese).

Ravetz, J. 2000. Integrated assessment for sustainability appraisal in cities and regions. *Environmental Impact Assessment Review* 20 (1): 31–64.

Saaty, T. L. 1977. Theory of analytical hierarchies applied to political candidacy. *Behavioral Science* 22 (4): 237.

Sha, J. 1997. Ecological culture is the only choice for mankind to lead a new life. *Forest & Humankind* 2: 44 (in Chinese).

Song, X. F. and Ruan, H. X. 2004. Ecological culture and city competence—the epoch connotation of city competence in the 21th century. *Ecological Economy* 12: 83–6 (in Chinese).

Su, M. R., Yang, Z. F., and Zhang, Y. 2007. Construction of urban ecological culture. *Environmental Science & Technology* 30 (9): 53–67 (in Chinese).

Su, M. R., Yang, Z. F., and Xu, L. Y. 2008. Building up urban ecological culture based on landscape ecology. *Environmental Science & Technology* 31 (4): 123–6 (in Chinese).

Sun, J., Han, Y. L., and Wang, R.S. 2003. Strategies for ecological culture construction of Yangzhou City. *Science & Technology and Economy* 16 (4): 42–5 (in Chinese).

Wu, L. Y. 2001. *Introduction to sciences of human settlements*. Beijing: China Architecture & Building Press (in Chinese).

Yan, X. Y., Meng, H., and Tang, M. X. 2006. Comparison and discussion about different weight methods of comprehensive evaluation. *Chinese Health Quality Management* 13 (4): 58–60.

Yang, D. W., Chen, Z. J., Chen, Y.J. et al. 2006. Biodiversity based on basic theories in landscape ecology. *Areal Research and Development* 25 (1): 111–5, 124 (in Chinese).

Yang, Z. F., He, M. C., Mao, X. Q. et al. 2004. *Programming for urban ecological sustainable development*. Beijing: Science Press.

Yi, W. J. 2001. From human centre view to ecological civilization view—Environmental ecology perspective of ecological culture. *Southeast Academic Research* 5: 38–42 (in Chinese).

Zeng, F. R. 2006. "Union of heaven and man" in ancient China and today's conception of an ecological aesthetics. *Journal of Literature, History and Philosophy* 4: 5–11 (in Chinese).

Zhang, L. Y., Zhou, Y. Z., Wen, C. Y. et al. 2005. Landscape ecological consideration on construction of ecocity. *Ecological Science* 24 (3): 273–7 (in Chinese).

wwwZhang, Q., Zhu, C., Liu, C. L. et al. 2005. Envrional change and its impacts on human settlement in the Yangtze Delta, PR China. *CATENA* 60 (3): 267–277.

8 Urban Ecological Planning Regulation

Bin Chen, Lixiao Zhang, and Zhifeng Yang

CONTENTS

8.1 ECO-ZONE UNIT–BASED ENVIRONMENTAL ELEMENTS REGULATION

As mentioned in the chapter of ecologically functional zoning, a region can be divided into four parts according to their functional orientation. The determination of orientation mainly depends on their limiting factors.

8.1.1 REGULATION FOR EXPLOITATION-OPTIMIZED ZONE

- Keep proper population density, sound regional construction density, and plot ratio.
- Strengthen landscape construction of the built area, emphasize the harmonious development of artificial ecology and natural ecology, keep reasonable layout for the public greenbelt.
- Construct the ecological industrial parks, taking "circular economy" and "clean production" as guidelines.

8.1.2   REGULATION FOR KEY EXPLOITATION ZONE

- Construct the ecological industrial parks, economic and technological development zone, labor-related industrial base, and so on, in line with the principle of "priority to ecology" and "circular economy."
- Keep proper population density, control the proportion of each land-use type, keep rational regional construction density, and plot ratio.
- Enhance landscape construction, emphasize the harmonious development of artificial ecology and natural ecology, and keep reasonable layout for the public greenbelt and ecological corridor.

8.1.3   REGULATION FOR EXPLOITATION-LIMITED ZONE

- Coordinate the relationship between regional development and ecological protection; develop eco-friendly agriculture and tourism.
- Control the developing scale of population, industry strictly; offer positive guidance and adjustment for reasonable industrial structure; develop eco-friendly industry, and forbid the enterprise characterized by heavy pollution and intensive energy.
- For areas under the national special protection, implement the national and local law or standard and forbid any kind of exploring activity harmful to the ecosystem. For the damaged ecosystem, restore them in a definite date for completion.
- Improve the environmental pollution of the industrial and mining enterprise having located there, move the heavily polluting enterprise gradually, and restore damaged ecosystem.
- Offer guidance for an integration of the basic farmland protection areas, forest land, orchard, drainage, and regional greenbelt.

8.1.4   REGULATION FOR EXPLOITATION-PROHIBITED ZONE

- Forbid variety of activities associated with exploration and construction.
- Accelerate eco-environmental protection and construction.
- Control the land use rigidly, maintain the balance of regional ecosystem, improve the structure of natural scenery, and protect the specialty of eco-environmental scenery.
- Improve the management system, protect the biodiversity, and at the same time rationally explore the land for ecotourism.

8.2   LIMITING FACTOR ANALYSIS AND REGULATION FOR URBAN ECOSYSTEM

8.2.1   INTRODUCTION

As a social-economic-natural complex ecosystem, urban ecosystem is confronted with various complicated problems when the energy and materials metabolism among a large amount of biophysical components is driving the integrated evolution. At the cost of environmental degradation resulted from the traditional development mode, many eco-environmental problems including air pollution, water scarcity, and energy shortage have emerged, lowering the human living standard and impeding the sustainable development of urban ecosystem. Generally, urbanization has caused increasing disturbances and pressures especially in the developing countries with expanding population. Particularly, more than 400 cities in China have water shortage. The annual economic cost brought by water shortage of urban industry has reached 120 billion yuan, amounting to 1% of gross domestic product of China.

Ecosystems are non-isolated and are in far most cases open systems that receive inputs and produce outputs (Patten et al. 1997). As for urban ecosystem, none has ever existed that did not depend on its hinterland for support, and even hunting and gathering tribes rely on inputs from a much larger area than that of the village itself (Costanza 2000). Lacking producer and decomposer, urban ecosystem is dominated by consumer, thus weakening the stability of the total ecosystem. High dependence on the environment and social metabolism may further lead to fragility of the urban ecosystem.

Regarding the dependence and complexity of the urban ecosystem, proper ecological regulation is needed to be conducted after necessary simulation and analysis. Yokohari et al. (2000) pointed out that the future of urban environment would be gloomy without effective regulation. To realize stable and sustainable urban development, it is necessary to explore the origin of eco-environmental problems and put forward corresponding ecological regulations for urban ecosystem. Most previous researches have focused on the urban development models (Huang 2001; Booth 2002; Rogerson 2003), regulative objectives (Yang et al. 2004), principles (Wang 2000; Wang et al. 2000; Rakodi 2001), and approaches (Wang 2000; Hogu and Roy 2001; Yi 2005). To probe into the cause of urban ecosystem degradation and set up the corresponding regulation scheme, we introduce a new indicator, urban ecological carrying capacity (UECC), representing the sustentation function of urban ecological supporting system (UESS), with the attempt to emphasize the fundamental utility of the urban biophysical components for the urban development mode.

8.2.2 METHODOLOGY

8.2.2.1 Urban Ecosystem Degradation in View of UECC

8.2.2.1.1 UECC

Urban ecosystem has the specific carrying capacity grounded on three subsystems including resource, environment, and human society that provide the required biophysical foundation to sustain the urban development without detrimental impacts. It has to rely on the UESS consisting of a series of factors with different supporting effects and functions to input and output energy and material flows. Here, we introduce the denotation and connotation of UECC to reveal the relationship and interaction between carrying media and object of the urban ecosystem. The sustaining functions for urban development mainly include the resource supply capacity determined by resource capacity (e.g., water, land, and energy resources), environmental bearing capacity, and environmental buffering capacity, which can thereby be quantified and termed as UECC.

Representing the sustaining functions of UESS, UECC is an integration of the supply of UESS and the demand of urban development. The relationship between the supply and demand determines the threshold of UECC, which is critical for the urban ecological regulation. Although infinite UECC is favorable for the flourishing and continuous urban development, there exists a fixed biophysical boundary for a given urban ecosystem because the structure and function of UESS within specified boundary tend to be steady, assuming the resource supply and environmental capacity to be constant during a certain period. Thus, the threshold of UECC is mainly concerned to the infinite demand and finite supply. When the demand exceeds the supply, the threshold will emerge and threaten the urban ecosystem stability, while under the threshold, when the supply and demand of UECC are in congruity and balance, the urban ecosystem can develop in a harmonious way.

8.2.2.1.2 Urban Ecosystem Degradation

It is certain that the Environment Kuznets Curve (Figure 8.1) depicts a wonderful future for the social development, indicating that eco-environment will be perfectly restored when the social economy can develop well.

However, it is not sufficient to understand the urban ecosystem evolution mechanism only from this curve because there is a hidden assumption that the environment quality degeneration is always under the threshold of UECC at the inflexion of associated development of socioeconomic and natural subsystems. As the threshold of UECC is considered, there will be four different modes describing the relationship between human and nature in the social evolution process as shown in Figure 8.2, which will help to analyze the origin of urban ecosystem degradation.

8.2.2.1.2.1 OKD Mode   As an ideal pattern, urban environment quality degeneration is under the warning line of UECC at the inflexion of relevant urban ecosystem. The warning line is assumed to be 80% of the UECC, representing a notable signal

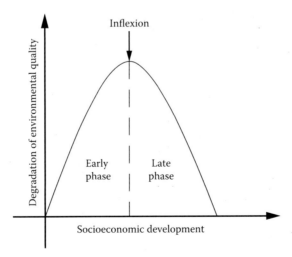

FIGURE 8.1 Kuznets Curve of social economy and environmental quality. (Modified from Yandle, B. et al., The Environmental Kuznets Curves: A review of findings, methods, and policy implications, PERC Research Study 02-1, 1–38, April 2004.)

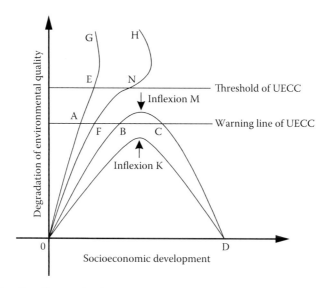

FIGURE 8.2 Coupling mechanism between urban economic development and environment based on urban ecosystem carrying capacity (UECC).

that certain regulations should be made to mitigate the deteriorated situation. On the right side of the inflexion, enhanced by the human environmental consciousness and corresponding environmental investment, the threat to eco-environment caused by urban development gradually decreases with the coordinated development of social economy and eco-environment.

8.2.2.1.2.2 OBMCD Mode The OBMCD mode can be described as a pattern of "developing first and then treating the pollution." At the inflexion of relevant development of socioeconomic and natural subsystems, the urban environmental quality degeneration has exceeded the warning line but still less than the threshold of UECC. After the inflexion, the environmental degeneration trend can be effectively held back by the proper ecological regulation. Subsequently, the environmental quality will turn gradually better and the coordinated development of human being and nature can be obtained.

8.2.2.1.2.3 OFNH Mode Characterized by the pattern of "developing first and then treating the pollution," OFNH mode is different from the OBMCD mode above because the urban environmental quality degeneration has not only exceeded the warning line but also broken through the threshold of UECC at the inflexion point. Thus, the environmental deterioration trend cannot be reversed through ecological regulation, thus leading to the stagnation of urban ecosystem development.

8.2.2.1.2.4 OAEG Mode It can be seen from Figure 8.2 that the influence of human activity on environment is not obvious due to the anthropocentric exploitation of natural resource and strong environmental self-purification capacity at the beginning. Afterwards, with the economical production and population expansion, not only the socioeconomic effect on environment breaks through the warning line of UECC but also the degeneration of urban ecosystem exceeds the limit of its self-regulation and self-restoration. Consequently, the urban ecosystem degradation emerges.

As for the OKD and OBMCD modes, the urban environmental quality degenerations are under the threshold of UECC at the inflexion of relevant urban ecosystem development. Given timely and proper regulation, eco-environmental quality can be improved gradually and harmonious development among society, economy, and environment of the urban ecosystem can be achieved. Considering the OFNH and OAEG modes, once the urban environmental quality degradation exceeds the threshold of UECC, the trend of eco-environmental deterioration cannot be held back to prevent the urban ecosystem degradation. Therefore, whether the urban ecosystem degradation will break out and can be eliminated relies on whether the relevant urban ecosystem development surpasses the threshold of UECC. Since the threshold of UECC is determined by the relationship between supply and demand of UECC, the urban ecosystem degradation originates from the imbalance between supply and demand of UECC, which is aggravated by the conflict between the infinite demand of socioeconomic development and the finite supply of UESS.

8.2.2.2 Limiting Factors of UECC

Although various ecological factors function simultaneously, only one or a few of them may dominate the urban ecosystem according to Liebig's law of the minimum. Similarly, the threshold of UECC can be represented with a few dominated limiting factors that can restrict the developing speed and scale of urban ecosystem in certain situations.

8.2.2.2.1 Supply and Demand of Limiting Factors

Supply and demand of limiting factors can be described by the virtue of economic analysis as shown in Figure 8.3, where the vertical axis expresses the socioeconomic scale and the horizontal axis represents the factors of UESS/UECC.

Three kinds of supply and demand relationships are, respectively, described by (a), (b), and (c) scenarios. S_1, S_2, and S_3, respectively, denote the supply curves of factors of UESS/UECC, whereas D_1, D_2, and D_3, respectively, represent the demand curves of factors. Q is the socioeconomic scale, of which Q_1, Q_2, and Q_3 are, respectively, the demand of factors of UESS/UECC determined by socioeconomic scale. Q_e expresses

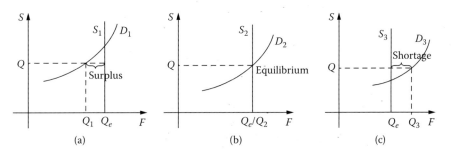

FIGURE 8.3 Supply and demand of the limiting factors of urban ecological supporting system (UESS)/UECC. (a) Supply exceeds demand, (b) supply equals demand, and (c) demand exceeds supply.

the quantity at the balance point determined by both supply and demand curves, which means the maximum sustainable supply of the factors of UESS/UECC.

In scenario (a), Q_1 is less than Q_e, and the surplus emerged. In scenario (b), there are no excess and shortage when Q_2 is equal to Q_e. In scenario (c), Q_3 is more than Q_e and then, the shortage emerged.

The development of urban ecosystem in scenarios (a) and (b) will not be greatly influenced by the excess and equivalent supply of factors of UESS/UECC, for their costs are much less than that of the scenario (c). In contrast, the development of urban ecosystem will be influenced when the supply of factors of UESS/UECC cannot satisfy the demand of social economy like scenario (c). Therefore, more attention should be paid to the limiting factors in scenario (c).

8.2.2.2.2　Assessment for Limiting Factors

8.2.2.2.2.1　Identification of Limiting Factors　A novel assessment method has been put forward by Hu (2005) to identify the limiting factors of UESS so as to implement effective urban ecological regulation. Firstly, factors of UESS are divided into resource factors and environment factors in view of the characteristics of supply and demand relationships. Then, different limiting indices concerning resource factors and environment factors are adopted respectively to quantify the gap between demand and supply, which can be calculated according to Equations 8.1 and 8.2:

$$\text{Resource factor: } R_i = \frac{L_i}{D_i} \times 100\% = \frac{D_i - B_i}{D_i} \times 100\% \tag{8.1}$$

$$\text{Environment factor: } R_i = \begin{cases} H_i - H_i^s & \text{(a)} \\ \dfrac{H_i - H_i^s}{H_i^s} & \text{(b)} \end{cases} \tag{8.2}$$

where R_i is the shortage index of the ith factor; B_i represents the amount of supply of the ith factor; D_i is the real demand of the ith factor determined by the socioeconomic system; L_i is the gap between the real and balanced demanding amount of the ith factor; H_i is the environmental quality status of the ith factor, and finally, H_i^s is the standard of environmental quality. The resource factors of water, land, and energy should be calculated by Equation 8.1, while the environmental factors by Equation 8.2a and b.

8.2.2.2.2.2　Grading Limiting Factors　Since there are various limiting factors of UESS, it is necessary to compare the supply and demand relationships of different factors and then decide the preferential levels of the regulated objects. Thus, the limiting factors of UESS can be graded into five levels, that is, very strong, strong, relative strong, weak, and none ones, as listed in Table 8.1, which are, respectively, denoted by the symbols "++++", "+++", "++", "++", and "=".

TABLE 8.1

Grading Standard of the Limiting Factors

		Grade of Limiting Factors (%)				
Item		Limiting Factors	Very Strong	Strong	Relative Strong	Weak	None
Limiting index	Major resources	Water, land, and energy	>30	15–30	5–15	<5	<1
		Forest acreage and green coverage	<−15	−10 to −15	−5 to −10	0 to −5	0
	Eco-environment	Energy structure (coal burning ratio)	>20	10–20	5–10	0–5	0
		Utilization ratio of atmosphere water environment capacity	>100	50–100	30–50	5–30	<5
	Mark		++++	+++	++	+	=

8.2.2.3 Regulations for Supply and Demand of Limiting Factors

The ecological regulation based on limiting factors of UECC is shown in Figure 8.4, where S_0 and D_0, respectively, stand for the supply and demand of the limiting factors before ecological regulation; S' and D', respectively, describe the situation after ecological regulation, whereas S_{F1} and S_{F2}, respectively, represent the lack of limiting factors after regulating supply and demand; and S_{F0} equals to the lacking amount before ecological regulation.

Considering the imbalance between supply and demand of UECC highlighted by the limiting factors, corresponding regulative measures can be carried out. According to the grades of the limiting factors, the higher level limiting factors are to be determined with priority. Then, the regulation of the selected limiting factors can be commenced from both the supply and the demand sides. On the one hand, exploiting new supply source may contribute to the supply capacity of the limiting factors by means of trade and technology, which can be sketched by the rightward movement of the supply curve (Figure 8.4a). On the other hand, the demand can be adjusted according to the following two different circumstances. In the first state, since the socioeconomic development goes beyond the supporting ability of UESS, the resource utilization efficiency should be improved to meet the demand and gradually balance the supply and demand. In the second state, although the socioeconomic scale is reasonable, the bottleneck still emerges due to the incomplete facility, laggard technique, and unreasonable utilization of resources, indicating

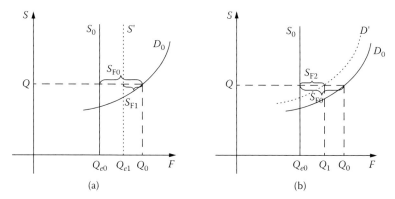

FIGURE 8.4 Ecological regulation based on the limiting factors of UECC. (a) State of supply and demand before/after ecological regulation (regulate supply); (b) state of supply and demand before/after ecological regulation (regulate demand).

that the effective supply ability of limiting factors should be heightened by engineering construction, technical renovation, and strengthened management. These two circumstances can be expressed by the upward movement of the demand curve (Figure 8.4b).

8.2.3 DISCUSSION

The exponential relationship illustrated by logistic curve shown in Figure 8.5a can be chosen to reveal the mechanism of urban ecosystem development and usage of UECC. In the early period, city develops in exponential way when most ecological factors are favorable for the urban development. Later, the development tends to be stable when some ecological factors grow into limiting factors when their ecological niches are gradually occupied (Alexander 1971; Sun et al. 1993).

The ecological regulation based on UECC aims to ensure the homeostasis between the supply and demand of the limiting factors, and control the demand of UECC required by urban development within the range of UECC. When the demand of UECC verges on the threshold of UECC, the regional interior structure is regulated and exterior surroundings is also improved to break down the bondage of limiting factors, and thereby, the supply level of UECC is promoted. Consequently, by the ecological regulation based on UECC, the urban ecosystem will develop step by step in a logistic-chain mode as shown in Figure 8.5b, in which P represents the developing status of the urban ecosystem, CC means the ecological carrying capacity, T is the time, T_i is the ith period, T_{ij} is the jth time point in the ith period, K_i is the UECC in the ith period, and K_{ij} is the corresponding carrying capacity level at the jth time point in the ith period. With this special developing mode, the total S-style evolution process will be divided into connected logistic units in each period that is convenient for monitoring, identification, and regulation, thus leading to well stabilization and balance mechanism of urban ecosystem evolution.

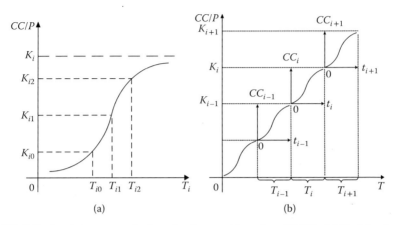

FIGURE 8.5   Logistic curve of urban development based on UECC. (a) Sketch of urban development based on UECC before ecological regulation; (b) combined logistic curve of urban development after ecological regulation based on UECC. (Modified from Cao, L. J., Science Press, Beijing, China [in Chinese], 78, 1999.)

8.2.4 Concluding Remarks

Urban ecosystem degradation and the dependence of urban ecosystem appeal for proper and timely urban ecological regulation. Meanwhile, the controllability of human-dominated urban ecosystem presents the opportunity to conduct effective regulation. UECC proposed in this chapter quantifies the biophysical capacity of urban ecosystem in both supply and demand aspects considering the conflict between infinite demand of socioeconomic development and finite supply of UESS as the essential reason of the urban ecosystem degradation.

Since the threshold of UECC is expressed by limiting factors, the urban ecosystem degradation can be mitigated by adjusting the imbalance of supply and demand of the limiting factors. Corresponding ecological regulation measures can thereby be taken to regulate the supply and demand of the limiting factors and realize the stable logistic-chain development during each period of the urban ecosystem evolution.

REFERENCES

Alexander, M., 1971. Microbial ecology. John Wiley, New York, NY, p. 95.
Booth, R. D., 2002. *Searching for paradise: economic development and environmental change in the Mountain West.* Littlefield Publishers, Lanham/Oxford, UK.
Cao, L. J., 1999. *Theory and method for evaluation of sustainable development.* Science Press, Beijing, China, p. 78 (in Chinese).
Costanza, R., 2000. The dynamics of the ecological footprint concept. *Ecol. Econ.* 32, 341–345.
Hogu, A. L., Roy, B. L., 2001. Theory of sustainable groundwater management: an urban case study. *Urban Water.* 3, 217–228.
Hu, T. L., 2005. *Study on urban ecological regulation rooted in ecological planning.* Beijing Normal University, Beijing, China, p. 57 (in Chinese).

Huang, G. Y., 2001. Evolution of eco-city in China. *Urban Environ. Urban Ecol.* 14(3), 6–8 (in Chinese).

Patten, B. C., Straškraba, M., Jørgensen, S. E., 1997. Ecosystems emerging: 1. Conservation. *Ecol. Model.* 96, 221–284.

Rakodi, C., 2001. Forget planning, put politics first? Priorities for urban management in developing countries. *Int. J. Appl. Earth Observ. Geoinf.* 3, 209–223.

Rogerson, C. M., 2003. Local economic development in Midrand, South Africa's Eco-city. *Urban Forum.* 14, 210–222.

Sun, R. Y., Li, B., Zhuge, Y., Shang, Y. C., 1993. *General ecology.* Higher Education Press, Beijing, China (in Chinese).

Wang, X. R., 2000. *Ecology and environment—new theory for urban sustainable development and ecological regulation.* Southeast University Press, Nanjing, China (in Chinese).

Wang, R. S., Zhou, Q. X., Hu, R., 2000. Method of urban ecological regulation. Meteorologic Press, Beijing, China (in Chinese).

Yandle, B., Bhattatai, M., Vijayaraghavan, M., 2004. The Environmental Kuznets Curves: A review of findings, methods, and policy implications. PERC Research Study 02-1, April, pp. 1–38.

Yang, Z. F., He, M. C., Mao, X. Q., Yu, J. S., Wu, Q. Z., 2004. Programming for sustainable development of urban ecology. Science Press, Beijing, China, pp. 3–4 (in Chinese).

Yi, W. J., 2005. Urban environmental problem and ecological regulation. *Safety Environ. Eng.* 12(1), pp. 1–4 (in Chinese).

Yokohari, M., Takeuchi, K., Watanabe, T., Yokota, S., 2000. Beyond greenbelts and zoning: A new planning concept for the environment of Asian mega-cities. *Landscape Urban Plann.* 47, 159–171.

Section II

Case Studies

9 Eco-City Guangzhou Plan

Linyu Xu and Zhifeng Yang

CONTENTS

The urban ecological planning of Guangzhou was drawn up from 1999 to 2003. The urban ecological planning work did not have constraints and guidelines because "The Guidelines for Construction Planning of Ecological Counties and Cities" had not been published at that time. The structures and content of Guangzhou urban ecological planning were therefore flexible, and although the ideas and methods used were immature in hindsight, they were cutting edge during that time. Generally, the Guangzhou urban ecological planning focused on hierarchical assessment, space controlling, and geographical information system and remote sensing technology, which had made important contributions to urban ecological planning and are excellent reference materials for the coming work. The spatial extent of this eco-plan can be seen in Figure 9.1.

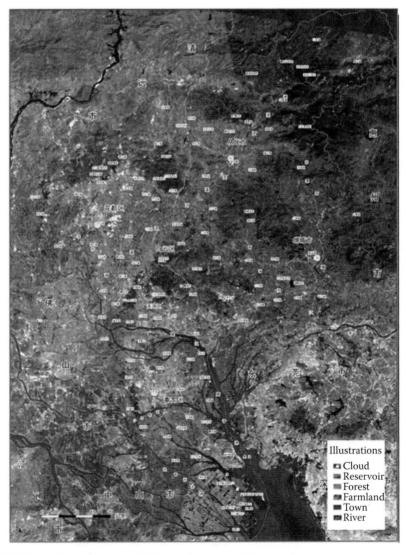

FIGURE 9.1 **(See color insert.)** The spatial extent of Guangzhou eco-plan.

9.1 PLANNING IDEAS AND THOUGHTS

9.1.1 IDEAS

Guangzhou urban ecological planning followed four basic planning ideas: health, security, vigor, and development.

1. Health is the overall condition of the external order and structure of the ecosystem, including the normal energy flow and substance circulation, the key ecological elements, resistance and resilience from natural or man-made disturbances, a steady structure, and self-controlling and ecological service supply for the demands of nature and humans.
2. Security refers to ecological security, including natural ecological security, economical ecological security, and social ecological security, all of which compose a complex artificial ecological security system. Additionally, a series of threshold or security levels exist in the urban ecological process, such as the carrying limitations of water and land resources as well as the constraints of forested and green areas. Corresponding to those thresholds, some key factors, local areas, or positional relationships exist in urban ecosystems, which form a certain potential security spatial pattern (i.e., the ecological security pattern) that plays a key part in maintaining and controlling the ecological process.
3. A healthy and secure ecological city is full of vigor. Ecosystem vigor is reflected in various social, economic, and ecological aspects. The vigor level of an urban ecosystem is improved through processes of capacity building for ecological support systems and interactive controlling between developmental and ecological support systems.
4. Development refers to urban sustainable development, including the systematic perspective, total beneficial perspective, and outlooks on the population, resources, and environment.

9.1.2 THOUGHTS

The Guangzhou City ecological planning adhered to the "top-down" and the "bottom-up" approaches in the macro- and the micro-perspectives, respectively. Restrictive factors were distinguished from advantages by ecological footprint (EF) estimation, ecosystem carrying capacity assessment, and urban ecosystem health analysis to establish a suitable aim for the sustainable development of the urban ecosystem. Research into the ecological support system made the quality, quantity, and spatiotemporal characteristics of all important resources and environmental elements clear. The study of the interactive demand and supply between the city development and the resource environment established the development scale and orientation of this moderate city. Based on the construction of an urban ecological security pattern, combined with different planning target year circumstances, landscape spatial pattern planning and ecological division planning were performed from macro-, medium-, and micro-levels. The construction, management, and maintenance of the urban ecosystem stimulated its vigor, improved its carrying capacity, and promoted its healthy and sustainable development. Through information integration of the ecological planning, visual management was realized, and the ecological planning program was refreshed according to a rolling mode. Above all, we put emphasis on the practical application and operability on the research achievement and attempted to make this planning program direct Guangzhou's eco-environment construction over a substantial period of time in the future.

9.2   ANALYSIS OF URBAN ECOLOGICAL SUSTAINABLE DEVELOPMENT TENDENCIES

9.2.1   Analysis of the State of Complex Ecosystems

We conducted ecological footprint analysis and ecological carrying capacity and ecological health assessments to discuss the present situation of the Guangzhou urban ecosystem.

9.2.1.1   Ecological Footprint Analysis

The calculation of an ecological footprint is based on two principles: (1) the quantity of most of the resources consumed by people and the waste produced by this consumption can be fixed and (2) the flow of resources and waste can be transformed into the corresponding biologically productive land area. Therefore, the ecological footprint of a given population equals the necessary biologically productive land area, including terrestrial and aquatic area, which produces the resources consumed and absorbs the waste from consumption (Guo et al. 2003). An EF can be defined by the following equation:

$$\text{EF} = Nef = N\sum(aa_i) = N\sum\left(\frac{c_i}{p_i}\right) \tag{9.1}$$

where i = types of consumption, p_i = the average productivity of type i, c_i = the consumption level of type i per capita, aa_i = the calculated biologically productive land area of type i per capita, N = the population, and ef = the ecological footprint per capita.

Biological resource accounts and energy resource accounts were opened according to the ecological footprint model. A biological resource account is composed of all types of agricultural products, and an energy resource account mainly includes coal, coke, fuel oil, crude oil, petrol, diesel oil, and electric power. It required a conversion for the energy resource account to fossil fuel production land area in calculation, in line with the standard of average unit area heating capacity of the global fossil fuel production land. According to the ecological footprint model, these two types of accounts can be converted into the corresponding necessary biologically productive land area, and the results concerning demands in the ecological footprint are shown in Tables 9.1 through 9.3. The equivalence factors of six types of land came from Chinese values from the World Ecological Footprint Report by Wackernagel (1997). We doubled those values for Guangzhou, and the permanent population of Guangzhou City was 9,943,000 in 2000.

TABLE 9.1
Ecological Footprint (EF) of Biological Resource Consumption in Guangzhou (2000)

Biological Resource	Gross Average Production (kg/hm²)	Total Consumption/t	EF (hm²)	hm²/capita	Biologically Productive Land Type
Grain	2,744	769,890.8	280,572.4	0.0400	Arable land
Vegetable oil	431	1,857.7	143,521.4	0.0205	Arable land
Vegetable	18,000	1,239,794	68,877.5	0.0098	Arable land
Pork	74	145,131.2	1,961,232	0.2799	Arable land
Beef and mutton	33	4,511	742,758.8	0.1060	Grassland
Egg	400	8,557.6	196,394	0.0280	Arable land
Meat birds	764	274,726.7	359,589.9	0.0513	Grassland
Aquatic products	29	283,121.3	9,762,803	1.3933	Water area
Sugar	4,997	18,352.4	3,672.7	0.0005	Arable land
Fruit	18,000	266,396.3	14,799.8	0.0021	Arable land
Milk	502	62,567.7	124,636.8	0.0178	Grassland
Wood	99 m³/hm²	147,625 m³	74,002.5	0.0074	Forest land

Source: Guangzhou Statistical Yearbook, China Statistics Press, Beijing, 2000 (in Chinese).

Biological resources include all types of agricultural products coming from arable land, grassland, aquatic areas, and forest land (Table 9.1). The ecological footprint of aquatic products and pork ranked first and second, 1.3933 and 0.2799 hm²/capita, respectively.

Energy resources include all types of fossil fuels from "energy land" and electric power from built-up land. Energy land is the portion of a land that should be set aside for the absorption of CO_2 released by the use of fossil fuels. The ecological footprint of crude oil and raw coal ranked first and second, 0.4267 and 0.3880 hm²/capita, respectively.

The demand of Ecological footprint in Table 9.3 is the summary of Tables 9.1 and 9.2. Because the biologically productive ability of units of arable land, energy land, grassland, and forest land varied from each other, each type of land area was multiplied by an equivalence factor to transform it into a comparable and uniform biologically productive area.

The Ecological footprint supply value of energy land equaled 0 because Guangzhou did not set any land aside to absorb CO_2. We also had to deduct 12% of the total supply land area to reserve biodiversity. The supply of Ecological footprint was larger than the demand for forest land and built-up land, and the ecological surpluses achieved were 0.039 and 0.073 hm²/capita, respectively. However, the supply of Ecological footprint was less than the demand for arable land, grassland, water area, and energy land; the ecological deficits were 1.037, 0.0927, 0.242, and 1.058 hm²/capita, respectively. The total Ecological footprint supply reached

TABLE 9.2

Ecological Footprint (EF) of Energy Resource Consumption in Guangzhou (2000)

Energy Resource	Consumption/ 10⁴ t standard coal	Conversion Coefficient (GJ/t)	Consumption (GJ/capital)	Gross Average Energy EF (GJ/hm²)	EF (hm²/ capita)	Biologically Productive Land Type
Raw coal	1013.57	20.93	21.340	55	0.3880	Energy land
Clean coal	28.62	20.93	0.603	55	0.0110	Energy land
Coke	17.41	28.74	0.499	55	0.0091	Energy land
Crude oil	942.08	41.87	39.699	93	0.4267	Energy land
Petrol	−83.83	43.12	−3.636	93	−0.0391	Energy land
Kerosene	23.05	43.12	1.000	93	0.0108	Energy land
Diesel oil	−142.74	42.71	−6.131	93	−0.0659	Energy land
Fuel oil	297.28	50.20	15.009	71	0.2114	Energy land
LPG	5.28	50.20	0.267	71	0.0038	Energy land
Natural gas	1.33	38.98	0.052	93	0.0006	Energy land
Other oil products	2.51	41.87	0.106	93	0.0011	Energy land
Electric power	20.46	11.84	0.244	1000	0.0002	Built-up land
Other energy	8.30	36.19	0.304	71	0.0043	Energy land

Source: Guangzhou Statistical Yearbook, China Statistics Press, Beijing, 2000 (in Chinese).
Note: LPG, liquefied petroleum gas.

TABLE 9.3

Demand and Supply of Ecological Footprint (EF) in Guangzhou (2000)

Land Type	EF Demand Area (hm²/ capita)	Equivalence Factors	Equivalence Area (hm²/ capita)	EF Supply Area (hm²/ capita)	Productivity Factors	Equivalence Area (hm²/capita)
Arable land	0.4074	2.8	1.141	0.01660	2.24	0.104
Grassland	0.1862	0.5	0.093	0.00002	3.29	0.00003
Forest land	0.0074	1.1	0.008	0.03580	1.20	0.047
Built-up land	0.0002	2.8	0.001	0.01180	2.24	0.074
Water area	1.2183	0.2	0.244	0.00890	1.00	0.002
Energy land	0.9614	1.1	1.058	0	0	0
Total EF demand			2.544	Total EF supply		0.227
Biodiversity reserve (12%)						0.027
Ecological deficit			−2.344			

0.2000 hm²/capita, but it could not meet the demand of the total Ecological footprint. Thus, Guangzhou had an ecological deficit of 2.344 hm²/capita in 2000.

The Ecological footprint in developed cities is often several times larger than their land area, even dozens of times larger for a few regions. In other words, in terms of domestic and international trade, developed countries need to import Ecological footprint from developing countries, and cities need to import Ecological footprint from rural areas or suburbs. Therefore, Guangzhou needs to import Ecological footprint from external areas to erase its ecological deficit to maintain the present consumption level. According to many reports, life patterns are a key factor influencing the urban Ecological footprint, and the consumption levels are proportional to the Ecological footprint. We should therefore promote domestic and international trade to erase Guangzhou's ecological deficit, use current resources more efficiently, change the present production and living model, and establish resource-saving and energy-saving systems to direct social consumption.

9.2.1.2 Urban Ecosystem Carrying Capacity Assessment

An urban area is an ecosystem composed of complex social, economic, and natural components. Urban carrying capacity is the balance between the supporting ability of ecological supporting subsystems and the developmental ability of socioeconomic subsystems. It can be defined by the following equation:

1. Supporting ability (N) model of the urban ecological supporting subsystem:

$$N = R \cdot \alpha_s^2 \cdot e^{\beta_s}$$

$$\begin{cases} R = k_1 \left(\sum_{i=1}^{n} S_i \cdot \log_2 S_i \right) \cdot \sum_{i=1}^{n} S_i \cdot P_i \\ \alpha_s = k_2 \left(\sum_{i=1}^{m} \frac{r_i}{G} \right) \\ \beta_s = \frac{1}{k} \sum_{j=1}^{k} \lambda_j k_j \end{cases} \qquad (9.2)$$

where R = the resilience of the ecosystem, α_s = the supply of resources, β_s = the environmental capacity, r_i = the supply of resource i, G = the GDP, S_i = the percentage of i's cover area in the whole city, P_i = the flexibility value of i, λ_j = the weight of pollutants, k_j = the emission standard of j, and k_1 and k_2 = constants.

2. Development ability model of an urban ecosystem:

$$F = \mu \cdot \delta \cdot E_{co}$$

$$E_{co} = \frac{\Delta G/G}{\Delta POP/POP} \qquad (9.3)$$

where F = the developmental ability of the socioeconomic subsystem, μ = the technical index, δ = the human resource index, E_{co} = the economic index, $\Delta G/G$ = the grow rate of the GDP, and $\Delta POP/POP$ = the change rate of the population.

3. Coupling model of an urban ecosystem:

$$UECC = f(N,F) = r \cdot N \cdot e^{F}$$
$$r = a\sum_{j=1}^{l}(\cos\frac{\pi}{2} \cdot \frac{M_j/POP}{M_{j0}/POP'}) + b \tag{9.4}$$

where $UECC$ = the carrying capacity of the urban ecosystem, r = a characteristic factor, M_j = the exploited amount of nonrenewable resource j, M_{j0} = the consumption amount of nonrenewable resource j, a and b = constants, and $a + b = 1$.

Carrying capacity always exists relative to pressure, and the pressure is rooted in the expulsion of the population and the intensification of the economy. The pressure can be defined by the following accounting identities:

$$EPIO = \alpha_u^2 \cdot e^{\beta_u}$$
$$\begin{cases} \alpha = k_3 \sum_{i=1}^{m}(\frac{POP \cdot s_i + G \cdot w_i}{G}) \\ \beta = \frac{1}{k}\sum_{j=1}^{k}\lambda_j(POP \cdot w_j + G \cdot \psi_j) \end{cases} \tag{9.5}$$

where $EPIO$ = the pressure of the ecosystem, α = the resource consumption index, β = the environmental pollution index, s_i = the consumption amount per capita of resource i, w_i = the resource consumption per million yuan GDP, λ = the weight of pollutants, w_j = the discharge per capita of pollutant j, ψ_j = pollutant discharge per million GDP, k_3 = constant, and k = the types of pollutants.

We chose 1992 as the starting point and calculated the carrying capacity and pressure of the Guangzhou urban ecosystem since then by analyzing the relevant changes in both to determine the urban ecology development trend. Relative carrying capacity and relative pressure can be defined with the following equations and are compared in Table 9.4:

$$UECC_{Ri} = \frac{UECC_{Ai}}{\min UECC_{Ai}}$$
$$EPIO_{Ri} = \frac{EPIO_{Ai}}{\min EPIO_{Ai}} \tag{9.6}$$

where $UECC_{Ri}$ = the relative carrying capacity over a certain period of time, $UECC_{Ai}$ = the absolute carrying capacity over a certain period of time, $\min UECC_{Ai}$ = the minimum absolute carrying capacity over a certain period of time, $EPIO_{Ri}$ = the relative pressure over a certain period of time, $EPIO_{Ai}$ = the absolute pressure over a certain period of time, and $\min EPIO_{Ai}$ = the minimum absolute pressure over a certain period of time.

Considering the United Nations Conference on Environment and Development (UNCED) in 1992, which globally affected the environment,

TABLE 9.4

Carrying Capacity and Pressure of the Guangzhou Urban Ecosystem

	Factor		
Year	$UECC_{Ri}$	$EPIO_{Ri}$	$1 = \dfrac{\Delta UECC_{Ri}}{\Delta EPIO_{Ri}}$
1992	1.00	1.00	—
1993	1.27	1.22	1.21
1994	1.63	1.25	14.45
1995	1.93	1.28	9.20
1996	2.19	1.15	1.99
1997	2.41	1.22	3.36
1998	2.75	1.31	3.71
1999	3.38	1.38	9.02
2000	3.93	1.40	30.74

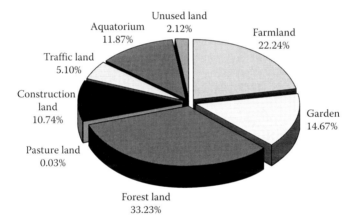

FIGURE 9.2 **(See color insert.)** The land use structure of Guangzhou in 2000.

the year 1992 was chosen as the initial research stage. The basic data for Guangzhou from 1992 to 2000 were collected and calculated by the UECCC evaluation model. Then, using the 1992 result as a benchmark, the comparable UECCC and UEPIO values for each subsequent year were obtained by the normalization method. The results are illustrated in Figure 9.2.

The UECCC and UEPIO variance ratios per year are expressed as $\Delta UECCC_{Ri}$ and $\Delta UEPIO_{Ri}$, respectively. When $\Delta UECCC_{Ri} > \Delta UEPIO_{Ri}$, $\gamma > 0°$, when $\Delta UECCC_{Ri} = \Delta UEPIO_{Ri}$, $\gamma = 0°$, and when $\Delta UECCC_{Ri} < \Delta UEPIO_{Ri}$, $\gamma < 0°$.

From the preceding results listed, we found that the UECCC of Guangzhou City has been steadily increasing since 1992. However, the UEPIO demonstrated a certain degree of fluctuation during this period, notably the sudden decreases in 1994 and 2000. This phenomenon may have been caused by the basic data of 1994 because this was when the statistic standards of industry data in China were amended. Furthermore, the concept of sustainable development was brought forward in the 1992 UNCED. Subsequently, related policies such as the "95 plan" (also known as the Ninth Five-Year Plan of China) were carried out, which caused rapid increases in economic development in Guangzhou City, especially in 1995. As a result, during 1992–1996 and 1996–2000, the UECCC of Guangzhou City harmoniously increased with economic development, and the developmental mode of Guangzhou City was sustainable (Xu et al. 2005).

9.2.1.3 Urban Ecosystem Health Assessment

Ecosystem health can be measured by its vigor, structure, resilience, service maintenance, people's health condition, and education level, and the most important elements are vigor, structure, and resilience for the health of an ecosystem. The urban ecosystem health assessment indicator system is shown in Table 9.5.

TABLE 9.5
Index System for Assessing the Health of an Urban Ecosystem

Element	Raw Indicator	Concrete Indicator
Vigor	Economic productivity	GDP per capita
	Material consumption efficiency	Material consumption/GDP
	Energy consumption efficiency	Energy consumption/GDP
Structure	Economical structure	R&D/GDP
		IT/GDP
	Social structure	Population density in the center of the city
		Gini coefficient
	Natural structure	Forest coverage
		Greening coverage of the urban area
		Nature reserve coverage
Resilience	Waste disposal index	Wastewater disposal rate
		Attainment rate of vehicle emission
		Utilization rate of industrial solid waste
	Utilization rate of circulated materials	Reutilization rate of industrial water
	Environment protection investment index	Environment protection input/GDP
Service maintenance	Environment condition commoditization	Environmental quality
		Per capita public green area
		Per capita floor space
		Per capita road area
People's health condition and education level	People's health	Engel's coefficient
		Average life expectancy
		Mortality rate of children aged 0–4
	Literacy	Average length of education

An ecosystem's health condition is a relative concept, which is a fuzzy problem. We established a health assessment model using the fuzzy mathematic method, and our results indicated that an urban ecosystem's structural characteristics and vigor hold the greatest potential for the development of the Guangzhou urban ecosystem. However, the ecosystem's resilience was inferior. Therefore, for Guangzhou to continue to maintain its developmental potential, something must be done to cover its shortcomings (Guo et al. 2002).

9.2.2 Ecological Supporting System Analysis

The carrying capacity of an urban ecological supporting system determines the future development of the urban area. Resource and environmental elements, which are the components of an urban ecological supporting system, support the development and expansion of population, economy, urban area, and so on. According to the Cask Effect, capacity is decided by the shortest board, and the key to improving capacity is defining the shortest board or bottleneck. We analyzed the water resources, water environment, land resources, urban greenbelt, energy resources, and urban atmospheric environment to find the bottleneck (shortest board).

9.2.2.1 Water Resources and Water Environment

Industrial wastewater and domestic sewage are the two principal sources of Guangzhou's urban sewage. The discharge amounts of industrial wastewater and domestic sewage are presented in Tables 9.6 and 9.7, respectively.

From 1995 to 2000 the amount of industrial wastewater and the discharge of principal pollutants tended to decline, but the amount of urban domestic water and domestic sewage increased.

Water consumption includes ecological water requirements, industrial water, and domestic water, and the water supply includes domestic water resources, cross-boundary water, recycled water, diversion water, and regulating water. We calculated

TABLE 9.6
Discharge Statistics of Industrial Wastewater and Principal Pollutants

Category	1995	1996	1997	1998	1999	2000
Industrial water in the city/million tons	309.30	278.36	267.64	254.67	238.67	246.22
Industrial water in the center of the city/million tons	246.52	221.88	21.348	200.71	181.05	210.58
COD amount/ton	56,038	54,441	53,067	57,114	44,585	41,332
Petroleum amount/ton	948	967	790	671	536	525
Heavy metal and its compounds/ton	8.05	7.86	11.83	9.50	5.04	5.215

Source: *Guangzhou Statistical Yearbook*, China Statistics Press, Beijing, 1995–2000 (in Chinese).
Note: COD, chemical oxygen demand.

TABLE 9.7
Discharge Statistics of Urban Domestic Sewage

Category	1995	1996	1997	1998	1999	2000
Domestic water in the center of the city/million tons	735.22	747.98	780.98	804.53	819.89	835.64
Domestic wastewater in the center of the city/million tons	588.18	598.38	624.78	643.62	655.91	668.51
Domestic wastewater in the entire city/million tons	742.28	756.47	785.63	826.15	829.80	831.99

Source: Guangzhou Statistical Yearbook, China Statistics Press, Beijing, 1995–2000 (in Chinese).

TABLE 9.8
Balance between Water Demand and Water Supply in the Base Year (100 Million Tons/Month)

District		Water Supply	Ecological Water Requirement	Domestic Water	Industrial Water	Gap
Conghua	Nonflood season	1.24	0.87	0.03	0.29	+0.05
	Flood season	2.10	1.73	0.03	0.29	+0.05
Zengcheng	Nonflood season	7.70	5.27	0.05	0.46	+1.92
	Flood season	28.67	23.08	0.05	0.46	+5.08
Huadu	Nonflood season	0.53	1.02	0.04	0.38	−0.91
	Flood season	1.12	1.02	0.04	0.38	−0.32
Panyu	Nonflood season	26.03	20.87	0.12	0.73	+4.31
	Flood season	98.72	88.90	0.12	0.73	+8.97
Center city	Nonflood season	20.19	22.83	0.7	1.52	−4.86
	Flood season	72.62	58.48	0.7	1.52	+11.89
Total	Nonflood season	46.22	43.42	0.94	3.38	−1.52
	Flood season	172.43	103.87	0.94	3.38	+64.24

Source: Guangzhou Statistical Yearbook, China Statistics Press, Beijing, 1995–2000 (in Chinese).
Notes: +, supply exceeds the demand; −, supply cannot satisfy the demand.

the relative data to analyze the balance between water demand and water supply for the year 2000, presented in Table 9.8, and predicted the future balance according to the present consumption level, presented in Table 9.9.

When we took the ecological water requirements into consideration, the total water supply could not satisfy the water demand of the entire city during the non-flood season, and the gap tended to increase. The water requirement in the process of improving water quality was the main reason for the large gap between water supply and water demand (Yang et al. 2004).

TABLE 9.9

Forecasted Balance between Water Demand and Water Supply in the Target Years (100 Million Tons/Month)

Item		Water Supply	Ecological Water Requirement	Domestic Water	Industrial Water	Gap
2005	Nonflood season	46.22	60.40	1.16	4.11	−19.45
	Flood season	172.43	103.80	1.16	4.11	+63.36
2010	Nonflood season	46.22	81.24	1.43	4.86	−41.31
	Flood season	172.43	103.80	1.43	4.86	+62.34
2020	Nonflood season	46.22	114.60	1.88	6.46	−76.72
	Flood season	172.43	114.60	1.88	6.46	+49.49

Notes: +, supply exceeds the demand; −, supply cannot satisfy the demand.

TABLE 9.10

Utilization Rate of Environmental Carrying Capacity

District	PCC-COD_{cr}/ (ton/d)	WLA-COD_{cr}/ (ton/d)	COD_{cr} discharge/ (ton/d)	Utilization Rate of PCC/%	Utilization Rate of WLA/%
Center city	439.53	160.68	487.64	110.95	303.5
Huadu	52.30	29.08	31.04	59.35	106.8
Panyu	465.75	173.78	98.98	21.25	57.0
Conghua	113.20	6.93	29.74	26.27	429.4
Zengcheng	132.95	50.73	50.48	37.97	99.5

Water environmental capacity is the maximum waste load allocation (WLA) of water under a certain condition, which is constrained by the water functional target. It is a variable concerned with objective natural conditions, such as temperature, flow, and the distribution and characteristics of waste, as well as with the subjective demands of citizens.

COD_{cr} is a common indicator to monitor pollution sources, but the COD_{Mn} indicator is applied in most environmental water monitoring. We converted COD_{Mn} into COD_{cr} to analyze the pollution carrying capacity (PCC) and WLA in Table 9.10.

The utilization of PCC was lower in the four districts, except for the center district, in which the rate exceeded 100%. When we took WLA into consideration, the city center and two other districts had overutilized their water carrying capacity, and Zengcheng district approached its limitation. Only Panyu district had a surplus (43%). In actuality, though the utilization rate of PCC was low in Huadu, Conghua, and Zengcheng, the results of water quality monitoring indicated that the water was in a poor condition in some sectors. PCC is only a theoretical indicator, which is much larger than WLA. We should control and regulate the waste discharge amount and distribution according to the WLA.

9.2.2.2 Land Resources

The area covered by the Guangzhou region measures 7434.4 km², among which there is 773 km² of aquatorium. The land use structure of Guangzhou in 2000 can be seen in Figure 9.2. Concerning the land area, 3432 km² is ≥50 m above sea level, and 3229.4 km² is <50 m above sea level (Figure 9.3). These data come from remote sensory pictures and 1:10,000-scale digital maps of Guangzhou in 2000. We analyzed the balance between the land demand and the supply from two points: whether the production from agricultural land meets the demand of agricultural products and whether the land area satisfies the demand of future urban construction.

According to the statistics from 1980 to 2000 of Guangzhou, we applied the trend extrapolation method to forecast agricultural production. Our results are shown in Table 9.11.

According to the forecasted data, domestic vegetables, fruit, and aquatic products can satisfy the corresponding demand, but grain production cannot meet this demand. Thus, Guangzhou will largely depend on imported grain in the future to meet the demand caused by the increase in population and the decrease of arable land.

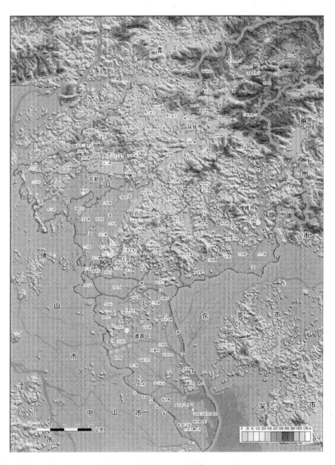

FIGURE 9.3 **(See color insert.)** The relief map of Guangzhou.

TABLE 9.11

Forecasted Balance between Agricultural Product Supply and Agricultural Product Demand

Item	Forecast	2000 (Base Year)	2005	2010	2020
	Yield	88.1	89.2	85.7	84.7
Grain	Demand	198.9	226.0	253.2	308.6
	Gap	−110.8	−136.8	−167.5	−223.9
	Yield	306.5	371.3	438.0	571.4
Vegetables	Demand		135.6	151.9	158.2
	Gap		+235.7	+286.1	+413.2
	Yield	33.0	38.3	40.2	43.4
Fruit	Demand		33.9	38.0	46.3
	Gap		+4.4	+2.2	−2.9
	Yield	32.5	45.3	46.8	49.1
Aquatic products	Demand		15.8	17.7	21.6
	Gap		+29.5	+29.1	+27.5

Source: *Guangzhou Statistical Yearbook*, China Statistics Press, Beijing, 2000 (in Chinese).

According to the average occupancy of nonagricultural construction on arable land from 1991 to 2000, it is estimated that the occupancy will reach 27,520 hm² from 2000 to 2020. In contrast, according to the General Plans for the Utilization of Land (1997–2010), the corresponding results are 4,523, 9,046, and 18,092 hm², respectively. To analyze how much area the arable land can supply for urban development, at the base amount of arable land area in 2000, we subtract the basic farmland protection area or first-grade basic farmland protection area. Thus, it was calculated that 33,639 or 80,600 hm² can be used for urban development.

From the preceding analysis, the required land area can be satisfied for urban expansion, and thus, Guangzhou possesses vast space for its modernization. However, urban construction will occupy a portion of the arable land, and mutual compensation is a feasible way to compensate for the decline of arable land.

9.2.2.3 Atmospheric Environment

Atmospheric environmental capacity means the maximum allowable amount of atmospheric pollutants that will still ensure the normal development of people and the ecosystem while achieving the standards of atmospheric environmental quality. We applied a modified box model, which is defined in Equation 9.7, to calculate the atmospheric environmental capacity of the districts in Guangzhou. The results are shown in Table 9.12.

$$Q_{si} = \frac{\sqrt{\pi}}{2} \bar{u} H \frac{S_i}{\sqrt{S}} C_{si} \tag{9.7}$$

where Q_{si} = the atmospheric capacity in district i, C_{si} = the atmospheric quality standard of district i, S_i = the area of district i, S = the total area of all districts, $\bar{u}$ = the average wind velocity in the mixing layer (2.3 m/s for Panyu district and 1.9 m/s for

TABLE 9.12

Distribution of Atmospheric Environmental Capacity in Guangzhou

District	Area/km²	Standard (Average Daily Concentration/mg/m³)			Capacity (t/a)
Dongshan	17.2	Second degree	SO_2	0.15	783
			NO_2	0.08	418
			TSP	0.30	1,567
Liwan	11.8	Second degree	SO_2	0.15	537
			NO_2	0.08	287
			TSP	0.30	1,075
Yuexiu	8.9	Second degree	SO_2	0.15	405
			NO_2	0.08	216
			TSP	0.30	811
Haizhu	90.4	Second degree	SO_2	0.15	4,117
			NO_2	0.08	2,196
			TSP	0.30	8,234
Tianhe	108.3	Second degree	SO_2	0.15	4,932
			NO_2	0.08	2,631
			TSP	0.30	9,865
Fangcun	42.6	Second degree	SO_2	0.15	1,940
			NO_2	0.08	1,035
			TSP	0.30	3,880
Baiyum	1,042.7	Second degree	SO_2	0.15	47,488
			NO_2	0.08	25,327
			TSP	0.30	94,975
Huangpu	121.7	Second degree	SO_2	0.15	5,543
			NO_2	0.08	6,050
			TSP	0.30	11,085
Panyu	1,313.8	Second degree	SO_2	0.15	107,838
			NO_2	0.08	57,514
			TSP	0.30	215,677
Huadu	961.1	Second degree	SO_2	0.15	43,771
			NO_2	0.08	42,298
			TSP	0.30	87,542
Zengcheng	1,741.4	Second degree	SO_2	0.15	79,308
			NO_2	0.08	52,872
			TSP	0.30	158,617
Conghua	1,974.5	Second degree	SO_2	0.15	89,924
			NO_2	0.08	47,960
			TSP	0.30	179,849
Total	7,434.4	Second degree	SO_2	0.15	392,396
			NO_2	0.08	209,274
			TSP	0.30	773,176

TABLE 9.13

Utilization Rate of SO$_2$ Environmental Capacity in Guangzhou in 2000

District	Industrial Source/t	Tertiary Industrial Source/t	Civil Source/t	Capacity/ (t/a)	Utilization Rate/%
Old part of town	105,516	864	1,565	71,546	150.88
Panyu	41,112	4	350	107,838	38.45
Huadu	5,190	10	227	43,771	12.40
Zengcheng	8,293	9	200	79,308	2.69
Conghua	2,217	3	200	89,924	2.69
Total	162,327	890	2,648	392,386	42.27

the other districts), and H = the average depth of the mixing layer (734 m for Panyu district and 551 m for the other districts).

As a result of the absence of actual data for NO$_x$ and total suspended particulates (TSP), we can only calculate the precise utilization rate of SO$_2$ environmental capacity according to its discharge amount in every district of Guangzhou. The results are shown in Table 9.13.

The utilization rate of the SO$_2$ capacity reached 150.88% in the old part of the town (including Dongshan, Liwan, Yuexiu, Haizhu, Tianhe, Fangcun, Baiyun, and Huangpu), which exceeds the environmental capacity. However, taking Guangzhou as a whole, SO$_2$ pollution was not yet evident. The SO$_2$ environmental capacity in Guangzhou remained in surplus.

9.2.2.4 Energy Resources

Guangzhou is poor in primary energy; only a few coal and hydroelectric power resources exist. According to the primary energy production and consumption in the recent years, the self-sufficient rate of energy tended to decrease, which has been below 1.00% since 1998 (Table 9.14).

TABLE 9.14

Energy Production and Consumption in Guangzhou (Unit: 10^4 Tons Standard Coal)

Year	Raw Coal	Hydroelectric Power	Total Production	Consumption	Self-Sufficient Rate (%)
1995	12.03	18.10	30.13	1,578.50	1.91
1996	5.37	18.13	23.50	1,716.36	1.37
1997	3.30	20.94	24.24	1,773.89	1.37
1998	0.91	16.63	17.54	1,801.09	0.97
1999	0	17.70	17.70	1,876.30	0.94
2000	0	20.46	20.46	2,133.38	0.96

Source: *Guangzhou Statistical Yearbook*, China Statistics Press, Beijing, 2000 (in Chinese).

According to the data on energy consumption and GDP increase rates from 1993 to 2000, we calculated the flexibility index between energy consumption and economic scale. Our results are shown in Table 9.15.

Based on the flexibility index from Table 9.15 and on the "General Scheme of Economic and Social Development from 2001 to 2020," we forecasted the future consumption of total energy and main point fuel. The results are shown in Table 9.16.

Based on the general developmental scheme, the total energy consumption in 2020 will increase by 123% of that in 2000. The large demand gap should be compensated for by diversity measures such as natural gas importation, electricity transmission, renewable energy development, and boiler improvement.

9.2.2.5 Urban Forest

According to the forest resource statistics in 2000, there were 30,8793.2 hm^2 forest land in Guangzhou, 0.044 hm^2 per capita, and forest coverage reached 41.3%. The forest area in Guangzhou tended to decrease from north to south, distributed mainly in Conghua and Zengcheng districts, where forest coverage reached 68.8% and 47.7%, respectively.

The forest stand volume (M_0) in 2000 was 7.4173 million m^3, the forest volume growth rate (P) was 6.4%, and forest consumption (M_1) was 0.147 million m^3. We applied three methods to forecast the future forest stand volume, and the results are shown in Table 9.17.

Program A: According to the present forest resource distribution, if the forest develops with the same growth rate, it can be defined by the following equation:

$$M_n = M_0(1+P)^n - M_1 \times \frac{(1+P)^n - 1}{P} \tag{9.8}$$

where M_n = the forest stand volume after n years, M_0 = the present forest stand volume, and M_1 = the forest consumption/year.

Program B: Forecast with the same equation as Program A, but P is increased by 0.05%.

Program C: Forecast by trend extrapolation, which can be defined by the following equation:

$$y = 523.56x^{0.1309} \tag{9.9}$$

where y = the total forest stand volume and x = D value between the target year and the year 1986.

The total forest growth production was 49.9 × 10^4 m^3 in 2000, and the total forest consumption in the same year was 14.7 × 10^4 m^3; that is, a surplus existed. The forest consumption tended to decrease since 1992, and the results shown in Table 9.17 suggest that the forest stand volume tended to increase, which indicated that the forest stand volume in Guangzhou can satisfy its demands.

TABLE 9.15
Flexibility Index of Main Energy Consumption from 1995 to 2000

Year	GDP Increase Rate/%	Fuel Consumption/10⁴ t/ SC/a	Increase Rate/%	Flexibility Index	Coal Consumption/10⁴ t/ SC/a	Increase Rate/%	Flexibility Index
1995	–	1578.5	–	–	888.7	–	–
1996	16.24	1716.4	8.73	0.54	938.5	5.61	0.35
1997	13.93	1773.9	3.35	0.24	916.4	–2.35	–0.17
1998	13.13	1801.1	1.53	0.12	893.9	–2.46	–0.19
1999	13.18	1876.3	4.18	0.32	892.5	–0.16	–0.01
2000	13.35	2133.4	13.70	1.03	1042.2	16.77	1.26
Average	–	–	–	0.45	–	–	0.25

Year	GDP Increase Rate/%	Petrol Consumption/10⁴ t/SC/a	Increase Rate/%	Flexibility Index	Electric Power/10⁴ t/SC/a	Increase Rate/%	Flexibility Index
1995	–	445.8	–	–	183.31	–	–
1996	16.24	516.2	15.80	0.97	195.84	6.84	0.42
1997	13.93	534.0	3.45	0.25	208.26	6.34	0.46
1998	13.13	550.9	3.16	0.24	225.19	8.13	0.62
1999	13.18	598.7	8.67	0.66	252.50	12.13	0.92
2000	13.35	671.4	12.15	0.91	292.98	16.03	1.20
Average	–	–	–	0.61	–	–	0.72

Source: *Guangzhou Statistical Yearbook*, China Statistics Press, Beijing, 1995–2000 (in Chinese).
Note: SC, standard coal.

TABLE 9.16

Forecasted Energy Consumption Based on the General Scheme (Unit: 10^4 Ton/SC/a)

Year	Total Energy	Coal	Petrol	Electric Power
2000	2133	1042	671	293
2001	2252.8	1074.3	722.1	319.5
2005	2717.6	1192.5	929.2	430.2
2010	3364.8	1343.1	1238.7	604.1
2020	4759.3	1628.3	1976.2	1049.2

Note: SC, standard coal.

TABLE 9.17

Forest Stand Volume in Different Target Years (Unit: 10^4 m³)

Program		2000	2005	2010	2020
A	The same growth rate	741.73	927.94	1181.72	2000.35
B	Growth rate increase by 0.05%	741.73	951.69	1243.36	2219.76
C	Trend extrapolation	774.90	774.90	797.90	833.90

9.2.2.6 Urban Greenbelt System

The area and the distribution of the greenbelt system in the current urban area are shown in Table 9.18.

Though Guangzhou urban green indicators are above the state standard for a garden city, in which the green coverage rate is 30% and public green area per capita is 6 m², when compared to the standard for a cosmopolitan city suggested by the WHO (green area rate = 40% and green area per capita = 40–60 m²), there is still a large gap. Thus, we should increase the green coverage rate and the green area per capita in Guangzhou.

9.2.2.7 Bottleneck Analysis of the Ecological Supporting System

The bottleneck factors of an urban ecological supporting system are the key elements that impede or constrain the development speed and scale of an urban socioeconomic system and lead the urban orientation. We used the shortage index to analyze bottleneck factors; the calculated model is as follows:

$$R_i = \min\left(\frac{D_i - B_i}{D_i} \times 100\%\right) \tag{9.10}$$

where R_i = the shortage index of factor i, B_i = the demand of i under the equilibrium of supply and demand, and D_i = the actual demand of i in the socioeconomic system.

The results indicated that water resources, water environment, energy resources, and energy structures are the four main bottleneck factors.

TABLE 9.18

Main Green Indicator Statistics in Current Guangzhou Urban Area (2000)

District	Regional area/hm²	Green Area/ hm²	Green Rate/%	Green Coverage/ hm²	Green Coverage Rate/%	Public Green Area/ hm²	Public Green Area Per Capita/m²
Guangzhou	29,750	8,797	29.57	9,400	31.60	2,704.93	7.87
Dongshan	1,720	386	22.44	481	27.97	196.86	3.30
Liwan	1,180	96	8.14	177	15.00	92.79	1.22
Yuexiu	890	189	21.24	251	28.20	164.79	3.81
Haizhu	4,827	559	11.58	569	11.79	275.60	3.78
Tianhe	7,230	2,007	27.76	2,167	29.97	995.85	20.92
Fangcun	1,820	456	25.05	527	28.96	109.86	7.52
Huangbu	3,898	1,283	32.91	1,335	34.25	207.10	13.48
Baiyun	8,185	3,821	46.68	3,893	47.56	692.30	17.56

9.3 INTERACTION BETWEEN URBAN DEVELOPMENT AND THE ECOLOGICAL SUPPORTING SYSTEM

9.3.1 MODERATE URBAN DEVELOPMENT SCALE

The population and the current urban area were used to represent the urban size during planning. Using the support of the resource subsystem, the constraint of the environment subsystem, and the interaction between economic intensity and population, the results for moderate economical intensity and population were obtained. We planned the construction land through two points: (1) the interaction between the population and the current urban area and (2) the constraint of land resources on the current urban area.

The system dynamics model and economic–population correlation model were applied to test the population target of the 10th Five-Year Program in Guangzhou. The results are shown in Table 9.19.

The results of system dynamics model and economic-population correlation model approximated to the target population of the 10th Five-Year Program. Therefore, the population target was reasonable, conforming to the conditions of the natural resources and economical scale.

The population forecasting and the development requirements of the socioeconomic system are the two principal foundations for determining the urban construction land area, which can be defined with the following equation:

$$U = \frac{P \times A}{10,000} \tag{9.11}$$

where U = the urban construction land area in the planned period, P = the urban population in the planned period, and A = the urban construction land area per capita.

According to the model, the urban construction land area in Guangzhou should be limited to 479, 608, and 780 km² for the years 2005, 2010, and 2020, respectively (Xu et al. 2003).

TABLE 9.19
Population Forecast in Guangzhou (Unit: 10⁴ People)

Method	2000	2005	2010	2020
The 10th Five-Year Program	994.3	1130	1266	1543
System dynamics	994.3	1102	1219	1522
Economic–population correlation	994.3	1069	1181	1227

9.3.2 URBAN SPATIAL STRUCTURE AND DISTRIBUTION

Resources and the environment expose either supports or constraints to urban development. We analyzed the natural conditions, natural resources, and ecological risks to determine their characteristics and influences. Combined with the regional strengths and weaknesses, the results indicated that the urban spatial development strategy of Guangzhou was reasonable and could be described by "advanced eastward, expanded southward, optimized northward, and united westward."

The population density in the southern and northern parts of Guangzhou will greatly increase in the future, and the dense population in the central part of the city will move to the new urban district. The population distribution forecasts of Guangzhou in different planned target years are shown in Table 9.20.

Based on the program of enlarging urban areas in eastern and southern regions of Guangzhou City, and according to the principles of circular economy, the future industry distribution in Guangzhou can be adjusted as follows:

1. Combine the factories in the central part of the city into an industrial park, and any new industrial land cannot be increased in the center of the city.
2. Utilize ecological planning during the industrial park planning to maintain the dynamic balance of the ecosystem.
3. Readjust the distribution of factory pollution in East Guangzhou.
4. Transfer the heavy industry to South Guangzhou.
5. Develop the tidal flats and establish industrial areas near the sea.

9.4 URBAN ECOLOGICAL SECURITY PATTERN

Guangzhou urban ecological space layout planning was linked to the overall objectives of ecological sustainable development planning and based on the ideology of an ecological security pattern. Through the assessment and evaluation of the spatial structures of Guangzhou's urban ecosystem (e.g., natural geographic features, eco-environmental vulnerability, ecosystem services, and eco-vigor), the urban ecological spatial pattern was categorized from the macro-, medium-, and micro-levels, as well as the established ecological security spatial pattern, to guarantee rapid and sustainable development of the urban area. This provides the foundation for the institution of ecosystem management measures and regulatory guidance on eco-units.

TABLE 9.20
Population Distribution in Various Target Years

District			Geographical Area	Population Density (/km²)		
				2005	2010	2020
Urban area	Old eight districts	Center city	Dongshan, Liwan, Yuexiu, Tianhe, old District of Haizhu	32,000	27,000	20,000
		Nearby suburban	Huangbu, Fangcun, a part of Haizhu, block of Baiyun	7,000	10,000	15,000
		Outer suburban	Baiyun except its block	1,000	2,000	3,000
		New urban	Zhujiang new urban	7,000	10,000	15,000
	Two new districts	Around urban	Xinhua, Shiqiao	5,000	7,000	10,000
		New urban	Nansha, Guangzhou new urban	4,000	6,000	10,000
			Dashi street, college town	2,000	4,000	7,000
Center town			Licheng, Jiekou	3,000	5,000	7,000

9.4.1 URBAN ECO-SPATIAL DIFFERENTIATION CHARACTERISTICS

There are five types of vulnerability in the Guangzhou urban ecosystem: (1) interface vulnerability on the coastal area exposed by rising sea levels under the influence of global climate change; (2) ecosystem vulnerability, including natural disasters, natural ecosystem degradation, and the decline of eco-environmental quality, caused by natural and man-made disturbances; (3) fluctuating vulnerability in the urban–rural ecotone; (4) substrate vulnerability of karst geology and the geological fault zone in the northwest urban and southern Guanghua basin; and (5) specific vulnerability of soil acidification, soil erosion, biodiversity loss, and human culture disappearance in the northern mountainous. According to the remote sensing data and digital maps, the Guangzhou natural ecosystem can be divided into four types (i.e., forest, grassland, farmland, and wetland ecosystems) to assess the ecosystem service value, which includes direct economic value (forest products and farming products) and indirect economic value (water and soil conservation, biodiversity, carbon storage and oxygen production, and air purification). These values were calculated in constant 1990 prices, and the results are shown in Table 9.21 (Xu et al. 2003).

The results indicate that different ecosystems contribute different ecosystem service values. If we take only the ecosystem service value into consideration, the order is wetland > forest > grassland > farmland.

Eco-vigor is a broad measurement of the comparative advantageous degree between economic and social function of eco-units in a regional socioeconomic-natural complex ecosystem. Natural resources and environmental functions, which include natural

TABLE 9.21

Ecosystem Service Values of Different Eco-Units (Unit: yuan/(hm²×a))

Type	Ecosystem Service Value								Direct Value
	Soil Conservation			Oxygen Production	Carbon Storage	Air Purification		Total	
	Water Conservation	Fertility	Reduce Sedimentation			SO$_2$ Absorption	Dust Retention		
Coniferous	357.50	863.80	61.68	6,669.96	5,419.77	128.18	5,644	19,144.89	840
Broad-leaved	418.02	863.31	61.64	4,622.40	5,024.43	96.63	1,718.7	12,805.13	840
Mixed forest	387.76	863.55	61.66	4,910.40	5,337.47	112.41	3,682.2	15,355.45	840
Shrub	328.30	864.39	61.72	5,227.20	5,674.35	96.63	1,718.7	13,971.29	120
Economic forest	307.46	864.16	61.71	4,416.00	4,800.06	96.63	1,718.7	12,264.72	10,872
Grassland	314.60	864.36	61.72	3,955.20	5,880.04	26.13	1,718.7	12,820.75	4,167
Farmland	243.20		61.78	3,955.20	4,299.17	26.13	1,718.7	10,304.18	10,872
Wetland	418.02	863.31	61.64	8,400.00	9,130.57	128.18	3,682.2	22,683.92	24,365

eco-vigor and socioeconomic eco vigor, can be defined by the following equations and Figure 9.4:

$$
\begin{aligned}
& N_i(A_{se}, A_N) \\
& N_i(l, \alpha) \\
& l = ABS(A_T) = \sqrt{A_N^2 + A_{SE}^2} \\
& \alpha = \arcsin[A_N/l] = \arcsin\left(A_N/\sqrt{A_N^2 + A_{SE}^2}\right)
\end{aligned}
\tag{9.12}
$$

where N_i = the eco-vigor of eco-unit i, A_{se} = the socioeconomic eco-vigor, A_N = the natural eco-vigor, l = the vector length (named as total eco-vigor), and α = the comparative advantageous degree between A_N and A_{se}. The larger the α is, the more advantageous the natural eco-vigor is.

Based on the measurable indicators presented in Table 9.22, the eco-vigor niche was calculated based on geographical information system (GIS), and the eco-vigor niche was analyzed by the ecosystem grid cells based on remote sensing

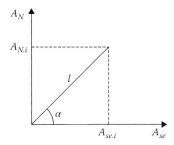

FIGURE 9.4 Eco-vigor.

TABLE 9.22
Measurable Indicators of the Eco-Vigor Niche

First-Level Indicator	Second-Level Indicator	Third-Level Indicator	Weight
Socioeconomic	Population	Population density	0.33
	Economy	Economy density	0.33
	Culture, education, and medical	Doctor/10^3	0.165
		Teacher/10^3	0.165
Natural	Environmental quality	Water environment	0.11
		Atmospheric environment	0.11
		Noise environment	0.11
	Urban thermal field (UTF)	Relative value of UTF distribution	0.33
	Natural ecological service	Ecosystem service value	0.33

or land patches. Then, spatial clustering was performed in two dimensions, and an assessment chart of absolute socioeconomic eco-vigor, absolute natural eco-vigor, and comparable eco-vigor niche was produced (Figures 9.5 and 9.6). We found that A_N-based area was mainly concentrated in Conghua, North Zengcheng, and North Huadu, while A_{se}-based area was mainly concentrated in the current urban area, South Zengcheng, South Huadu, and Panyu.

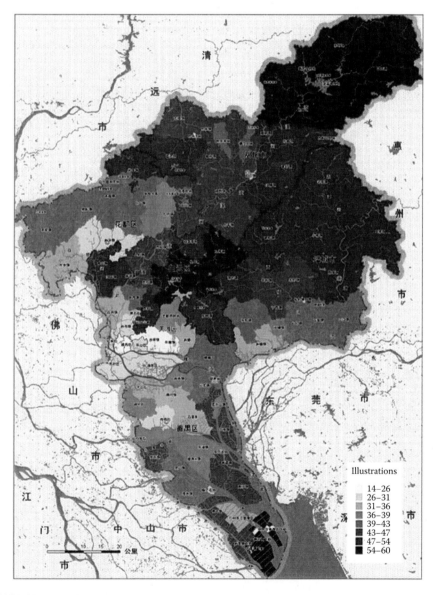

FIGURE 9.5   **(See color insert.)** Levels of natural eco-vigor in Guangzhou.

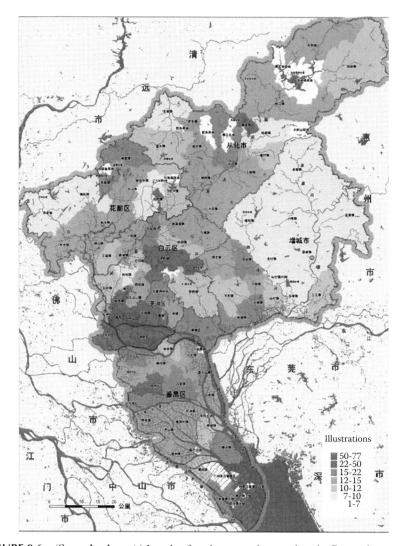

FIGURE 9.6 (**See color insert.**) Levels of socioeconomic eco-vigor in Guangzhou.

9.4.2 URBAN LANDSCAPE ECOLOGICAL SECURITY PATTERN

Based on the theory of the landscape ecological security pattern, the identification of Guangzhou's potential landscape ecological security pattern includes sources, buffer zones, radiating routes, intersource linkages, and strategic points. The key sources are mainly distributed in North Conghua, North Huadu, Northwest Zengcheng, East Baiyun District, North Panyu District, and Nansha. The orchard and farmland surrounding the forested area is a type of important buffer zone. Strips of external radiation from the source to the ridge lines form radiating routes. Intersource linkage facilitates the formation of ecological corridors, which are efficient channels for

FIGURE 9.7 (**See color insert.**) Potential landscape eco-security of Guangzhou.

ecological flow. The vegetation in Hualong Town becomes a linkage for connecting the sources between the north and the south (Figure 9.7).

9.4.3 URBAN ECO-REGIONAL PLANNING

According to the Guangzhou spatial development strategy, the urban eco-regional planning mainly includes constructing three green corridors along the northern region, opening up ecological channels between the north and the south, establishing intergroup ecological isolation zones, and placing emphasis on protecting the vegetation along the southern water networks (Xu et al. 2004).

Based on urban eco-regional planning, we performed eco-unit management assessment from macro-, medium-, and micro-scales. The urban ecosystem was classified into three types: the ecological protection and management zone, ecological control zone, and ecological reconstruction zone. Furthermore, 22 ecological subzones (e.g., the urban center, east industrial zone, and Panyu urban zone) and 66 ecological control units (e.g., the Liwan eco-control unit, Luogang academic city, and Guangzhou Higher Education Mega Centre) were plotted in detail.

9.5 ECOSYSTEM MANAGEMENT

9.5.1 MACRO-MANAGEMENT OF THE ECOSYSTEM

1. Ecological security measures should be taken to protect important ecological assets, including land, water, food, biodiversity, and energy.
2. Three indexes (i.e., building density, hardened ground area percent, and road network density) were applied to constrain the land development intensity in the current Guangzhou urban area, and the optimal targets are shown in Tables 9.23 through 9.25.

TABLE 9.23
Target of Average Building Density (Unit: %)

District	Year			
	2000	2005	2010	2020
Dongshan	41	40	39	38
Yuexiu	54	53	52	50
Liwan	49	48	47	45
Baiyun	37	35	33	30
Tianhe	29	29	29	29
Huangbu	34	33	32	30
Haizhu	39	38	37	35
Fangcun	41	40	38	35
Xinhua	36	35	33	30
Licheng	32	31	31	30
Shiqiao	38	36	34	30
Jiekou	38	36	34	30
New urban	—	29	29	29

TABLE 9.24
Target of Hardened Ground Area Percent (Unit: %)

Year	Haizhu	Liwan	Dongshan	Yuexiu	Fangcun	Tianhe	Baiyun	Huangbu	New Urban
2000	73.28	75.00	61.95	68.87	57.98	68.07	47.77	60.26	—
2005	67.83	70.92	62.10	68.52	59.59	64.57	48.27	59.74	46.71
2010	61.56	67.00	59.72	66.21	58.79	62.87	47.78	57.89	46.71
2020	48.63	57.11	55.96	57.11	57.69	56.45	46.71	54.68	46.71

TABLE 9.25
Target of Road Network Density (km/km²)

Year	Huadu	Panyu	Zengcheng	Conghua	Baiyun	Tianhe	OSD[a]	New Urban
2000	076	0.57	0.73	0.49	1.06	3.68	1.90	—
2005	1.02	1.44	0.99	0.66	1.31	4.50	2.06	9.0
2010	1.27	2.31	1.26	0.83	1.56	5.31	2.22	9.0
2020	1.62	3.05	1.62	1.06	1.77	6.40	2.44	9.0

Note:
[a] OSD, the six old districts: Dongshan, Liwan, Yuexiu, Huangbu, Haizhu, and Fangcun.

TABLE 9.26

Total Emission Control of Main Pollutants in Guangzhou

Item	2000	2005	2010	2020
COD_{cr}/(t/a)	254,727	229,254	203,782	178,309
SO_2/(t/a)	191,881	153,505	134,317	115,129

3. Adjustment of environmental function zone includes atmospheric quality function regionalization, functional adjustment of water environment, and regional adjustment of noise function.
4. According to the indicator requirements in the National Total Discharge of Pollutants and the present environmental pollution characteristics in Guangzhou and due to the limited availability of data we present only COD and SO_2 in Table 9.26.
5. Adopt economic incentives to promote the construction of ecological forests and pollution control.

9.5.2 MICRO-MANAGEMENT OF ECOLOGICAL CONTROLLING UNITS

We analyzed the resource strength, ecological problems, and hidden ecological dangers based on the management and control of eco-units and constructed an eco-controlling guild rule from, for example, resource utilization, pollution treatment, population control, and eco-construction, to make the planning results more feasible and workable. Liuxihe Reservoir is used as an example to demonstrate the eco-controlling guide rule in Table 9.27.

9.5.3 ECO-CONSTRUCTION MEASURES

Due to the prospective feasibility of eco-construction planning, the measures we put forward are aimed at the period from 2000 to 2010, which includes ecological pattern construction, pollution treatment and environmental protection, a clean energy program, traffic construction, a population development scheme, industry distribution, and the urban appearance.

9.6 EVALUATION OF THE PLANNING PROGRAM

Cost-effectiveness analysis was performed to evaluate the planning program, which was specific to the project level to estimate the input required to complete the planning program or achieve planning targets. The estimated costs are shown in Table 9.28.

The effectiveness of the planning program was also tested through urban ecosystem health assessment according to the fuzzy forecasting method. The results are shown in the following identities:

$$H_{2005} = (0\ 0.12\ 0.33\ 0.48\ 0.07),$$
$$H_{2010} = (0\ 0.048\ 0.166\ 0.603\ 0.183),\ and$$
$$H_{2020} = (0\ 0\ 0.126\ 0.398\ 0.476).$$

TABLE 9.27

Eco-Controlling Index of Eco-Units in the Liuxihe Reservoir

Controlling Indicators		2005	2010	2020
Pollution control	Industrial waste water discharge rate up to standard/%	100	100	100
	Vehicle emission rate up to standard/%	>90	>95	>98
	Repeat rate of industrial solid waste/%	>90	>95	>96
	Safe disposal rate of domestic waste/%	100	100	100
	Treatment rate of hazardous waste/%	100	100	100
	Environmental investment/GDP(%)	>2.5	>3	>3
Environmental quality	Annual concentration of NO_x/(mg/m³)	I	I	I
	Annual concentration of SO_2/(mg/m³)	I	I	I
	Annual concentration of TSP/(mg/m³)	I	I	I
	Coverage rate of noise up to standard/%	95	98	100
	Regional average value of noise/[dB(A)]	<50	<50	<50
	Water quality rate of drinking water source up to standard/%	100	100	100
	Water quality rate up to standard/%	100	100	100
Population	Population density/(capital/km²)	100	100	100
Eco-construction	Forest coverage/%	80	80	80

TABLE 9.28

Cost Estimation of the Planning Program

Item	Engineer or Compensation	Cost Estimation/10⁹ yuan		
		2001–2005	2006–2010	2011–2020
Water resources and water environment	Water pollution control	67.966	116.529	189.696
	Urban river comprehensive improvement	60.000	100.000	40.000
	Lake water environmental improvement	0.177		
	Forest construction to save soil and water	0.500	0.650	1.650
	Liuxihe protection	0.6	0.5	
	Rainwater and waste water pipe system	16.671	20.375	
	Urban water supply pipe system	17.044	17.044	
	Rural tap water	10		
	Water factory	27		
Atmosphere	Atmospheric pollution control	10.310	24.870	
Solid waste	Solid waste disposal	22.644	7.090	
Energy	LNG project	24.500	10.500	
	Urban fuel gas engineer	25		

(*Continued*)

TABLE 9.28 (*Continued*)
Cost Estimation of the Planning Program

Item	Engineer or Compensation	Cost Estimation/10^9 yuan		
		2001–2005	2006–2010	2011–2020
Ecological	Eco-forest compensation	2.130	1.775	
construction	Eco-forest construction	5.863	4.886	
	Eco-agricultural town construction	0.1		
	Eco-agriculture demonstration zone in Baiyun	0.917		
	Biogas ecosystem demonstration	0.1		
	Pollution-free agricultural products demonstration engineer	0.1		
	Forest park and nature reserves	14.53	10.47	
	Liuxihe ecological corridors	4.8	7.2	
	Forest virtualization improvement at both sides of urban exit	0.800		
	Urban garden	15		
Traffic	Airport	100.00	88.624	
	Roads	112.38		
	Urban rail transportation	106.76	160.0	200
	Port	33.1	17.8	
Environmental protection	Environment protection management building	5.355		
Urban, industry, and human resources	Education, scientific research	244	436.6	1323
	Intellectuals import	0.051		
	Job training	10		
	Park construction of Guangzhou university	13		
	International biological island	100		
	New and high technology industry	25		
	Company informationization	3.9355		
	Interactive graphic service system	4.15		
Total investment		1084.48	1024.55	1754.35
Total investment/ GDP (/%)		4.44	2.77	1.67

These data given in brackets, from left to right, represent the degree for morbid, ill healthy, sub-healthy, healthy, and very healthy, respectively. According to the maximum membership degree principal, the Guangzhou urban ecosystem appeared healthy in 2005, for which the membership degree was 0.48 and the membership degree for morbid decreased to 0. Guangzhou would have been still healthy in 2010, with a membership degree increase to 0.603. Moreover, Guangzhou will be very healthy in 2020, with a membership degree of 0.476.

As the ultimate purpose, development permeated the entire planning program, in which eco-management and eco-maintenance served to improve the urban eco-vigor. The implementation of the planning program was also focused to effectively and healthily drive the urban ecosystem development. Therefore, the Guangzhou urban ecological planning was reasonable, feasible, and evolutive.

REFERENCES

Guangzhou Statistical Yearbook, 1995, Beijing: China Statistics Press (in Chinese).
Guangzhou Statistical Yearbook, 1996, Beijing: China Statistics Press (in Chinese).
Guangzhou Statistical Yearbook, 1997, Beijing: China Statistics Press (in Chinese).
Guangzhou Statistical Yearbook, 1998, Beijing: China Statistics Press (in Chinese).
Guangzhou Statistical Yearbook, 1999, Beijing: China Statistics Press (in Chinese).
Guangzhou Statistical Yearbook, 2000, Beijing: China Statistics Press (in Chinese).
Guo X. R., Yang J. R., Mao X. Q., 2002, Primary studies on urban ecosystem health assessment, *China Environmental Science*. 22(6): 525–529 (in Chinese with English abstract).
Guo X. R., Yang J. R., Mao X. Q., 2003, Calculation and analysis of urban ecological footprint: A case study of Guangzhou, *Geographical Research*. 22(5): 654–662 (in Chinese with English abstract).
Wackernagel M. et al. 1997, Ecological footprint of nations: How much nature do they use?—How much nature do they have? The Earth Council, San Jose, Costa Rica, March.
Xu L. Y., Yang Z. F., Mao X. Q., 2003, Seeking optimal urban population: A case study in Guangzhou, *Acta Scientiae Circumstantiae*. 23(3): 355–359 (in Chinese with English abstract).
Xu L. Y., Yang Z. F., Li W., 2004, Urban environmental protection plan based on eco-priority rule, *China Population, Resources and Environment*. 14(3): 57–62 (in Chinese with English abstract).
Xu L. Y., Yang Z. F., Li W., 2005, Theory and evaluation of urban ecosystem carrying capacity, *Acta Ecologica Sinica*. 25(4): 771–777 (in Chinese with English abstract).
Xu Q., He M. C., Yang Z. F., Yu J. S., Mao X. Q., 2003, Assessment on urban ecosystem services of Guangzhou city. *Journal of Beijing Normal University* (*Natural Science*). 39(2): 268–272 (in Hu T. L., Yang Z. F., He M. C., 2004, An analytical method on limiting factors of urban ecological supporting system and its application to Guangzhou city, *Acta Ecologica Sinica*. 24(7): 1493–1499 [in Chinese with English abstract].
Yang Z. F., Zhao Y. W., Cui B.S., Hu T. L., 2004, Ecocity-oriented water resources supply-demand balance analysis, *China Environmental Science*. 24(5):636–640 (in Chinese with English abstract).

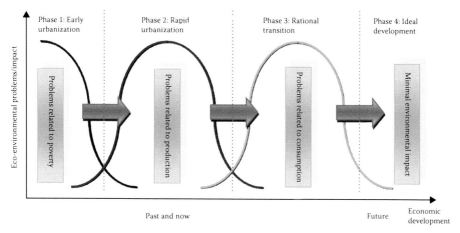

FIGURE 1.1 Stages of urban development.

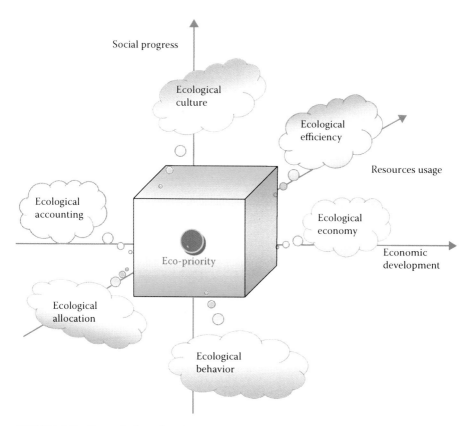

FIGURE 1.3 Eco-priority cube of eco-city planning.

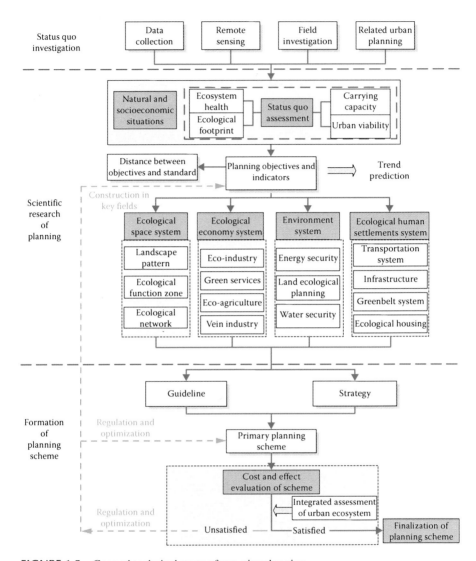

FIGURE 1.5 General technical route of eco-city planning.

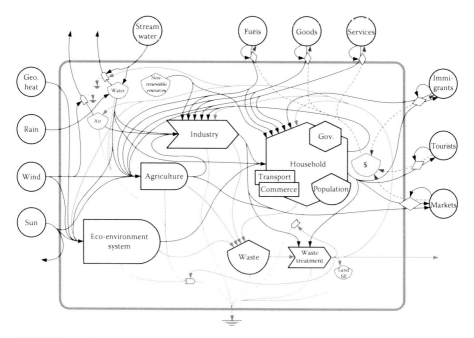

FIGURE 2.4 Emergy diagram of Beijing ecosystem.

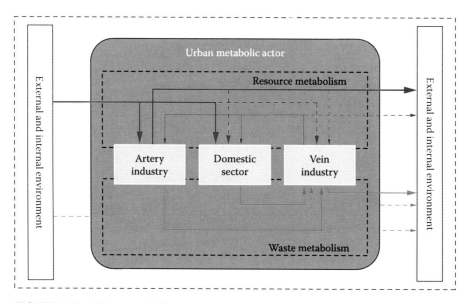

FIGURE 2.15 Urban metabolic process based on two venation.

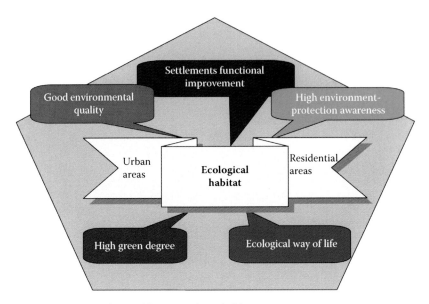

FIGURE 7.1 Connotation and features of eco-habitat.

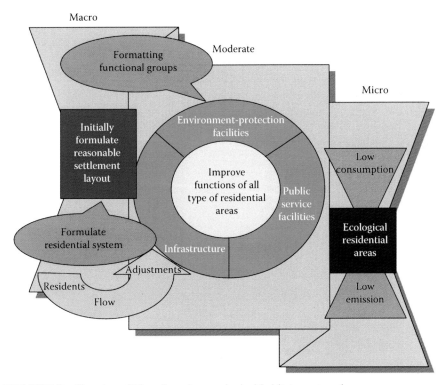

FIGURE 7.2 Planning of ideas for urban ecological habitat construction.

FIGURE 7.3   Schematic diagram of road green.

FIGURE 7.4   Schematic diagram of vertical green wall.

FIGURE 7.5   Green roof diagram.

FIGURE 7.6 Landscape design diagram of settlement water feature.

FIGURE 7.7 Landscape design diagram of settlement simulation.

FIGURE 9.1 The spatial extent of Guangzhou eco-plan.

Illustrations
- Cloud
- Reservoir
- Forest
- Farmland
- Town
- River

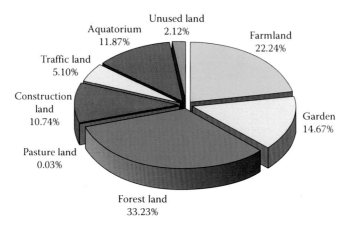

FIGURE 9.2 The land use structure of Guangzhou in 2000.

FIGURE 9.3 The relief map of Guangzhou.

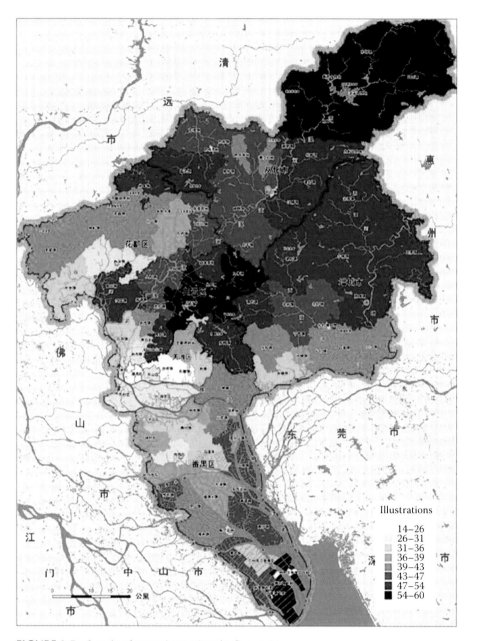

Illustrations

14–26
26–31
31–36
36–39
39–43
43–47
47–54
54–60

FIGURE 9.5 Levels of natural eco-vigor in Guangzhou.

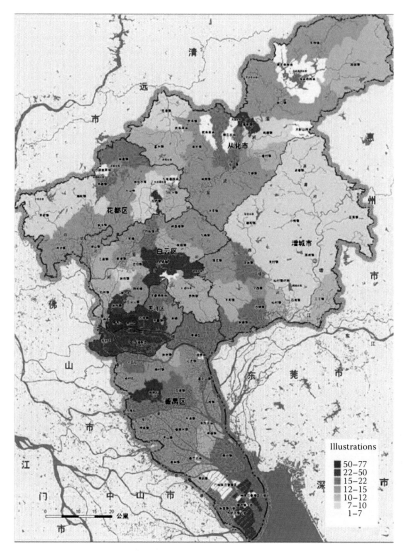

FIGURE 9.6 Levels of socioeconomic eco-vigor in Guangzhou.

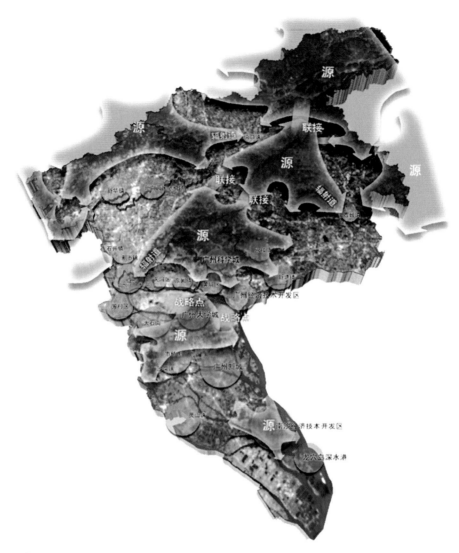

FIGURE 9.7 Potential landscape eco-security of Guangzhou.

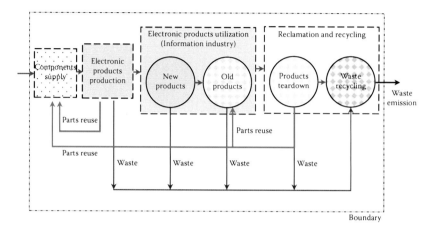

FIGURE 10.3 Electronic and information service ecological industry system conceptual model.

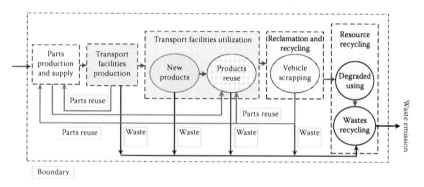

FIGURE 10.5 Traffic and transport engineering industry conceptual model.

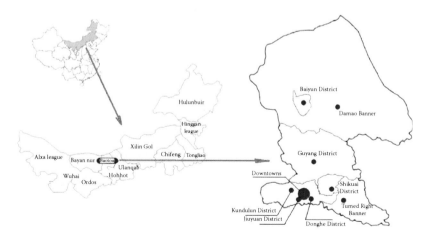

FIGURE 11.1 Geographic location of Baotou.

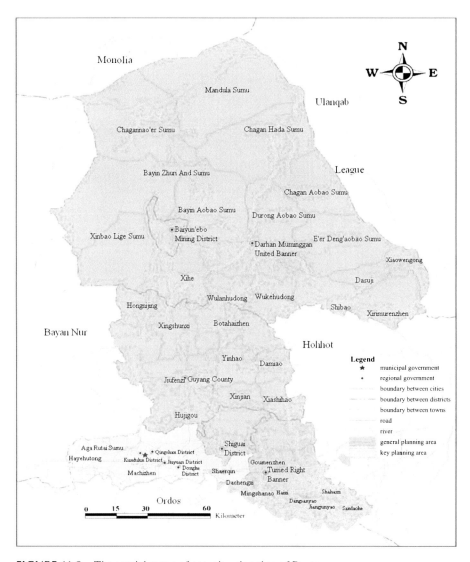

FIGURE 11.2 The spatial range of eco-city planning of Baotou.

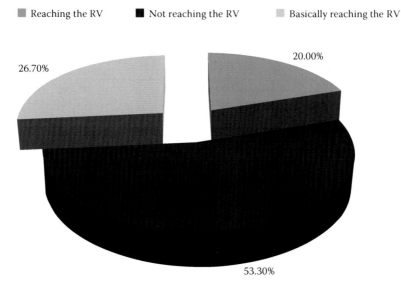

26.70%

20.00%

53.30%

FIGURE 11.4 The ratio between the status value and the reference value (RV) of the index of eco-city construction of Baotou.

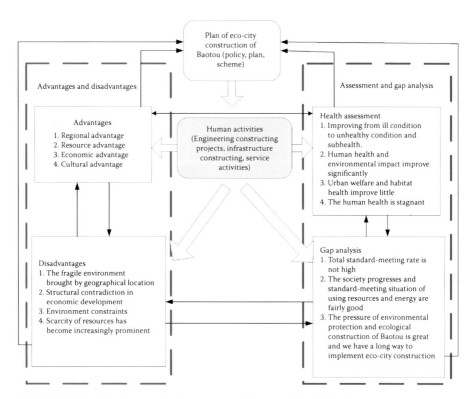

FIGURE 11.5 Advantages, disadvantages, and gap analysis of Baotou.

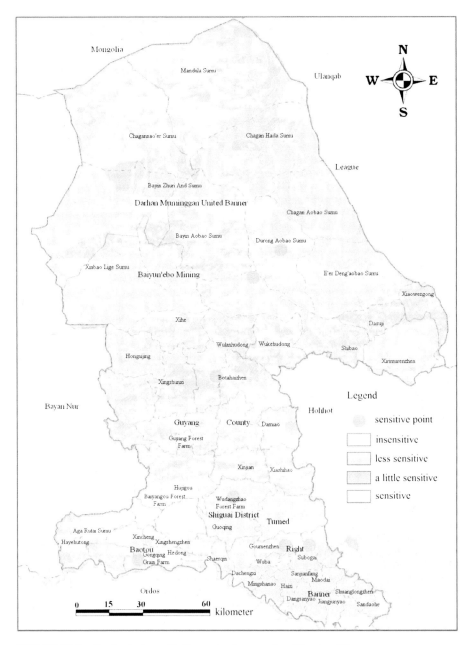

FIGURE 11.6 Ecological sensitivity evaluation results of Baotou City.

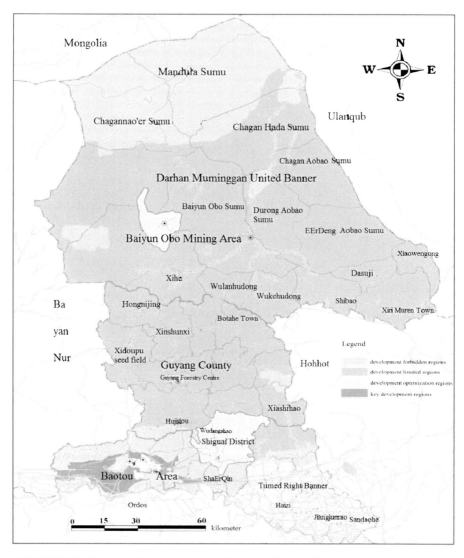

FIGURE 11.7 Ecological function regionalization of Baotou City.

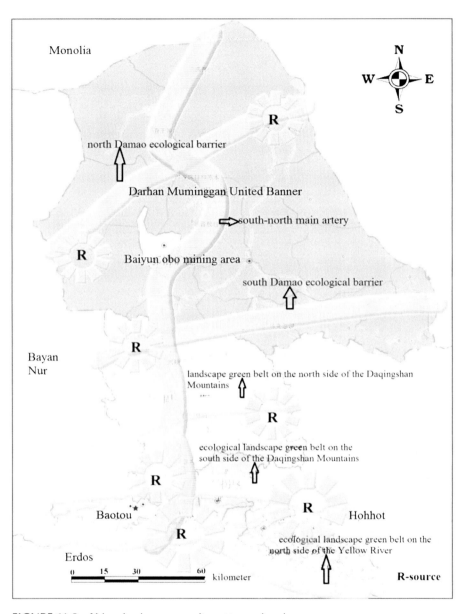

FIGURE 11.8　Urban landscape security patterns planning.

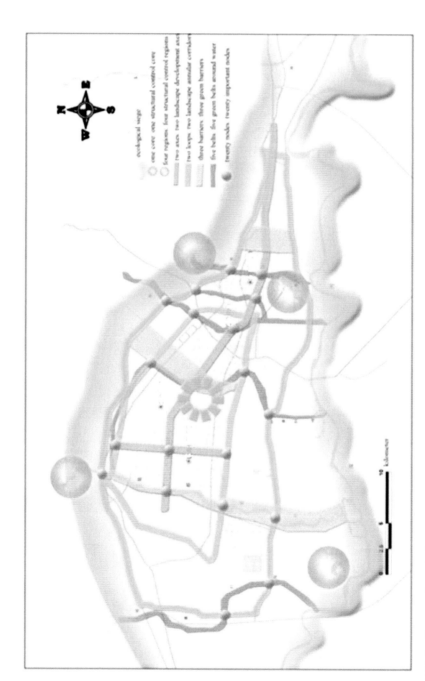

FIGURE 11.9 Construction of urban areas landscape security patterns.

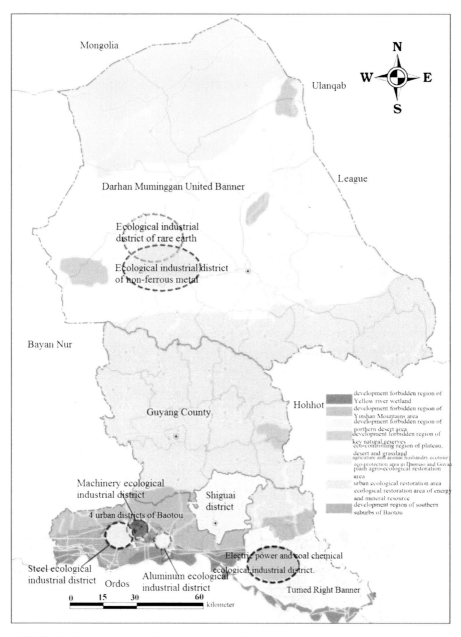

FIGURE 11.10 Integration planning results of Baotou's industrial layout.

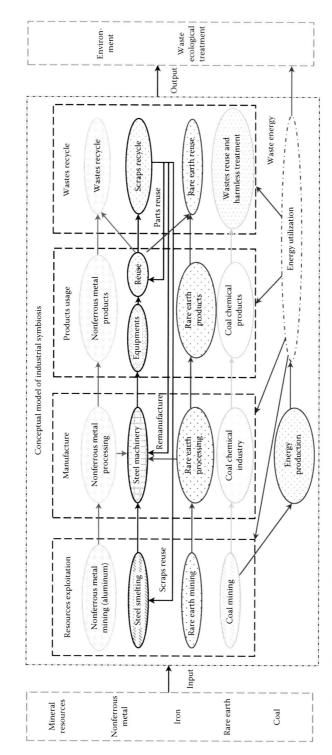

FIGURE 11.11 Conceptual model of Baotou's industrial symbiosis system.

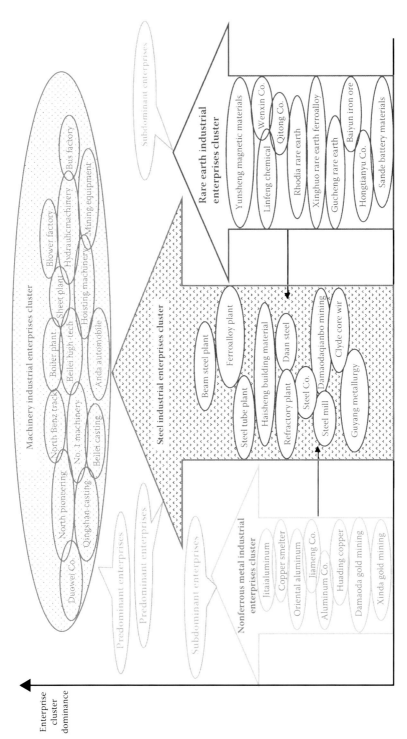

FIGURE 11.12 Framework of Baotou's industrial enterprises clusters.

FIGURE 11.15 Baotou's ecological protection projects.

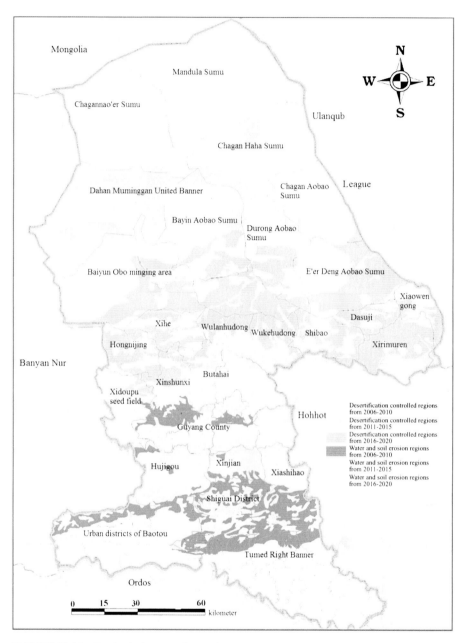

FIGURE 11.16 Ecological restoration and management projects in Baotou City.

FIGURE 12.1 Location map of Wuyishan City.

FIGURE 12.2 Ecological environmental sensitivity evaluation map of Wuyishan City.

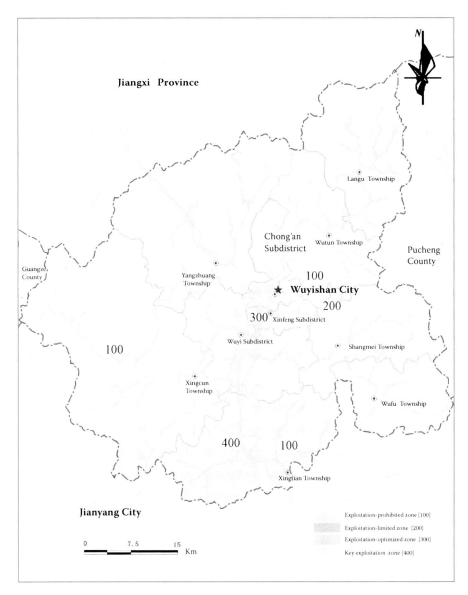

FIGURE 12.3 First-grade ecological functional zoning map of Wuyishan City.

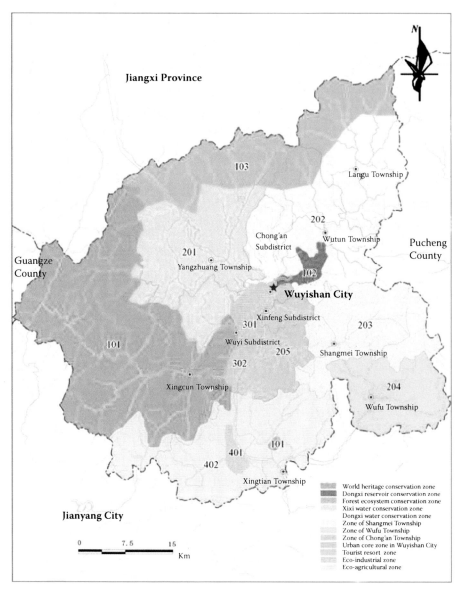

FIGURE 12.4 Ecological functional regulating district map of Wuyishan City.

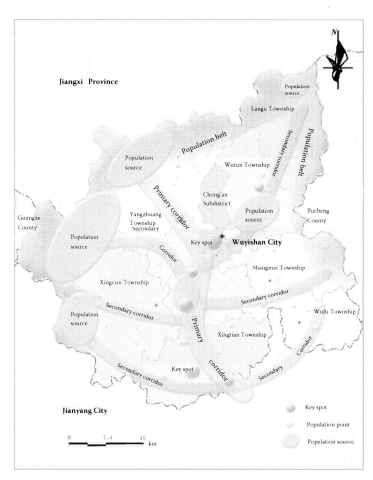

FIGURE 12.5 Landscape pattern in Wuyishan City.

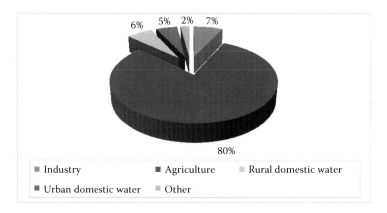

FIGURE 12.8 Water consumption situation of Wuyishan City in 2004.

FIGURE 12.10 Zoning map of water resource in Wuyishan City.

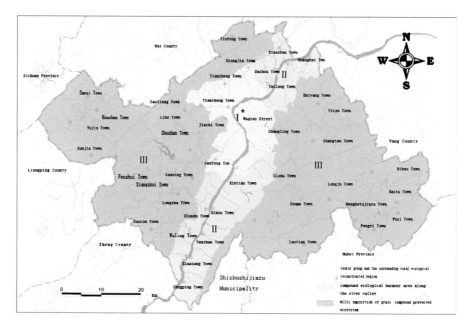

FIGURE 13.1 First-level ecological function zoning of Wanzhou.

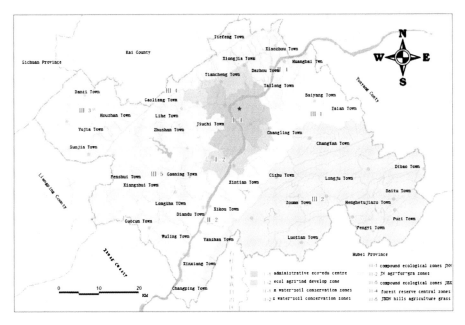

FIGURE 13.2 Second-level ecological function zoning of Wanzhou.

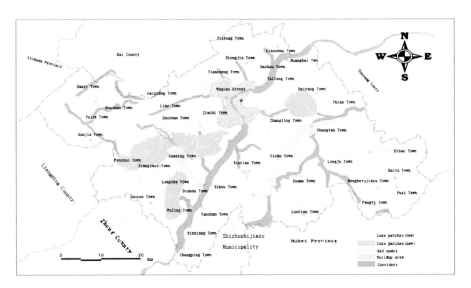

FIGURE 13.3 Wanzhou butterfly-pattern landscape ecological network system.

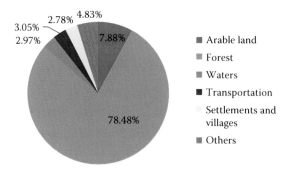

FIGURE 14.1　The percentage of various types of land utilization in Jingdezhen.

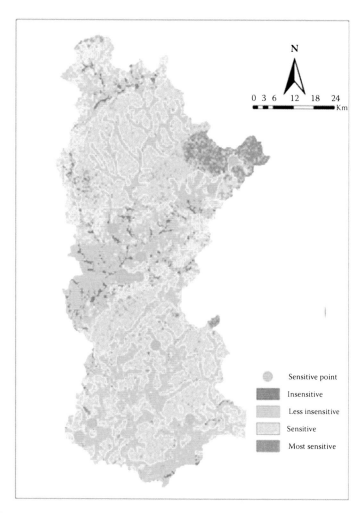

FIGURE 14.2　Ecological sensitivity evaluation map of Jingdezhen.

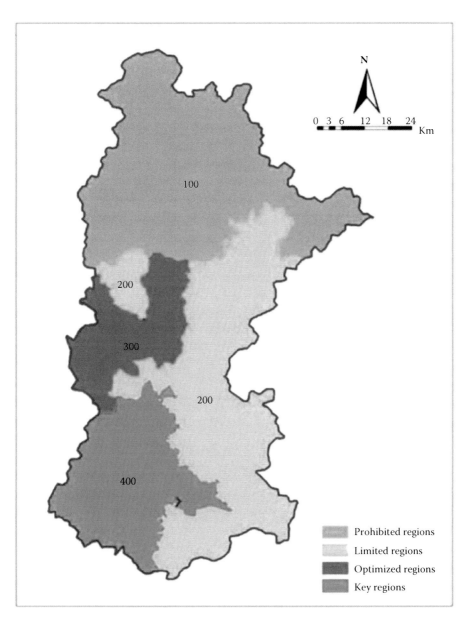

FIGURE 14.3 Map of ecological functional regions in Jingdezhen.

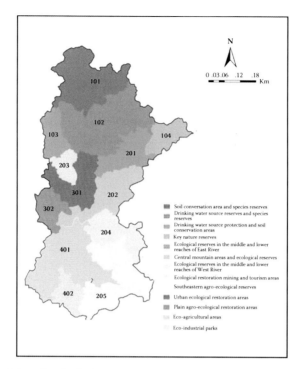

FIGURE 14.4 Map of ecological regulatory zones in Jingdezhen.

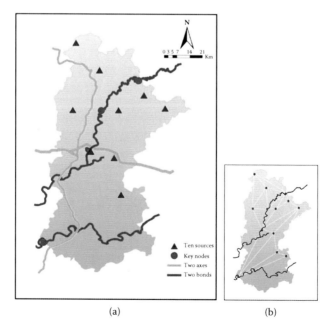

FIGURE 14.6 Map of ecological network. (a) Main landscape framework construction, and (b) The point-connection network construction.

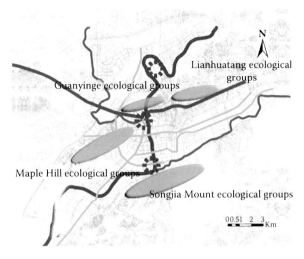

FIGURE 14.7 The ecological network of Jingdezhen City District.

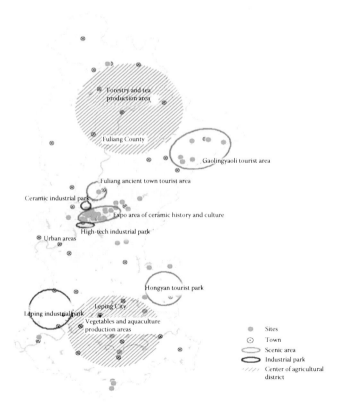

FIGURE 14.8 The current industrial layout of Jingdezhen City.

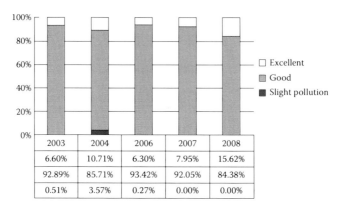

	2003	2004	2006	2007	2008
	6.60%	10.71%	6.30%	7.95%	15.62%
	92.89%	85.71%	93.42%	92.05%	84.38%
	0.51%	3.57%	0.27%	0.00%	0.00%

FIGURE 14.11   The proportion of air quality levels in Jingdezhen from 2002 to 2008.

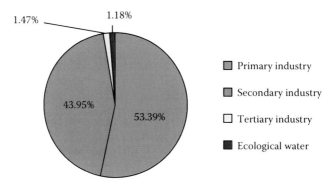

FIGURE 14.12   Structure of production water in Jingdezhen in 2007.

FIGURE 15.1   Solar panel used for heating. The district heating station is shown in the background.

FIGURE 15.2 Wind mills off-shore, Samsø. Wind mills on land have also been erected on Samsø.

FIGURE 15.3 Storage of straw applied for district heating.

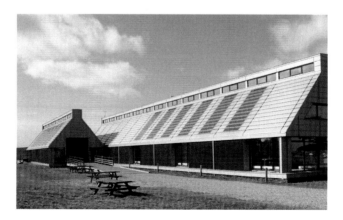

FIGURE 15.4 The Samsø Energy Academy building, Strandengen 1, Ballen, Samsø.

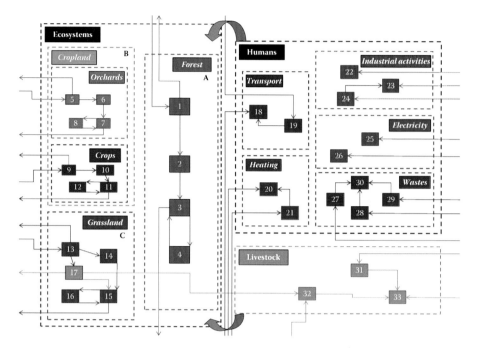

FIGURE 15.6 General conceptual diagram of the carbon cycle model of Siena Province: boxes denote a summary of the state variables (indicated by different numbers), arrows between boxes denote the processes, and arrows entering or leaving the system denote the forcing functions.

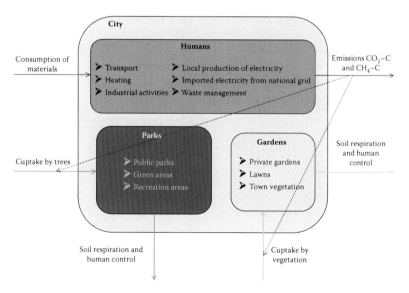

FIGURE 15.7 Submodels and landscape elements of a carbon cycle model of a city. The arrows indicate the general forcing functions involved in the urban processes.

10 Eco-City Xiamen Plan

Linyu Xu, Zhifeng Yang, and Yanwei Zhao

CONTENTS

The urban ecological planning of Xiamen was compiled from 2003 to 2004. During this period, the ecological city planning guidelines were drafted and issued by the state Environmental Protection Administration; thus, some departments had already obtained practical experience in urban ecological planning. The content and structure of Xiamen urban ecological planning were relatively standardized, and the schemes for eco-city planning are considered to be relatively mature. The indicators and reference values are determined by the state ecological city appraisal standards. Our work here will focus on the planning operability, ecological function regionalization, and ecological socioeconomic problems associated with urban ecological planning.

10.1 STATUS ANALYSIS AND TREND ANALYSIS OF XIAMEN'S URBAN ECOLOGICAL PLANNING

According to the open nature of urban ecosystems, SWOT analysis (SWOT analysis is a kind of assessment method in terms of strengths, weaknesses, opportunities, and threats) should be expanded and applied on the different types of ecological analysis, including urban alliance ecosphere analysis, Fujian Province ecosphere analysis, and the cross-strait ecosphere analysis. Based on Xiamen's ecosphere analysis and its urbanization development profile, Xiamen's ecological environment is considered to be satisfactory on the whole. It possesses abundant ocean resources, great tourism resources, and rare wildlife resources. The temporal and spatial variation of the resources will determine the ecological pattern of Xiamen's future development (Tian, 2008).

10.1.1   XIAMEN URBAN DEVELOPMENT PROFILE

Xiamen, the second largest city in the Fujian Province following the capital Fuzhou, covers a total area of 1569 km². The six districts in the Xiamen municipality are Xiangan, Siming, Haicang, Huli, Jimei, and Tongan. By 2003, there were 2.17 million people residing on the Xiamen Island, of which 1.42 million were registered as permanent residents. The tourism and service industries are the main contributors to the gross domestic product (GDP) of Xiamen. The overall length of the Xiamen Island is 234 km. Xiamen has a subtropical oceanic climate and an annual average temperature of 21°C. From July to September, the summer is warm with temperatures reaching 37°C. Winter is cool with average temperatures around 11°C. The whole city is widely populated with southern tropical evergreen coniferous forests, broad-leaved evergreen forests, saline herbs, and mangrove communities.

In 1981, the Chinese government designated Xiamen as a special economic zone. Since Xiamen's special economic zone was first established, it has adhered to the state's basic policy of reform and opening-up and absorbed investment from foreign countries and overseas Chinese. In 2003, the GDP of Xiamen reached 76 billion yuan, a rise of 17.2% over the previous year. The GDP per capita is 53,621 yuan, 13.4% more than the previous year.

Xiamen is a picturesque coastal city in southeast China, having earned the nickname "Pearl on the East China Sea." Famous for its beautiful scenery, unique location, and mixed culture, Xiamen has enjoyed its role as the "garden on the sea" for decades. Being the nearest city to Taiwan, Xiamen is now shining at the west coast of Taiwan Strait with its booming economy. Enjoying its exceptional location, Xiamen is a port for economic cooperation and cultural diffusion, bringing aspiration to this young city.

10.1.2   ANALYSIS OF XIAMEN ECOSPHERE DEVELOPMENT

Based on the geographic characteristics of Xiamen, we designate the ecosphere into four levels: (1) the cross-strait ecosphere, (2) the Fujian Province ecosphere, (3) the Minnan ecosphere, and (4) the Xiamen City ecosphere.

As Xiamen is the nearest central city to Taiwan, it is in the special position to strengthen ties and cooperation between the mainland and Taiwan. As such, Xiamen plays an important role in Fujian's economic development. However, even with its exciting advantages, Xiamen's economy is small and its natural resources are relatively scarce. Now, with the rapid development of Fuzhou, Quanzhou, and Zhangzhou, the role of Xiamen is still undecided.

Xiamen's future development will adhere to the following principles:

1. Form the Xia-zhang-quan economic circle.
2. Focus on an export-oriented economy due to its lack of natural resources.
3. Develop the tourism and port industries.
4. Maintain the "ecology first" strategy.

The status assessment of Xiamen's urban ecological planning is mainly based on the method of urban ecosystem health assessment and the analytical method for the limiting factors of an urban ecological supporting system (UESS).

10.1.3 EVALUATION OF URBAN ECOSYSTEM CARRYING CAPACITY IN XIAMEN CITY

The urban ecosystem is a social–economic–natural complex consisting of human beings and their surroundings. The urban population and their activities jointly form the core component of the urban system, which is then interlinked with the urban eco-environment. Resources and environmental quality are considered crucial to the existence and development of an urban ecosystem; as such, the development of an urban ecosystem is built from interactions among environmental carrying capacity (ECC), resource carrying capacity (RCC), and social–economic development capacity (SEDC). A single component's capacity should not be studied without first considering the integrity of the whole system. The concept of "compound carrying capacity" (CCC) is introduced and studied as an index of the interactions among ECC, RCC, and SEDC and has been adopted as the basis for meeting the challenges of urban sustainable development and eco-city building. The extent of these services is dependent on both the natural carrying capacity and the acquired carrying capacity of the urban ecosystem, in addition to the human beings who behave as the receptor of the services. This theoretic model shows the carrier, the carried target, and the carrying mechanism of the urban ecosystem's CCC. The methodology for both calculating and obtaining the adjustment mechanism of the CCC is outlined in Table 10.1 and Figure 10.1, with reference to the urban ecosystem health index and evaluation models of sustainable development.

Selecting 2000 as the basic year and collecting data through 2003 (Xiamen statistical yearbook, 2003), we calculated the development trend angle between CCC and pressure (γ).

$\Delta UECCC_{Ri}$ represents the changing ratio of the carrying capacity between years, and $\Delta UEPIO_{Ri}$ represents the pressure's annual changing ratio. If $\Delta UECCC_{Ri} > \Delta UEPIO_{Ri}$, then $\gamma > 0°$. If $\Delta UECCC_{Ri} < \Delta UEPIO_{Ri}$, then $\gamma < 0°$. If $\Delta UECCC_{Ri} = \Delta UEPIO_{Ri}$, then $\gamma = 0°$.

TABLE 10.1
Xiamen Urban Ecosystem CCC and Pressure Comparing Results

	Factor		
Time	$\Delta UECCC_{Ri}$	$\Delta UEPIO_{Ri}$	Γ
2000	——	——	——
2001	0.85	0.67	+
2002	0.15	0.33	−
2003	−0.02	−0.34	+

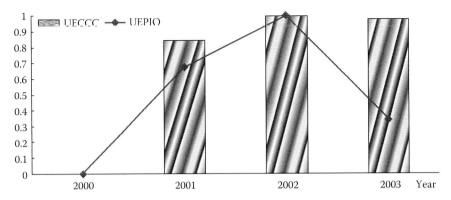

FIGURE 10.1 The relationship between Xiamen urban ecosystem CCC and pressure.

According to the results, we can conclude that the carrying capacity and pressure on the ecosystem fluctuated. In 2001 and 2002 these values increased, while in 2003 they declined. Because the γ is greater than zero, the urban ecosystem can be said to be basically healthy. Since 2000, Xiamen's γ's are always in a positive–negative state; as such, Xiamen's ecosystem can be said to be developing toward an S-shape curve. In our planning, we will strive to maintain healthy development for Xiamen.

10.1.4 ANALYTICAL METHOD FOR LIMITING FACTORS OF XIAMEN'S URBAN ECOLOGICAL SUPPORTING SYSTEM

A city is a complex ecosystem comprised of social, economic, and natural subsystems. Its development depends on the integrity of its UESS. However, traditional urban planning has tended to either neglect the ecological research or isolate the individual factors of the UESS. These practices prevent the consideration of the UESS's whole state in setting development goals and resource allocation. Sustainable development theory requires the careful consideration of the earth's life supporting system capacity. It is important to attempt to harmonize the relationships between man and nature by using a system-based approach to identify the carrying capacity of the UESS. Limiting indices for resource and environmental factors were established accordingly as they frequently have different supply and demand attributes. The effects of resource factors to urban ecosystem can be expressed directly in terms of the factors' quantity. Thus, the limiting indices for resource factors were established by establishing the contrast of scarcity to demand. As some environmental factors, such as forestland, urban vegetation, energy structure, and so on, have national, regional, or industrial guidance standards, the corresponding limiting indices were established by finding the difference of the factor's state subtracted from its standard value. For air and water specifically, we developed the limiting indices through establishing the contrast between the residual ECC and the total ECC. Also, the limiting effect was quantified with a scale of five grades ranging from grade I through grade V, with grade I indicating the strongest limiting effect and grade V indicating none (Tables 10.2 through 10.4). The grade

TABLE 10.2

Estimating Standards for Limiting Factors of Resources and Environmental Components

		Standards of Limiting Factor Identification and Its Blocking Effects Grades (Limiting Grades)				
		I Stronger (%)	**II** Strong (%)	**III** Less Strong (%)	**IV** Weak (%)	**V** None (%)
Assessing Content	Indicator					
Bottleneck indicator	Key resources	Water, land, marine resources	>30	15 to 30	5 to 15	<5	<1
	Ecological environment	Forest land urban vegetation	<−15	−10 to −15	−5 to −10	0 to −5	0
		Atmospheric environment, water environment, marine environment	<−100	−100 to −50	−50 to −30	−30 to 5	>5
	Indicating symbols		▲	▲	▲	▲	▲

		Ecological Capacity Grades of Factors (Supporting Capacity)				
		I Stronger (%)	**II** Strong (%)	**III** Less Strong (%)	**IV** Weak (%)	**V** None (%)
Assessing Content	Indicator					
Abundance Index	Key resources	Water, land, marine resources	<−100	−100 to −50	−50 to −30	−30 to −10	>−10
	Ecological environment	Forest land urban vegetation	>15	10 to 15	5 to 10	0 to 5	0
		Atmospheric environment, water environment, marine environment	>50	30 to 50	20 to 30	5 to 10	<5
	Indicating symbols		●	●	●	●	●

TABLE 10.3
Xiamen Urban Ecosystem Resource Limitation Identification

| Assessing Factors | Resource Limitation | | | Limiting Factors | Limiting Index |
	Water	Land	Marine (Port Capacity)		
Base year	●	▲	●	Land resources	▲
2010	▲	▲	●	Water resources, land resources	▲
2015	▲	▲	●	Water resources, land resources	▲
2020	▲	▲	●	Water resources, land resources	▲

Note: limiting index: ▲ stronger, ▲ strong, ▲ less strong, ▲ weak, ▲ none; abundance index: ● stronger, ● strong, ● less strong, ● weak, ● none

of each factor was established based on the Delphiwmethod. Finally, two methods for controlling and regulating limiting factors were proposed, one to control the demand and the other to expand the supply.

Based on the analytical method for determining the limiting factors of a UESS, the situation is not optimistic in terms of resources and the overall environmental aspect of Xiamen. As such, planning efforts should pay close attention to water resources and environment, land resources, energy, the oceanic environment, and the atmospheric environment. According to our assessment of the urban ecosystem health in Xiamen, we have discovered that the ecosystem is unhealthy. This indicates that corresponding methods and countermeasures should be exercised to regulate the limiting factors of the ecosystem. Efforts should be made to slow the ever-increasing pressure placed on resources and the environment from high-speed urbanization. Only in this way can we meet the demand of regional socioeconomic development, and ensure that Xiamen remains healthy. Finally, these efforts will allow us to achieve the strategic objectives of sustainable urban development (Lu et al., 2004).

10.2 GOAL AND OUTLINE OF XIAMEN'S URBAN ECOLOGICAL PLANNING

10.2.1 GENERAL OBJECTIVES AND STAGE TARGETS

Based on our status assessment of Xiamen's urban ecosystem, we should adhere to the urban development strategy of "ecology first." This strategy will promote the concept of building a society with an advanced ecological economy, prosperous ecological culture, and beautiful ecological environment in the frame of a

TABLE 10.4
Xiamen Urban Ecosystem Environment Limitation Identification

Ecological Environmental Status	Assessing Factors Year	Ecological Environmental Limitation					Limiting Factor	Limiting Index
		Water	Atmosphere	Marine	Forest	Vegetation		
Remain existing environmental protection measures	Base year	◀	●	◀	●	●	Water environment, marine environment	◀
	2010	◀	◀	◀	●	●	Water environment, marine environment, atmospheric environment	◀
	2015	◀	◀	◀	●	●	Water environment, marine environment, atmospheric environment	◀
	2020	◀	◀	◀	●	●	Water environment, marine environment, atmospheric environment	◀

Note: limiting index: ◀ stronger, ◀ strong, ◀ less strong, ◀ weak, ◀ none; abundance index: ● stronger, ● strong, ● less strong, ● weak, ● none

bay-type city. According to the local resources and environmental conditions, a reasonable input of overseas resources will be essential. Additionally, we should coordinate the relationship among urban development, resource utilization, and environment protection in urban ecological planning. We should also regulate the overall urban activity while providing effective ecological strategies and operational control measures. From the perspective of ecological sustainability, we will try our best to enhance the macroscopic strategic planning required for reasonable resource utilization, ecological risk prevention, and ecological compensation.

The reconstruction of Xiamen as an ecological city will require spending approximately 15 years working to transform it into a gulf-type ecologically friendly city with an ecologically prosperous economy, culture, and environment.

Xiamen's ecological city construction planning can be divided into three main stages.

Stage 1 (2006–2010): In this initial stage, efforts are concentrated on controlling existing environmental pollution and restoring the damaged terrestrial and marine ecosystems. The authorities should advocate the implementation treatment and restoration projects while providing the education needed to promote ecological urban construction. In this stage, the city's environmental quality should be brought up to the environmental standards, and pollutants should be adequately controlled as required by the standards. The proposed solid waste disposal centers will be built and put into operation. The overall implementation of Xiamen's ecological city reconstruction will be triggered during this phase. We will strive to achieve the ecological urban construction standards and acceptance criteria established by the Natural Environmental Protection Agency within the proposed timeframe.

Stage 2 (2011–2015): In the second stage, we will work to strengthen the construction and management of the major projects proposed to address key issues in Xiamen's ecological planning. In order to improve the city's ecological construction in a comprehensive and in-depth way, the government will strive to obtain breakthroughs on a few crucial links. Based on the theory of circular economy and efficient resource utilization, an ecological economic system having specific regional characteristics will be built. This will allow Xiamen's urban living environment to be further improved while the urban ecological landscape and security patterns are formed. The city will issue a comprehensive river basin management system for the Jiulong River. The total amount of chemical oxygen demand (COD) in effluent where rivers discharge into the sea should meet the established control requirements. The ecological function of wetlands will be effectively restored. The attainment rate of the environment quality in offshore areas will exceed 95%. One hundred percent of villages and towns will meet the qualifications established for cities perceived as environmental elegant towns. The construction of the ecological urban city will be in full swing.

Stage 3 (2016–2020): The third stage will involve the construction of an eco-logical city reaching the national advanced level. By this stage, we will have educated the public to promote the viewpoint necessary to form the economic structure, environmental system, and public administration system able to support the demands of sustainable development requirements. Based on the completion of major engineering design and projects, this stage will allow us to achieve superior performance and improve the overall efficiency needed to meet the planning projects' requirements. In this stage, the environmental management system will be operating effectively while the damaged ecosystem will be undergoing a comprehensive restoration process. The output value of the ecological industry will contribute more than 65% of the GDP growth. The attainment rate of the environment quality in offshore areas will exceed 95%. The terrestrial and marine ecosystems will be healthy and stable. The concept of ecology and environmental protection will been accepted among the Xiamen citizens. The authorities should be advocating the ideas of green consumption while encouraging the establishment of green schools across the city. The comprehensive decision-making system considering the environmental development will be fully implemented in government works.

10.2.2 Indicator Systems

The indicator system was established by way of theory analysis, hierarchy analysis, expert counseling method, and the Delphi method. The indicators of ecological city planning usually fall into three categories: assessment, research, and supervision.

The assessment indicators are established by the state Environmental Protection Administration to evaluate the process of ecological city construction. The indicator establishment should give prominence to the local characteristics. The selection of research indicators should comply with the following principles:

1. The principle of systematicness and representativeness
2. Guidance and operability principles
3. The principles of classification monitoring and step-by-step assessment

The assessment indicators and research indicators should be integrated according to the following five systems: (1) ecological landscape and security pattern system, (2) ecological industry and circular economy system, (3) resources guarantee and efficient utilization system, (4) environmental improvement and pollution control system, and (5) ecological culture and human settlement system. These systems will eventually define the indicator systems of Xiamen's ecological city planning. Based on Xiamen's ecological city planning indicator systems, some relatively poor indicators were selected based on gap analysis of these five systems. These indicators were selected to establish the supervision indicator systems in the Xiamen ecological city construction (Zhao et al., 2009).

The Xiamen City was used as an example to develop a suit of eco-city indicator systems with five layers, including 67 indicators (Table 10.5).

TABLE 10.5
Eco-City Construction Planning Indicators

System	No.	Indicator	Unit	Standard Value	Status Value	2010	2015	2020	Supervision Agency	Note
(A) Ecological landscape and security pattern system	A1	Ecological construction land ratio	%	—	35.71	37	38	39	Municipal Land Bureau, Municipal Forestry Bureau	*
	A2	Per unit area output value	10,000 yuan/km^2	10,000	4,000	10,000	15,000	19,000	Municipal Forestry Bureau	
	A3	Forest coverage rate	%	40	42.6	43	44	45	Municipal Forestry Bureau	*
	A4	Protected land area ratio	%	17	11.82[a]	17	18	19	Municipal Forestry Bureau	*
	A5	Urban per capita vegetation area	m^2/person	≥11	13.35	15	16	17	Bureau of Urban Utilities and Landscaping	*
	A6	Bare mountain area	ha	—	46.7	20	10	5	Municipal Forestry Bureau	
(B) Ecological industry and circular economy system	B1	GDP per capita	Yuan/person	≥33,000	53,586	67,000	93,000	110,000	Municipal Economy and Development Bureau	*
	B2	Annual per capita financial revenue	Yuan/person	≥5,000	10,526	17,000	23,500	29,000	Municipal Economy and Development Bureau	*
	B3	Farmer annual per capita net income	Yuan/person	≥8,000	5,152	7,000	8,000	9,000	City Planning Commission, Municipal Economy and Development Bureau	*

(Continued)

TABLE 10.5 (Continued)
Eco-City Construction Planning Indicators

System	No.	Indicator	Unit	Standard value	Status value	2010	2015	2020	Supervision Agency	Note
	B4	Urban annual per capita disposable income	Yuan/person	≥16,000	12,915	16,000	20,000	25,000	Municipal Economy and Development Bureau	*
	B5	The tertiary industry of GDP	%	≥45	39.1	41.00	43.00	45.00	Municipal Economy and Development bureau	*
	B6	The rate of enterprises carrying out clean production	%	100	10	80	90	100	Environmental Protection Bureau	*
	B7	The rate of enterprises passing ISO14000	%	≥20	10	20	65	80	Environmental Protection Bureau	*
	B8	The percentage of organic product and green component of main agriculture product	%	≥20	—	45	55	60	Municipal Agriculture Bureau, Environmental Protection Bureau	
	B9	Solid waste emissions per industry output value	t/100 million yuan	—	0.52	0.29	0.14	0.09	Environmental Protection Bureau	
	B10	Material circulation rate	%	≥80[b]	36[c]	50	70	80	City TDC	

B11	The ecological industry output of the GDP	%	≥65[b]	20[c]	35	50	65		Environmental Protection Bureau, Municipal Economy and Development Bureau
(C) Resources guarantee and efficient utilization system									
C1	Energy effective use rate	%	43[b]	33[d]	37	40	45		City Planning Commission, Municipal Economy and Development Bureau
C2	Clean energy utilization rate	%	—	40	65	80	95		Environmental Protection Bureau
C3	Energy consumption per GDP	t/10,000 yuan	≤1.4	0.76	0.61	0.53	0.47	*	City Planning Commission, Municipal Economy and Development Bureau
C4	Water consumption per GDP	t/10,000 yuan	≤150	74[e]	42	32	28	*	Municipal Water Resources Bureau
C5	The rate of industrial water reuse	%	≥50	78[f]	85	88	90	*	Municipal Water-saving Office
C6	The animal waste comprehensive utilization rate from livestock and poultry farm	%	≥90	—	90	95	100		Environmental Protection Bureau
C7	Crop straw comprehensive utilization rate	%	≥95	>90	100	100	100		Municipal Agriculture Bureau, Environmental Protection Bureau

(Continued)

TABLE 10.5 (*Continued*)
Eco-City Construction Planning Indicators

System	No.	Indicator	Unit	Standard value	Status value	2010	2015	2020	Supervision agency	Note
	C8	The recycling rate of urban domestic sewage	%	—	0.83	10	20	25	Environmental Protection Bureau, Bureau of Urban Utilities and Landscaping	
	C9	Urban rainwater collection use rate	%	—	—	3	10	15	Bureau of Urban Utilities and Landscaping	
	C10	Increasing value of marine industry per GDP	%	44.7	23.4	26.3	31.2	44.7	Municipal Oceanic Administration, Municipal Statistics Bureau	
(D) Environmental improvement and pollution control system	D1	Urban water function zone water qualified rate	%	100	—	90	100	100	Environmental Protection Bureau	*
	D2	Coastal environment water qualified rate	%	100	23.1	75	85	95	Environmental Protection Bureau, Municipal Oceanic Administration	*
	D3	Major atmospheric pollution emission intensity SO_2	kg/10,000 yuan (GDP)	<5.0, no more than the country's main pollutant total control index	5.90^{g}	1.9	1.5	1.4	Environmental Protection Bureau	*
		NO_2			3.15[7]	2.7	2.6	2.3	Environmental Protection Bureau	
	D4	Major water pollutant emission intensity (COD)			7.04	5.0	3.0	2.0	Environmental Protection Bureau	*

	Indicator	Unit						Responsible Department	
D5	Centralized drinking source water qualified rate	%	100	99	100	100	100	Environmental Protection Bureau	*
D6	Urban domestic sewage centralized treatment rate	%	≥70	50.87	70	90	98	Environmental Protection Bureau, Bureau of Urban Utilities and Landscaping	*
D7	Village drinking water qualified rate	%	100	—	100	100	100	Municipal Public Health Bureau, Environmental Protection Bureau	
D8	Sewage emission qualified rate from livestock and poultry farm	%	≥75	—	90	95	100	Environmental Protection Bureau	
D9	Intensity of fertilizer use	kg/hm²	≤280	458.6	280	250	225	Municipal Agriculture Bureau, Environmental Protection Bureau	
D10	Total COD into the sea	t/a	—	43,123	35,000	28,000	25,000	Environmental Protection Bureau	
D11	Industrial wastewater qualified rate	%	—	97.57	98.5	99.5	100	Environmental Protection Bureau	
D12	Urban air quality	The number of days better or equal to the secondary standard/year	≥330	364	Up to the standard	Up to the standard	Up to the standard	Environmental Protection Bureau	*

(Continued)

TABLE 10.5 (*Continued*)
Eco-City Construction Planning Indicators

System	No.	Indicator	Unit	Standard value	Status value	2010	2015	2020	Supervision agency	Note
	D13	Coverage rate of smoke and dust control area	%	>90	95.47	100	100	100	Environmental Protection Bureau	
	D14	Noise qualified coverage rate	%	≥95	83.06	90	100	100	Environmental Protection Bureau	*
	D15	Urban domestic waste disposal rate	%	100	93.95	100	100	100	Environmental Protection Bureau	*
	D16	Industrial solid waste use rate	%	≥80; no hazardous waste emission	77.06	90	92	95	Environmental Protection Bureau	*
	D17	Tourist site environmental qualified rate	%	100	100	100	100	100	Municipal Tourism Administration	
	D18	Restoration rate of degraded land	%	—	Up to the standard	Up to the standard	Up to the standard	Up to the standard	Municipal Forestry Bureau	*
	D19	The amount of phytoplankton	×103 cell/L	<300	785.3	<500	<300	<300	Environmental Protection Bureau, Municipal Oceanic Administration	
	D20	The restoration area of damaged mangrove	hm²	140	20.8	80	120	140	Environmental Protection Bureau, Municipal Oceanic Administration, Municipal Forestry Bureau	

Category	Code	Indicator	Unit						Responsible department
	D21	Tide volume of semiclosed bay	×10⁸ m³	7.6	5.5	6.5	7.2	7.6	Municipal Oceanic Administration
	D22	The environment protection investment of GDP	%	≥3.5	2.2	3.5	3.6	3.8	Municipal Government *
(E) Ecological culture and human settlement system	E1	Urbanization level	%	≥55	59.1/65]	72	80	88	Municipal Construction and Administration Bureau *
	E2	Average life expectancy	Age	77	78.23	78.5	78.8	79	Municipal Government, Municipal Public Health Bureau
	E3	Urban gasification rate	%	≥92	99.2	100	100	100	Bureau of Urban Utilities and Landscaping
	E4	Population density of built-up area in the island	10,000 people/km²	—	—	1.2	1.1	1	Municipal Government, Municipal Construction and Administration Bureau, Department of City Planning
		Population density of built-up area outside the island		—	—	0.6	0.7	0.75	
	E5	The rate of green community	%	≥50	2.4	50	65	80	Environmental Protection Bureau, Municipal Construction and Administration Bureau

(Continued)

TABLE 10.5 (*Continued*)
Eco-City Construction Planning Indicators

System	No.	Indicator	Unit	Standard value	Status value	2010	2015	2020	Supervision Agency	Note
	E6	Bus number every 10,000 people	1	—	24.15	27	29	30	Municipal Government, City Traffic Commission	
	E7	Vehicle emission qualified rate	%	—	64.74	90	100	100	City Traffic Commission, Environmental Protection Bureau	
	E8	Engel's coefficient	%	<40	41.28	38	35	30		*
	E9	Gini coefficient		0.3–0.4	0.41	0.35	0.32	0.30	City Planning Commission, Municipal Economy and Development Bureau	*
	E10	Higher education enrollment rate	%	≥30	14	35	45	60	Municipal Education Bureau	*
	E11	Environmental protection education and propagation rate	%	>85	—	90	95	98	Environmental Protection Bureau, Bureau of Culture	*
	E12	Public satisfaction regarding the environment	%	>90	86.3	95	96	99	Environmental Protection Bureau	
	E13	Implementation rate of planning EIA	%	>95	—	95	96	98	Municipal Government, Municipal Legislation Bureau	*
	E14	Active enterprises with environmental information	%	>90	—	90	93	96	Environmental Protection Bureau, Bureau of Culture	

E15	Green school rate	%	>80	4	50	80	90	Municipal Education Bureau, Environmental Protection Bureau
E16	The operating condition of urban ecosystem	%	≥80	—	Up to the standard	Up to the standard	Up to the standard	City construction, transportation, fire-fighting, sanitation, earthquake
E17	Land for public facilities per capita	m^2	—	10.75	13	15	16	Municipal Government, Municipal Construction and Administration Bureau, Department of City Planning
E18	Beds with every 10,000 sick people	1	—	46	60	70	80	Municipal Public Health Bureau

Notes: Indicators appearing in the last column with the asterisk "*" are instructive ones; all remaining are research indicators.

a Data is obtained from the nature reserve area ratios from the Xiamen statistics yearbook. The actual protected land area ratio shall be verified.

b The value is determined by the situation of developed countries.

c The number is estimated.

d Data is from 2001 (Xiamen statistical yearbook, 2001).

e From Xiamen water communique of 2003 (Xiamen statistical yearbook, 2003).

f From Xiamen Water Saving Office.

g The value of 2002 (Xiamen statistical yearbook, 2002).

h From the Xiamen target-complied planning of environmental quality, excluding data from the Jiulong River's.

i In this planning, semiclosed gulf refers to Tongan Gulf and the western sea area.

j The former data is calculated by census register population, and the latter is calculated by the actual population, including the temporary residential population.

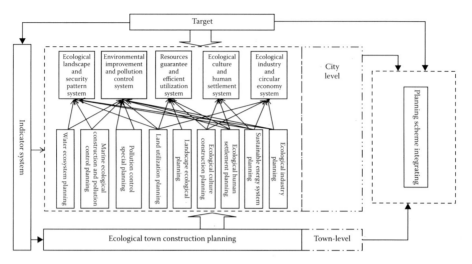

FIGURE 10.2   Xiamen eco-city planning diagram.

10.2.3   OUTLINE OF XIAMEN'S URBAN ECOLOGICAL PLANNING

Five systems of Xiamen eco-city planning were established: (1) ecological landscape and security pattern system, (2) ecological industry and circular economy system, (3) resources guarantee and efficient utilization system, (4) environmental improvement and pollution control system, and (5) ecological culture and human settlement system. The ecological landscape and security pattern system contain the planning for the forest and green area. The ecological industry and circular economy system focus on the agriculture, industry, and service industry. In the resources guarantee and efficient utilization system, we aim to efficiently use water resources, land resources, and energy. The environmental improvement and pollution control system comprise the water environmental protection plans, atmospheric environmental protection plans, noise pollution control measures, and solid waste pollution control rules. The ecological culture and human settlement system include ecological culture, ecological consumption, and ecological residential environment planning.

It is imperative for us to grasp the regional ecological relationship between the ecosystem and the human beings and intensify the self-repairing and supporting capacity of nature. Eco-city planning has important significance and ecological impact for building a regional ecological security pattern. The research and exploration of urban ecological function areas is important if the urban ecological management and protection is to be improved (Figure 10.2).

10.3   ECOLOGICAL FUNCTION REGIONALIZATION OF XIAMEN

Ecological function regionalization aims to apply sound ecological methods and principles to reveal the similarities and differences existing in natural ecosystems, and to quantify the environmental disturbances caused by human activities.

Ecological function regionalization provides the foundation for the management of our natural and social resources. It is different from natural regionalization because it takes the natural environment's characteristics and processes into account, as well as the influence of human activity.

The ecological function regionalization was reconstructed by analyzing ecological environmental fragile degrees, the evaluation on ecosystem service function value, and the preexisting ecological function regionalization results. The regionalization process is managed in three levels: the first level includes three ecological function regions, the second level includes eight ecological function subregions, and the third level includes 25 ecological function units (Tables 10.6 through 10.8).

TABLE 10.6
Ecological Function Zone Hierarchical Structure

No.	Function Zone	No.	Function Subzone	No.	Ecological Function Units
01	Island ecological function zone	0101	Ecological management and protection zone	010101	Gulangyu ecological unit
				010102	WanShiShan ecological unit
		0102	Ecological reconstruction zone	010201	Eastern island ecological unit
				010202	Western island ecological unit
02	Outside ecological function zone	0201	Ecological management and protection zone	020101	Caijianwei mountain scenic forest ecological unit
				020102	Western hill soil and water conservation and scenic forest ecological unit
				020103	Shidou–Bantou Reservoir drinking water source protection ecological unit
				020104	Northwestern hill soil and water conservation and scenic forest ecological unit
				020105	Tingxi Reservoir drinking water source protection ecological unit
				020106	Northeastern hill soil and water conservation and scenic forest ecological unit

(Continued)

TABLE 10.6   (Continued)
Ecological Function Zone Hierarchical Structure

No.	Function Zone	No.	Function Subzone	No.	Ecological Function Units
		0202	Ecological buffer zone	020201	Tongan district urban and rural coordinated construction ecological unit
				020202	Xiamen eastern city and industrial environment ecological unit
		0203	ecological reconstruction zone	020301	Haicang southern port and industrial environment ecological unit
				020302	Maluan Bay and Xinglin Bay industrial environment ecological unit
				020303	Dadeng Island commercial and travel ecological unit
03	Marine ecological function zone	0301	Ecological management and protection zone	030101	Xiamen southern seas landscape ecological unit
				030102	Xiamen eastern seas landscape ecological unit
				030103	Tongan Bay port environment and rare marine creature protection ecological unit
		0302	Ecological buffer zone	030201	Dacheng Island aquaculture and travel ecological unit
				030202	Dongju port aquaculture pollution control ecological unit
				030203	Xun River port shipping ecological unit
		0303	Ecological reconstruction zone	030301	Jiunlong River estuary environment ecological unit
				030302	Western sea south port environment and rare marine creature protection ecological unit

TABLE 10.6 (*Continued*)
Ecological Function Zone Hierarchical Structure

No.	Function Zone	No.	Function Subzone	No.	Ecological Function Units
				030303	Western sea north travel environment and rare marine creature protection ecological unit
				030304	Maluan Bay wetland and travel environment ecological unit

TABLE 10.7
Function Description of Ecological Control Units

Ecological Function Unit	Main Function
Gulangyu ecological unit	Travel, ecological environment
WanShiShan ecological unit	Travel, ecological environment
Eastern island ecological unit	Urban commercial and trade, living ecological environment
Western island ecological unit	Urban commercial and trade, living ecological environment
Caijianwei mountain scenic forest ecological unit	Travel, ecological environment
Western hill soil and water conservation and scenic forest ecological unit	Soil and water conservation
Shidou–Bantou Reservoir drinking water source protection ecological unit	Drinking water sources
Northwestern hill soil and water conservation and scenic forest ecological unit	Soil and water conservation, drinking water sources
Tingxi Reservoir drinking water source protection ecological unit	Drinking water sources
Northeastern hill soil and water conservation and scenic forest ecological unit	Soil and water conservation and water source conservation
Tongan district urban and rural coordinated construction ecological unit	Urban commercial and trade, ecological environment
Xiamen eastern city and industrial environment ecological unit	Urban commercial and trade, industrial ecological environment
Haicang southern port and industrial environment ecological unit	Port and industry, ecological environment
Maluan Bay and Xinglin Bay industrial environment ecological unit	Urban commercial and trade, industrial ecological environment protection and pollution absorption

(Continued)

TABLE 10.7 (*Continued*)
Function Description of Ecological Control Units

Ecological Function Unit	Main Function
Dadeng Island commercial and travel ecological unit	Travel, commercial, trade, and ecological environment protection
Xiamen southern seas landscape ecological unit	Bathing beach, travel, and ecological environment protection
Xiamen eastern seas landscape ecological unit	Bathing beach, travel, and ecological environment protection
Tongan Bay port environment and rare marine creature protection ecological unit	Rare marine species, ecological environment, port environment
Dacheng Island aquaculture and travel ecological unit	Aquaculture and travel development
Dongju port aquaculture pollution control ecological unit	Short time: marine aquaculture Long term: port environment
Xun River port shipping ecological unit	Short time: marine aquaculture Long term: port environment
Jiunlong River estuary environment ecological unit	Port and pollution absorption
Western sea south port environment and rare marine creature protection ecological unit	Port and pollution absorption
Western sea north travel environment and rare marine creature protection ecological unit	Port and pollution absorption
Maluan Bay wetland and travel environment ecological unit	Travel and ecological environment

TABLE 10.8
Regulating Strategy of Function Subzones

No.	Function Subzone	Strategy
0101	Ecological management and protection zone	Scientifically develop the tourism economy, protect and cultivate the urban ecological landscape, strictly control pollution, mountain landscape restoration coupled with soil and water conservation, improve the ecological service value, achieve balance in the development of mountain, city, island, and oceanic areas
0102	Ecological reconstruction zone	Adjust the relationship between urban construction and ecological protection to increase green and sustainable development, improve land use efficiency, optimize urban ecological space, build ecological channels based on parks, create an urban vegetation and transport network, impel the urban pollution treatment infrastructure, restore the water environment and improve the water quality, control the city's development intensity

TABLE 10.8 (*Continued*)
Regulating Strategy of Function Subzones

No.	Function Subzone	Strategy
0201	Ecological management and protection zone	Determine the development intensity and scale of tourism activities, strengthen the construction of ecological and water conservation forests, improve soil and water erosion controls, address the livestock pollution problems in water source protection areas, promote the overall ecological service value, provide guarantees for urban ecological environment optimization
0202	Ecological buffer zone	Coordinate the relationship between industry production and ecological environment protection, develop more low-emission industries, reconstruct the ecological industry chain, improve the ability of pollution treatment by the means of infrastructure construction, protect and improve rural ecological environment and coastal ecological environment
0203	Ecological reconstruction zone	Reconstruct the layout of urban functional zones, promote rural urbanization, complete the basic framework of urban pollution prevention and control, protect the shoreline and tidal resources, integrate the industrial park (as dictated by port efficacy), build a modern port area and an ecological industrial park
0301	Ecological management and protection zone	Strictly implement national and local laws, regulations, and the relevant standards to protect the natural reserve and tourism landscape reserve, prohibit all the activities with the potential to damage the ecosystem (such as establishing drainage outlets, encroaching on beaches, and sand mining), timely restoration efforts should be planned and executed for the damaged ecosystem, as well as the establishment of wetland parks and special reserves
0302	Ecological buffer zone	Strictly control land reclamation and coastal engineering construction activities, actively guide and adjust the industrial structure while developing the ecological industry, construct the coastal shelter forest system to combat shoreline erosion
0303	Ecological reconstruction zone	Restore the damaged ecosystem as soon as possible, stress the importance of coordinated development amongst the natural and human-made ecosystems, implement an environment functional zone, strictly carry out environmental quality standards and pollution emission standards

10.4 KEY AREAS AND CRUCIAL TASK OF XIAMEN'S URBAN ECOLOGICAL PLANNING

10.4.1 Ecological Landscape and Security Pattern Construction

In accordance with the theory of safety patterns in landscape ecology, the best potential ecological security pattern should be identified for Xiamen's urban ecological planning efforts. Xiamen's ecological security pattern consists of a core region that is subdivided further into 4 districts, 8 residential groups, 3 levels corridors, 7 ecological isolation belts, and 10 landscape strategy points (Su et al., 2006).

The landscape of green space is the most accessible feature to urban inhabitants and also plays the most important role in the cityscape. The green space landscape planning is the key to the development of the regional safety ecological pattern. Green space is an important part of complex urban ecosystems and provides significant services. The green space landscape planning aims to promote both healthy and comfortable ecological human settlements. For example, the green space will involve the planting of trees in local factories, companies, and around offices as part of the common landscaping and installing lawns along streets or in urban areas. The green space landscaping will be designed on the basis of the spatial characteristics.

The greenland in the Xiamen Island should be carefully protected. Areas of special interest include the Zhongzhai Reservoir scenic woodlands, Xianyue Mountains, Huweishan Park, Wanshi Mountain, and Yuanshan Forest region. Outside of the Xiamen Island, we should protect the greenland of Caijianwei Mountain, Jingkouyan Mountain, Dalun Mountain, and Mei Mountain. Swamp restoration and planting efforts should also be prioritized in Xinglin Bay, Maluan Bay, Tongan Bay, and Weitou Bay. Tourist attractions to include the Guanxun Fruit Forest Sightseeing Park, the Xike Ecological Park, and Sightseeing region in Dong Stream, and Xi Stream will be developed. We will conduct research on mountain ecological reconstruction techniques and create a suitable environment. The channel between the mountains and the important visual corridor shall be retained, and neither damage to the mountain nor filling the sea will be permitted.

We should construct and protect an ecological isolation zone extending from the forest ecological background zone to Tongan Bay. We should plan green spaces along Maluan Bay and arrange the Houpu, Xiyuan, and Wencuo shelter forest passing through Maluan Bay's western region. The external greenbelt in the Canghai district contains the green space in Maluan Bay, oriental golf course, coastal road, and Shugang Road. Focusing on the intersection of Jiu River and Neitian River, we should establish a radiated permanent green channel including areas such as Tongkeng Bay, Dadeng Bay, and Ma Port. The distribution of afforestation should be consistent with the distribution of population. In other words, the larger the flow of the population, the higher the quality of the greening standards. In the island, Huangdao Road and its adjacent sea area will form the coastal green area, which can serve as a buffer zone against the outside disturbances. The main vegetation in this area consists of public green space, protection forest, and scenic woods. The river regulation project should consider the present water quality issues of Xiamen's river and lake ecosystems, while striving to preserve the traditions and local characteristics of the coastal city. The Yuandang Lake, Hubian Reservoir, Liliao Reservoir, Zhongzhai Reservoir, Xinglin Bay, Jiutian Lake, Maluan Bay, and Shidou Reservoir are candidates for water storage. The coastal greenbelt includes the green fields circling the Xiamen Island and gulf coast. This greenbelt can prevent seawater erosion while filtering the salt air.

The green spaces located in the park of Xiamen are presently unequally distributed; as such, we should control its layout by calculating the service radius of public green space. It is stipulated that the service radius on a municipal level should be no less than 2 km, while the service radius on a district level should be no less than 1 km. To establish these radii, we will construct 2 municipal parks, 7 regional parks, 29 community parks, and 3 theme parks. With the development of the city, the economic center will migrate from the island to the outside (Su et al., 2006). We estimate that it will take 3–5 years to establish some large-scale public green engineering projects in the Hubian Reservoir, Zhongzhai Bay, Xinglin Bay, and Maluan Bay. The distance between community parks should be less than 1 km. Green shelters should be installed in garage and solid waste disposal locations to help minimize dust emissions and material leakage. Finally, we hope to see that there will be a small, community garden maintained in each city block, a green island in each square, and that roof plantation practices will be widely utilized in Xiamen.

10.4.2 Ecological Industry Construction

The current industry development trends indicate that Xiamen's economic growth foundation is relatively weak. The performance of different industrial sectors in the Xiamen economy has been seen to vary significantly. The processing industry accounts for the largest proportion of the industrial economy. However, there are many economical operation problems; primarily, the industry chain is fairly short with a weak industry relevance degree. To address these issues, the theory of material circulation should be applied to our ecological industry planning (which presently consists of ecological agriculture planning, ecological industry planning, ecological third industry planning, and the interregional cooperation planning).

10.4.2.1 Ecological Agriculture

In addition to our other important goals, it is also important to promote agriculture industrialization, to enforce private economy development, and to transfer the surplus work force from the countryside to the secondary and tertiary industry.

However, Xiamen's primary industry is going to shrink. In order to meet the demands of the local population, we should vigorously develop our agriculture industry in our planning efforts. We should also increase the proportion of fisheries in our primary industry, which will highlight the unique marine characteristics of Xiamen.

It is essential to make an adjustment in the relationship between economic development and the preservation of water resources, a shift in which the conservation of water allotted for agricultural purposes is a key element. The following measures should be implemented:

1. Improve the agricultural water supply facilities to reduce the water loss in irrigation.
2. Promote the stereo-agriculture model to increase irrigation efficiency.

3. Increase the proportion of rainwater utilization and the use of urban sewage in the agriculture irrigation process.
4. Implement a three-dimensional culture model such as "pig-swamp-grass-pig" or "pig-swamp-fruit-forest," and establish an ecological agriculture demonstration area having high yield, high quality, low pollution, and low emissions.
5. Spread the technology of straw mature accumulation, bioconversion for energy regeneration, technology of the agriculture film reutilization, and comprehensive utilization of animals' metabolite.
6. Promote the technology of animal waste utilization in agriculture, reduce the use of chemical fertilizers, and cut the source of emissions from daily life.

Pesticides (fertilizer application) are a major contributor to Xiamen's water pollution. As such, ecologic pest control techniques should be implemented to control the application of pesticides and fertilizer.

The three-dimensional model can save 10%–20% of agricultural land. Through the improvement of water supply facilities, 10% of the water consumption can be saved and the water utilization efficiency will improve by 15%.

10.4.2.2   Ecological Industry

Industrial goals should include optimization of the industrial structures, coordination of the industrial spaces, integration of the city's industrial groups, construction of an ecological industry, increasing the proportion of the secondary industry, and eventually forming the economies of scale for Xiamen's industry.

10.4.2.2.1   Industry Structure Adjustment

Based on the current statistics, the proportion of industry in Xiamen's secondary industry should be increased until it reaches 94% in 2020. Until 2020, the pillar industry will continue to perform at a high level, providing 90% of the total industrial output.

10.4.2.2.2   Ecological Industry System Construction

We will develop the following industries to help construct the ecological industry of Xiamen: electronics and information service, petrochemical, transportation, and machinery.

10.4.2.2.3   Electronics and Information Service Ecological Industry System

Based on examples from the existing electronics industry, such as Dell China Co., Ltd, service providers within the information industry should be combined. Additionally, we should strive to extend the maintenance service abilities during

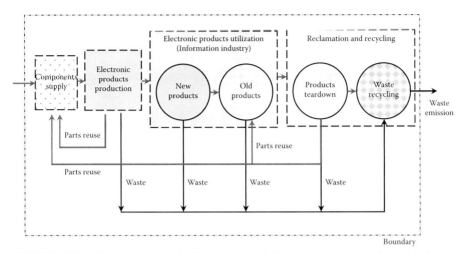

Boundary

FIGURE 10.3 **(See color insert.)** Electronic and information service ecological industry system conceptual model.

a product life cycle. The recycling of waste from electrical equipment, electronics, and waste treatment comprehensively makes up the "half product life cycle" industry system. In the ecological industry system, the electronics and information service industries will balance the creation of product, utilization, and maintenance throughout a product lifetime and the recycling process. Throughout this cycle up to the final process of recycling, material utilization efficiency will be largely improved (see Figure 10.3).

Disadvantages exist in Xiamen's current electronics and information service industry. Primarily, most electronic components are imported from outside locations. Secondly, current waste recycling processes need to be improved. Third, the reuse of the old electronic products is not being maximized and also requires further improvement. Finally, the availability of electronic products is limited and Xiamen needs to establish a stronger electronic product community.

10.4.2.2.4 Petrochemical Ecological Industry System

Based on the production of PTA (purified terephthalic acid) and spinning fibers associated with Xianglu's petrochemical industry, the proposed ecological industry should involve a waste recovery industry focusing on garment processing and petrochemical waste (see Figure 10.4).

As shown in Figure 10.4, the petrochemical ecological industry system will incorporate chains devoted to textile, rubber, and plastic products. The rubber and plastic products should be given priority in identifying opportunities for reuse and then considered for recycling and hazard-free treatments. Conversely, textile products can be considered immediately for recycling and treatment at the end of their life cycle.

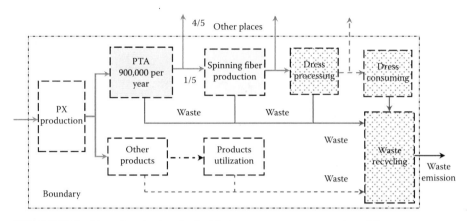

FIGURE 10.4   Petrochemical ecological industry system conceptual model.

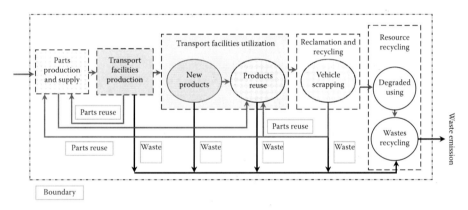

FIGURE 10.5   **(See color insert.)** Traffic and transport engineering industry conceptual model.

10.4.2.2.5   *Transportation and Machinery Ecological Industry System*

- The transportation and machinery components of the ecological industry system should include the vehicle manufacturers, maintenance, and utilization groups, and companies involved in the recycling, reuse, and reclamation process (see Figure 10.5).
- Issues are apparent with these existing industries. Primarily, the authorities lack a systematic approach to management. Secondly, the material recycling rate is fairly low with only 54.5% steel able to be reclaimed. Third, the reuse of waste is presently minimal and also requires further improvement. Finally, used products and their associated components are not recycled in an efficient way.

10.4.2.2.6   Eco-Industrial Park Construction

The Eco-Industrial Park (EIP) will consist of telecom, petrochemical transportation, and machinery industries. Based on the example of Dell China Co., Ltd, we will combine the telecom service industry with the Oasis Sources Company, an environmental protection limited company, to establish a "Half Product Life Cycle" ecological demonstration park. The main goals of this effort include the following:

1. Launch new enterprises to supplement for the functional defect of the industry system.
2. Upgrade the electronic products with more advanced technology.
3. Develop the production of auxiliary products, while fashioning IT industry products group in the form of Dell and Cankun.

The "Half Life Cycle" ecological demonstration part can be constructed at the Jinlong Co., LTD and Taigu Plane Engineering Co., LTD, while also making use of the roads, traffic authority, and shipping administration bureau. From the transportation aspect, these parks should be designed on the basis of "ecological design," "green manufacturing," and "green transportation."

In the construction of the EIP, water recycling and the cascade utilization of energy present key issues. To promote water recycling, water resource conservation and water reuse management will be the focus. In terms of the energy consumption, attention should be given to the large energy consumers. We should constantly promote the use of renewable energy resources.

10.4.2.2.7   Life Cycle Assessment

Life cycle assessment will be adopted as the main goal of the pillar industry (see Tables 10.9 and 10.10), to include "design for environment" the concept that product design must address the product's impact upon the environment), "enterprise resources planning," green manufacturing, green package and transportation, reclamation, recycling, and reuse.

10.4.2.2.8   Implementation of Cleaner Production and ISO14000

Based on the audits documenting the relationships among enterprise production, energy consumption, and environmental pollution, we have observed the disadvantages in Xiamen's industry development and will strive to eliminate the outdated technology.

With the planned adjustments to industrial structure, Xiamen's material recycling efficiency can be improved by 8%, while the ecological efficiency improves by 10%–20%. Through increasing product reuse within each industry, the associated resource consumption intensity will be reduced by 20%–40%. Simultaneously, with the construction of EIP, the ecological efficiency will be increased by 30%–50%. With product life cycle assessment and related improvements, the ecological efficiency will be increased further by 20%–40% (Zhao et al. 2007).

TABLE 10.9
Scenario Analysis of the Pillar Industry Ecological Indicators

	Indicators	Unit	Status Value	2010	2015	2020
Scenario 1 Traditional industry	Material circulation rate	%	36	36	36	36
	Solid waste emission per industrial added output	t/100 million	0.63	0.63	0.63	0.63
	Solid waste discharge value	t	200	510	800	1030
Scenario 2 Ecological industry	Material circulation rate	%	—	70	78	85
	Among them: Industrial structure adjustment		—	8	9	10
	Construction of industrial system		—	8	9	10
	LC management and LCA		—	8	10	12
	EIP construction		—	8	10	12
	Clean production and ISO14000		—	6	7	8
	Solid waste emission per industrial added output	t/100 million	—	0.25	0.17	0.12
	Ecological efficiency		—	2.5	3.7	5.35
	Among them: Industrial structure adjustment		—	1.1	1.2	1.3
	Construction of industrial system		—	1.2	1.3	1.4
	LC management and LCA		—	1.2	1.3	1.4
	EIP construction		—	1.3	1.4	1.5
	Clean production and ISO14000		—	1.2	1.3	1.4
	Solid waste discharge of pillar industry	t	200	202.5	217.6	196.8
Profit	Material circulation rate	%	—	34	42	49
	Reduction of solid waste emission	t	—	307.5	582.4	833.2

TABLE 10.10
Trend Analysis with Certain Index of Pillar Ecological Industry

Names of Index	Unit	2010	2015	2020
Ratio of pillar industry to cleaner production enterprise	%	80	90	100
Ratio of scale enterprise through the ISO14000 certification[a]	%	45	65	80
Among them: The electronic industry	%	40	70	90
Petrochemical industry	%	30	50	70
Machinery industry	%	60	70	80
Material circulation rate[a]	%	70	78	85
Among them: Electronic industry	%	60	80	90
Petrochemical industry	%	50	60	70
Rubber/plastic/water	%	55/30/60	70/40/70	80/50/80
Machinery industry	%	60	70	80
Steel	%	80	85	90
Solid waste emissions of pillar industry unit value added of industry	t/100 million yuan	0.25	0.17	0.12

Note:
[a] The two indexes are the weighted composite index of Xiamen's pillar industry, with the weight coefficients being weight taken in the pillar industry.

10.4.2.3   Ecological Service Industry

Primary goals of the ecological service industry include speeding up the development of the tertiary industries and new technology industry, while promoting the growth and development of the service sector, especially the tourism industry. Due to its progressively improving relationship with Taiwan and other trade districts, Xiamen stands to benefit significantly by increasing their open and global trade system. Due to the geographic advantages of its location, Xiamen will be constructed into an ecological port and modernized bay. Additionally, the development of Xiamen's tourism industry will grow, and we will strengthen the centrality of the third industry.

In order to enhance the prosperity of Xiamen's tourism industry, it should be established with the perspective of offering whole-course services. Simultaneously, Xiamen's tourism will effectively protect the ecological resources and human environment and help to promote the coordination of environmental protection with economic development.

A whole-course service pattern is possible for the tourism industry if we strengthen the integration of tourism with communications, catering, and other related industries to establish a comprehensive level of service. In terms of passenger flow, we should advocate for the reduction of related energy consumption and waste emissions. In terms of material flow, we should apply a green package and environmentally friendly transportation. In terms of management, we should improve the information level of

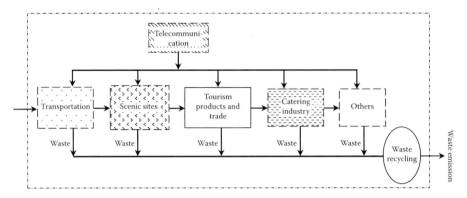

FIGURE 10.6 Xiamen tourism ecological industry system conceptual model.

tourism and enhance it with the latest electronic information technology. Finally, we will also bring the catering industry into the ecological tourism industry system. Our goal is to ensure that as the tourism industry expands, the energy consumption and waste emissions will be maintained at acceptable levels (see Figure 10.6).

10.4.2.4 Ecological Marine Industry

Xiamen has rich marine resources mainly used for biomedical, fine chemical engineering, aquaculture, and the conservation of fresh water resources. The proposed measures for preserving and enhancing this industry are as follows:

1. Apply existing technologies pertaining to the transgene and active material extraction of marine plants and animals to form the marine biological pharmaceutical products community.
2. Take the EIP mode in the northern area of Shandong as an example and focus on the production of halogen and enzyme.
3. Develop the sea water breeding industry and accelerate Xiamen's role as a distributing center of offspring seed, develop a nuisance-free aquaculture product, and produce green marine products consistent with the guidelines imposed by national and international standards.

Taking the Dadeng salt field and Liuwudian seawater seeding field as examples of quality chemical engineering and aquaculture, respectively, a marine ecological demonstration park will be built thanks to joint efforts of scientific research institutes such as Beidazhilu, Tebao biology, and Haiyangsansuo (see Figure 10.7).

Table 10.11 (Xiamen statistical yearbook, 2003) demonstrates that the marine ecological output of 1.2%, 2.35%, and 3.42% of the GDP is forecasted in 2010, 2015, and 2020, respectively. Should this forecasted increase take place, Xiamen's traditional situation involving excessive dependence on outside resources will be changed. Additionally, the products of marine chemical engineering can help to improve Xiamen's environmental quality. Caustic soda, for example, has the ability to neutralize acidic waste gas. Associated enzyme technologies have the ability to improve the efficiency of both industrial production and waste water disposal.

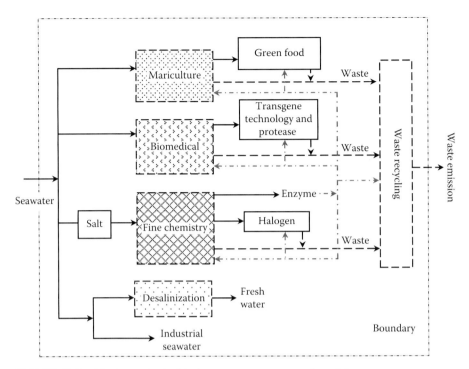

FIGURE 10.7   Marine ecological industry system conceptual model.

TABLE 10.11
Trend Analysis with Certain Index of Xiamen Marine Ecological Industry System

Names of Index	Unit	Status Value (2003)	2010	2015	2020
GDP	100 million yuan	759.69	1800	2800	3600
The added value of biological medicine[a]	100 million yuan	—	17	56	109.5
The added value of marine ecological breeding[b]	100 million yuan	—	4.68	9.84	13.76
The added value of marine industry system	100 million yuan	—	21.68	65.84	123.26
Ratio of marine industry system to GDP	%	—	1.20	2.35	3.42

Notes:
[a] Refers to the biological medicine and fine chemicals industry.
[b] Refers to ocean green seedlings and food industry.

10.4.2.5   Cross-Region Cooperation

Xiaquanzhang (Xiamen, Quanzhou, and Zhangzhou) City Combination is proposed to effectively raise the competitive power of the economic zone in Xiamen. The combination of these cities is believed to have the potential to accelerate industry transfer, increase industry scale, and integrate and optimizing the structure of industry. Thus, this proposal should be adopted and implemented with all strength. Meanwhile, the cooperation between the Pear and the Yangtze River deltas should be advocated. Beneficial to both areas are mutual harbor construction and ecological protection measures, the conservation of water resources, and waste treatment. These efforts will comprise the working emphasis of our government and will be embraced as an effective way to support Xiamen's sustainable development.

The strategy for Xiamen's cross-region-cooperation development is as follows. Initially, we construct the Xiaquanzhang (Xiamen, Quanzhou, and Zhangzhou) City Combination through the development of a multilevel platform involving the resources of marine, port, tourism, water resources, and energy industries. Additionally, we will complete the ecological construction of the dominant industry while promoting equal use of the urban infrastructure. Secondly, the cooperation between the Pear River delta and the Yangtze River delta should be advocated and prioritized. The two regions should be educated on the benefits of this cooperation, to include further improvements to their resource utilization efficiencies.

We will transfer a part of Xiamen's industry, which has demonstrated a relatively low level of consistency with the pillar industry to Quanzhou and Zhangzhou, thus establishing a new ecological industry there. This will not only ease land resources shortage in Xiamen but also fortify the industry potential of both Quanzhou and Zhangzhou. Due to the limited carrying capacity of Xiamen's port, it would be reasonable to combine Xiamen's port resources with those of Quanzhou and Zhangzhou. Only through such mergers can the ports' comprehensive competitive strength be improved, while also effectively relieving conflict between harbor construction and environmental protection. Measures should be taken to strengthen the water resource conservation and environmental protection efforts at the Jiulong and Ji Rivers. Additionally, through the development of the tourism industry in the Xiaquanzhang City Combination area, the development of commerce and trade will also be stimulated accordingly.

Finally, according to the consistency analysis of industry distribution and functional regionalization, the EIP devoted to the mechanical and telecom industries will be located in the industrial area of the northern island. The petrochemical EIP will be located to the south of the Haicang district. The marine EIP will be seated to the south of the Maxiang Town in Xiangan Area. The agricultural EIP will locate in the agricultural region of the Tongan district.

10.4.3   Effective Utilization of Resources

This section points out some important options to promote the effective utilization of energy, water resources, and land resources.

10.4.3.1   Energy System Planning

The energy planning strategy should be determined by the energy demand analysis of Xiamen. This analysis will involve an energy-saving priority strategy, improvements oriented towards effective energy utilization, measures for optimizing the energy structure, and the implementation of an environment-friendly energy policy.

Before 2010, Xiamen's energy system planning efforts should mainly focus on energy conservation. The government should introduce methods such as liquefied natural gas (LNG) to improve the energy efficiency and meet both residential and commercial demand. By 2015, the key issue will shift to optimization of the energy structure and transitioning to the use of LNG for the industrial fuel supply. By 2020, the environmentally friendly energy policy will be widely implemented, and the utilization of clean energy will be promoted more significantly.

10.4.3.1.1   Improve the Energy Utilization Efficiency

By aiming for reductions in overall unit energy consumption, for the step utilization of energy, combined heat and power generation, and for the recovery of industry waste heat, Xiamen's industry energy utilization efficiency can dramatically improve. In a specific effort, power plants will transit from their original vertical classifiers to horizontal classifiers, allowing smoke preheat the air, recycle and reuse the condensation water, and reduce the leakage of sealed steam and sealed compressed air. The unit energy consumption of major industrial products in 2015 will approach or reach internationally established levels of the 1990s, and in 2020 it will reach the currently established internationally advanced level. We forecast that energy utilization efficiency will be improved by 6%–7% in 2015 and by 11% in the year 2020.

In terms of the mechanical and oil refining industries, some effective measures should be taken to eliminate outdated equipment and technologies, thus increasing the investment potential of projects either recently built or expanded. This effort will also help to develop efficient and energy-saving products and increase the design level and manufacturing ability of energy-saving mechanical products. In the process of oil refining, we will strive to improve operating capacity and heat exchange efficiency. We should also decrease the processing losses, optimize cooling systems, and redesign key components such as ethylene-cracking furnace.

In terms of building energy conservation, Xiamen strived to reduce its building energy consumption by 30% in 2005 and 50% in 2010 (since this book has been finished in the year 2008, it was hoped to happen in 2010.), and by 2020 overall building energy consumption will be reduced by another 50%. Efforts toward energy conservation will involve:

1. Application of ice-cold storage air condition technology
2. Active promotion of cooperating heat and cooling units
3. Decreasing the indoor thermal load, specifically by improving window shades and the insulation capacity of building materials
4. Implementation of green lighting project
5. Construction of green buildings and provide demonstrations to communities

In terms of energy conservation associated with transportation, the regulative authorities should increase rates for public parking and strive to limit the annual increase of vehicles from 9.36%, 7.15%, and 5.31%. At the same time, the use of automobiles powered by cleaner fuels, fuel cell vehicles, electric vehicles, or other types of low-emission vehicles will be encouraged. The service life of cars should be encouraged to be more than 10 years. To improve the urban traffic system as well as the public transportation, the use of motor scooters in Xiamen should be prohibited by 2010. This change will dramatically reduce the emissions of hydrocarbon pollution.

10.4.3.1.2 Optimize the Energy Structure

Xiamen's industry will work to continuously enhance the resource utilization ratio in addition to the production of cleaner fuel in order to promote sustainable and rapid development. In 2010, 70% of the energy resources will be replaced by LNG, and diesel consumption will be reduced by 60%. By 2015, all diesel fuel will be replaced in the industry. In the public transportation department, we suggest the introduction of dual-fuel and natural gas buses. By 2010, the number of LNG buses and taxis will equal 180 and 600, respectively, subsequently leading to an increase in the number of LNG automobiles. The number of CNG buses in active service will be 200 in 2010, 500 in 2015, and 800 in 2020. There will be 300 CNG taxis in 2010, 750 in 2015, and 1200 in 2020 (Zhao et al., 2005).

The use of solar energy will be widely applied in Xiamen, to include the use of solar water heaters, passive solar houses, and solar energy appliances (such as stoves). In our planning, we have found it necessary to establish a solar water heater utilization system to provide hot water to the rural areas. We also recommend demonstration projects for promoting solar products and gradually extending the scope of solar energy utilization.

The northwestern region of Xiamen has tremendous hydropower potential with the ability to increase the proportion of hydropower in Xiamen's energy consumption structure. Other strategies involve establishing an agricultural circular economy system to increase the use of methane, and the construction of more household biogas pools in the Tongan and Xiangan districts. The number of household biogas pools in Xiamen was 2000 in 2010, should be 2500 in 2015 and 3000 in 2020. By 2020, Xiamen will also make efforts to harness the power of wind in order to reduce fossil fuel consumption in the electric power industry.

10.4.3.1.3 The Implementation of Environmentally Friendly Energy Policy

It is imperative to replace fossil fuels with clean energy and also to control dust sources. The sulfur content in coal and oil-burning fuel sources should be reduced to less than 0.7% by 2015. By 2020, the sulfur content in oil should be less than 0.65%. The usage rate of industrial clean energy was 85% in 2010, should be 90% in 2015 and 95% in 2020.

10.4.3.2 Water Resource Sustainable Utilization Planning

Through the analysis of water supply and demand equilibrium in Xiamen, we present reasonable countermeasures against Xiamen's current water policies. The water diversion quantity should be increased to promote unified allocation, optimal operation, and comprehensive utilization. It is also necessary to consider the role of ecological restoration in improving the overall water quality in the reservoir ecosystem. Reclaimed water resources including seawater, rainwater, and recycling water should be considered to reduce strain to Xiamen's water resources.

To alleviate strain on its internal water sources, Xiamen should import water from the Bei Stream Diversion Project, thus allowing diverted water to flow in volumes dictated by the region's actual demand. Xiamen should also depend on local water resources such as the Shidou–Bantou reservoir to minimize water transfer from external sources. The first phase of Longjin-Fangyang Water Control Project should be initiated for use in 2010. Long-term strategies include enhancement of the networking water supply of the Xiamen and Tongan subsystems. In 2015, the second phase of Fangyang Water Control Project should be executed, while construction of the Lianhua Reservoir is considered based on water resources supplies and related demand.

The water resource sustainable utilization planning includes two parts: (1) water supply project planning and (2) risk supply planning.

10.4.3.2.1 Water Supply Project Planning

In terms of the water supply project planning, Longji Stream Diversion Project and Linahua Reservoir will be constructed. The Longji Stream Diversion Project will involve two phases: (1) construction of river barrage and artificial water channels to allow a diversion discharge volume of 10 m^3/s and (2) construction of a reservoir in the upstream with a holding capacity in excess of 113 million m^3/s of water. The Lianhua Reservoir can be constructed with a total holding capacity of 49 million m^3/s of water to supplement the Tingxi Reservoir and meet the water demands of the Tongan district. Due to its huge costs, however, it will serve as a perspective control and will be constructed only following completion of the Longji Stream Diversion Project. In addition to these projects, Xiamen's seawater utilization increased to 900 million m^3/s in 2010, will increase to 1 billion m^3/s in 2015 and to 1.2 billion m^3/s in 2020. With the decreasing cost of producing desalinated water, it can be widely used as a future drinking water resource for Xiamen.

Urban wastewater reuse can be implemented at three levels: regional, community, and individual. At the regional level, advanced waste water treatment plant should be constructed, whereas at the community level, a water reuse project will be designed. On the individual level, we will encourage local companies and industries to construct their own waste water treatment equipment (Table 10.12).The rainwater resources utilization planning is provided in Table 10.13.

To summarize Table 10.14, we anticipate that seawater utilization, waste water reuse, and rainwater resource utilization will account for 9%, 13%, and 16% of the total water consumption by 2020.

10.4.3.2.2 Risk Water Supply Planning

In times of drought or water shortage, the lake reservoir has the ability to maintain Xiamen's water supply for up to 36 days in 2010 and 26 days in 2020. Some small and scattered water sources can be used to meet the Xiamen Island's emergency water supply for up to 17.5 days in 2010 and 12.5 days in 2020 (Xu et al. 2006). The

TABLE 10.12
Plan of Deep Sewage Treatment System (10,000 t/d)

Name	Deep Processing Scale	Construction Period	Utilization Method of Recycling Water	Remarks
Secondary sewage treatment factory	1	Recent	Water for municipal administration, domestic use, and landscape use	Centralized processing and recycling
Songyu sewage treatment plant	3	Recent	Water for municipal administration, domestic use, and landscape use	Centralized processing and recycling
Tongan sewage treatment plant	5	Medium term	Water for agricultural production, low-quality industry, municipal administration, domestic use, and landscape use	Centralized processing and recycling
Xinglin sewage treatment plant	3	Medium term	Water for municipal administration, domestic use, and landscape use	Centralized processing and recycling
Jimei sewage treatment plant	1	Long term	Water for municipal administration, domestic use, and landscape use	Centralized processing and recycling
Xike sewage treatment plant	2	Medium term	Water for municipal administration, domestic use, and landscape use	Centralized processing and recycling
Aotou sewage treatment plant	1	Long term	Water for agricultural production, municipal administration, domestic use, and landscape use	Centralized processing and recycling
Xiangan sewage treatment plant	5	Long term	Water for municipal administration, agricultural production, domestic use, and landscape use	Centralized processing and recycling
Sewage treatment system of some towns (including Gulangyu)	2	Gradually building	Water for agricultural production, municipal administration, domestic use, and landscape use	Decentralized processing and on-site recycling

TABLE 10.13
Project Plan for Rainfall Utilization

Region	Catchment Area (km²)			Collected Water Quantity (10,000 t/a)			Purpose
	2010	2015	2020	2010	2015	2020	
Bendao	1.5	3	5	42	84	140	Water for municipal administration and domestic use
Gulangyu	0.5	1	1	14	28	28	Water for municipal administration and domestic use
Jimei	0	0.5	1	0	14	28	Water for municipal administration, domestic use, and industrial use
Haicang	0	0.5	1	0	14	28	Water for municipal administration, domestic use, and industrial use
Tongan	0	0.5	1	0	14	28	Water for municipal administration and domestic use
Xiangan	0	0.5	1	0	14	28	Water for municipal administration and domestic use
Total	2	6	10	56	168	280	

TABLE 10.14
Reclaimed Water Planning of Each Target Year

	Seawater Utilization		Sewage Water Reuse			
Target Year	Direct	Desalinization	Recycled Water	Deep Treatment of Municipal Administration	Rainfall Utilization	Total
2010	3000	600	600	2920	56	7176
2015	3500	950	900	6935	168	12,453
2020	4000	1200	1500	9855	280	16,835

Shidou and the Bantou Reservoirs also have the capacity to provide water for up to 11 days in 2010 and 7 days in 2020. In order to improve the reliability of Xiamen's water supply, a cross-ocean channel should be constructed near Canghai Bridge. Additionally, we propose an increase in rainwater storage efforts and an upgrade to the seawater desalination equipment.

10.4.3.3 Land Utilization Planning

Xiamen's objective in the following 10 years is to become a center of technological development, port shipping and logistics, international finance, education, and culture. While we work toward these goals, we will strive to adjust and optimize our approach toward land resources utilization. Strict management efforts will be in place to protect the environment of the eastern island, and regions located outside of the island will be designated for development. In this way, a framework will be established for creation of a land utilization pattern.

10.4.3.3.1 Land Resources Sustainable Development Planning

Thirdly, due to the existing problems in Xiamen's land resources utilization, future land utilization planning will focus on function transformation. Large-scale exploration of the eastern island's land resources will be closely monitored with the eastern and western regions outside of the Xiamen Island designated as an important development area. The developmental framework of the eastern region should be gradually established in conjunction with its land use pattern to support the ecological environment of the whole city.

The new city will mandate planned and necessary construction practices as opposed to miscellaneous or petty construction. Construction within residential areas should mainly focus on middle income households, with newly built residential areas complete in infrastructure. According to our planning, Xiamen will consist of eight industrial districts designated as follows:

- Electronic and aviation industries: Northern Island
- Petrochemical industry: Southern Haicang district
- Electronic and machinery industries: Xinyang–Dongfu industry parks
- Machinery and chemical industries: Xingxi–Guankou industry parks
- Light and food industries: Xike industry park
- High-tech and light industries: Southern Tongan and Magang districts
- High-tech R&D manufacturing enterprises: Northern Jimei district

The water source protection zone, key cultivation land area, and ecological forest zone should also be involved in the land resources sustainable development plan.

10.4.3.3.2 Water and Soil Conservation Planning

Considering a small watershed as a single unit, an ecological demonstration area should be established. Measures taken to explore the new approach toward ecological environment reconstruction should include both the shelter forest system and slope farmlands and implementation of integral management planning for soil and water erosion. Adjustments to the land utilization structure over the next 15 years will increase the potential to guard against disasters and mitigate resulting damages, while soil erosion will account for only 2% of the total land mass.

There are three water and soil conservation regions in Xiamen. The first is the northwest mountain conservation region, where efforts should concentrate on protecting existent vegetation and constructing water conservation forests upstream.

The second is central plain control region, where the priority should involve taking various effective measures to conserve the soil and water resources. Considering the small watershed as a whole unit, protection against water and soil erosion is necessary to improve land productivity. The third region consists of the southern coastal regions and islands. In this area, the supervision of water and soil erosion should be intensified to improve the land resources protection efficiency and improve the coordinated development of economy and environment.

Dimensions for the total cultivated area should be fixed at 140 km². To limit the construction of a new industrial park will also help to implement the policy of "grain for green." These efforts will result in an increase of the forest area from 497 to 500 km², and by 2010 the total area of shelter forest was no less than 300 km² and will be 300 km² by 2015.

10.4.3.3.3   Green Space Security Planning

Implementation of scenic spot protection shall be engineered to protect the constituents of Xiamen's greenbelt such as the Gulangyu-Wanshshan State-level Scenic Resort, Lianhua National Forest Park, Beichenshan scenic spot, and so on. We will also form an urban green space system comprised of an urban protective greenbelt, productive plantation area, and road afforesting. The urban natural reserve will consist of forest parks, scenic spots, and green shelters. The regulative authorities should prioritize the protection of natural reserve and wildlife habitat to help maintain the biodiversity in the Xiamen Island.

10.4.4   Environmental Pollution Control Planning

The environmental pollution control planning efforts will aim to improve overall environmental quality through increasing the ecological efficiency in pollution treatment strategies and by reducing the discharge of pollutants. These planning efforts will focus on four categories of pollution control: water, atmosphere, solid waste, and noise.

10.4.4.1   Water Environment Pollution Control Planning

The current water quality status of Xiamen's resources is considered to be relatively poor. We can classify surface waters into three categories requiring them to be placed into special protection, key protection, and general protection, respectively. The regulative authorities should implement corresponding measures to control pollutants derived from industrial, domestic, livestock and poultry, nonpoint source, and mariculture sources. A strategy involving the treatment of urban and rural sewage treatment, nonpoint source pollution control, key functional area control, and water environment comprehensive treatment will also be formulated. Specifically, the water quality of drinking water supplies will be protected through the prohibition of industrial pollution emission sources near drinking water source protection areas. The factories located within drinking water source protection areas will be relocated outside of these zones. In terms of the main city areas, the urban sewage collection system will be improved as we enhance the construction of large wastewater treatment plants. Outlying villages

TABLE 10.15

Key Towns Sewage Treatment System Planning

No.	Towns	Treating Method	Treating Scale (10,000 t/d)	Usage
1	Lianhua	Land ecological treatment	0.22	Greening and irrigation
2	Tingxi	Land ecological treatment	0.35	Greening and irrigation
3	Xinwei	Land ecological treatment	1.1	Greening and irrigation
4	Neicuo	Land ecological treatment	0.9	Greening and irrigation
5	Dadeng	Land ecological treatment	1.1	Greening and irrigation

found to severely threaten the drinking water sources will be encouraged to incorporate ecological management techniques allowing them to minimize water pollution.

Green agricultural practices, such as ecological and pollution-free agriculture, will be advocated. The percentage of green and ecological farmland increased to 30% in 2010, will increase to 50% in 2015 and 70% in 2020. Specifically, strategies will involve adjusting the fertilization structure; promoting water-saving irrigation technology; and constructing features such as artificial wetlands, stabilization ponds, and vegetated buffer zones. The lengths of the vegetated buffer zones will increase from 10 to 20 km. To control and manage urban nonpoint source pollution sources, engineering measures will include vegetation control, filtration system construction, and the design of the sewers and wetland retention system. Additionally, regulations shall prohibit aquaculture activities in sensitive marine zone while reducing aquaculture density, improving feed efficiencies, and developing a mode of three-dimensional ecological farming.

A main priority will involve the gradual upgrade of the sewage collection network, thus enhancing its ability to provide advanced treatment and reclamation. The urban sewage treatment rates for each target year are 70%, 85%, and 95%, while corresponding processing powers are 157.29, 230.37, and 309.05 million tons annually.

Outlying villages and towns are the location for many small lakes, irrigation canals, and ditches. These features can be rebuilt into stabilization pools and manmade wetlands. Additionally, the ecological treatment efficiencies for rural domestic sewage will reach 20%, 45%, and 60% in each target year. Table 10.15 shows the planning approach for sewage treatment in Xiamen's key towns.

According to ecological function regionalization, it is prohibited to discharge polluted water into the Bantou, Shangli, or Zhuba Reservoir, as well as other upstream reservoirs. Measures will be taken to decrease the water contamination. The detailed measures and reduction amounts are provided in Table 10.16.

10.4.4.2 Atmospheric Environment Pollution Control Planning

The air quality of Xiamen is fairly good, ranking among the highest of key cities in the country. In the light of analyzing the ECC and air quality functional regionalization, we were able to approach planning from macrolevel, medium level, and microlevel.

At the macrolevel, we should implement our sustainable energy strategy and implement a clean production strategy. This will involve optimization of the energy supply structure and consumption structure, to include the use of LNG as opposed to fossil

TABLE 10.16
Main Surface Water Pollution Controlling Strategies

| Water Body | Cut-Down Amount (t/a) | | Scheme | |
	COD	Ammonia Nitrogen	2004–2007	2007–2010
Bantou Reservoir	23.37	4.45	Control large-scale livestock breeding industry	Control pollution emissions from rural domestic emission and agricultural runoff
Shangli Reservoir	35.59	7.12	Remove and close high-polluting enterprises, control large-scale livestock breeding industry	
Zhuba Reservoir	109.5	21.90	Control large-scale livestock breeding industry	
Upstream of Xixiyingbian bridge and Tingxiguoxi bridge	1188.49	221.52	Control large-scale livestock breeding industry and urban domestic emissions	Control pollution emissions from rural domestic emission and agricultural runoff
Hubian Reservoir	728.98	84.62	Central treatment of sewage from rural areas and large-scale livestock breeding industry	Control pollution emissions from rural domestic emission and agricultural runoff
Xinglin Bay	5573.71	920.54	Control the sewage from industry, rural areas, and large-scale livestock breeding industry	Control pollution emissions from rural domestic emission and agricultural runoff
Downstream of Tongwanxixiyingbian bridge	2199.92	355.86	Promote large-scale breeding and ecological breeding, centralize treatment of sewage from rural and large-scale livestock breeding industry	

fuel, and exploring terrain in Xiamen's northwest region for hydroelectric resources. A clean production strategy will be executed on multiple levels. Through source reduction and process control, material and energy utilization efficiency will be improved. The regulatory authorities should incorporate their environmental management techniques with clean production to assure that clean production methods are rolled out.

At the medium level, control policy and control of major pollution sources will be fully implemented. Without having a discernable boundary, atmospheric pollution diffusion requires a total balance control strategy and optimization of the industrial layout and structure. Emphasis will be placed on developing energy savings and environmentally friendly industry.

At the microlevel, we will work to increase pollution control investment. Through upgrades made to production equipment, adjustments to the product structure and implementation of clean production technology energy use efficiency can be improved while and reductions are made to industrial pollution emissions. Specifically, we will improve motor vehicle emission standards and work to minimize both production and sales of high-emission vehicles. Simultaneously, we will work to promote the development of low-emission cars. By these means, Xiamen's urban air quality will come up to excellent standards and will continue to improve annually.

10.4.4.3 Solid Waste Pollution Control Planning

Based on the analysis of Xiamen's current solid waste conditions, we plan to propose countermeasures for the management of domestic, ordinary industrial solid, and hazardous waste. In terms of dealing with industrial solid wastes, the following measures should be considered: explore new ways of safe disposal and industrial zone disposal; urban industrial parks can reuse their own slag and coal ash; in the long run, industrial solid waste use and disposal center should be constructed to absorb much of the industrial solid waste. In terms of hazardous wastes, by 2005 a regional medical waste processing center was constructed. By 2010, all medical wastes are treated safely and the construction of Xiamen's hazardous wastes disposal center was completed. By 2020, a city-level hazardous wastes disposal base will have been constructed to where all hazardous wastes can be safely treated and disposed.

10.4.4.4 Noise Pollution Control Planning

According to the status and trend analysis of Xiamen's sound environment, establishing a law-supporting system for noise control is essential to sound environmental planning. The planning should consider four aspects: transportation, social activities, building operation, and industrial production. Specifically, cars will be prohibited to blow their horns in sensitive areas. Commercial streets will be made into one-way and walking streets to reduce traffic. We will eliminate and upgrade the old cars in excess of established noise criteria. Roadwork projects, to include alterations and extensions to existing roads as well as new roads, should adopt reasonable noise reduction technology and apply a low noise material. Simultaneously, these road projects will enhance road greenbelt construction. The rate of road green land will exceed 15%, and the rate of newly built green space will meet planting requirements established for urban roads.

The authorities will strictly regulated construction activities within quiet residential areas, and noise control will be focused on in both the catering and entertainment industries. For some sensitive regions, construction will be prohibited at night. Regulatory authorities will strictly implement construction permits to control construction, and construction projects may also be prohibited in sensitive regions.

10.4.5 Ocean Ecological Construction and Environment Protection Planning

For the purpose of establishing a healthy and live marine ecosystem, Xiamen's oceanic ecological planning will focus on coastal tourism industry development, marine environmental protection, and the restoration of destroyed marine ecosystem.

Based on the advantage of its geographical location and resource characteristics, Xiamen should be positioned as a tourism center and also a coastal tourist city. In the future, Xiamen will be transformed from a sight-seeing tourism city into a multifunctional city supplying travel, relaxation, tours, holiday, and business. The awareness of environmental protection should be also be strengthened in the development of the tourist resources. The coastal tourism industry planning efforts will emphasize the protection of our natural environment as well as cultural and historic heritage. Promoting the ecological and cultural tourism industries can help Xiamen to balance relations between economic development and environmental protection, also helping to achieve sustainable development in the long run. Xiamen has rich tourism resources allowing for the integration of culture, tourism, and other related industries. We will strive to promote Xiamen's tourism industrialization.

Xiamen's sea pollution mainly comes from the land, so initial efforts will focus on controlling the emissions from land sources. A large amount of these pollutants come from nonpoint sources; therefore, apart from the infrastructure construction, we aim to adjust the industry structure and improve the existing marine functional regionalization. Meanwhile, focus will also be given to the protection of marine biodiversity in Xiamen. The marine ecosystem of Xiamen is now adversely affected by economic development. As such, the government should propose resolutions to include the adjustment of industrial structures and ecological restoration in impaired marine ecosystems such as the western sea area, Majian Bay, and the coastal wetlands.

10.4.5.1 Ocean Resources Development and Utilization Planning

Leisure fishery had become a highlight of marine tourism as well as a new industry of island economy in Xiamen. As such, focus should shift from traditional aquaculture to leisure fishery. The key area of Xiamen's marine industry is tourism and port shipping. Efforts will be made to foster the marine high-tech industry and optimize the coastal industrial base. The following measures should be considered for the development of the coastal tourism industry:

1. Establish business ideas, promote the integration of tourism industry, and develop a number of powerful enterprises
2. Highlight the development of leisure and ecological tourism products
3. Pay attention to the development of Jinmen City
4. Construct ecological protection park; for example, wetland park and urban park

The Xiamen port can be regarded as the shipping center for the Fujian Province, South Jiangxi, and East Guangdong. We should integrate Xiamen's port shipping resources and continue to strengthen its infrastructure development. The fishery industry of Xiamen is high quality, and it is also a fry cultivation and supply base. We should improve the aquaculture fry industry, promote the intensive land-based, pollution-free aquaculture industry, and optimize the aquaculture structure while vigorously developing the leisure fishery. We will strive to strengthen the establishment of the ocean technology innovation system, the construction of ocean high-tech industry parks, and the implementation of high-tech marine projects.

10.4.5.2   Ocean Environmental Protection Planning

10.4.5.2.1   Marine Pollution Control

Xiamen's ocean pollutants are mainly derived from nonpoint source pollutants on land. Therefore, apart from some basic measures such as infrastructure construction, the industry structure should be adjusted, allowing for the formulation of large-scale, offshore functional divisions and plan comprehensive marine environment protection. In the eastern ocean area, it is important to reduce the pollutants migrating with farmland runoff. Control measures such as the reuse of livestock and poultry manure, the mode of ecological farming, and organic fertilizer projects should be considered. In the estuary areas, we should stress the control of rural domestic refuse while improving the water supply and drainage systems in rural areas. In the western sea area, we should control the pollution sources in the cities and the tail water from the urban sewage treatment plants. For the southern marine area, the regulatory authorities should collect and dispose of urban domestic sewage, while intercepting polluted water from Xiagangbifengwu. Through the comprehensive coordination of major issues with the concept of ecological compensation, the concept of basin comprehensive renovation should be implemented so as to protect the ecological environment and achieve the sustainable development of water resources in the JiuLongJiang basin.

10.4.5.2.2   Islands Exploitation and Protection

There are 17 uninhabited islands and 27 key rocks scattered in Xiamen's sea area. Xiamen will work toward a policy of "putting prevention first, while enhancing management and minimizing exploitation." This means that we will establish special marine natural reserves, protect the cultural heritage on the island, and plant more trees to protect and rebuild the island landscape. The authorities will also implement coastline erosion protection and implement island vegetation restoration.

10.4.5.2.3   Marine Biodiversity Protection

There is a concern that the ecological restoration of Xiamen's natural reserves is not an optimistic goal. The following measures should be considered to improve the chances of restoration:

1. Promote public participation and natural reserve management.
2. Reform the management system while intensifying coordination ability.
3. Improve the existent protection legislation.
4. Intensify publicity and education while improving scientific monitoring.
5. Formulate policies regarding sustainable use of natural resources.

In order to protect Xiamen's marine biology resources, strict measures should be taken to limit the constructions of coastal engineering projects. We will work to adjust the fishery structure and control fishing intensities while adopting a logical strategy for fishing. It is also important to release fish fries into the sea, which can help to restore the damaged marine ecosystem more effectively. Xiamen is vulnerable to species invasion such as *Spartina alterniflora* Loisel invasion, and *Mytilopsis sallei* Reeluz invasion. Some guidelines should be issued concerning activities that could lead to the introduction of invasive species. Establishing risk assessments and early warning mechanisms are also very necessary.

10.4.5.3 Damaged Marine Ecological System Restoration and Construction Planning

Maluan Bay is the most seriously damaged area in Xiamen's marine ecological system. In this bay, the blind expansion of aquaculture and intensive breeding practices has led to significant organic pollution. Our objective is to support the development of the Maluan region, restore the ecological service function required by its regional development, and maintain the health of the marine ecosystem.

The main priorities of coastal wetland restoration and construction efforts include constructing coastal wetland reserve, prohibiting illegal reclamation, conserving the region's biodiversity, and reducing the high-concentrated pollutants emission (see Table 10.17).

The detailed plan for restoration includes two parts: habitat and marine organism restoration. We should implement the mud-cleaning project, forbidden "fishing project" and "source cut & pollution control project." In terms of marine organism restoration, we should regulate exotic species, construct large-scale mangrove plantation, and restore the structure and function of the aquatic ecosystem.

10.4.6 ECOLOGICAL HUMAN SETTLEMENT PLANNING

Based on the analysis of Xiamen's population situation, social status, and human living conditions, its infrastructure should be upgraded and constantly optimized to allow for comprehensive service capabilities. It is important to promote the

TABLE 10.17
Coastal Wetland Ecosystem Restoration Indicators

Restoration Index (unit)	Status Value	Target Value		
		2010	2015	2020
Restoration of damaged mangrove (hm²)	4.5	>8	>9	>10
Tide volume of semiclosed gulf (×10⁸ m³)	5.5	6.5	7.2	7.6
Water high-quality rate of bathing place (%)	46.3	60	70	80

TABLE 10.18

Maluan Bay Ecosystem Restoration Indicator

Restoration Rate (unit)	Status Value	Target Value	
		2010	2020
Sea area (km²)	4.5	>8	>10
Water quality	IV or V	III	III
Phytoplankton (×10³ cell/L)	1620 eutrophic	<800	<500
Bentonic organism biodiversity index H'	2.87	>3	>3
Sediment sulfide (mg/kg)	>1000	<600	<300

importance of creating environmentally elegant towns and building environmentally friendly residential areas. Xiamen's population forecast is based on the method of trend extrapolation while excluding some restrictive conditions (such as ecological ECC and economical development). The results are given in Table 10.18.

10.4.6.1 Optimize Urban Public Utility Services

The urban public utility services contain cultural centers, health care facilities, leisure and entertainment facilities, and commercial/financial facilities. The focus of our planning is to improve the medical security network while constructing some leisure areas to enhance efficiency and convenience for the people. We can optimize the urban service structure at different regional levels: city, district, and community. At the city level, we aim to construct the Yundang, Jimei, and Haicang central areas. According to their regional positions and geographic features, the public service facilities should be designed in a highly centralized pattern, thus promoting its commercial operational efficiency and attracting more customers.

10.4.6.2 Infrastructure Construction and Environmental Pollution Control

An effective public transportation system should be developed. This system will improve traffic conditions and minimize commuting time. To accomplish this, we should introduce more bus-only lanes to encourage more people to use public transport during rush hours. We propose the improvement of bus speed by 10% and will control the bus service radius within 500 m. The total number of buses should be increased, and by 2020 there will be more than 18 buses for every 10,000 people. To support this growth, we will construct the bus routes and railway stations simultaneously. Recently, we learned that we can connect the Xiamen Island, Jimei, Xinglin, and the Haicang district by subway. In the future, Gulangyu and Tongan will also be added to the system. In order to reduce the inflow of traffic onto the island, we propose the construction of large-scale subway stations and bus terminals across the Xiamen Island. Environmental sanitation infrastructure also plays an important role in urban construction. We propose the sorting of rubbish to improve the garbage classification, collection, transportation, warehouse, utilization, and disposal systems.

It is predicted that the rate of recyclable rubbish will increase in the future, with 35% of rubbish classified as recyclable by 2020. Along the main roads of populated areas, waste disposal containers should be positioned every 100 m. We propose four types of public disposal containers: one for toxic (such as electronic) waste, one for degradable organic waste, one for recyclable garbage (such as paper, metal, and plastic), and the last for hard-degradation waste. In the future, garbage recovery rates will reach 100%. Through the adoption of integrated waste treatment methods such as waste comprehensive classification screening, waste-to-energy power generation technology, and sanitary land fill technology, we can optimize Xiamen's production of solid wastes.

10.4.6.3 Construction of Beautiful and Environmentally Friendly Towns

The precondition of eco-city construction requires that 80% of the county attain the indicators for building ecological counties. Similarly, the precondition of building an ecological county requires that 80% of the towns attain the indicators for constructing beautiful, environmentally friendly towns. To model this, a central town, such as Guankou, Dadeng, Xinwei, and Dongfu, should be developed so as to highlight the characteristics of that particular town. The concrete process for constructing Xiamen's ecological towns is provided in Figure 10.8.

10.4.6.4 Construction of a Safe and Environmentally Friendly Ecological Community

Through the adoption of sound technology and policy measures, we will reasonably improve the residential sector's comfort, thereby improving overall living conditions and the quality of life. To accomplish this, we must first implement the reuse of

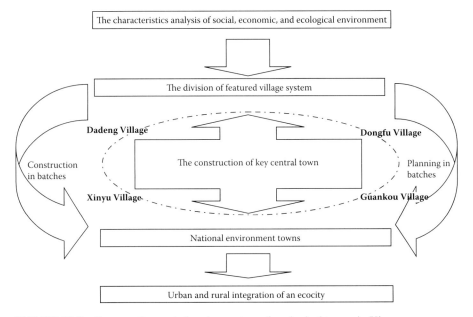

FIGURE 10.8   Construction and planning system of ecological towns in Xiamen.

residential waste and then largely reduce the impacts of anthropogenic environmental pollution and destruction.

A portion of the island's population should be transplanted to the mainland to help ease the island's population density and also expand the scale of economic development outside of the island. One benefit of this approach is the acceleration of infrastructure construction. Additionally, the mainland area will also benefit in that we can build ecological residential sectors outside of the Xiamen Island. In the future, this approach will cause the population density of the island to decline by 10,000 people/km², and outside the island, the density will increase to 7,500 people/km². In this way, Xiamen will transform from an island-type to harbor-type city. The construction of the ecological residential area should follow three main principles. First, the buildings should be designed to be energy efficient and make use of renewable resources such as solar and wind energy. Secondly, measures should be implemented to conserve water resources and dispose of domestic wastes. Thirdly, we will "green" our buildings by roof, wall, and outdoor planting.

10.4.7 Ecological Culture System Construction

Presently, Xiamen's traditional and distinguishing culture is in a decline. Publicity and educational concerns surrounding ecological improvements and environmental protection have not been widely communicated or embraced in rural areas. There are other problems, as well, such as the shortage of educational capital investments and the lack of local residents willing to participate. We should adopt the mode of establishing green production and consumption while increasing public awareness surrounding environmental protection. Based on the concept of circular economy, the overconsumption of resources, and heavy pollution, unsustainable methods of production should be altered to make way for a resource saving and environmentally friendly consumption structure. Xiamen's eco-city construction will make great progress through initiating environmental protection activities, advocating ecologically friendly lifestyles, promoting the education for sustainable development, building an interaction platform, impelling public participation, and building a harmonious society together.

10.4.7.1 Green Production and Green Consumption

10.4.7.1.1 Industry Culture

Based on the essence of traditional agriculture, modern science, and technological achievements, it is possible to cultivate awareness of ecological agriculture by means of introducing an efficient agricultural producing model and establishing sound ecological agriculture bases. We will foster the construction of an industrial ecological culture through strengthening of technical reformation, spreading clean production policy, adjusting industrial structure, and extending the chains of environmental industry. In the process of constructing industrial parks, we will focus on the development of Xiamen's pillar industries to include the electronic, machinery, and chemical industries. Based on the existing industrial development zones, Xiamen will take advantage of its own characteristics to construct a new industrial park able to achieve the target goal of "resource sharing, unambiguous responsibility, and combined operations, where all sides benefit and can commonly develop."

10.4.7.1.2 Community Culture

To begin with, the regulatory authorities will encourage citizens to select green products for their properties, which promote safety, energy savings, sanitation, minimal pollutants, and convenience. Secondly, consumption activities should be revised to produce less pollution and protect the environment. Thirdly, the existing consumption concepts should be changed, and we should save our energy resources to achieve sustainable development.

To support the concepts and characteristics of ecological consumption, Xiamen's residents should choose "green clothes" and optimize their everyday diet structure. The construction of ecological residential areas will be actively promoted. We will develop a three-dimensional traffic network including public transport and will encourage the use of environmentally friendly transportation such as clean energy cars and bikes. We will suggest that an ecotype consumption safeguard mechanism be established to prohibit the products of unhealthy consumption, such as cigarettes and drinks, via tax revenue. We will also set up a price system to internalize the costs of environment and resource conservation while improving the legal and administrative systems charged with managing ecological consumption step by step. The regulatory authorities launched a campaign to establish a "green community," and by 2010, 20% of the community met this definition, which will allow the number to rise to 80% in 2020.

10.4.7.2 Popularization of Ecological Environment Protection Knowledge

It is recommended that the government carry out various activities intended to propagate the knowledge of environment protection. The environmental education efforts reached 90% in 2010, should reach 95% in 2015 and 100% by 2020. These efforts will involve introducing a development course for agriculture and industry and showing how they can contribute toward an ecological civilization in a flexible and diversified way. These efforts must make the public understand that ecological civilization will be the inevitable choice for long-term health and human comfort. Large-scale publicity activities should be organized throughout the city every year on pertinent days of recognition such as World Environment Day, Earth Day, and so on. To enhance the population's personal initiative and sense of responsibility, these events will include advertising design, signature activities, and some thematic song and dance shows. Guided by the principles of reduction, recycling, and safe treatment, we will introduce the concepts of ecological environmental protection and technological development. Other topics may convey the importance and necessity to evolve from upscale consumption toward ecological consumption, the categories of environmentally friendly products, the significance of green products in the concept of environmental protection, food safety, and the differences and categories of safe food.

It is also important to intensify the level of environmental publicity and education. Policy makers must be taught of the importance and necessity of strategic environmental assessments. Additionally, a supervisory institution and expert panel responsible for major decisions should be established. These government officials should actively take part in symposiums where they can learn and then share their experiences pertaining to environmental protection. They should subscribe to the environmental protection newspapers and study new findings in environmental research regularly. In their daily work, these officials should set a good example and be thrifty with the consumption of resources in their daily life. With regards to entrepreneurs, we will propagate the environmental law, ISO14000, for environmental management and ecological industry design. It will also be necessary to issue incentive policies to reward enterprises that distinguish themselves in the area of environmental protection. The public will receive knowledge on environment laws regularly. Before taking up their posts, Xiamen workers should be trained on the pollutions and environmental hazards that can be produced as a result of their work. We can also popularize the concept of environment protection via mass media such as TV, newspaper, and so on.

Finally, green schools should be established in Xiamen to improve the effectiveness of public participation. Xiamen should adopt schooling as a focus in developing the environmental education base and launch a campaign to promote "green school" construction. By 2010, the percentage of green schools in Xiamen reached 50%, and will reach 80% in 2015 and 100% by 2020. Similar campaigns involving the construction of green hotels, restaurants, and houses should also be encouraged.

10.4.7.3 Improving Public Participation

To arouse public awareness surrounding the need for environmental protection, and to educate citizens on how to protect the environment, regulatory authorities should take steps to improve the effectiveness of public participation. Public participation will occur in diverse forms to include public gathering, seminars, legal notices, public polls, and written comments. It is recommended that different public participation methods be adopted according to participants' different backgrounds (see Tables 10.19 and 10.20).

TABLE 10.19

Population Size Prediction for Xiamen's Districts (Unit: Per 10,000 people)

District	Status Value	2010	2015	2020
Siming	46.88	55.74/33.44	62.15/34.18	68.61/34.31
Huli	13.88	16.50/9.90	18.40/10.12	20.31/10.16
Haicang	8.66	10.30/6.18	11.48/6.31	12.67/6.34
Jimei	17.15	20.39/12.23	22.74/12.50	25.10/12.55
Tongan	29.40	34.96/20.97	38.98/21.44	43.03/21.52
Xiangan	25.80	30.68/18.41	34.20/18.81	37.76/18.88

Note: Status value represents registered permanent residence in Xiamen. In each planning column, the former represents the registered population while the latter represents transient-floating population. Based on Xiamen's comprehensive planning, we take the natural growth rate of the registered population as 5%, and the mechanical growth rate of the registered population as 20% in 2010, 17% in 2015, and 15% in 2020. The mechanical growth rate of transient-floating population was 60% in 2010, will be 55% in 2015 and 50% in 2020.

TABLE 10.20
Types of Public Participation and Qualified Participants

Types of Participants	Public Gathering	Seminar	Law Announcement	Information Announcement Meeting	Citizen Consultancy	People Will Test	Seek Opinion on the Spot	Written Comment
Well educated			●					●
Ill educated	●	●			●		●	
Staff	●			●		●	●	
Mid-manager		●	●	●	●	●	●	●
Manager			●			●		
Job loser			●	●	●	●	●	
Highly involved participated	●	●	●	●	●	●	●	●
Multiple identity	◇		◇	○	○		◇	○

Note: ● high, ○ medium, ◇ low.

10.5 PLANNING SCHEME AND BENEFITS EVALUATION OF XIAMEN'S URBAN ECOLOGICAL PLANNING

10.5.1 PLANNING SCHEME

Our planning scheme is comprised of five parts: the ecological landscape and security pattern system, the ecological industry and circular economy system, the resources guarantee and efficient utilization system, the environmental improvement and pollution control system, and the ecological culture and human settlement system. The project of eco-city construction is the foundation for each of the planning measures and is comprised of three parts. To date, the project is proceeding as planned. We will attempt to spend 15 years making Xiamen into a gulf-type ecological city with a prosperous ecological economy, a thriving ecological culture, and a beautiful ecological environment (see Figure 10.9).

There are a total of 132 ecological construction projects in Xiamen's ecological planning strategy. These projects span ecological industry, resources security, urban

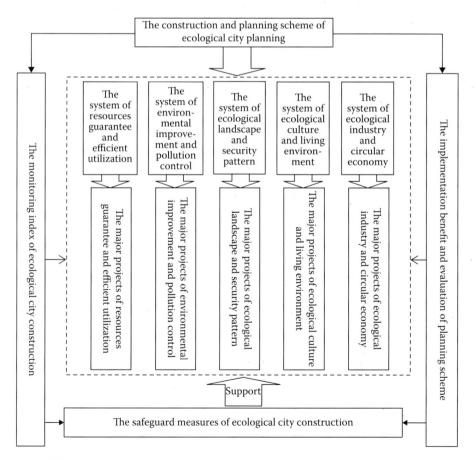

FIGURE 10.9 Xiamen eco-city construction planning system diagram.

eco-residential construction, and urban ecological culture planning. These planning efforts include five types of engineering strategies: blue sky, green water, green, clean, and peace engineering. The total investment needed for the project amounts is 84 billion yuan.

The individual construction projects can be classified as three types. One type is currently under construction or already built up, meaning that in 2003 the project has already been completed or is currently under construction. A second type is a planned project, meaning that the project is scheduled for construction at some point in the national "11th Five-Year Plan." The third type of project has been designed in this plan, with specific statistics given below.

According to its current status, Xiamen's existing planning effort has already resulted in the design and drafting of a few plans and projects. However, this existing effort has focused only on severe problems that have arisen recently and can only be expected to produce short-term and/or regional results. Based on a deep analysis of the restrictive factors facing Xiamen's development, our proposed eco-city construction provides five systems and many specific assessment indicators, the use of which will produce global benefits in addition to Xiamen's continuous development.

10.5.2 BENEFITS EVALUATION OF XIAMEN'S URBAN ECOLOGICAL PLANNING

The effect of Xiamen's ecological planning can be expressed by its ecosystem service value. Through the assessment of this ecosystem service value, we can provide scientific and rational methods for an urban ecological planning assessment. Apart from having a natural value, the urban ecosystem service value will also include social and economic value. We will be able to obtain the results of the plan's benefits, for the local environment, society, and economy. In our ecological landscape and security pattern system, we recognize the need for the rational use of land resources, to increase forest coverage, to regulate the climate, and to preserve biodiversity. In an ecological industry and circular economy system, planning efforts will reduce overall material consumption, conserve energy resources, promote the economic development, and create more job opportunities.

In a resources guarantee and efficient utilization system, both air quality and the energy use efficiency can be improved. In an environmental improvement and pollution control system, we can expect to see improvements in environmental quality when implementing measures such as power plant desulphurization, the blue sky project, solid waste center construction, and so on. In an ecological culture and human settlement system, more ecological service values will be produced through the construction of ecological demonstration residential areas, green government, green schools, and ecological culture projects.

The environment benefits will increase slowly over time, while the economic benefit will increase more quickly. Meanwhile, Xiamen's ecosystem service value will also rise gradually. The input will become smaller and smaller; however, the output will be larger and larger. In the early stage, benefit effects of the plan may not seem notable; however, with the plan's further implementation, resulting economic improvements will be tangible as profit margins continue to increase.

10.6   GUARANTEE MEASURES OF XIAMEN'S URBAN ECOLOGICAL PLANNING

The construction of an ecological city is a long and difficult task; as such, an effective and long-term management plan is necessary to ensure the smooth and ongoing implementation of our plan.

First, we should strengthen the leadership of different organizations and organize in the comprehensive coordination to address major issues. We will incorporate eco-city construction into the chief executive target responsibility system. The division of our vision into several achievable targets and goals will allow our progress to be routinely assessed and documented.

Secondly, we should enhance the information supporting system from the perspective of an innovative administrative mechanism. The government will give priority to the key projects and items pertinent to the eco-city construction. Measures should also be taken to improve the economic compensation system for natural resources utilization. An ecological compensation system will also be established. In order to form a dynamic feedback process between programming and implementation, we will develop indicators to establish the eco-city supervision indicator system.

Finally, we will encourage the expansion of external economic collaboration to spread advanced technology. We will establish a "green" national accounting system, which takes the environment and natural resources into account. We will accelerate the formation of the framework for multi-investment opportunities, to include personal individual proprietorship, joint venture, financing cooperation, Build-Operate-Transfer (BOT), and so on. We will explore the method of mortgage loan, which leaves franchise and waste water treatment toll rights, and commits to the access of forest, mineral, and marine resources as a pledge. We will formulate ecological industry and environmentally friendly standards, seeing that ecological agriculture be measured by these standards. We will improve the existing incentive mechanisms to encourage venture investments. Finally, we will partner with experts to work in Xiamen and will establish long-term cooperation with foreign research institutes and colleges.

REFERENCES

Lu G. X., Yang Z. F., and He M. C., 2004, Urban space characteristics of self-organizing of Xiamen, *Journal of Beijing Normal University (Natural Science)*. 40(6): 825–830 (in Chinese with English abstract).
Su M. R., Xu L. Y., and Yang Z. F., 2006, Study on the urban eco-economic environmental management system, *Journal of Safety and Environment*. 6(4): 42–45 (in Chinese with English abstract).
Su M. R., Yang Z. F., Wang H. R., and Yang X. H., 2006, A kind of method and its application for urban ecosystem health assessment, *Acta Scientiae Circumstantiae*. 26(12): 2072–2080 (in Chinese with English abstract).
Tian G. J., 2008, Urban spatial-temporal dynamic pattern in Xiamen multi-center metropolitan area, *Tropical Geography*. 28(5): 423–427 (in Chinese with English abstract).
Xiamen statistical yearbook, 2001, Beijing: China statistics press (in Chinese).
Xiamen statistical yearbook, 2002, Beijing: China statistics press (in Chinese).
Xiamen statistical yearbook, 2003, Beijing: China statistics press (in Chinese).
Xiamen statistical yearbook, 2004, Beijing: China statistics press (in Chinese).

Xu L. Y., Yang Z. F., Shuai L., Yu J. S., Liu S. L., 2006, Eco-compensation of reservoir project based on ecosystem service function value, *China Population, Resources and Environment.* 16(4): 125–128 (in Chinese with English abstract).

Zhao Q., Yang Z. F., Zhang L. P., Chen Z. T., 2007, Eco-city characteristics index discrimination and its application, *Journal of Safety and Environment.* 7(2): 86–90 (in Chinese with English abstract).

Zhao Q., Zhang L. P., Chen Z. T., 2009, Studies on indicator system of eco-city: A case of Xiamen City, *Marine Environmental Science.* 28(1): 93–112 (in Chinese with English abstract).

Zhao W., Yang Z. F., Li W., 2005, Optimization of energy structure in eco-urban construction, *Journal of Safety and Environment.* 5(6): 54–59 (in Chinese with English abstract).

11 Eco-City Baotou Plan

Yan Zhang, Yanwei Zhang, Meirong Su,
Jiansu Mao, Gengyuan Liu, and Zhifeng Yang

CONTENTS

Situated in the western part of the Inner Mongolia Autonomous Region of China, Baotou is the largest industrial city in this region. It is an important industrial base for metallurgy, machinery, chemical production, and energy generation and enjoys world fame as the capital of rare earth. At present, the industrialization of Baotou City is still at a low level. However, in the future, the city will face huge environmental pressure and challenges arising from the reconstruction of old industrial bases and the treatment of pollutants. These include grassland destruction and extinction of grass species, serious damage to forests and vegetation, a decrease in forest coverage, and serious eco-environmental destruction due to water erosion, wind erosion, soil and water losses, desertification, and salinization. Therefore, the government has proposed that the development of Baotou should follow the eco-city model. This could make Baotou a model city in northern China for the harmonious coexistence of ecological protection and industrial development.

11.1 GENERAL REMARKS

11.1.1 BACKGROUND AND MEANING OF THE ECO-CITY PLAN

Baotou is located in the rich "golden triangle" zone of the great bend of the Yellow River in the central part of Inner Mongolia (Figure 11.1). The geographical coordinates are east longitudes 109° 15′ ~ 11° 26′ and north latitudes 40° 15′ ~ 42° 44′. It borders Togtoh County, Tumed Left Banner, and Wuchuan County of Hohhot in the east, faces the Dalad Banner and Jungar Banner of the Alliance of Yikezhao

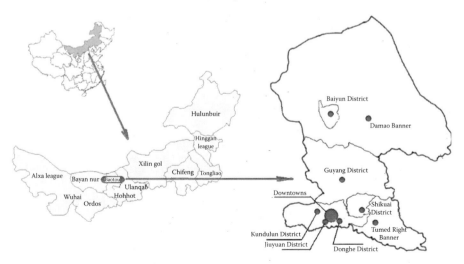

FIGURE 11.1   **(See color insert.)** Geographic location of Baotou.

across the Yellow River in the south, borders Urat Middle and Front Banners of the Alliance of Bayanzhuoer in the west, and borders Mongolia in the north. The maximum length of Baotou City is 270 km from north to south, and the maximum width is 180 km from east to west. The total area is 27,768 km^2 including 160.6 km^2 of built-up area. Baotou governs nine districts (or banners or counties) including Kundulun District, Qingshan District, Donghe District, Jiuyuan District, Tumed Right Banner, Guyang County, Shiguai District, Baiyun District, and Damao Banner. Among them, Kundulun, Qingshan, Donghe, and Jiuyuan Districts are downtown areas, while Shiguai and Baiyun Districts are mining areas, and Tumed Right Banner, Guyang County, and Damao Banner are agriculture and animal husbandry–centric districts.

Baotou was constructed in the early days of Chinese independence as an industrial city and occupies an important position in the national industrial economic system. With the new economic reforms and industrial open door polices introduced by the government, the position of Baotou as a regional core city of economic development has been strengthened further. However, in recent years, Baotou is facing a series of urban development problems affected by historical practices and ecological imbalance. For example, the infrastructure for environmental protection is lagging behind the high-intensity high-speed economic development that is taking place, the living environment is declining, the circular economy system of urban industrial enterprises is poorly set up, the urban aquatic ecology of Baotou is deteriorating, there is a water shortage phenomenon due to limited water resources and poor water quality coexists, energy utilization efficiency and the use of renewable energy sources is limited, the grasslands and hinterlands in Baotou are facing serious degradation, and the city has not formed a reasonable ecological landscape pattern. Therefore, seeking a scientific and rational development mode especially in the context of urban construction is a key issue during the urban development of Baotou. The eco-city model promotes ecological harmony and emphasizes efficient allocation of natural resources, at the same time leading to urban development; it is imperative to implement this model for the construction of Baotou.

11.1.2 IDEOLOGY GUIDING THE PLANNING

1. To achieve ecological sustainable development and to construct a harmonious Baotou living environment, the population, resources, and socioeconomic development of Baotou should be coordinated; the development between built-up area and ecological hinterland of Baotou be balanced; and the development of agriculture, primary, secondary, and tertiary industries be distributed on an even scale. The coexistence and continuous symbiosis between humans and nature should be guaranteed to realize "rich, civilized, ecological and safe Baotou."

2. The two key goals of Baotou's development are (1) constructing Baotou to be the most important urban area in China's northern and western regions and (2) constructing Baotou to be the core city in the Jing-Jin-Hu-Bao-Yin economic belt. To accomplish this, we should be guided by the scientific development concept, which considers innovation as the key driving force

and results in an economy that strengthens the city, a science and education infrastructure that revitalizes the city, environmentally protects the city, promotes the Baotou society to develop harmoniously, and on the whole, realizes the eco-city construction of Baotou. This will be the first demonstration of the eco-city model of construction in northwestern China and can act as a blueprint for the development of other cities, nationally and globally.

3. We should consider the following points while designing policies and guidelines for Baotou's development: harmony between humans and nature, symbiosis between urban construction and ecological environment, control of environmental pollution, ecological conservation and recovery for the comprehensive improvement of the urban environment, protection of agriculture and animal husbandry to maintain a balanced ecotone, recovery and regeneration of degraded grasslands, and construct a stable and safe urban landscape pattern relying on the water areas, wetlands, and greenbelt.

4. With the industrial ecology principle and circular economy theory as guidance, we should use cleaner production methods and operation processes that include industrial-ecological transformation to adjust the industrial structure and layout of Baotou. In turn, this will successfully construct a symbiotic system between industry and ecology, as well as the urban environment and ecology, enhance the efficiency of using energy and natural resources such as water, soil, and grasslands, and implement a comprehensive strategy of sustainable development in Baotou.

11.1.3   Thoughts behind the Plan

The eco-city plan for Baotou must be supported by the reasonable use of natural resources and energy and the optimization of environmental quality. Combined with the effective remediation of environmental pollution, this can lay the foundation for constructing an ecological industrial system, aquatic system, and grassland ecological system, thus resulting in an overall improvement of Baotou's living environment and realizing sustainable development.

11.1.4   Spatial and Temporal Range of the Plan

11.1.4.1   Temporal Range

The base year of planning was 2005, with the short-term, medium-term, and long-term planning phases as 2006–2010, 2011–2015, and 2016–2020 respectively. It is important to note that the time required to tackle major ecological environment problems should be extended or contracted appropriately.

11.1.4.2   Spatial Range

The plan encompasses two scales, namely administrative and downtown regions. The administrative regions are "general" planning areas and include all nine districts with a total area of 27,768 km². The downtown regions include Kundulun, Qingshan, Donghe, and Jiuyuan Districts and are the "key" planning regions.

The spatial planning scale is shown in Figure 11.2.

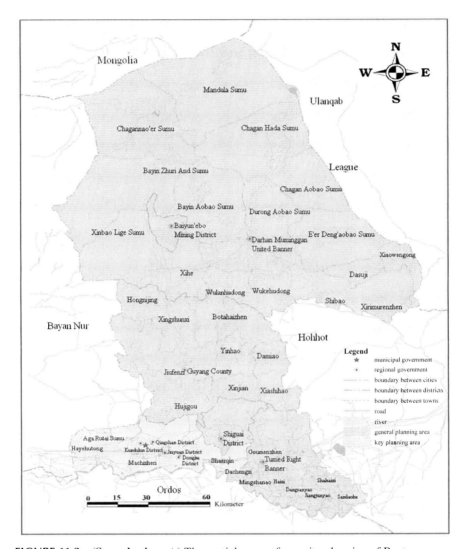

FIGURE 11.2   (See color insert.) The spatial range of eco-city planning of Baotou.

11.1.5   Technical Route of the Plan

The technical route of the eco-city plan for Baotou is shown in Figure 11.3.

11.2   ANALYSIS OF THE CURRENT STATUS OF THE URBAN ECOLOGICAL SYSTEM OF BAOTOU

11.2.1   Health Assessment of the Urban Ecological System of Baotou

A delineated picture of the current ecological health of Baotou's urban environment will help isolate the main problems that could arise during the eco-city

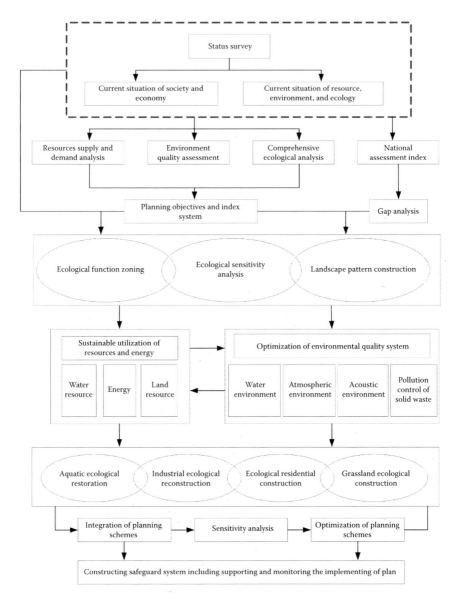

FIGURE 11.3   The technical route of eco-city planning for Baotou.

development process of Baotou and enable the design of a reasonable eco-city model. Although the standards for health assessment of an ecosystem are multiple and dynamic (Rapport, 1995), the following eight aspects are commonly used to indicate ecological health: vigor, organizational structure, resilience, function maintenance, management strategy, external inputs, influence on a neighboring ecological system, and human health. Of the eight basic standards, the first three are primarily applied to a natural ecosystem (Odum, 1983). However, to obtain a systematic understanding of the health of Baotou City's urban ecosystem (Ma and Wang, 1984), four assessment factors were chosen, including vigor (V), organizational structure (O), resilience (R), and function maintenance (F).

Three indicators for vigor included ratio of emergy investment, ratio of electricity to total emergy, and ratio of net emergy yield. Ratio of electricity to total emergy is used to describe the industrialization degree of an urban ecosystem, while ratio of net emergy yield can be regarded as an indicator of the environmental impacts of emergy consumption to estimate the depletion of emergy feedback. Next, nonrenewable emergy ratio, which describes the resource utilization structure, and energy exchange rate, which characterizes the coupling input and output structures of a commercial economy, were used as indicators of organizational structure. A third organizational structure indicator, per capita emergy usage, was used to determine the intensity of energy utilization. The input indicators for resilience included environmental load rate and population carrying capacity, which respectively represent the environmental and social imported pressure. The output indicator for resilience was the waste generation rate, which reflects the ability of the ecosystem to recycle waste and reduce the usage of nonrenewable resources. Thus, this indicator can quantify environmental potential and restoring power under specific conditions of environmental and population pressure. Finally, function maintenance of urban ecosystem services, which dominates the resource supply chain of the ecosystem, was determined by emergy self-sufficiency, ratio of emergy to money, and emergy density.

Data for environmental protection, industry, agriculture, import, and export during the 5 years of the study were collected from the standard annual statistical yearbooks compiled by the local government and sorted into the emergy flow table of the Baotou urban ecosystem (Baotou Statistical Office, 2001–2005). Furthermore, 12 dynamic emergy indicators, for example, emergy investment ratio, ratio of electricity to total emergy use, net emergy yield ratio (NEYR), and so on, were calculated to evaluate the urban ecosystem health (Table 11.1) (Brandt-Williams, 2001; Brown and Bardi, 2001).

11.2.1.1 Vigor

The rapid economic growth of Baotou depends heavily on the import of resources, with a high environmental price being paid for industrial growth. During 2000–2004, emergy investment ratio of Baotou declined, indicating the increasing competitive power (Table 11.1) with regard to economy. Meanwhile, the ratio of electricity to total emergy use increased, indicating that Baotou had enhanced new industrialization and the vitality of the urban ecosystem. NEYR followed a slow fluctuating upward trend, implying that the feedback process of urban socioeconomic systems can satisfy the energy and resource demands of the growing economic activities. Overall, the economic efficiency and vigor of Baotou's urban ecosystem increased during the study period.

11.2.1.2 Organizational Structure

Considering the inputs of the urban ecosystem, the nonrenewable emergy ratio increased from 2000 to 2004, showing that the energy structure of Baotou was unreasonable, which was caused by the dominated consumption of nonrenewable resources rather than alternative renewable sources. Emergy exchange ratio (EER) steadily

TABLE 11.1

Emergy Evaluation Table for Urban Ecosystem Health of Baotou (2000–2004)

Index	Expression	Unit	2000	2001	2002	2003	2004
Vigor							
Emergy investment ratio	IMP/(R + N)	—	4.16	4.34	3.49	3.75	3.26
Electricity to total energy use	Elc/U	—	1.09E−2	1.08E−2	1.12E−2	1.11E−2	4.85E−2
Net emergy yield ratio	(R + N + IMP)/IMP	—	1.24	1.23	1.29	1.27	1.31
Organizational Structure							
Nonrenewable emergy ratio	N/U	—	0.19	0.18	0.22	0.21	0.23
Emergy exchange ratio	IMP/EXP	—	107.28	101.24	88.03	75.00	60.30
Emergy per capita	U/pop	sej/p	1.32E+17	1.44E+17	1.52E+17	1.61E+17	1.72E+17
Resilience							
Environmental loading ratio	(U − R)/R	—	521.12	570.83	611.31	649.23	696.02
Population carrying capacity	8 × (R/U) × pop	—	31,305	28,842	27,178	25,755	24,130
Waste producing rate	W/U	—	4.07E−2	3.95E−2	4.10E−2	4.16E−2	6.84E−2
Function Maintenance							
Emergy self-sufficiency	(N + R − EXP)/U	%	18.29	17.63	21.18	19.84	22.17
Emergy money ratio	U/GDP	sej/$	9.49E+13	9.52E+13	7.62E+13	5.98E+13	4.75E+13
Emergy density	U/area	sej/m²	9.74E+12	1.07E+13	1.14E+13	1.21E+13	1.30E+13
Three Indices							
Emergy Sustainability Index (ESI)	NEYR/ELR	—	9.65E−3	8.74E−3	8.53E−3	7.91E−3	7.61E−3
Emergy Index of Sustainable Development (EISD)	NEYR × EER/ELR	—	1.04	0.88	0.75	0.59	0.46
Emergy-Based Urban Ecosystem Health Index (EUEHI)	(NEYR × EER × ED)/(ELR × EMR)	—	0.106	0.099	0.113	0.120	0.126

Note: IMP, imported emergy; U, total emergy use; R, renewable resource; N, nonrenewable resource; Elc, electricity; EXP, exported emergy; pop, population of the city; W, waste producing volume; NEYR, net emergy yield ratio; ELR, environmental loading ratio; EER, emergy exchange ratio; ED, emergy density; EMR, emergy money ratio.

declined, which illuminated an energy trade disadvantage of the urban ecosystem due to the local emergy loss. The emergy per capita, as the output that is used to decide the living standard of the resident by the possession of emergy use per capita, had increased from 1.32×10^{17} sej/pop in 2000 to 1.72×10^{17} sej/pop in 2004, manifesting that the studied site had been intensively explored with increasing economic activities and high emergy use per capita. Together, the three indicators of urban organizational structure showed that the urban ecosystem of Baotou is an unreasonable state.

11.2.1.3   Resilience

Like every industrialized city in the world, the Baotou ecosystem has already affected its surrounding environment in a deleterious manner. This effect increased during the 5 years of the study (Table 11.1), probably due to the import of large amounts of resources and the overexploitation of local resources. Meanwhile, the carrying capacity of the urban ecosystem decreased, despite an annual increase in the population. For example, according to emergy analysis, the population carrying capacity of Baotou in 2004 was only 11.5% of the total population (Table 11.1). With regard to the outputs from energy utilization and other industrial processes, the waste producing rate (WPR) of Baotou gradually increased from 6.84% in 2004 compared with 4.07% in 2000, suggesting that the ecological stress could be due to the accretion of waste gas emission.

11.2.1.4   Function Maintenance

Emergy self-sufficiency, which is an input indicator of function maintenance, showed an upward trend, albeit with some fluctuations, implying that Baotou relied on external trade, imported material, and energy flow; this promoted local economic development from 2000 to 2004. A second indicator, emergy to money ratio, dropped from 9.49×10^{13} sej/\$ in 2000 to 4.75×10^{13} sej/\$ in 2004, revealing that most of the resources required for the economic activities of Baotou were extracted from the natural environment. However, the rapidity of Baotou's urban expansion and industrialization has put great pressure on the environment and diminished its capacity as a biophysical support for the urban ecosystem. Meanwhile, a third indicator, emergy density, increased significantly, indicating that the Baotou socioeconomic and industrial region had been intensively explored and developed. On the whole, the three indicators of urban function maintenance showed that the well-being and economic growth of Baotou increased from 2000 to 2004, and this can be primarily attributed to commodity trade instead of the local consumer environment.

11.2.1.5   Urban Ecosystem Assessment Indices

Three different indices, Emergy Sustainability Index (ESI) (Brown and Ulgiati, 1997), Emergy Index for Sustainable Development (EISD), and Emergy-Based Urban Ecosystem Health Index (EUEHI), were employed to assess the urban ecosystem of Baotou (Table 11.1) (Zhang and Yang, 2007; Zhang et al., 2009). Our analysis showed that EISD and ESI were constantly on the decline from 2000 to 2004 while EUEHI was increasing. In fact, the NEYR, which is used to calculate each of these indices, decreased each year. This is because, with the economic development of Baotou, the growth of the total input emergy invested in the city was higher than that of the industrial emergy output. Therefore, the high consumption of nonrenewable fossil fuels and raw material as well as waste emissions applied significant pressure on the urban ecosystem and lowered sustainability. In addition, external trade of Baotou City led to higher emergy expenses compared to emergy gain, detrimentally affecting sustainability. In contrast, the health of Baotou's urban ecosystem steadily increased from 2000 to 2004, except in the year 2001. In spite of a relatively low competitive advantage and resilience (Table 11.1), improved resource structure, higher utilization efficiency, and better ecosystem services contributed to the improved health level of Baotou's urban ecosystem during the study period.

Taken together, despite high environment pressure and low ecosystem recovery capacity, the integrated Baotou urban ecosystem obtained a better organizational structure and ecosystem service function during 2000–2004, due to an emphasis on resource structure adjustment and utilization efficiency.

11.2.2 IDENTIFICATION OF COMPREHENSIVE PROBLEMS

Based on the analysis of Baotou's current urban ecological system in Section 11.2.1, we can see that the economy in Baotou has developed preferably and that the degree of health of the ecosystem has grown steadily. Given that this region has abundant minerals and biological resources, the spatial distribution and time-varying characteristics of these resources will determine the eco-city pattern in the future. However, there are still some problems that need to be addressed; these are shown in Table 11.2.

We can see from the preceding comprehensive analysis that environmental quality, utilization of resources and energy, ecological protection, and construction are the factors that should be considered in the urban development of Baotou. Planning should focus on the contents of these factors and the key human practices that affect each factor by carrying out focused research.

11.3 PLANNING OBJECTIVES AND INDEX SYSTEM

11.3.1 PLANNING OBJECTIVES

11.3.1.1 General Objectives

To construct the regional core of Baotou City, we should focus on sustainable and coordinated development between the social economy and the ecological environment, tackle the requirements of constructing a livable city in the harsh climatic conditions of the middle and western regions, and control regional environmental pollution. Furthermore, we should introduce industrial ecological reforms that serve as driving forces for the construction of an ecologically sound industrial pattern, design healthy and safe ecological landscape security patterns to serve as a strong support for healthy urban life, and sustainably utilize natural and energy resources to safeguard the environment for future generations. A key priority should be placed on the protection, restoration, rebuilding, and maintenance of grassland, wetland, and urban ecological system, simultaneously optimizing the industrial symbiosis structure in order to reduce the overexploitation of resources and energy; this will result in an eco-friendly transformation of the economic growth mode. Overall, we should

TABLE 11.2
Analysis of the Comprehensive Problems of Baotou

Essential Factors	Specific Problems
Utilization of resources and energy	1. The usage rate of urban renewable water is still low and water efficiency indices such as GDP per 10,000 have the potential to improve. 2. Multiple industrial types and traditional economic growth modes lead to high energy consumption and severe waste—the level of energy usage needs to be much lower. 3. The demand of land resources increases with urban development and its scarcity becomes a bottleneck for further development. At present, the used lands are in state of obvious waste and degradation with a greater representation of industrial land versus ecological land.
Environmental quality	1. The air quality has a typical "soot-type" pollution characteristic with serious pollution in winter and spring. Air-borne fluoride pollution is also a serious concern. 2. Water quality problems are also prominent because industries emit huge quantities of effluents in a water-based form—the larger the emission load, the greater is the harmfulness of effluents such as fluorides and other inorganic contaminants. The sewage treatment rate needs to be further improved. 3. The volume of industrial solid waste produced is huge and its treatment rate is low. The lack of recycling and comprehensive utilization of such waste negatively impacts the surrounding environment. Hazardous radioactive and nonradioactive wastes require science-based effective solutions. 4. Traffic noise issues are prominent.
Ecological protection and construction	1. Industrial ecology: Heavy industries constitute a large part of the industrial development program of Baotou. However, the industrial chain connecting these enterprises to an effective industrial ecological system is incomplete leading to an inconsistency between industrial development and ecological environment construction. 2. Aquatic ecology: Overexploitation has lead to serious ecological damage of rivers—urban rivers are extremely polluted, the wetland area is shrinking, and the ecological restoration work is inadequate. In addition, the situation of groundwater is very grim—overexploitation by the city has resulted in the formation of large area of groundwater funnels and this area is constantly increasing. 3. Grassland ecology: The phenomena of desertification, alkalization, and degradation of grasslands in and around Baotou City are key issues that are on a continued downward trend. 4. Ecological living: While there are several advantages to ecological living, this can be achieved only if the urban environment, green space, public transportation, and other services are improved. Moreover, the current ecological residential district has not formed into scale.

construct Baotou to be an eco-city with a reasonable spatial pattern, fully equipped infrastructure, powerful ecosystem service function, developed economy, rich environment, harmonious living, prosperous culture, and developed industrial system, all within the framework of ecological equilibrium.

11.3.1.2 Staging Objectives

The eco-city model of Baotou City cannot be accomplished overnight. Realistically, the work should be divided into short-term, medium-term, and long-term stages that focus on different aspects from 2006 to 2020. Each stage will target specific objectives that are relevant to the problems faced at that time.

1. *Short Term (2006–2010):* The short-term phase from 2006 to 2010 is the start-up stage where several of the current issues faced by Baotou were tackled. These include the treatment of environmental pollution in the built-up regions of Baotou, high-intensity control of air and water pollution, the promotion of ecological utilization of resources and energy, initial measures for the restoration of damaged grassland and wetland ecological systems, and propaganda and education work describing the Baotou eco-city model to ensure that it satisfies the requirements of the national environmental protection model city introduced in 2007 by the State Environmental Protection Administration (SEPA). During this stage, through the preliminary construction of a rational industrial ecological system, through the installation of infrastructure to control environmental pollution, and through the introduction of ecological protection measures, the basic framework of the eco-city model will be established.

2. *Medium Term (2010–2015):* The medium-term phase is the period during which the overall standard of the city will be brought up to par, particularly with regard to improving the ecological sustainability and health of the Baotou ecosystem. Importantly, the construction of Baotou as a model city for national environmental protection will be realized with breakthroughs such as the creation of an improved natural landscape pattern; control of urban pollution and building the infrastructure for ecological protection; effective protection of water areas, improved air environment; proper disposal and comprehensive utilization of industrial solid waste; restoration and reconstruction of damaged ecosystems; improved service function of the natural ecology; enhanced utilization efficiency of land, water, and energy resources; a better ecological living system with the development of "green schools" and ecological communities; adopting a low consumption, low pollution, and high benefit mode of economic growth, which is ecologically safe; and finally, all the indices satisfy the eco-city assessment criteria established by the SEPA.

3. *Long Term (2015–2020):* The long-term phase is the consolidating and perfecting stage of the eco-city plan. During this stage, the ideal eco-friendly landscape pattern and natural ecological network will be completely set up; the environmental pollution control system will be comprehensively

established; clean industry and circular economy practices will become the basic requirements for industrial management; the quality of every environmental factor will be optimal; the utilization efficiency of land, water, and energy resources will attain the national advanced level; the damaged grassland and aquatic ecosystems within the administrative regions of Baotou City will be fully repaired; urban groundwater funnel areas will recover to a certain degree; green consumption and green economy will become deeply rooted in society; and the high-quality ecological model of economic growth will be instituted in its entirety.

11.3.2 PLANNING INDEX SYSTEM

In the eco-city model proposed by SEPA, 32 indices are selected using the correlation analysis method and sorted according to their weight through the entropy method—these indices are the standards that determine whether a given city is an eco-city or not. In the planning index system for Baotou City, we selected the first 24 indices and added 6 characteristic indices, which are based on practical problems in the ecological environmental protection and construction of Baotou. These 30 indices can be divided into three categories: utilization of resources and energy, improvement of environmental quality, and ecological protection and construction. Their desired values in each planning period are detailed in Table 11.3.

11.3.3 GAP ANALYSIS OF THE CURRENT SITUATION

Gap analysis was performed to analyze the gap between the current ecological situation in Baotou City and the SEPA standards—such an analysis sheds light on the problems that should be addressed during eco-city construction and provides positive development strategies. Among the 24 evaluating indices of SEPA and 6 characteristic indices of Baotou, there are only 6 indices that are currently in line with the reference standards: 5 evaluating indices of SEPA and 1 Baotou characteristic index (Table 11.3, year 2005). Therefore, the total standards are reached at a rate of just 20%, with the evaluating indices of SEPA at 20.8% and the characteristic indices at 16.7%. Of the indices not meeting the reference value, there are eight indices that are close to the reference value (i.e., the exponential value is >0.9) and will meet the standard requirements in the short-term stage of the eco-city plan.

Of the 10 indices in the category of improving environmental quality, only 1 index, that is, environmental protection investment, completely measures up to the reference value. Four indices including urban air quality, standard-meeting rate of centralized drinking water, emission intensity of chemical oxygen demand (COD), and treatment rate of urban waste are close to the reference value, while five other indices do not meet the standards.

There are five indices in the category of utilization of resources and energy—four of these are evaluating indices of SEPA and one is a characteristic index; that is, the utilization rate of municipal reclaimed water. Two SEPA indices meet the reference value while three do not. By reducing the energy consumption per GDP, improving the comprehensive utilization of industrial solid waste, and enhancing the

TABLE 11.3

Planning Index System and Desired Values of the Eco-City Construction of Baotou

Category	NO.	Name of Index	Unit	2005 (Year)	2010 (Year)	2015 (Year)	2020 (Year)	Reference Value	Note
Improvement of environmental quality	A1	Urban air is better or equal to the days of secondary standard	Days	256	292	315	>329	≥280	Data of urban area
	A2	Standard-meeting rate of water quality in urban water function areas	%	80	100	100	100	100	Data of whole city
	A3	Standard-meeting rate of centralized drinking water	%	97.7	100	100	100	100	Data of whole city
	A4	Standard-meeting rate of fluoride in urban air	%	79.2[a]	90	95	100	100	Data of whole city
	A5	Discharge of fluoride in urban water	t/year	785.1	700	600	500	500	Data of whole city
	A6	Emission intensity of SO_2	kg/10,000 yuan	21.1	≤10.0	≤6.5	<5.0	<5.0	Data of whole city
	A7	Emission intensity of COD	kg/10,000 yuan	5.2	<2.5	<2.0	<1.5	<5.0	Data of whole city
	A8	Centralized treatment rate of urban sewage	%	60.5	70	85	95	≥70	Data of whole city
	A9	Treatment rate of urban consumption waste	%	97	99	100	100	100	Data of whole city
	A10	Index of environmental protection investment	%	3.16	3.4	3.7	4.0	≥1.7	Data of whole city
Utilization of resources and energy	B1	Energy consumption per GDP	tce/10,000 yuan	2.58	1.9	1.6	1.4	≤1.4	Data of whole city
	B2	Water consumption per GDP	m^3/10,000 yuan	123	65	45	30	≤150	Data of whole city
	B3	Repetition rate of industrial water	%	86.5	90	93	95	≥50	Data of whole city
	B4	Utilization rate of municipal reclaimed water	%	23	60	65	70	40	Data of built-up area
	B5	Disposal and utilization rate of industrial solid waste	%	18.27	30	60	80	≥60	Data of whole city

Category	Code	Indicator	2004	Unit				Target	Data source
Ecological protection and construction	C1	GDP per capita	33,100	Yuan per capita	67,000	90,000	11,0000	≥33,000	Data of whole city
	C2	Net income of farmers per capita per year	4,667	Yuan per capita	10,000	15,000	19,000	≥6,000	Data of whole city
	C3	Disposable income of urban residents per capita per year	13,218	Yuan per capita	26,400	39,000	52,000	≥14,000	Data of whole city
	C4	The ratio of scaled enterprises approved by ISO14000	0	%	10	20	30	≥20	Data of whole city
	C5	The proportion of tertiary industry accounted for GDP	42.1	%	44	45.5	47	≥45	Data of whole city
	C6	Gini coefficient	0.41[a]	—	0.39	0.37	0.35	Between 0.3 and 0.4	Data of whole city
	C7	Recovery rate of urban groundwater funnel area	0	%	10	30	50	100	Data of built-up area
	C8	The proportion of protected regions accounting for land area	14.8	%	17.0	17.5	18.3	≥17	Data of whole city
	C9	The ratio of high-covering grassland area	41.5	%	50	65	70	65	Data of whole city
	C10	The ratio of degraded land	39.2	%	55	75	90	>90	Data of whole city
	C11	The level of urbanization	72	%	77	82	90	≥55	Data of whole city
	C12	Urban public green land area per capita	10.6	m²/pop	11	12	14	≥11	Data of built-up area
	C13	Coverage rate of standard-meeting region of noise	83	%	90	100	100	≥95	Data of built-up area
	C14	Coverage factor of urban gas	80	%	95	100	100	≥92	Data of built-up area
	C15	Coverage factor of centralized heat supply	70	%	90	95	100	≥65	Data of built-up area

Note:

[a] Data from the year 2004.

TABLE 11.4

Comparison between the Index of Baotou Eco-City Construction and the Reference Value

	Index of National Eco-City Construction			Characteristic Index of Baotou		
	Reaching the RV	Not Reaching the RV	Close to Reaching the RV	Reaching the RV	Not Reaching the RV	Close to Reaching the RV
Improvement of environmental quality	0	3	4	1	2	0
Utilization of resources and energy	2	2	0	0	1	0
Ecological protection and construction	3	6	4	0	2	0
Total	5	11	8	1	5	0

Note: RV, reference value.

regeneration of urban sewage, these three indices will be improved from becoming bottlenecks of Baotou eco-city construction.

There are 15 indices in the category of ecological protection and construction. Three indices including GDP per capita, the level of urbanization, and coverage factor of centralized heat supply reach the standards, while four indices including disposable income of urban residents per capita per year, the proportion of tertiary industry accounting for GDP, Gini coefficient, and urban public green land area per capita are close to the standard. The remaining eight indices do not meet the reference value.

The various indices and their approximation to the reference value in Baotou eco-city construction are shown in Table 11.4, and the overall condition is depicted in Figure 11.4.

11.3.4 POTENTIAL DEVELOPING TREND ANALYSIS

According to the health assessment of the urban ecosystem of Baotou (Table 11.1) and the gap analysis of the present index value of Baotou to the reference value (Table 11.4), we can obtain a clear understanding of the "advantages," "disadvantages," and "gaps" that have affected Baotou's development in the years 2000–2004 (Figure 11.5). Generally speaking, in the 5 years of the study period, the economy of Baotou developed well, the ecosystem health steadily improved, and the environmental quality picked up. This region has rich mineral and biological resources, and the spatial distribution and time-varying characteristics of these resources have laid the foundation for the eco-city plan in the future. However, the current degree of development is still far from the reference index values of the SEPA eco-city model,

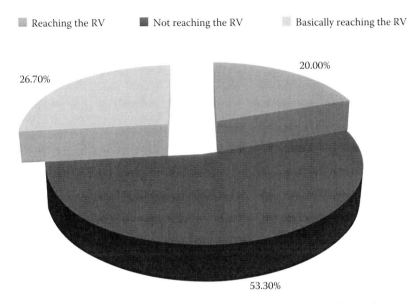

FIGURE 11.4   (See color insert.) The ratio between the status value and the reference value (RV) of the index of eco-city construction of Baotou.

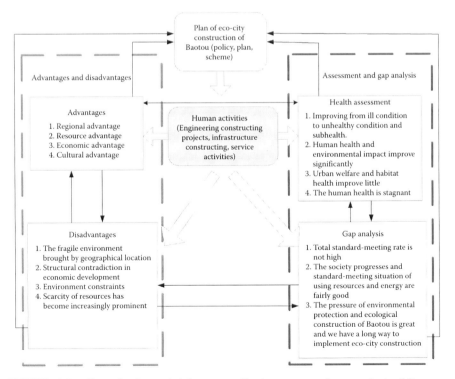

FIGURE 11.5   (See color insert.) Advantages, disadvantages, and gap analysis of Baotou.

and the pressure on environmental protection and ecological construction of Baotou is enormous. We have a long way to go toward comprehensively implementing eco-city construction of Baotou.

Based on the gap analysis and health assessment results, we identified the environmental impact of key factors on Baotou's urban ecosystem: resource factors such as water, land, and mineral resources and environmental factors such as desertification of soil, salinization of soil, soil erosion, water environment, air environment, forest, and grassland environment and wetland environment. The current and predicted (i.e., upon the implementation of the eco-city model) environmental impacts of these factors are shown in Table 11.5.

Upon analyzing the results in Table 11.5, we conclude that problems still remain with regard to Baotou's ecosystem and that the eco-city construction of Baotou faces constraints from specific ecological elements. For example, scare water resource is the current bottleneck among all the natural resources and will remain so in the future. In contrast, while the current situation of land and mineral resources is fairly good, in the future, these will become the bottlenecks, primarily due to urban development and an increase in energy demand. Furthermore, the status of aquatic, forest, and grassland environments, which are the main ecological environmental bottlenecks, are not optimistic. Also, degradation, desertification, and salinization of soil and soil erosion will remain serious issues in several parts of Baotou, especially as urban development increases.

11.4 ECOLOGICAL FUNCTION REGIONALIZATION OF BAOTOU CITY

11.4.1 Ecological Sensitivity Analysis

Ecological sensitivity analysis was performed to evaluate the extent to which Baotou's ecological environmental change is influenced by human activities. The resulting sensitivity evaluation factors are summarized in Table 11.6 and depicted in Figure 11.6. Note that the values range from 0 to 4, and the larger the value, the higher the sensitivity. As shown in Table 11.6, Baotou's ecological sensitivity to various factors lies between 1.12 and 3.61 with an average of 1.72. The results reflect that Baotou's overall ecological sensitivity is high although spatial differences are obvious. For example, the wetlands near the Yellow River, the Daqingshan Mountains, and the deserts in the Inner Mongolian Plateau region are the most sensitive regions, while the alluvial plains and basin valleys that have good water and nutrient conditions are the least sensitive.

1. *Sensitive Regions:* Sensitive regions account for 21.06% of the total area of Baotou City, mainly distributed in the southern, central, and northern parts including the wetlands near the Yellow River, the Daqingshan Mountains, and the deserts in the Inner Mongolian Plateau. These regions are important ecological barriers and indicate and control Baotou's ecological environment. Thus, these regions are the key areas where ecological regulations have to be implemented.

TABLE 11.5
Trend Matrix of Ecological Development of Baotou (According to the Current Situation)

	Water Resource		Land Resource		Mineral Resource		Soil Desertification		Soil Salinization	
	Present	Future	Present	Future	Present	Future	Present	Future	Present	Future
All the regions of Baotou	▲	++	▲	+		++	▲	+	▲	+
Four downtown areas	▲	++	▲	++		++	▲	−	▲	−
Guyang county	▲	+	▲	0	▲	0	▲	+	▲	−
Damao Banner	▲	0		− −	\	0	▲	++	▲	−
Tumed Right Banner		−		+	\	\	▲	0	▲	+
Baiyun mineral region	▲	++	▲	0	▲	++	▲	0	\	\

	Soil Erosion		Water Environment and Water Ecology		Air Environment		Forest and Grassland Environment		Wetland Environment	
	Present	Future	Present	Future	Present	Future	Present	Future	Present	Future
All the regions of Baotou	▲	+	▲	++	▲	+	▲	++		+
Four downtown areas	▲	0	▲	++	▲	+	▲	−		++
Guyang County	▲	+	▲	0		−	▲	+	\	\
Damao Banner	▲	+	▲	0		− −	▲	++	▲	−
Tumed Right Banner	▲	+	▲	+		−	▲	0	▲	0
Baiyun mineral region	\	\	▲	++	▲	+	▲	0	\	\

Note: ◁, Slight environmental problems; ◁, environmental problems of medium intensity; ▲, serious environmental problems; \, no such problems; +, increase of the severity in future; 0, no obvious change; −, reduction of the severity.

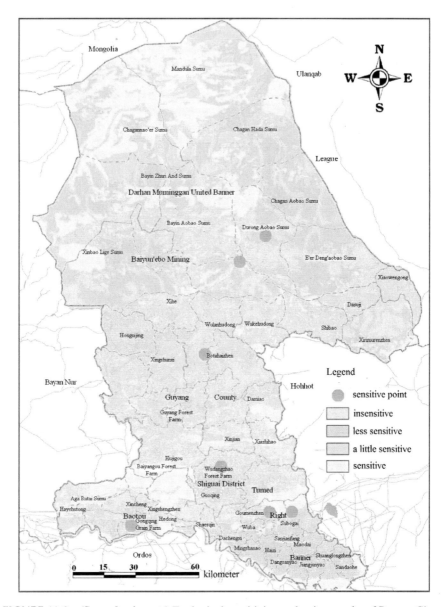

FIGURE 11.6   **(See color insert.)** Ecological sensitivity evaluation results of Baotou City.

2. *Moderately Sensitive Regions:* Moderately sensitive regions account for
   19.35% of the total area of Baotou City and are mainly distributed in the
   northern Guyang and Damao Counties. Continuous land degradation has
   led to the ecological sensitivity in these regions. Conditions such as dry and
   windy climate, fragile soil matrix, low vegetation coverage, together with
   unreasonable human activities such as cultivation of slope farmland and

TABLE 11.6
Ecological Sensitivity Evaluation Factors of Baotou City Environment

Weights	Sensitive Issues	Weights	Factors	Grading Standard			
				Sensitive	Moderately Sensitive	Less Sensitive	Insensitive
0.2	Water loss and soil erosion	0.2	Surface runoff (10⁴ m³)	>40	20~40	10~20	<10
		0.4	Slope (°)	>15	10~15	5~10	<5
		0.2	Soil types	Brown calcic soil, aeolian sandy soil, gray cinnamonic soil	Lithosol, skeleton soil, chestnut soil	Chernozem	Solonchak, moisture soil
		0.2	Land covers	Bare sandy land, sparse woodland, low coverage grassland	Cultivated land	Construction land	Forest, grassland, water areas
0.2	Desertification	0.3	Rainfall (mm)	<200	200–300	300–400	>400
		0.3	Wind speed (m/s)	>5.2	4.4~5.2	3.2~4.4	<3.2
		0.2	Soil types	Brown calcic soil, chestnut soil, aeolian sandy soil	Skeleton soil	Chernozem, lithosol, gray cinnamonic soil	Solonchak, alluvial, moisture soil
		0.2	Land covers	Dry land in hills, overgrazing areas in grassland	Dry land, grazing areas in grassland	Forest, high coverage grassland	Towns, water areas
0.1	Salinization	0.6	Terrain	River valley, plain	Fluvial plain	Diluvial plain	Plateau, hills, mountain
		0.4	Tillage forms	Irrigable land	Ordinary farmland	Hilly farmland	Other land
0.5	Wetland, natural reserves	1	Wetland, natural reserves	√ Cultural relics			
	Sensitive points						
Grading assignment				4	3	2	1
Grading range				>1.75	1.65~1.75	1.55~1.65	<1.55

grassland overgrazing have led to land degradation in these regions. At present, the desert land is increasing consistently, resulting in a rise in ecological sensitivity. This region of Baotou has also become the source of sandstorms in Inner Mongolia as well as North China. In fact, the regions north of the Yinshan Mountains have already been classified as restricted development zones in the National 11th Five-Year Plan.

3. *Less Sensitive Regions:* The area of less sensitive regions accounts for 36.23% of the total area and is mainly distributed in the Inner Mongolia High Plain and the Tumote Plain in the administrative districts of Damao, Tumed Right, and Jiuyuan Counties. The dominant industries in these regions are grassland animal husbandry and irrigation farming. Based on field surveys, although the grassland has degraded to some extent, the overall vegetation coverage is high and is resistant to wind erosion and desertification. In addition, long-term irrational irrigation patterns have caused the soil in the Tumote Plain to undergo secondary salinization. However, this pattern will gradually change since the irrigation water quota cannot increase any further. Therefore, the ecological sensitivity of these regions is not high, and in the future, these regions can become important commodity grain production bases for Baotou City.

4. *Insensitive Regions:* The area of insensitive regions accounts for 23.36% of the total area, mainly distributed in alluvial plains and basin valleys that have good water and nutrient conditions. In these regions, vegetation coverage is high, soil condition is good, and the ability to resist erosion is strong. Nevertheless, the distribution of insensitive regions is broken and scattered and results in difficulties of development.

From the preceding analysis, we find that the overall ecological sensitivity of Baotou City is high. The urban areas in Baotou City belong to regions that are moderately ecologically sensitive, while the southern and northern parts belong to the Yellow River wetland region and the Daqingshan Mountain region, respectively, both of which are ecologically sensitive regions. If we only consider ecological sensitivity, we can conclude that Baotou City should expand along the east–west direction. However, the Baogang industrial park and Tailing dam are situated to the west of Baotou making the eastern direction an important expansion direction. Overall, the future focus should be on developing the Donghe and Jiuyuan Districts in the east of Baotou.

11.4.2 REGIONALIZATION OF ECOLOGICAL FUNCTION

Based on the principles and requirements of ecological function regionalization (Ministry of Environmental Protection of the PRC, 2008), we determine the basic boundaries of ecological function regions with the aid of roads, rivers, and administrative boundaries. Accordingly, these regions are classified into four categories: type I red line regions or development forbidden regions, type II yellow line regions or development limited regions, type III green line regions or development optimization regions, and type IV blue line regions or key development regions (Figure 11.7).

1. *Development Forbidden Regions:* Development forbidden regions in Baotou City mainly include the Yellow River wetland, the Yinshan Mountain area, the northern desert area, and the Bayinhanggai and Tenggezuoer wetland natural reserves.
2. *Development Limited Regions:* Development limited regions mainly include the northern grassland; that is, the agriculture pasture ecotone on the northern side of the Yinshan Mountains. These regions belong administratively

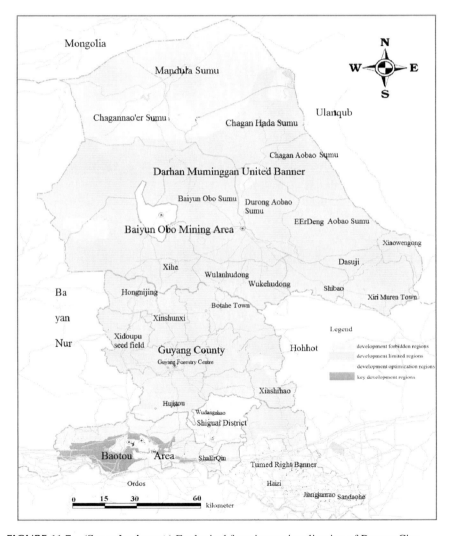

FIGURE 11.7   (See color insert.) Ecological function regionalization of Baotou City.

to central and South Damao District and the North Gubei District. Factors in these regions that are incompatible with enhanced development comprise harsh natural environment, sparse and damaged vegetation, limited and overloaded native grass, unreasonable cultivation, water loss and soil erosion, severe grassland degradation, low population carrying capacity, overgrazing in the hilly agriculture pasture ecotone, and low economic benefit of agriculture and grasslands.

3. *Development Optimization Regions:* Development optimization regions include the urban areas of Baotou City, the Tumote plain agricultural region, and key regions of energy and mineral resource development that belong administratively to the urban areas, Tumed Right district, and the Baiyunebo and Shiguai mining areas. These regions have a lot of industrial enterprises and are already highly developed—consequently, they face a suite of ecological environmental problems. Compared with other regions, development optimization regions have relatively low vegetation coverage, simple ecosystem structure, abundant consumption of resources, and severe pollution of the local environment.

4. *Key Development Regions:* Key development regions include the planning expansion suburbs in Baotou City, which administratively belong to the Donghe and Jiuyuan Districts. These regions have large populations, smooth terrain, fertile land, rich water resources, and good irrigation conditions. These regions produce key commodities such as grains, vegetables, and fruits for consumption in Baotou City; and the land is predominantly used for agriculture.

11.4.3 PLANNING OF URBAN LANDSCAPE PATTERNS

11.4.3.1 Construction of Urban Landscape Ecological Security Patterns

On the urban scale, we propose that the framework of Baotou's landscape ecological security pattern should consist of seven sources, one artery, two barriers and three belts. Urban landscape ecological security patterns should be based on the following principles: (1) protection of important sources of biodiversity, (2) breaking through the main arterial roads, (3) building green barriers, and (4) construction of several ecological isolation belts (Figure 11.8).

The seven sources include Jiufengshan, Bayinhanggai, Meiligeng, Chunkunshan, Nanhaizi wetland, Tenggenaoer wetland, and Honghuaaobao natural reserves.

The north to south main artery along provincial road 211, which spans the whole Baotou City, will form the artery of the urban landscape security pattern. This artery will be converted into a green corridor.

The two barriers are the North Damao and South Damao ecological barriers for wind prevention and sand fixation.

The three isolation belts will consist of the greenbelt on the north side of the Daqingshan Mountains for water and soil conservation, ecological landscape greenbelt on the south side of the Daqingshan Mountains, and ecological landscape greenbelt on the north side of the Yellow River.

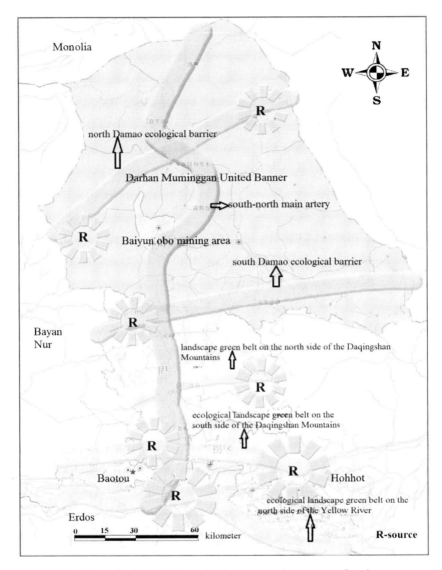

FIGURE 11.8 (See color insert.) Urban landscape security patterns planning.

11.4.3.2 Construction of Suburban Area Landscape Ecological Security Patterns

Based on the existing greening basis, urban space development strategy, and potential landscape ecological patterns, we recommend the following landscape ecological development pattern for the suburban areas surrounding Baotou City. The first step will involve the building of ecological barriers outside the suburban areas, incorporating the ecological landscape greenbelt on the south side of the Daqingshan Mountains in North Baotou and the ecological landscape greenbelt on the north side

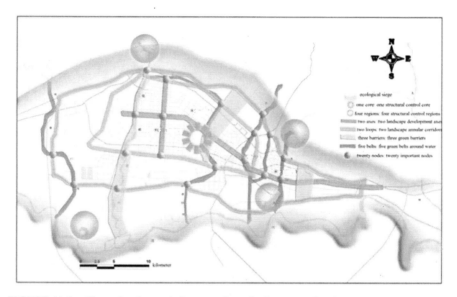

FIGURE 11.9   **(See color insert.)** Construction of urban areas landscape security patterns.

of the Yellow River in South Baotou. These barriers will wrap up the Baotou urban area and form a solid ecological city.

Inside the suburban areas, the eco-city plan will result in the formation of an ecology-incorporated city that has the Ghengis Khan Ecological Park as the "green heart"; four regions around the suburban areas as the "green lung"; isolation belts, rivers, and roads as the "green skeleton"; and important green nodes as the "articulation." In short, this will be a "one core, four regions, two axes, two loops, three barriers, five belts and twenty nodes" green natural system (Figure 11.9).

The Ghengis Khan Ecological Park will form the core of the landscape pattern.

The four regions covered will include the Nanhaizi wetland, Kundulun reservoir, Zhaojundao wetland, and Liubaoyao reservoir.

The two axes will be formed by the 20–100-m-wide green landscape axis along Aerding Avenue and the 20–100-m-wide green landscape axis along Gangtie Avenue, Jianshe Road, and Bayantala Avenue.

The two loops are the 40–100-m-wide annular green corridor along the peripheral railway around the city and the 40–100-m-wide annular green corridor on both sides of the peripheral highway around the city.

The three barriers will consist of the ecological barrier in West Baotou on both sides of the Kundulun River between Baogang industrial park and the urban areas, the ecological barrier in Center Baotou between the new and the old districts including Ghengis Khan Ecological Park, and the ecological barrier in East Baotou between Donghe old district and Baolv industrial park.

The five isolation belts include the greenbelts on both sides of the Hademen Channel, Sidaosha River, Erdaosha River, West River Channel, and East River Channel.

The junctions of green isolation belts, river corridors, and road corridors will form nodes in the landscape pattern. These nodes are very important for animal migration; therefore, we plan to design these nodes within woodlands to increase the probability of species migration.

11.5 ECOLOGICAL PROTECTION AND CONSTRUCTION OF BAOTOU CITY

11.5.1 CONSTRUCTION OF INDUSTRIAL ECOLOGICAL SYSTEM

11.5.1.1 Integration of Layout

1. Consistency Analysis of Industrial Layout and Ecological Function Regionalization
   Based on the distribution of natural resources, the existing industrial layout, the outline of land use in the eco-city plan, and the plan's ecological function regionalization of Baotou City (Section 11.4.2), we analyzed the consistency of the industrial layout with the ecological function regionalization of the nine districts of Baotou City as follows:
   a. *The urban areas:* Ecological function regionalization of Baotou City considers Kundulun, Qingshan, and Donghe Districts as key development and development optimization regions and Jiuyuan District as a key development region. Based on the current industrial layout, Kundulun, Qingshan, and Donghe Districts are built urban areas, while Jiuyuan District is an urban area that is being developed. Thus, the industrial development and ecological functions of the four urban districts of Baotou City are consistent.
   b. *The mining districts:* Based on the ecological function regionalization of Baotou City, Baiyunebo mining area and Shiguai Districts are planned as development optimization regions. Therefore, these districts can be planned to develop industries and optimize the current industries.
   c. *Damao, Guyang, and Tumed Right districts:* According to the ecological function regionalization of Baotou City, most areas of Damao and Guyang Districts are either development limited or development forbidden regions, while most areas of Tumed Right district are either development optimization, development forbidden, or development limited regions. However, in the current industrial layout, some industries have arisen, such as steel and rare earth and nonferrous metal and coal chemical industries. It is thus clear that these existing industries in Damao and Guyang Districts are inconsistent and conflicting with the ecological function realization and should be adjusted. These districts should not develop primary and secondary industries on a large scale, but can develop tertiary ecological industries such as the information service industry.
2. *Integration Planning of Industrial Layout:* The proposed integration planning results of Baotou's industrial layout are shown in Figure 11.10. Compared with the original industrial layout, industrial enterprises in the

eco-city model will form ecological groups and be under the unified management of ecological parks. This will increase the resource utilization efficiency and promote waste exchange and reuse. At the same time, to realize ecological recovery, the number and scale of industrial enterprises will be reduced in a stepwise manner in regions where development is either forbidden or limited.

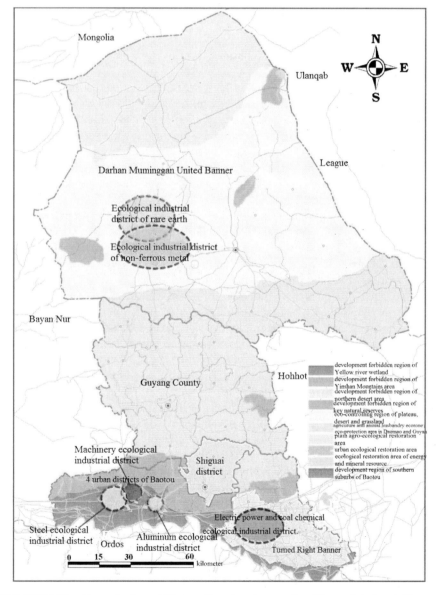

FIGURE 11.10   (**See color insert.**) Integration planning results of Baotou's industrial layout.

a. *Ecological industrial layout of the urban areas:* The ecological industrial layout results for the four urban districts are as follows: Kundulun District is planned as the steel ecological industrial district, Qingshan District as the machinery ecological industrial district, Donghe District as the aluminum ecological industrial district, and Jiuyuan District as reserve land for industrial development, high-tech development district, and national rare earth district.

b. *Ecological industrial layout of the mining districts:* Considering the advantages of the natural resources found in the mining districts, Baiyunebo is planned as the rare earth ecological industrial district and Shiguai District as the electric power and coal-based chemical ecological industrial district.

c. *Damao, Guyang, and Tumed Right districts:* Most areas of Damao, Guyang, and Tumed Right districts belong to development forbidden and development limited regions, according to the ecological function regionalization. Therefore, existing industries should either move out or become consolidated to reduce the damage to resources and environment.

11.5.1.2 Construction of Industrial Clusters

To construct industrial clusters, Baotou should make the best use of its mineral resources and existing industries. The eco-city plan proposes to build two main industrial chains (steel and rare earth) and two supplementary industrial chains (nonferrous metal and coal chemical industry), and an energy cascade utilization system. The resulting industrial symbiosis system of Baotou is shown as a conceptual model in Figure 11.11. This plan focuses on developing the steel industry and its related chain of secondary industries, enhances the research sector of the rare earth industry, develops a logistics chain for the rare earth industry based on technical and economic appraisal and risk prevention, and builds an effective resource and energy utilization cascade. The development of two supplemental industrial chains should meet the demand of steel industry logistics and enhance corresponding support services.

Using the preceding strategies, the guiding aim is to develop the steel industry and related secondary industries as the core enterprises cluster, with the nonferrous metal, rare earth, and related secondary industries as the subdominant enterprises cluster. This is because the subdominant enterprises cluster is mainly involved in designing steel-dependent products. The conceptual framework of Baotou's industrial enterprises clusters is shown in Figure 11.12.

Figure 11.13 shows the life cycle of steel, nonferrous metal, and rare earth industries from resource exploitation, smelting, processing, and utilization/production to waste discharge and treatment. The first step involves the exploitation of iron, copper, and rare earth ores. These mineral resources are then processed to produce several primary intermediate products such as pig iron powder, blister copper, and aluminum ingot. Finally, these primary intermediate products are processed by machinery manufacturing industry (i.e., a secondary steel-dependent industry) to produce building materials, vehicle parts, and appliance

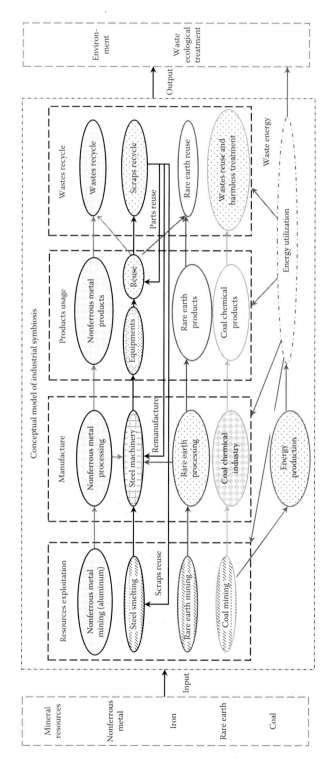

FIGURE 11.11 (See color insert.) Conceptual model of Baotou's industrial symbiosis system.

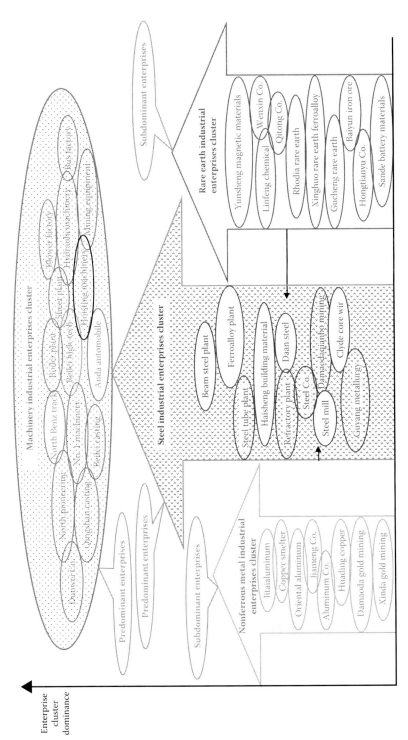

FIGURE 11.12 (See color insert.) Framework of Baotou's industrial enterprises clusters.

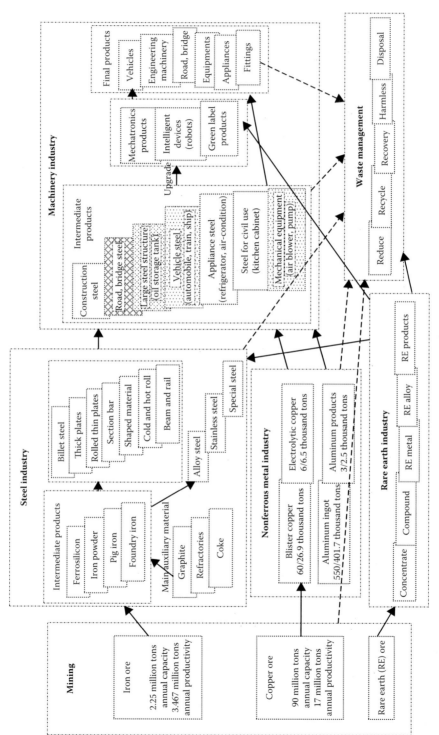

FIGURE 11.13 Conceptual model of Baotou's industrial products clusters.

materials. For example, steel can be used in the manufacture of plates, pipes, rails, wires, and other kinds of products. Similar to the steel industry, the nonferrous metal and rare earth industries can produce intermediate products, which link a variety of secondary enterprises and result in the formation of reticular products cluster. However, industrial finished products with direct use value can also be produced.

Figure 11.14 shows the direct and indirect relationships between Baotou's industrial chains from the perspective of their inputs and products. In Baotou, the two main industrial chains are the steel and equipment manufacturing industries, while the two supplementary industrial chains are nonferrous metal and rare earth industries. From Figure 11.14, it is clear that the development of the main and supplementary industrial chains relies heavily on sufficient energy supply. Thus, the thermal power industry and its industrial chain are the primary bases to support the development of the other dominant industries. This also highlights the need for an efficient energy generation and utilization cascade for Baotou eco-city.

Taken together, we hypothesize that Baotou will become the core of economic growth, regional industrial cluster center, and the scientific and technological innovation base of the Hohhot-Baotou-Yinchuan and Hohhot-Baotou-Erdos regions by 2010 (the plan was made in 2007, so the predicting year is 2010). The overall aim is to develop Baotou into regional capital of the economy of Midwestern China.

11.5.2 GRASSLAND ECOLOGICAL PROTECTION AND RESTORATION

The types of ecosystem in Baotou are complex and diverse. The northern foot of the Yinshan Mountains is a particularly fragile region consisting of typical desert grasslands. It is important for Baotou's green space construction (see Sections 11.4.2 and 11.4.3) and has multiple functions such as production, protection, and ecological regulation. However, for a long time, only the production function of this region has been emphasized, which is grassland animal husbandry. Its ecological function has been ignored, leading to continuous ecosystem degradation. To rectify this, we outline the following grassland ecological protection and restoration plan.

11.5.2.1 Grassland Ecological Protection Scheme

1. *Returning Grazing Land to Grassland:* The process of changing grazing land back to grassland includes two measures: forbidden grazing and rest grazing. In ecologically fragile natural grasslands with vegetation cover lesser than 50% or in areas that are more than 4500 m above sea level, grazing should be forbidden. Moderately degraded grasslands with vegetation cover between 50% and 70% and slightly degraded grasslands with vegetation cover between 70% and 80% can be ecologically recovered by rest grazing.

   In Baotou City, the area of grazing land that will be restored to grassland from 2006 to 2020 is planned to be 1.2 million hm^2 (see Table 11.7 and Figure 11.15). Of these, the forbidden grazing area comprises 325,000 hm^2,

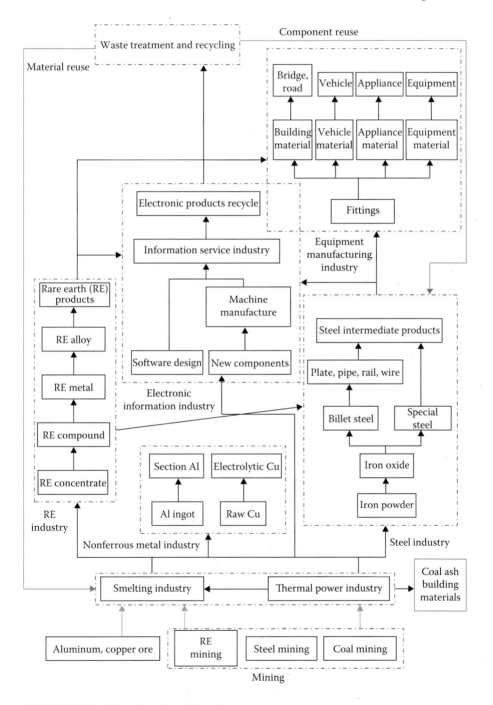

FIGURE 11.14 Baotou's industrial chains.

TABLE 11.7

Area, in Thousand hm², of Returning Grazing Land to Grassland in Baotou

Area of Returning Grazing Land to Grassland in Baotou		1200	
Forbidden Grazing	By Reseeding	Rest Grazing	By Reseeding
325	98	875	262

of which the reseeding area is 98 hm², and the rest of the grazing area comprises 875,000 hm², of which the reseeding area is 262 hm². In the eco-city plan, during the short-term stage, the aim is to curb northern grassland degradation. During the medium-term stage, that is, by 2015, the goal is to recover desertified and degraded grasslands. For a comprehensive recovery of grasslands, along with a coordinated development of grassland ecology and animal husbandry, the target is the long-term stage of the plan, that is, by 2020.

The first step in the process of project implementation is to compile a work design based on regional requirements and distribute the tasks into three levels: villages, families, and meadows. The next step is to use global positioning system (GPS) to pinpoint meadow position by longitude and latitude and compile each village's operation map. To coordinate the electronic management system, technical measures should be standardized.

2. *Natural Forest Protection Project:* Baotou City launched its first Natural Forest Protection Project in 2000. After several years of efforts, the Jiufengshan natural forest protection region in the western part of the city has developed into a green barrier that protects Baotou's ecological environment. The area of this protected region is 130,000 hm² and encompasses 5 districts, 44 villages, and 9 national forest farms. If the current project construction zones are taken into consideration, the actual area is 170,000 hm². However, because Baotou is located in arid and semiarid regions, forest development is under tremendous stress. Moreover, the afforestation-driven ecological restoration idea is being questioned by many people. For these reasons, the construction area for natural forest protection is planned to be within 300,000 hm². Implementation measures include the abolition of natural forest harvesting, effective management of natural and other forests, active development of fast-growing and high-yielding timber, and the strengthening of the forest cover of barren mountains and land. The total area of afforestation from 2006 to 2020 is planned to be 80,000 hm².

3. *Returning Farmland to Grassland/Forest:* The key regions of farmland that will be returned to grassland/forest include sloping farmlands with gradients over 15° and farmlands suffering from desertification and/or degradation. Specifically, sloping farmlands with gradients over 25° and severely desertified farmlands will be returned to grassland/forest by 2010 (the plan

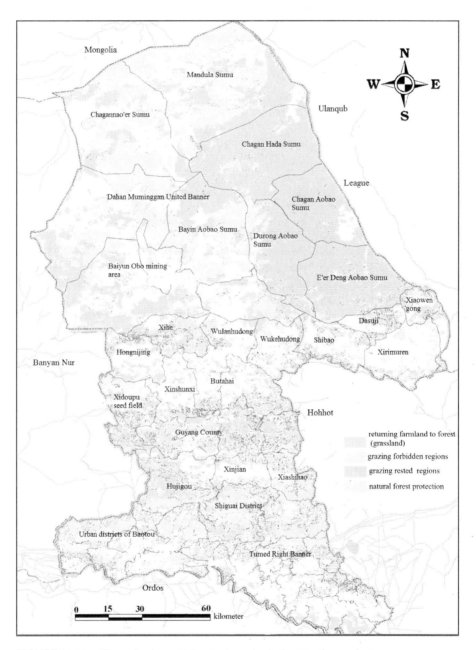

FIGURE 11.15   (See color insert.) Baotou's ecological protection projects.

was made in 2007, so the predicting year is 2010); they cover 67,000 hm². Sloping farmlands with gradients over 15° and severely degraded farmlands with poor water and nutrient conditions will be returned to grassland/forest by 2020; they cover an area of 120,000 hm². However, Baotou is not suitable for large area forestation, given its arid and semiarid climate. Therefore, the south mountainous region, the shady slopes of mountains, and the banks of rivers will be planted with forests, while farmlands will be mainly restored to grasslands. Combined with Natural Forest Protection Project mentioned earlier, a small portion of the farmlands could be returned to forests using artificial or natural afforestation methods.

4. *Ecological Migration Project:* Theoretically, ecological pressure that is caused by overpopulation can be resolved by transferring population from ecologically degraded regions to the outside. The resulting principle of ecological migration aims to rationally distribute ecological pressure in larger space. To achieve this in Baotou, we propose the movement of people who live in villages with poor ecological condition, scarce water resources, and severe desertification to regions with improved traffic and water resource conditions. The target population for ecological migration includes 15,000 people from Damao District and 10,000 people from Guyang District. Three different arrangements will be in place for these people. Firstly, the construction of new ecological migrant villages around an already developed town—activities such as drylot feeding and vegetable growing will be encouraged to provide the residents with agricultural and sideline products. Secondly, encourage the movement of ecological migrants to regions with good water and nutrient conditions, where efficient ecological agriculture and husbandry practices can be developed. Thirdly, encouraging them to work in the towns—this will not only increase their employment chances but also alleviate the ecological pressure on the land.

11.5.2.2 Grassland Ecological Construction Scheme

1. *Desertification Grassland and Wandering Dune Comprehensive Treatment:* The severely desertified area of Baotou is 19,008 km^2, while the moderately or above wind erosion desertification area is 12,746 km^2; these are mainly located in north and central Baotou's agricultural pastoral ecotone and North Baotou's grazing region. The current plan mainly aims at overgrazing-induced desertification. Also, since the desertification area of Baotou is too large, it is not economically feasible to treat the entire area. Therefore, this plan emphasizes the treatment of desertified land in Damao and Guyang Districts. Chief treatment measures include forbidden grazing, returning grazing land to grassland, and restoring farmland to grassland. The total treatment area is planned to be 1620 km^2, which consists of 90 km^2 of wandering dunes and 1530 km^2 of desertified grasslands (Figure 11.16).

Given the rate of development of Baotou in recent years as well as the projected rate of future development, it is imperative to utilize desertified grassland in an ecological manner and improve land quality. Rational use, which in itself is a kind of scientific management, and improvement of desertified grassland is a complex problem that cannot be solved by a single approach. For fixed and semifixed desert grasslands, building fences and reseeding appropriate plants, growing these plants into pioneering species, fixing the desert grassland, and then developing a stable vegetation community will have a strong sand stabilization effect. For wandering dunes, stabilization of

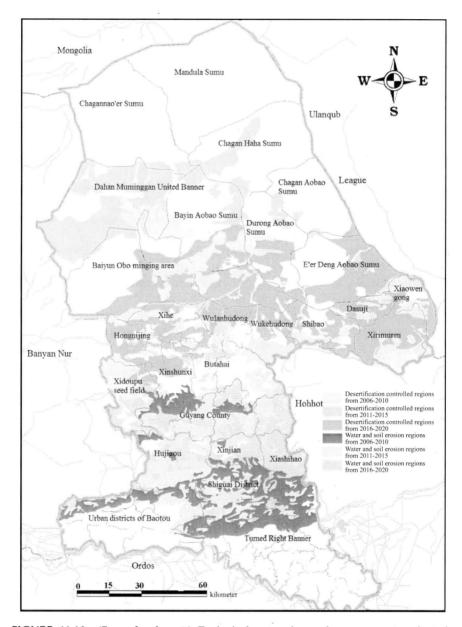

FIGURE 11.16 (See color insert.) Ecological restoration and management projects in Baotou City.

sand should be engineered by building fences and grass square sand barriers, by planting shrubs in the sand barrier, and by reseeding local pastures. This stabilization will in turn help in the recovery of a vegetation canopy.

2. *Comprehensive Treatment of Water Loss and Soil Erosion:* The area of water loss and soil erosion in Baotou is 8402 km², accounting for 30.26% of the

total area of Baotou City. In particular, the area of severe water loss and soil erosion is 5500 km², mainly in the medium and low hills of Jiuyuan and Guyang Districts. The primary measures to treat the water loss and soil erosion area are outlined in the Natural Forest Protection Project, Returning Grazing Land to Grassland Project, and Returning Farmland to Grassland (Forest) Project. By 2010, in the short-term stage of the eco-city plan, the treatment area of water loss and soil erosion is planned to be 3534 km2; this is 42.06% of the total deteriorated area (the plan was made in 2007, so the predicting year is 2010). In the medium-term stage, the treatment area is planned to be 5000 km², that is, 59.5% of the total deteriorated area will be treated by 2015. Finally, the treated area will reach 80% by 2020, in the long-term stage of the eco-city plan. This will be achieved by adopting comprehensive treatment measures, which are a combination of biological and engineering measures. For example, on barren slopes and abandoned farmlands with a gradient above 25°, grass will be artificially grown and quality of natural grasslands will be improved. On barren mountains and slopes with a gradient between 10 and 25°, trees will be planted for water and soil conservation—shrubs if the soil layer is thin, tall forests or mixed forests of trees and shrubs if the soil is thick. On thick fertile soil layers with a gradient between 5° and 10°, economic forests that cooperate with cave-shaped soil preparation will be planted. On sloping fertile lands with a gradient less than 5°, basic farmlands will be developed. Additional measures include the planting of "shelter" forests around farmlands, grasslands, and on the sides of roads; construction of roads in traffic-deficient districts; and digging of wells in districts where there is no direct irrigation source.

- *Perspective:* The ultimate objective of the eco-city model is to realize sustainable development. Accordingly, the development of Baotou into an eco-city will be based on ecological function regionalization and landscape layout and will be specifically implemented by constructing ecological industrial and grassland systems in a systematic and long-lasting manner. Because the project will involve the hard work of the Chinese government and Baotou's citizens, effective measures should be taken to promote comprehensive cooperation between the participating governing agencies, institutions, enterprises, and residents. This will ensure the sound implementation of the planning objectives and realize the coordinated development of the economy, society, and environment of Baotou, converting it into a model eco-city in China.

REFERENCES

Baotou Statistical Office, 2001–2005. *Baotou yearbook of statistics* (2001–2005). Beijing: Chinese Statistic Publisher (in Chinese).

Brandt-Williams, S. 2001. *Handbook of emergy evaluation: A compendium of data for emergy computation issued in a series of folios. Folio 4, emergy of Florida agriculture.* Gainesville, FL: Center for Environmental Policy, Environmental Engineering Sciences, University of Florida.

Brown, M.T. and Bardi, E. 2001. *Handbook of emergy evaluation: A compendium of data for emergy computation issued in a series of folios. Folio 3, emergy of ecosystems.* Gainesville, FL: Center for Environmental Policy, Environmental Engineering Sciences, University of Florida.

Brown, M.T. and Ulgiati, S. 1997. Emergy-based indices and rations to evaluate sustain ability: Monitoring economies and technology toward environmentally sound innovation. *Ecological Engineering* 9: 51–69.

Ma, S.J. and Wang, R.S. 1984. The social-economic-natural complex ecosystem. *Acta Ecologica Sinica* 4(2): 27–33 (in Chinese).

Ministry of Environmental Protection of the PRC. 2008. National ecological function regionalization. Beijing (in Chinese).

Odum, H.T. 1983. *Systems ecology.* New York: John Wiley & Sons.

Rapport, D.J. 1995. Ecosystem health: Exploring the territory. *Ecosystem Health* 1(1): 5–13.

Zhang, Y. and Yang, Z.F. 2007. Emergy analysis of urban material metabolism and evaluation of eco-efficiency in Beijing. *Acta Scientiae Circumstantiae* 27(11): 1892–1900 (in Chinese).

Zhang, Y., Yang, Z.F., and Yu, X.Y. 2009. Evaluation of urban metabolism based on emergy synthesis: A case study for Beijing (China). *Ecological Modelling* 220(13–14): 1690–1696.

12 Eco-City Wuyishan Plan

Lixiao Zhang, Linyu Xu, Yan Zhang,
Meirong Su, and Zhifeng Yang

CONTENTS

12.1 INTRODUCTION OF WUYISHAN CITY

Wuyishan City is a country-level city located in the northwestern part of Fujian Province within Nanping Prefecture. It is a unique tourist city named after the famous mountain in Fujian Province. Owing to the grand and beautiful natural scenery, rich historical and cultural relics, a mild climate with distinctive seasons, and abundant local products, Wuyishan City has been confirmed as a key city for national tourism and a historical city by the State Council. Benefiting from the geographic features, Wuyishan is renowned for its rich water and mineral resources, and the agricultural production such as tea, grain, and timber. It is also a key area for biodiversity conservation given its rich biodiversity (Gao et al., 2010).

12.1.1 NATURAL CONDITION

Wuyishan City is located near the southeast coast of China, on the northwest of Fujian Province and bordering Jiangxi Province, at 27°27′–28°05′ N, 117°37′–118°19′ E. (Figure 12.1). It covers an area of 2813.91 km², and is 382 km away from the provincial capital Fuzhou.

Wuyishan is a hilly city; mountains and hills cover 88.9% of the total area. The landform consists of mountains, hills, plains, and intermountain basin, overall the terrain slopes from the northwest to the southeast. Wuyishan City lies in the subtropical humid monsoon climate zone with a significant regional climate variations. The dominant wind direction in winter and summer are northerly and southerly, respectively.

In general, Wuyishan has a healthy natural environment and is classified as a water source protection zone. The average quantity of water resource per year is 3.61 billion cubic meters (including surface water and groundwater) and the per capita availability is more than 20,624 m³. The theoretical potential of hydropower is estimated at 215 thousand kilowatts. The quality of the air and water greatly exceeds the state's standard. It is also rich in forest resources, with the forest covering 79.9%. In addition, the subtropical forest ecosystem of the Wuyi Mountain is a key area for biodiversity conservation given its rich biodiversity (Zhao et al. 2008).

FIGURE 12.1   **(See color insert.)** Location map of Wuyishan City.

12.1.2 Economic Condition

Wuyishan City mainly relies on an export-oriented economy, while tourism is its pillar industry. The city's gross domestic product (GDP) in 2005 was 3.02 billion yuan and GDP per capita was 13,571 yuan. Primary, secondary, and tertiary industries accounted for 23.8%, 23.9% and 52.3%, respectively. The main industries in Wuyishan City include manufacturing; the production and distribution of electricity, gas, and water; and mining. The tertiary industry mainly includes wholesale trade and retail trade, real estate trade, hotel and restaurant, and tenancy and commercial services. The gross annual value of industrial and agricultural products was 2137.47 billion yuan; of this, 978.01 million yuan came from industrial production and 1159.46 million yuan from agricultural production. By relying on its rich bamboo, timber, rock tea, and granite resources, it has formed a series of sturdy industries mainly serving tourism, including the production of tourist arts and crafts, processing of bamboo and forest products, and making fine rock tea and granite. Wuyishan City is also a commodity grain base and key forest area, as well as a famous tea producer in China. Its agriculture has constantly been optimized along with the development of tourism. Overall, the economic gross of Wuyishan City is still small, with a weak industrial foundation, emphasizing the necessity of economic restructuring (Yang et al. 2011).

12.1.3 Social Condition

Wuyishan City has the relatively small urban population and large minority. The total population reached 223.7 thousand in 2005, of which urban population accounted for 44.9%. At the stage of rapid urbanization, the spatial distribution of Wuyishan City is seriously unbalanced, with the urban population density of 79.5 persons per square kilometer. Generally, 14 nationalities distribute over Wuyishan, including the Han, She, Mongolian, Miao, and Hui. Of these, the Han population accounts for 99.6% of the total.

Wuyishan City retains the rich ancient culture and human landscape resources, such as rock tea culture, Zhuxi's culture, and Ancient Minyue Royal City. Wuyishan City has been named as "Top Tourist City in China," owning to its unique natural and cultural resources. Wuyi Mountain, the most popular tourist destination in Fujian, and a top ranked national scenic spot, was denoted as both natural and cultural UNESCO World Heritage Sites in 1999.

12.2 ECO-CITY WUYISHAN PLANNING OBJECTIVES AND THOUGHTS

12.2.1 Planning Contents and Thoughts

12.2.1.1 Planning Scope

The time range for this proposal will use 2005 as the planning base year and 2010, 2015, and 2020 as the short-term, mid-term, and long-term planning years. In contrast, the planned period for ecological risk assessment is 2005–2050.

The region under consideration for this plan will cover the entire 2813.91 km² of Wuyishan City, including three streets (Chongan, Wuyi, Xinfeng) and seven towns (Xingcun, Xingtian, Wufu, Shangmei, Wutun, Langu, Yangzhuang), which is coordinated with Wuyishan's overall plan.

12.2.1.2 Key Planning Content

According to the investigation and assessment results of the ecosystem plan in Wuyishan City, the content will be divided into the following four systems:

1. Planning for ecological zoning, including ecologically functional zoning, landscape construction planning, and biodiversity conservation
2. Planning for the ecological industry system, including eco-tourism, eco-agriculture, and eco-industrial planning
3. Systemic planning for the ecological environment, including water use, aquatic environmental protection, planning for sustainable use energy, air pollution prevention and control, and managing solid waste pollution
4. Planning for an ecologically sensitive cultural system, including ecological awareness and literacy, a culture that prioritizes ecology and ecological habitat planning

In addition to the above four systems, ecological risk analysis on the effects of rapid urbanization will also be performed.

12.2.1.3 Planning Considerations

In this study, we have used the theory of sustainable development, setting the construction of an eco-city as the core goal, determining ecosystem value accounting as the main evaluation method by using geographical information system (GIS). Therefore, the plan for the eco-city Wuyishan focuses on developing four systems including ecological zoning, ecological industry system, ecological environment system, and ecological culture system, while an analysis of the ecological risk in the face of rapid urbanization is also performed (Liu 2011; Yang et al. 2004; Ouyang and Wang 2005; Zhang 2009). The plan highlights Wuyishan's features and explores cultural influence, and finally forms a planning scheme of "One core, two securities, four systems, and four cards." The aim is to refine the key points of eco-city construction for a typical southeast coastal tourist city and to establish technical guidelines of eco-city construction and planning in China.

12.2.2 Planning Objectives and Indicators

12.2.2.1 Planning Objectives

One objective of this plan is to establish the market-oriented brand of the "World's Natural and Cultural Heritage." A second objective is to use this brand to drive the development of other related industries with the pillar industry of tourism, while maintaining the integrity of natural ecosystems at the same time. Wuyishan City will be built into an ecological, modern, international tourist city, and the main provincial historical and cultural city.

12.2.2.2 Planning Indicators

12.2.2.2.1 Characteristic Indicators of Wuyishan City

Referring to the eco-city assessment indicators, 13 characteristic indicators were chosen according to the requirements of constructing an eco-city and its own ecological characteristics (Table 12.1).

TABLE 12.1
Characteristic Indicators of Wuyishan City

No.	Indicator	Unit	Reference Value	Status Value	2010	2015	2020
1	The proportion of organic and green products in agricultural products	%	≥20[a]	10.2[b]	15	20	25
2	Comprehensive utilization rate of crop stalks	%	100[a]	99[b]	100	100	100
3	Availability of quality drinking water in town	%	100[a]	96	100	100	100
4	The use of fertilizer (pure)	kg/ha	<250[a]	464	300	200	150
5	Utilization rate of clean energy	%	43[c]	50	70	85	98
6	Protection rate of historical and cultural legacy	%	100[d]	100[e]	100	100	100
7	The popularization rate of secondary education[f]	%	≥99[a]	—	92	96	99
8	Hospital beds per million persons[f]	—	—	7,600	7,800	8,000	8,300
9	Buses per million persons[f]	—	—	900[b]	1,100	1,400	1,600
10	Protection rate of rare and endangered species	%	100[g]	—	95	100	100
11	The area of bare mountains	ha^2	—	680[b]	660	620	580
12	Utilization coefficient of irrigation water	—	—	0.5[h]	0.65	0.75	0.8
13	Solid waste emission per unit industrial added value	t/billion yuan	—	5,000[b]	500	100	10

Notes:

[a] Standard of eco-county.

[b] Data in 2004.

[c] The national average of the United States in 2005.

[d] Reference to other cities and requirements of the plan.

[e] On-site consultation.

[f] Only rural area.

[g] Standard of eco-province.

[h] Planning report of primary irrigation county of Wuyishan City, Fujian.

12.2.2.2.2 Planning Index System and Target Values of Eco-City Wuyishan

The planning index system of eco-city Wuyishan is mainly derived from Wuyishan's characteristic indicators and eco-city assessment indicators, integrating the ecological zoning, ecological industry system, ecological environmental system, and ecological culture system (Table 12.2).

12.3 PLANNING OF ECOLOGICAL ZONING IN WUYISHAN CITY

12.3.1 ECOLOGICALLY FUNCTIONAL ZONING

Ecologically functional zoning can completely avoid the disadvantages of traditional spatial planning, which does not consider the resources and environmental carrying capacity. In this case, the zoning plan will be built rationally and scientifically based on the premise of the resource and environment carrying capacity (Cai et al. 2010).

12.3.1.1 Analysis of the Ecological Factors Affecting Ecologically Functional Zoning

1. *Soil erosion:* With the natural features of high mountains and steep slopes, shallow soil and heavy rainfalls, Wuyishan City has an innate ecological fragility, which will result in serious soil erosion if there is damage to the surface vegetation. In 1999, the soil erosion area in Wuyishan reached 24.28 thousand square hectometers, accounting for 8.68% of the total.

2. *The ecological environmental problems caused by tourism:* The ecological environment in Wuyishan City will face tremendous pressure in the development of eco-tourism, mainly shown in the following respects: (a) the pollution caused by waste and sewage; (b) the interference with activities of organisms; (c) the hunting of wild animals seriously impacts biodiversity.

3. *The environmental pressure facing the World Heritage Sites:* Wuyishan City is under the increasing environmental pressure, including the pressure to develop and provide resources for environmental and tourism purposes.

4. *The protection of the drinking water source:* Preventing the direct discharge of domestic sewage is the main measure to protect the drinking water source. In addition, monitoring the protection of forest vegetation surrounding the Chongyang Stream is also essential to further improve its water storage and purification functions. Avoiding nonpoint source pollution caused by the use of fertilizers and pesticides will also be key to protecting drinking water.

   Industrial park construction and eco-industrial problems: In accordance with the development plan, Wuyishan City has launched construction of an industrial park in the Economic Development Zone of North Fujian. Considering that Wuyishan has determined tourism and other related tertiary industries as an economic pillar, developing the production of tourist craft, food or high-tech products must be done without pollution in the construction area. Projects that cause pollution must be forbidden.

TABLE 12.2

Planning Index System and Target Values of Eco-City Wuyishan

System	No.	Indicator	Unit	Reference value	Status Value	2010	2015	2020
A. The ecological zoning	A1	Forest coverage rate	%	≥75[a]	77.9[b]	78	78.5	79
	A2	The proportion of the protected area	%	≥20[a]	35.21[c]	35.21	35.21	35.21
	A3	Protection rate of rare and endangered species	%	100[d]	—	95	100	100
	A4	The area of bare mountains	ha	—	680[e]	660	620	580
B. Ecological industry system	B1	GDP per capita	yuan	≥25,000[a]	13,573	16,182	18,581	21,336
	B2	Financial income per capita	yuan	≥3,800[a]	656	1,500	3,800	4,200
	B3	Rural per capita net income	yuan	≥4,500[a]	4,516	6,790	9,980	14,660
	B4	Urban per capita disposable income	yuan	≥12,000[a]	9,321	13,073	19,208	28,224
	B5	The proportion of organic and green products in agricultural products	%	≥20[a]	10.2[c]	15	20	25
	B6	The proportion of the tertiary industry	%	≥45[e]	52.1	52.2	53.4	54.6
	B7	Water consumption per GDP	t/thousand yuan	≤15[a]	80[f]	60	30	15
	B8	Repeated utilization rate of industrial water	%	≥40[a]	30.06[f]	50	70	80
	B9	Energy consumption per GDP	tce/million yuan	≤120[a]	18	16	16	16
	B10	Environmental compliance rate in tourist district	%	100[a]	—	100	100	100
C. Ecological environment system	C1	Comprehensive utilization rate of crop stalks	%	100[a]	99[c]	100	100	100
	C2	Utilization rate of clean energy	%	43[g]	50	70	85	98
	C3	Utilization coefficient of irrigation water	—	—	0.5[h]	0.65	0.75	0.8

		Units					
C4	Water quality compliance of centralized drinking water source	%	100[a]	100	100	100	100
C5	Availability of quality drinking water in town	%	100[a]	96	100	100	100
C6	Centralized urban household sewage treatment rate	%	≥60[a]	12.04	50	70	100
C7	Standardized days of urban air quality	Days better than or equal to Grade II standard per year	≥333[i]	365	365	365	365
C8	Treatment rate of harmless urban household garbage	%	95[i]	50[c]	65	80	95
C9	Comprehensive utilization rate of industrial solid waste	%	95[i]	94.4[c]	95	95	95
C10	The use of fertilizer (pure)	kg/ha.	<250[a]	464[a]	300	200	150
C11	Solid waste emission per unit industrial added value	t/billion yuan	—	5,000[c]	500	100	10
D. Ecological culture system							
D1	Urbanization rate	%	≥50[a]	44.91	50	53	55
D2	Protection rate of historical and cultural legacy	%	100[i]	100[k]	100	100	100
D3	The popularization rate of secondary education[l]	%	≥99[a]	—	92	96	99
D4	Engel's coefficient[l]	%	<40[a]	46.7	44	42	40
D5	Gini coefficient[m]		0.3–0.4[a]	0.27	0.29	0.30	0.31
D6	The popularization rate of environmental publicity and education	%	>85[a]	28[c]	50	85	95
D7	Public satisfaction rate over environment	%	>95[a]	99[c]	99	100	100

(Continued)

TABLE 12.2 (*Continued*)
Planning Index System and Target Values of Eco-City Wuyishan

System	No.	Indicator	Unit	Reference value	Status Value	2010	2015	2020
	D8	Hospital beds per million persons[l]	—	—	7600	7800	8000	8300
	D9	Buses per million persons[l]	—	—	900[c]	1100	1400	

Notes:

a　Standard of eco-county.
b　Average value, not divided into mountains and hills.
c　Data in 2004.
e　Standard of eco-city.
d　Standard of eco-province.
f　Estimated.
g　The national average of United States in 2005.
h　Planning report of primary irrigation county of Wuyishan City, Fujian.
i　Reference to related indicators of eco-city Xiamen plan
j　Reference to other cities and requirements of the plan.
k　On-site consultation.
l　Only rural area.
m　Only urban area.

5. *Problems of eco-agriculture construction:* Tea, tobacco, and other agricultural production increased soil erosion; one-sided pursuit of high yields led to the excessive usage of fertilizer, resulting in increasingly serious nonpoint source pollution; the ability of water to support agricultural development will be declining because of the resultant water pollution and its distribution, both in time and space.

12.3.1.2 Sensitivity Analysis of the Ecological Environment

From the analysis of the ecosystem status, it was concluded that soil erosion, biodiversity conservation, and the protection of historical and cultural sites are the major environmentally sensitive issues in Wuyishan City.

The impact factors of the above sensitive environmental issues will be evaluated and calculated into four assessment grades (Table 12.3). Results show that the ecological sensitivity values in Wuyishan are between 1.18 and 4, with an average of 2.71 (the larger the value, the higher the sensitivity). According to the single-factor grading

TABLE 12.3

Assessment Factors of Ecological Sensitivity in Wuyishan

Weight	Sensitive Environmental Issues	Weight	Impact Factor	Grading Standard			
				More Sensitive	Sensitive	Less Sensitive	Insensitive
0.6	Soil erosion	0.2	Rainfall	>2,200	2,000–2,200	1,800–2,000	<1,800
		0.3	Gradient	>30	20–30	10–20	<10
		0.3	Soil type	Meadow soil, yellowish red soil	Red earth, purple soil	Paddy soil	Moisture soil
		0.2	Land use pattern	Slope farmland, tea plantation, building site, sparse woodland, unused land	Cultivated land	Woodland, grassland	Water area
0.4	Biodiversity conservation	1	Natural reserves	√			
0.4	Cultural relic protection	1	Heritage sites	√			
	Sensitive spots						
Grading assignment of evaluation factors				4	3	2	1
Grading standard of evaluation results				>3	2.5–3	2–2.5	<2

Note: As the biodiversity conservation and cultural relic protection are two aspects of the world's natural and cultural heritage protection in Wuyishan, which are parallel in space, they share the same weight (0.4).

standards, Wuyishan has a high ecological sensitivity on the whole, but with an obvious spatial difference (Figure 12.1).

The results of the sensitivity analysis suggest that the development of Wuyishan should go southbound, especially south of railway line, avoiding the ancient sites of Han Dynasty.

12.3.1.3 Ecologically Functional Regionalization

Based on the classification method of the 11th National Five-Year Plan (FYP), starting with current urban spatial distributions and regional environmental characteristics in Wuyishan City, and then following the principle of integrating, areal differentiation, landscape heterogeneity and environmental capacity, the ecologically functional zones of Wuyishan City will be divided into the following four categories with the help of roads, rivers, administrative boundaries, and other status elements: (1) red line zone, exploitation-prohibited zone, which requires comprehensive ecological protection and will strictly prohibits all types of construction and development activities; (2) yellow line zone, the exploitation-limited zone, which gives priority to the protection of important ecological functional areas, and allows for restricted development mainly of ecological construction for other regions; (3) green line zone, the exploitation-optimized zone, which focuses on the development of ecological construction; (4) blue line zone, the key exploitation zone, which carries out priority development mainly on ecological construction (Fu et al. 1999; Liu and Fu 1998). After that, each functional zone is further subdivided into ecological functional districts for the regulation of ecological function (Figures 12.2 and 12.3).

12.3.1.3.1 Ecologically Functional Zoning

1. *Exploitation-prohibited zone:* This zone includes World Heritage Site of Mount Wuyi, the Dongxi reservoir conservation area, the northwest and northern mountain area at the highest altitude areas of Wuyishan City, with an average altitude of more than 1000 m. It is endowed with abundant natural resources, an integrated ecosystem, rich biodiversity, and a well-preserved, broad-leaved evergreen forest community. In addition, it is the cradle of the Minjiang River and the key zone that embodies the characteristics of "natural and cultural heritage sites" and achieves a good ecological environment for Wuyishan City. Currently, the major ecological and environmental problems are bamboo cutting, tea cultivation, and serious soil erosion caused by mining. Long-term reclamation of bamboo and deforestation of the broad-leaved forests will result in pure forests of bamboo and a decrease in species diversity. Meanwhile, the phenomenon of hunting wild animals still exists, which also adds great pressure on biodiversity conservation. The contradiction between the protection of World Natural Heritage Site and the development of community residents becomes more and more incongruous.

2. *Exploitation-limited zone:* This zone includes the existing water source (Xixi), the catchment area of the alternate water source (Dongxi), and the eastern hilly and ecologically fragile area. The major environmental problems are deforestation pressure, severe soil erosion, and nonpoint source

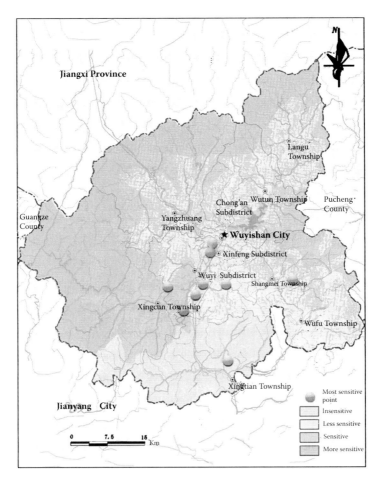

FIGURE 12.2 (**See color insert.**) Ecological environmental sensitivity evaluation map of Wuyishan City.

pollution caused by aquaculture wastewater and fertilizer use. Both the urban and the rural ecological environments need to be improved.

3. *Exploitation-optimized zone:* This zone includes the urban and tourist area of Wuyishan City, with a high level of development, dense population, large amounts of industrial enterprises, and rich cultural heritage. The major environmental issues are the low forest coverage rate, simple ecosystem structure, tea plantation resulting in deforestation, and environmental pollution, especially severe water pollution. Meanwhile, the disorder of construction projects has some effects on the landscape of scenic spots. In general, the overall urban landscape quality is not high.

4. *Key exploitation zone:* This zone includes parts of Xingtian and Wufu Townships with relatively flat terrain, convenient transportation, and a high level of agricultural development. It is also the location of the Wuyi New Area, where the Xiandian startup hub has begun to take shape. The future

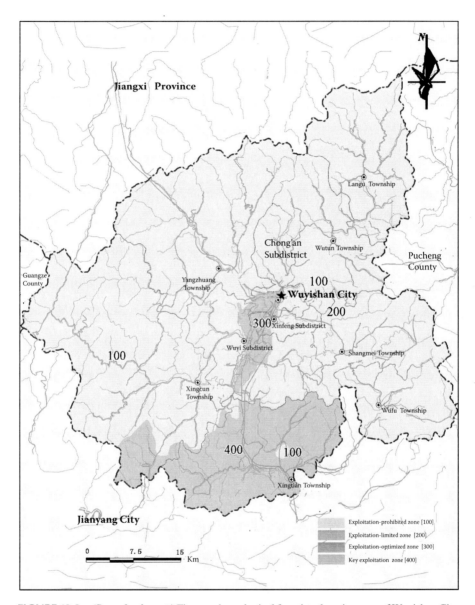

FIGURE 12.3   (See color insert.) First-grade ecological functional zoning map of Wuyishan City.

development of this zone will focus on the concept of "Ecology," namely eco-industry and eco-agriculture.

12.3.1.3.2   Ecological Functional Districts and Regulating Guidelines

In accordance with the structure and function of each ecologically functional district, the formulation of the regulation guidelines and protection measures is shown in Table 12.4 (Figure 12.4).

TABLE 12.4
Division Types of the Ecological Functional Zones in Wuyishan City

Ecological Functional Zoning	Ecological Regulating District	Position of Main Ecological Service Function	Regulating Guidelines
Exploitation-prohibited zone (100)	World heritage conservation zone (101)	Biodiversity conservation, protection of the world cultural and natural heritage	Strictly implement protection; strengthen the management and protection of soil erosion area; control the planting scale of tea and the deforestation of bamboo; protect and repair cultural sites; develop ecotourism moderately in the ecological buffer zone
	Dongxi Reservoir conservation zone (102)	Water conservation, protection of water quality	Prohibit damage to the reservoir environment and any activities against water-source including forests, river bank protective belts, and water source protections; eliminate the use of highly toxic and persistent pesticides; control and prevent soil erosion; protect water quality of reservoir
	Water and forest ecosystem conservation zone (103)	Maintenance and improvement of the capacity of source runoff and water conservation, conservation of soil and water, biodiversity conservation	Strictly implement protection; prohibit deforestation; strengthen the construction of ecological forest
Exploitation-limited zone (200)	Xixi water conservation zone (201)	Xixi's water conservation, conservation of soil and water, landscape corridor in sight of traffic	Control and prevent soil erosion; protect the quality of drinking water source; gradually move the Yangzhuang industrial zone; encourage the development of characteristic agriculture and ecological agriculture; strengthen landscape construction along the roadways
	Dongxi water conservation zone (202)	The conservation of soil and water in Dongxi valley; moderate forestry and eco-agriculture	Overall enclosure to mountains; strengthen the management of ecological forest and soil erosion areas; prevent nonpoint source pollution from the farmland; develop eco-agriculture mainly of grain and tobacco, and moderate bamboo forestry industry
	Zone of Shangmei Township (203)	The conservation of soil and water in the Meixi valley; moderate eco-agriculture	Strengthen the construction of ecological forest; use forest resources rationally and improve the comprehensive utilization efficiency of timber; develop eco-agricultural mainly on grain, lotus seeds, and edible mushrooms

(Continued)

TABLE 12.4 (*Continued*)
Division Types of the Ecological Functional Zones in Wuyishan City

Ecological Functional Zoning	Ecological Regulating District	Position of Main Ecological Service Function	Regulating Guidelines
exploitation-optimized zone (300)	Zone of Wufu Township (204)	Eco-agriculture	Strengthen basic farm construction and control the use of fertilizer and pesticides; develop eco-agricultural mainly on grain, tea, and edible mushrooms;
	Zone of Chong'an Township (205)	Eco-agriculture	Strengthen the comprehensive regulation of the urban and rural environments; develop eco-agricultural mainly on tea and vegetable cultivation; prevent nonpoint source pollution of the Chongyang stream; develop ecotourism
	Urban core zone in Wuyishan City (301)	Ecotourism and green residential environment	Initiate regulatory planning and comprehensive treatment of environmental pollution, especially for water pollution and restoration and construction of water ecology; strengthen reforestation of urban area and main road, realize ecological network construction of the whole space and improve the overall landscape quality; improve the infrastructure of tourism to achieve the goal of being recognized as an international tourist city worldwide
	Tourist resort zone (302)	Travel service	Strengthen the management of tourism and catering industry, domestic sewage and garbage treatment; improve service quality and level; strengthen reforestation, and improve landscape quality of holiday resorts; build an integrated economic and service area involving tourism and commerce
Key exploitation zone (400)	Eco-industrial zone (401)	Eco-industry	Insist on the processing of high value-added products mainly on the deep-processing of tea, tourist food processing (including tea-related beverages and food), and bamboo products; strictly prohibit enterprises of high pollution and low energy consumption; strengthen the recycling economy and the construction of an eco-industrial park; accelerate the construction of environmental infrastructure
	Eco-agricultural zone (402)	Eco-agriculture	Strictly enforce provisions of relevant laws and regulations and strengthen the protection of basic farmlands; reduce the use of fertilizers and pesticides; increase the usage of organic fertilizer; develop a production base construction of tea, organic vegetable, lotus seeds, and edible mushrooms; expand green food base; and establish stereoscopic agricultural and ecological agricultural model integrating various related industries

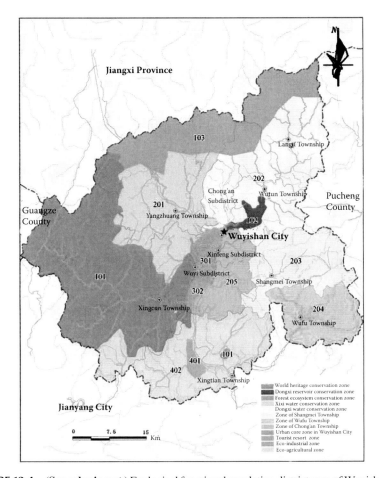

Jiangxi Province

Lanxi Township

103

202

Chong'an
Subdistrict
Wutun Township

Pucheng
County

201
Yangzhuang Township

102

★Wuyishan City

Guangze
County

Xinfeng Subdistrict
301
Wuyi Subdistrict
205
205
203
Shangmei Township

101

302

204
Wufu Township

Xingcun Township

401
101

402

Xingtian Township

World heritage conservation zone
Dongxi reservoir conservation zone
Forest ecosystem conservation zone
Xixi water conservation zone
Dongxi water conservation zone
Zone of Shangmei Township
Zone of Wufu Township
Zone of Chong'an Township
Urban core zone in Wuyishan City
Tourist resort zone
Eco-industrial zone
Eco-agricultural zone

Jianyang City

0   7.5   15
          Km

FIGURE 12.4 **(See color insert.)** Ecological functional regulating district map of Wuyishan City.

12.3.2 Landscape Construction Planning

Wuyishan City is located in the important water source conservation area in the upper Minjiang River. It is one of the world's natural and cultural heritage sites with good conservation of natural resources, rich species, and high forest coverage rate of 77.9%. Through the dynamic analysis of the landscape patterns in 1995, 2000, and 2005, our results reveal that Wuyishan has suffered a serious degradation of the landscape, showing a degraded mode of "forest-sparse woodland-grassland," which has serious adverse effects on the conservation of biodiversity. Therefore, the construction of the landscape pattern in Wuyishan should be based on landscape restoration.

12.3.2.1 Landscape Construction Planning Objectives

Based on the restoration of the degraded landscape, the objective of the construction plan will be to build the landscape pattern of "Six sources and one belt, one trunk and six branches." This means constructing and protecting the six population sources,

TABLE 12.5

Planning Index System of Landscape Construction

Indicator	Unit	Index Standards	Status Value	2010	2015	2020
Forest coverage rate	%	≥75[a]	77.9[b]	78	78.5	79
The proportion of the protected area	%	≥20[a]	35.21[b]	35.21	35.21	35.21
Protection rate of rare and endangered species	%	100	—	95	100	100
Recovery rate of degraded land	%	≥90	—	30	60	90
The area of bare mountains	ha	—	680[b]	660	620	580

Notes:

[a] Reference to the construction index of eco-county and eco-city.

[b] Data in 2004.

building annular biodiversity conservation belt consisting of the eastern, northern, and western mountain, opening up the north–south artery (Xi Stream–Chongyang Stream), and finally achieving the ecological pattern system of "going through north to south, radiating east to west." The objective is to build the landscape pattern with a reasonable layout, complete structure, and perfect function, which integrates mountains, water, city and fields, and blending towns and nature (Wu 2000; Zhang 1999). The planning index system of landscape construction is shown in Table 12.5.

The overall objective of landscape restoration in Wuyishan is as follows: by 2020, restore the landscape, which has degraded between 1995 and 2005 to its original landscape type. The total area for restoration is 28.01 thousand hectares, including 1.33 thousand hectares of bushes to be restored to forests, 1.33 thousand hectares of grassland to be restored to forests, 10.67 thousand hectares of grassland to be restored to sparse woodlands, and 13.34 thousand hectares of grassland to be restored to bushes.

12.3.2.2 Landscape Planning Contents

12.3.2.2.1 Design of Population Source

A population source is the distribution area of existing or potential native species, constituting the source point for species expansion and maintenance. It is the main protection target in the landscape construction planning. This plan will construct and improve six population sources of the Wuyishan area (Table 12.6) and 50 biodiversity conservation sites. An annular conservation belt of population sources is composed of the population source of the national nature reserve, nature reserves of the upper Jiuqu Stream, and drinking water source protection areas of the upper Xi Stream and Dong Stream. In addition to the six population sources in the table, we should focus on the population sources potential development of the upper Mei Stream.

12.3.2.2.2 Corridor Construction

According to the river distribution of Wuyishan City, build the secondary corridor of "One trunk and six branches" (Figure 12.5).

TABLE 12.6
Population Source of Wuyishan City

Name	Location	Area (hm²)	Main Protection Objects
Wuyi Mount nature reserve	Huanggang Mount	565,270	Forest ecology, rare animals and plants
Nature reserves of the upper Jiuqu Stream	Xing Village to the both sides of Pikeng, Liyuan, and Sixin Streams	3,345.33	Forest ecology, rare animals and plants
Drinking water source conservation area	Watershed of Shixiong to Da'an	2,096.3	Vegetation
Macaque protection area of the upper Dong Stream in Zhang Village	Langu Town, Zhang Village	4,57.4	Macaque, silver pheasant, and pangolin
Dong Stream reservoir nature reserve	Dam sites to Wutun reservoir and both sides	2,166.67	Forest ecology, rare animals and plants
Scenic spot	In scenic spot	6,000	Forest vegetation, flora and fauna park

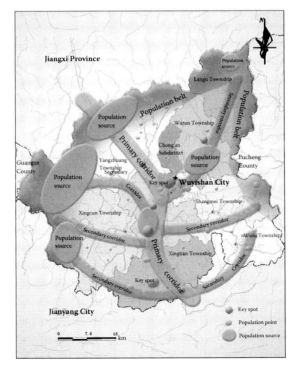

FIGURE 12.5 **(See color insert.)** Landscape pattern in Wuyishan City.

One trunk: Xi Stream–Chongyang Stream go through from north to south in Wuyishan City, constituting the central axis of the corridor. The upper Xi Stream is rich in vegetation and species, and also the drinking water source for the conservation area, requiring a 500 m buffer zone at the unilateral bank where human activities are limited. Both sides of the river should retain the necessary width of 30 m. On both sides of the Chongyang Stream, more than 30 m green belt of the ecological landscape will be built, and a riverside leisure park is designed, highlighting the theme of Wuyi tea culture, and historical and cultural landscapes.

Six branches: the **Dong Stream** flows through the reservoir, and is the outward radiant channel of the macaque population source in the Zhang Village-reservoir population source, requiring a 300 m buffer zone with a 30 m reserved width. **Huangbai Stream and Jiuqu Stream** are the outward radiant channels of national nature reserves and nature reserves of the upper Jiuqu Stream, requiring a corridor width of 60 m, 600–1200 m if the conditions permit, in order to meet the function of plant and animal migration and biodiversity conservation. **Tan Stream and Mei Stream** basins include the important cultural heritage protection areas, guaranteeing the 25 m green belt of historical and cultural heritage sites to improve the ornamental value of the former residences and relics. Meanwhile, the construction of the Mei Stream corridor will play an important role in the development and protection of upstream species. The **Chenghu Stream** flows through the eco-agricultural economic zone in the southern part of Wuyishan City, requiring 10–20 m corridor along the river, which not only helps the improvement of the water's self-purification capacity, but also prepares this system for the increasing pollutant emissions from the industrial park construction. Further, this barrier can help achieve the barriers between the agricultural patches and reduce soil erosion in the region.

12.3.2.2.3   Seven Key Points

The key point is the location, which has an important role in the maintenance of ecological process continuity. These seven key points include (Figure 12.5): the downtown of Wuyishan City, Wuyi Town, Yangzhuang Village, Wutun Village, Xingtian Town, Provincial Huangtu farm, Wuyi Palace—Residence (the confluence of Jiuqu Stream and Chongyang Stream). The most critical point is the downtown of Wuyishan City.

The downtown of Wuyishan City is located in the confluence of Dong Stream, Xi Stream, and Chongyang Stream, a high-emergy zone through which the ecological corridor goes, and a vital strategic point in the landscape construction. The planning will adopt the basic mode of "Ring, Belt, District, Park." The peripheral green ring will be built to rely on the green belts on both sides of Chongyang Stream, Small Wuyi Park, and Duba ecotourism economic park. In the inner city, the ecosystem that combines the block and the mesh green space integrates shrubbery, grass, and flowers, and will be built to blend the urban and rural areas. The aim is to reduce the urban resistance to population source radiation channel and to improve the ecological carrying capacity of Wuyishan City.

Other key points are all located at the confluence of river corridors, with a variety of habitats. Planning will designate construction areas of ecological nodes, where the ecological construction should be closely combined with the construction of water corridors, and kept up with the construction requirements of ecological nodes in terms of the widths of corridors and species mix. This will also prohibit cultivation, hunting, mining, and construction of any production facilities.

12.4 PLANNING OF ECOLOGICAL INDUSTRY SYSTEM IN WUYISHAN CITY

12.4.1 INDUSTRY STATUS AND THE SWOT ANALYSIS

12.4.1.1 Analysis of Industry Status

Figure 12.6 and Table 12.7 reflect the economic development status of Wuyishan City between 1995 and 2005. The notable feature is that the proportion of the service industry is relatively high, which verifies that the Wuyishan City has strictly enforced the industrial access system after establishing tourism as the pillar industry, and gradually affected the industrial structure.

According to the characteristics of Wuyishan City, several indicators, as shown in Table 12.8, were selected and compared with the national level. Results show that the specialization ratio of tourism revenue is the highest, which is in accordance with the pillar industry in Wuyishan City. Meanwhile, the specialization ratios of primary and tertiary industry are greater than 1, indicating that both of them are likely to become the pillar industry in this region. The specialization ratio of the secondary industry is less than 1, demonstrating that the industry remains in development. Based on the above analysis, the related industries of tourism in Wuyishan City are agriculture and industry.

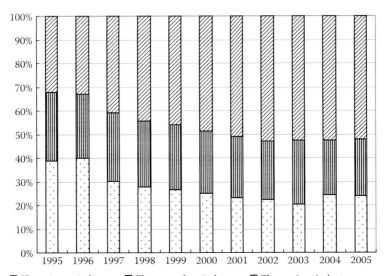

FIGURE 12.6 The industrial structure in Wuyishan City.

TABLE 12.7
Industrial Added Value and per Capita GDP in Wuyishan City, 1995–2005

	1995	1996	1997	1998	1999	2000	2001	2002	2003	2004	2005
Primary industry (10^4 yuan)	34,617	42,697	42,101	42,476	43,240	43,670	44,456	47,094	48,140	64,607	71,743
Secondary industry (10^4 yuan)	25,958	28,886	40,684	42,089	44,500	45,175	48,680	52,383	63,885	60,882	72,286
Tertiary industry (10^4 yuan)	28,847	35,404	57,243	67,130	74,660	84,685	97,464	111,117	123,815	138,882	157,786
Per capita GDP yuan	4,293	5,112	6,639	7,165	7,649	8,109	8,820	9,687	10,779	12,009	13,571

TABLE 12.8
Specialization Ratio of Added Value in Each Sector

Indicator	National Level (billion yuan)	Wuyishan City (million yuan)	Specialization Ratio of Added Value (%)
GDP	18,232.1	3,028.5	—
Primary industry	2,271.8	717.4	1.901172
Secondary industry	8,620.8	722.9	0.504799
Industry	7,619.0	432.4	0.341633
Construction	1,001.8	290.5	1.74573
Tertiary industry	7,339.5	1,577.9	1.294262
Tourism revenue	768.6	1,551.0	12.14852

12.4.1.2   SWOT Analysis

Various factors influencing the industry development were identified by the method of SWOT analysis, so as to provide direction for more rational planning. The analysis results are shown in Table 12.9.

12.4.2   ECOLOGICAL INDUSTRY PLANNING OBJECTIVES

12.4.2.1   Planning Objectives and Indicators

The overall objective of industrial planning in Wuyishan City is industrial restructuring, the construction of industrial system, research and development, construction of ecological parks and infrastructure, ecological environment protection, and so on. This plan will promote the common development of eco-industry, eco-agriculture, and ecotourism, finally formulating the interrelated and stable industrial structure.

The economic gross indicators in Wuyishan City are shown in Table 12.10. The planning index system of eco-industry is divided into three categories: ecotourism, eco-agriculture, and eco-industry, containing 18 indicators totally. They will each be described in detail separately.

12.4.2.2   Planning Thoughts

The eco-industrial planning for Wuyishan City focuses on industrial restructuring. Based on the indicators of eco-city construction, this restructuring should increase the proportion of the tertiary industry, reaching 54.61% in 2020. Restructuring will also accelerate the development of the primary and secondary industries by the development of the tertiary industry, with the development mode of the eco-industry and eco-agriculture.

12.4.3   ECOLOGICAL INDUSTRY PLANNING CONTENT

12.4.3.1   Ecotourism Planning

12.4.3.1.1   Tourism Status in Wuyishan City

The tourism revenue in Wuyishan City has been increasing in recent years. In 2005, the total tourism income reached 1.551 billion yuan, and 3.75 million visitors have travelled here (Figure 12.7).

TABLE 12.9
Industrial Development Measures Based on SWOT Analysis

SWOT Analysis	Disadvantage (Threat)		Advantage (Opportunity)		
	Intensive Regional Competition	Deterioration of Ecological Environment	Tourism under the Development Stage	Location Advantage	Policy for Industry Development
Measure	Innovate tourist routes and excavate ecological and cultural connotation	The protection of ecological environment	Establish tourism as the pillar industry	Strengthen regional cooperation	Guide the development of ecological industry

SWOT Analysis	Industrial Advantages		Industry Disadvantage			
	Rich Tourist Resources	Mineral Resources	Weak Economic Foundation	Scattered Distribution of Tea Processing Enterprise	The Shortage of Talents	Tourism at the Low Level
Measure	Tourism drives the development of other related industries	Scientific development and reasonable usage	Develop the ecological industrial park based on the advantage	Industrialization of tea	Strengthen the cultivation and introduction of talents	Regulate tourism management, develop tourist products and activities, tourist infrastructure construction

TABLE 12.10
Economic Gross Indicators in Wuyishan City

No.	Indicator	Unit	Reference Value	Status Value	2010	2015	2020
1	GDP	billion yuan	—	3.028	5.319	8.566	13.795
2	Primary industry	billion yuan	—	7.17	9.15	12	16.55
3	Secondary industry	billion yuan	—	7.23	14.84	24	41.38
4	Tertiary industry	billion yuan	—	15.78	29.2	49.66	80.02
5	Industrial production value	billion yuan		9.78	12.48	17.51	28.2
6	Industrial structure	—	—	24:24:52	17.2:27.9:54.9	14:28:56	12:30:58
7	GDP per capita	yuan	≥25,000[a]	13,571	22,000	32,000	49,000
8	Annual financial income per capita	yuan	≥3,800[a]	656	1,500	3,800	4,200
9	Urban per capita disposal income	yuan	≥12,000[a]	9,321	13,073	19,208	28,224

Note:
[a] Reference to the standard of eco-county.

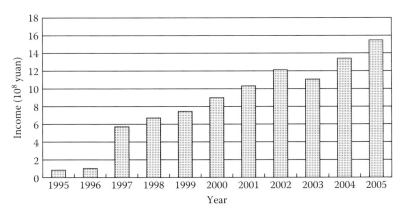

FIGURE 12.7 Increase of tourism revenue, 1995–2005.

After the calculation of the environmental capacity for tourists, we can determine that the maximum number of daily visitors in Wuyi Mount scenic spot is 16,844, and then the maximum number annually will be 6.14 million. The number of visitors in 2005 was 3.75 million. If the development of tourism maintains the status quo, with the growth rate of 12.23%, Wuyi Mount scenic spot will reach saturation in 5 years.

12.4.3.1.2 Planning Indicators and Focus

12.4.3.1.2.1 Planning Indicators   The indicators of the proportion of the tertiary industry and the public satisfaction rate over environment have exceeded the national standards for an eco-city. The popularization of environmental publicity and education will be the focus in the ecotourism planning in Wuyishan City (Table 12.11).

12.4.3.1.2.2 Planning Focus   According to the gap analysis of tourism, planning should focus on the following points: excavate and innovate tourist routes against the danger of saturation in the scenic spot at Wuyi Mount; ecological environment protection is the focus of ecotourism planning; carrying out industrial tourism and agricultural tourism to enrich tourism development; and continue developing the tourist infrastructure and related service industry.

12.4.3.1.3 Development Measures of Ecotourism

a. The plan must take care of the relationship between ecological tourist stakeholders. First, carry out effective management to the tourists entering the ecotourism zone. Second, focus on satisfying the legitimate interests of local residents. Finally, develop environmental education for tourist managers.
b. Environmental impact assessment and environmental audit results are of great significance for the implementation of ecotourism, environmental protection, and sustainable development.

TABLE 12.11
Planning Indicators of Ecotourism

No.	Indicator	Unit	Reference Value	Status Value	2010	2015	2020
1	Popularization rate of environmental publicity and education	%	>85[a]	28	62	86	90
2	Public satisfaction rate over environment	%	>95[a]	99	99	100	100
3	Proportion of the tertiary industry	%	≥45[b]	52.1	52.2	53.4	54.6

Notes:
[a] Reference to the standard of eco-county.
[b] Reference to the standard of eco-city.

c. Innovate tourist routes and excavate historical and cultural connotation: highlight the theme of "World Heritage Sites" and open up new tourist routes.

d. Develop tourist products and activities with Wuyishan characteristics, specifically leveraging the development potential of Wuyishan tea.

e. Carry out ecological environmental protection. First, pay attention to the environmental protection publicity and education. Second, build the coordination mechanism of tourism development and environmental protection. Third, establish and strengthen the management agencies for tourist assets and ecological environment. Finally, formulate the tourism management regulation to build a good management structure.

f. Regulate tourism management, focusing on building a supervision center for visitor regulation and strengthen the control functions of the tourism supervision. Meanwhile, consider strengthening the integration and restructuring of Wuyishan tourism industry, making the tourist enterprises bigger and stronger.

g. Accelerate the construction of tourist facilities including attractions, three-dimensional traffic network, hydropower, and communications. In addition, try to make attempts of MICE (Meeting, Incentive, Convention, and Exhibition) tourism, industrial tourism, and agricultural tourism.

h. While vigorously developing the tourism industry, improve the proportion of service industry, optimize the service structure, and promote the comprehensive and rapid development of service.

12.4.3.2 Eco-Agriculture Planning

12.4.3.2.1 Agriculture Status in Wuyishan City

The output value of the primary industry in 2005 was 1.16 billion yuan, in which agriculture accounted for 48.3%. At present, Wuyishan City has formed an agricultural structure featuring three industries including rock tea, flue-cured tobacco, and high-quality rice. Other industries are flourishing such as mushroom, dry taro, flowers, fruits, and vegetables.

Livestock pollution is the major contributor to agriculture pollution. In addition, fertilizer loss has become the main source of nitrogen and phosphorus pollution in water, accelerating eutrophication. Further, residual plastic film mulches reduce the infiltration function and water content of the soil, all of which have negative effects on agricultural production and our daily life.

Soil erosion in Wuyishan City cannot be ignored. The phenomenon of disorderly digging and spoiling has occurred commonly, while, in contrast, projects for protective engineering and vegetation restoration are rare. And there still exists 300.13 hectares of cultivated land with the slope of more than 25 degrees.

12.4.3.2.2 Planning Indicators and Focus

12.4.3.2.2.1 Planning Indicators
The indicator of rural per capita net income is up to the standard, while urban per capita disposal income falls, indicating that peasants' livelihood is relatively well-off, and employment in agriculture has more

TABLE 12.12
Planning Indicators of Eco-Agriculture

Indicator	Unit	Reference Value	Status Value	2010	2015	2020
Rural per capita net income	yuan	≥4500[a]	4516	6790	9,980	14,660
Engel's coefficient	%	<40[a]	46.7	44	42	40
Proportion of organic and green products in agricultural products	%	≥20[a]	10.2[b]	15	22	30
Comprehensive utilization rate of excrement in the large-scale livestock and poultry farms	%	≥90[a]	—	80	91	95
Use of fertilizer (pure)	kg/ha	<250[a]	464[c]	300	200	150
Comprehensive utilization rate of crop stalks	%	100[a]	99[b]	100	100	100
Proportion of new energy in the rural household energy consumption	%	≥30[a]	—	20	40	60
Popularization rate of secondary education	%	≥99[a]	—	95	100	100

Notes:
[a] Standard of eco-county.
[b] Data in 2004.
[c] Calculation of the 2005 statistics yearbook.

advantages over industry. The indicators of Engel's coefficient and the proportion of organic and green products in agricultural products are substandard, demonstrating that organic and green products in crops are far from the standards of ecological agriculture. This provides the direction for the next step of eco-agriculture construction. Fertilizer intensity is also far from the standard, correlating with the low proportion of organic green products in Wuyishan City. Engel's coefficient with a gap of 17% indicates the need to further improve farmers' income (Table 12.12).

12.4.3.2.2.2 Planning Focus According to the gap analysis of agriculture, planning should focus on the following points:

1. Adjust and optimize the agricultural structure, minimizing the investment of resources and waste to the lowest and maximizing both output and economic benefits.
2. Vigorously promote the industrialization of agriculture, and set up leading enterprises of processing and circulation.

3. Strengthen scientific and technological agriculture, and improve the business skills of technical staff in rural areas.
4. Ecological environmental management and restoration. Livestock pollution control is the focus of eco-agriculture planning in Wuyishan City.

12.4.3.2.3 Development Measures of Eco-agriculture

12.4.3.2.3.1 Countermeasure of Eco-agriculture Development Given the agricultural characteristics in Wuyishan City, the following suggestions are made:

1. Emphasis on the development of new agricultural talents.
2. Considering the principle that agriculture services tourism, several ecological agriculture development modes can be considered: soil and water conservation, integral agroforestry system, ecological food chain, and the integral agricultural and industrial system.
3. Expand acquisition of agricultural products, and establish both large-scale agro-processing factories and market channels, to expand the development pathway for agriculture in Wuyishan City.
4. Build and expand biogas digesters to both solve the fuel shortage of rural life and play a positive role in environmental management and ecological restoration.
5. Promote harmless processing methods of agricultural products, and establish leading processing industries, finally realizing agricultural industrialization.
6. Protect and improve transportation routes of agricultural products, and strive to make the number of roads through the administrative village achieve 100% in 2010.
7. To open up agricultural markets and increase farmers' income, the following marketing strategies should be made: cultivate the circulation subject of agricultural products to move toward systematization and mass production; promote an agricultural supermarket chain; enhance the brand marketing of agricultural products.

12.4.3.2.3.2 Develop Recreational Agriculture Recreational agriculture has changed the traditional agricultural development mode based on cultivation and load capacity, making agriculture become repeated selling of resources, with higher economic benefits than traditional agriculture. Wuyishan City is a tourism-oriented city. With its beautiful scenery and rich cultural landscape, the development of recreational agriculture not only saves resources and energy, but also meets the demand of urban residents for rural agricultural life experience. Wuyishan's recreational agriculture can include the following types: sightseeing plantations, sightseeing farms, and agricultural parks.

12.4.3.2.3.3 Agricultural Pollution Control Through the above-mentioned description of the agricultural pollution status in Wuyishan, the largest source of agricultural pollution is from livestock pollution, followed by pesticides, fertilizers, mulches, and soil erosions.

1. *Livestock manure and fertilizer pollution control:* Apply the ecological farming model of "source control and recycling" to prevent livestock pollution. First, control the source, focusing on approvals of the construction projects and the implementation of pollutant permit system under the total quantity control. Second, promote the establishment of biogas digester in each village, forming a variety of ecological chains that link to biogas.
2. *Pesticide pollution control:* Adjust the pesticide structure and control the amount of pesticide application. Encourage the combination of biological control, physical control, and chemical control, and make full use of natural enemies to reduce the pesticide use. Promote the use of biopesticides and new pesticides with high efficiency, low toxicity, and low residue.
3. *Mulch pollution control:* Source control is an effective way for the difficult degradable pollutants. Encourage farmers not to use plastic film, and promote biodegradable plastic films. In addition, set relevant policies for plastic film recycling based on the principle that the "polluter pays."
4. *Soil erosion control:* Improve farmland capital construction and reform cultivating sloped land, combining both biological and engineering measures. Enhance the protection of the existing woodland, and soil and water conservation facilities. Destruction of vegetation caused by mining, repairing roads, and building factories should be restored as soon as possible.

12.4.3.3 Eco-Industry Planning

12.4.3.3.1 Industry Status in Wuyishan City

The industrial added value in Wuyishan City accounted for 52% of GDP, increasing at an average annual rate of 3.35% from 1995 to 2005. The gross industrial output ratio of light industry to heavy industry was 2.74:1 in 2000 from 1.1:1 in 1995, indicating that the light industry development is faster than the heavy industry development. The features of the industry in Wuyishan City are that the manufacturing sector has the largest share, including wood processing industry, textile and garment manufacturing, plastics processing, and beverage manufacturing industry. In general, most enterprises are small scale and decentralized. In addition, the township enterprises in Wuyishan City are gradually developing, with the output value of 536.34 million yuan in 2000, accounting for 72% of the total industrial output value.

The overall environmental quality of Wuyishan City is good. Industrial pollution has declined with the structure adjustment in recent years. In particular, as a result of the "World Heritage Sites" declaration, large-scale environmental renovation works have been carried out, bringing significant improvement in environmental quality.

12.4.3.3.2 Planning Indicators and Focus

12.4.3.3.2.1 Planning Indicators   In addition to the two indicators without statistics, all the indicators were substandard in the base year (2005). This shows that clean production and industrial environmental management need further improvement. Meanwhile, there is still a big gap between the current situation and eco-city standard in particular in terms of the indicator of annual per capita income (Table 12.13).

TABLE 12.13
Planning Indicators of Eco-Industry

No.	Indicator	Unit	Reference Value	Status Value	2010	2015	2020
1	GDP per capita	yuan	≥25,000[a]	13,571	16,182	18,581	21,336
2	Annual financial income per capita	yuan	≥3,800[a]	656	1,500	3,800	4,200
3	Urban per capita disposable income	yuan	≥12,000[a]	9,321	13,073	19,208	28,224
4	The proportion of investment in environmental protection to GDP	%	≥3.5[a]	3	3.5	3.6	3.8
5	The proportion of enterprises implementing cleaning production	%	100[a]	—	15	40	60
6	The ratio of large-scale enterprises passing the ISO14000 certification	%	≥20[b]	—	5	10	20

Notes:
[a] Reference to the standard of eco-county.
[b] Reference to the standard of eco-city.

12.4.3.3.2.2 Planning Focus According to the gap analysis for industry, the focuses for planning are described as follows:

1. Eco-industry development should be combined with ecotourism and eco-agriculture, focusing on deep processing of agricultural products (such as tea) and further development of manufacturing of tourist products (such as tourist souvenirs).
2. Develop high-tech industrial system with low or even no pollution. Industrial lands should be distributed rationally, and located in areas that have less impact on the urban environment. Construction should follow the principles of garden style and eco-parks. In addition, new industrial projects should implement a strict environmental impact assessment system, and pollution prevention projects, as well as ecological and environmental protection facilities should be designed, constructed, and put into operation simultaneously with the principal part of new projects.
3. Promote clean production and centralized control of industrial pollution. Implement the total quantity control of pollutants, and strengthen the comprehensive management of pollution sources.
4. Establish eco-industrial parks, mainly on light industry.

12.4.3.3.3   Development Measures of Eco-Industry

1. Arrangement of industrial parks should be considered broadly, concentrating industries and sectors that are suitable for industrial development in Wuyishan City, to achieve optimal efficiency. In 2006, Wuyishan City began the construction of industrial parks, as part of a comprehensive tourism development project integrating industry (mainly manufacturing), tourism and agriculture—Mount Wuyi "World Cultural and Natural Heritage."
2. Develop clean production, and strengthen the regulation of industrial pollution.
3. Develop Wuyishan industries based on local advantages. Consummate the wood processing industry, which has formed a certain scale under the premise of environmental protection. Strengthen industrialization of tea industry in Wuyishan. The textile and garment industries will take a new road with greater use of new technologies, which allow for high performance and green environmental protection, relying on the development of tourism.
4. Place emphasis on talents and introduce high-tech industries to enhance industrial competitiveness.

12.5   PLANNING OF ECOLOGICAL ENVIRONMENT SYSTEM IN WUYISHAN CITY

12.5.1   PLANNING OF RIVER ECOSYSTEM

12.5.1.1   Planning Objectives and Indicators

12.5.1.1.1   Planning Objectives

The main objective of the plan for the river ecosystem is to keep the river ecosystem healthy in Wuyishan City. The main rivers can reach the water quality standards, and have plenty of water to meet the production needs of both domestic and ecological water use. Aquatic animals and riparian vegetation biodiversity will get effective conservation and restoration. This will further require a reduction of the impact of human activities on the river, and broadening of the utility of rivers. The main object also requires the following: use water economically; establish and improve the urban and rural drainage system; control and prevent agricultural nonpoint source pollution, and reduce soil erosion; enhance the landscape, cultural, entertainment functions of the urban waterfront; broaden the riparian vegetation ecosystems; strengthen the protection of aquatic animals; and protect important rivers where drinking water sources are located.

12.5.1.1.2   Planning Indicators

The planning indicators of river ecosystem construction in Wuyishan City are shown in Table 12.14.

TABLE 12.14
Planning Indicators of River Ecosystem in Wuyishan City

Indicator	Unit	Reference Value	Status Value	2010	2015	2020
Repeated utilization rate of industrial water	%	$\geq 40^a$	30.06^b	50	70	80
Water consumption per GDP	t/thousand yuan	$\leq 15^a$	80^b	60	30	15
Utilization coefficient of irrigation water	—	—	0.5^c	0.65	0.75	0.8
Collection and utilization rate of urban rainfall	%	—	—	10	15	25
Discharge attainment rate of industrial waste water	%	—	99.17^d	100	100	100
Water quality compliance of centralized drinking water resource	%	100^a	—	100	100	100
Waste water emissions per thousand yuan	t/thousand yuan	—	—	≤ 1.1	≤ 0.9	≤ 0.6
Rural sewage irrigation attainment rate	%	—	—	100	100	100
Hygienic qualification rate of drinking water in town	%	100^a	96^e	100	100	100
Centralized urban household sewage treatment rate	%	$\geq 60^a$	12.04^f	50	70	100

Notes:
[a] Standard of eco-county.
[b] Estimated.
[c] Planning report of primary irrigation county of Wuyishan City, Fujian.
[d] Urban environment comprehensive improvement quantitative assessment of Wuyishan City in 2005.
[e] Provided by the local.
[f] Environmental planning of Wuyishan City.

12.5.1.2 Health Condition Analysis of River Ecosystem in Wuyishan City

12.5.1.2.1 Volume of Water

The total length of the river in Wuyishan City is 820 km, with a total basin area of 2,861.4 km². The city's average volume of surface water resources is 3.597 billion cubic meters, and the per capita one is 16,026 m³, which is richer than the provincial level where the per capita surface water resources is 3,522 m³.

The total volume of water supply in 2005 was 405 million cubic meters. Owing to the lack of data in 2005, the water usage numbers come from 2004 instead. The total water consumption in 2004 was 237.5 million cubic meters, and the specific situation is shown in Figure 12.8. The above-mentioned analysis shows that the current water consumption situation of each department is fully guaranteed, with a sufficient

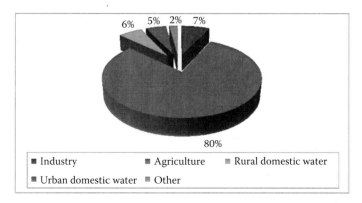

FIGURE 12.8 **(See color insert.)** Water consumption situation of Wuyishan City in 2004.

surplus of water supply capacity. The amount of agricultural irrigation water is the largest use of water, indicating that there is still potential savings to be had in agricultural water consumption.

12.5.1.2.2 Quality of Water

12.5.1.2.2.1 Main River   After years of sampling and analysis, results indicate that regionally the quality of surface water is good in general. However, with the increased population and the development of industry, agriculture, and tourism in recent years, water pollution has shown an increasing trend. However, the water quality has degraded to Class III from Class II in the most residential areas.

Four surface water monitoring sites have been set in Wuyishan City. According to the monitoring data of environmental monitoring stations in 2005, the average water quality of all the monitoring sections in different periods has reached the standard. The water quality monitoring data of the two monitoring sites was used to evaluate the water quality change in the main stream of Chongyang River from 2000 to 2005 (Figure 12.9). Dissolved oxygen and petroleum were basically steady, but with a slight increase in dissolved oxygen. Ammonia nitrogen experienced a greater decline in 2003, and rose back to the level of 2002 in 2005. BOD_5 showed a slow rise in the 6 years. All statistics pollutants reached the Class I of "Surface Water Quality Standard" (GB3838-2002), indicating water quality of Chongyang Stream has remained at a good level in recent years.

12.5.1.2.2.2 Drinking Water Source   The water quality of drinking water sources in Wuyishan City remains good. The water quality compliance rate of centralized drinking water source is 100%, and the hygienic qualification ratio of drinking water in town is 96%. According to the monitoring data of Shixiong water plant throughout the year of 2005, these evaluation results showed that the indicators of centralized drinking water sources all met the standards, with a good water quality (Table 12.15). Using the monitoring data from 2000 to 2005 to evaluate the water quality change, the result is that all statistics pollutants reached the Class I of "Surface Water Quality Standard" (GB3838-2002).

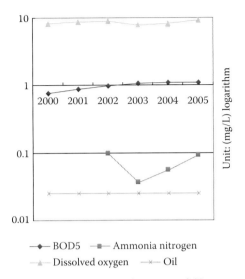

FIGURE 12.9   Water quality change in the main stream of Chongyang River.

TABLE 12.15
Water Quality Assessment of Centralized Drinking Water Source (Unit mg/L)

Assessment Factor	Standard	Wet Season	Medium Season	Dry Season	Average
PH	6–9	0.06	0.05	0.02	0.04
Dissolved oxygen $\geq$	6	0.18	0.3	0.3	0.26
$BOD_5 \leq$	3	0.34	0.36	0.38	0.36
Permanganate index $\leq$	4	0.29	0.27	0.31	0.29
Volatile phenol $\leq$	0.002	<1	<1	<1	<1
Fluoride $\leq$	1	0.25	0.22	0.21	0.23
Cyanide $\leq$	0.05	<0.001	<0.001	<0.001	<0.001
Hg $\leq$	0.00005	<1	<1	<1	<1
As $\leq$	0.05	0.1	<0.08	<0.14	<0.1
Pb $\leq$	0.01	<0.1	<0.1	0.2	<0.1
Ammonia nitrogen $\leq$	<0.5	<0.1	<0.1	<0.1	<0.1
Fecal coliforms per litre $\leq$	2,000	0.65	0.06	0.04	0.49

12.5.1.2.3   Present Situation of Aquatic Organisms

Wuyi Mount National Nature Reserve has the largest area and the best-preserved subtropical forest ecosystems in southeast China. It is also the World Biosphere Reserve. There exist many rare species of fish and amphibians in the river system, accounting for 17% and 63% of the total in Fujian Province, respectively.

12.5.1.3   Disturbance Analysis of Human Activities on the River Ecological System

12.5.1.3.1   Analysis of the Balance of the Water Supply and Demand

According to water distribution characteristics, the integrity of administrative divisions, and related planning reports, the Wuyishan City will be divided into five districts (Figure 12.10).

The forecasts of water demand are shown in Table 12.16. Results indicate that based on the assumption that river base flow remains unchanged, domestic and landscaping water demand will keep increasing, while industrial and agricultural water demands will continue declining.

When $P = 90\%$, supply and demand forecasts of each district are shown in Table 12.17. By comparison, we find that the ecological water demands in all districts have been fully met, but industrial and domestic water demands of No. 3, 4, and 5 districts are slightly insufficient. With the implementation of various water-saving irrigation measures, agricultural water demand of No. 4 district will fall 9.49 million cubic meters from 2010 to 2020. In addition, the No. 4 district is located in the downstream of Dongxi Reservoir and it will almost achieve the balance of supply and demand in 2020. The problem of industrial and domestic water supply in No. 3 and 5 districts needs to be solved firstly. The reasons for the shortage

FIGURE 12.10   (See color insert.) Zoning map of water resource in Wuyishan City.

TABLE 12.16

$P = 90\%$ **Water Demand of Different Types in Wuyishan City (10^4 m^3)**

Type of Water Demand		Base Year	2010	2015	2020
Ecological water demand	River base flow	157,600	157,600	157,600	157,600
	Landscaping water demand	106	154	214	263
Industrial water demand	Industrial water demand	1,956	1,731	1,412	1,471
	Agricultural water demand	33,644	27,049	24,033	23,101
Domestic water demand	Residents and tourists water demand	831	1,237	1,634	2,038
	Livestock water demand	98	109	120	133
Total		194,235	187,880	185,013	184,606

Note: *P* refers to probability of water supply.

of water supply of the above-mentioned two districts are different, which should be handled separately.

12.5.1.3.2 Analysis of Point-Source and Nonpoint Pollutant Discharge

Point-source pollution is just a handful of the total river pollution in Wuyishan City, but it lacks the necessary monitoring data. The emission trend is that industrial pollution gradually reduces, while domestic pollution sources keep increasing. The primary cause of nonpoint source is agricultural nonpoint source pollution. Since nitrogen is the major element of nonpoint source loss, and the amount of nitrogen fertilizer usage has increased in recent years, this plan calculated the total nitrogen loss of the agricultural nonpoint source pollution. Using the method of the output coefficient model represented by Johnnes, we can get that the nitrogen loss per unit area is 0.02 t/km². Comparing this result with the nitrogen loss of southeastern rivers of 0.5t/km², the agricultural nonpoint source pollution in Wuyishan contributes less, and therefore does not pose a greater threat to the water quality.

12.5.1.3.3 Analysis of Drainage System

Urban drainage system is imperfect in Wuyishan City. The confluence of untreated rain and sewage is spilled directly into the river through roadside channels or pipes. Drainage pipe network is insufficient, and there are no sewage treatment facilities in Wuyishan City.

12.5.1.3.4 Soil Erosion and Cause Analysis

The area of soil erosion is 242.83 km² in Wuyishan City, accounting for 8.68% of the total land area. Natural and man-made disturbances are the main causes of soil erosion in Wuyishan City. There are four main factors of natural disturbances to the soil: rainfall, slope, soil type, and vegetation cover. The area of soil erosion caused

TABLE 12.17
Supply and Demand Forecasts of Each District (10^8 m³/a)

District	Base Year		2010		2015		2020	
	Total Supply and Demand	Industrial and Domestic Supply and Demand	Total Supply and Demand	Industrial and Domestic Supply and Demand	Total Supply Domestic Supply and Demand	Industrial and Domestic Supply and Demand	Total Supply and Demand	Industrial and Domestic Supply and Demand
No. 1	−0.08	+1.13	+0.09	+1.31	+0.17	—	+0.18	—
No. 2	+2.43	−0.12	+2.56	+0.02	2.64	—	+2.64	—
No. 3	+0.75	−0.18	+0.84	−0.09	0.87	—	+0.88	—
No. 4	+0.75	−0.26	+0.92	−0.09	0.99	—	+1.01	—
No. 5	+0.34	−0.17	+0.41	−0.10	0.44	—	+0.44	—
Total	+4.19	0.40	+4.82	+1.05	+5.11	—	+5.15	—

Note: "+"indicates supply exceeding demand; "−"indicates demand exceeding supply.

by man-made disturbance such as excessive logging, unreasonable mountain and sloping field development, and disorderly digging occupies 92% of the total erosion.

12.5.1.3.5 Analysis of Disturbance on Fish

Wuyishan City has promulgated a series of policies and regulations to ensure the protection of aquatic life. Due to the good water quality, it is suitable for the survival of fish, amphibians, and other aquatic organisms. Currently, the existing negative impacts include the phenomenon of killing wild animals for food, the threat of over-fishing, tourist disturbance on water, and the influence of water storage projects.

12.5.1.3.6 Analysis of the Influence on the Riparian Ecological System

The main influences on the riparian ecosystems include the following: poor urban waterfront landscape; water culture has not been prominent; riparian zones are occupied by farmlands and buildings; and garbage pollution. The lack of ecological revetment will affect normal material metabolism between water and lands.

12.5.1.4 Planning Content

Based on the above analysis and combined with the characteristics of Wuyishan City, this plan will put forward several protective measures for the river ecosystem from the following aspects: drinking water source protection, drainage system planning, nonpoint source pollution prevention and control, soil and water conservation, water landscape and riparian zone management, and aquatic ecosystem conservation.

12.5.1.4.1 Planning of Water Supply and Drainage

12.5.1.4.1.1 Planning for the Water Supply   Based on the analysis of the balance of the water supply and demand, the focus areas of the water supply are No. 3 and 5 districts. The type of water supply is meant to meet the agricultural water demand and guarantees the basic industrial and domestic water at the same time.

No. 1, 2, and 4 districts can solve the problem of production and domestic water demand in 2010, with no trouble in the future as well. The No. 3 district has insufficient industrial and domestic water supplies due to too much agricultural irrigation water consumption. But with many nature reserves and scenic spots, the No. 3 district should prohibit the construction of water storage projects to avoid disturbance on the river ecosystem. Therefore, plans to increase investments in a diversion project add 5 million cubic meters of water to the No. 3 district and 4 million cubic meters to the No. 2 district. The No.5 district will use comprehensive measures to address the problem of inadequate water resources by planning to establish small reservoir in 2010 with the capacity of 1 million cubic meter. In addition, water-saving irrigation measures should be strictly implemented. Plan to add 5 million cubic meters to the water supply by 2015 and 8 million cubic meters in 2020 will solve the problem of water supply.

Implementation of water supply measures of sustainable development include: encouraging the collection and use of rainwater; collecting surface runoff from rainwater, which will be used as cooling water, irrigation water, and groundwater sources (the collection and utilization rate of urban rainfall will reach 25% in 2020); developing water plants and improving the construction of diversion projects (the water supply rate of administrative villages will reach 100% in 2009); and carrying out centralized water supply in rural areas.

12.5.1.4.1.2   Planning for Water Conservation   Owing to the characteristics of abundant water resources, industries dominated by agriculture, a weak economic foundation in Wuyishan City, planning should focus on agricultural and rural water conservation while encouraging industrial and urban water conservation at the same time.

1. *Agricultural irrigation water planning:* The irrigation and drainage projects in Wuyishan City are seriously aging and ineffective, meaning there is an opportunity for great water-saving potential. The current water-saving irrigation projects in Wuyishan City mainly are seepage control of canal, pipe irrigation, and sprinkling irrigation projects. The area of water-saving irrigation is 5.99 thousand hectares, accounting for 37.32% of the total effective irrigation area.

2. *Industrial water planning:* Mining; manufacturing; and water, electrical, and the gas supply industries are the main water-consuming industries. Currently, industrial water recycling rate is only 30.06%. There is a plan to increase these levels to 50%, 70%, and 80% in 2010, 2015, and 2020 respectively, and the industrial water consumption per thousand yuan will reach 1400 m^3, 800 m^3, 500 m^3 in 2010, 2015, and 2020 respectively. In addition, 30% of the enterprises whose water consumption is more than 300 thousand cubic meters per year will reach national standards for water-saving enterprises. These measures are further tap the water-saving potential of large enterprises and focus on rectification of small business; vigorously develop the recycling economy and clean production; and pay attention to the introduction and development of industrial water reuse technology.

3. *Domestic water planning:* According to the characteristics of decentralized location of water use, simple agricultural products processing, inefficient rural water use, shabby water supply equipment, and a lack of safe drinking water sources, the plan should develop village water-saving technology, especially the development of centralized water supplies. Specific water-saving measures are as follows: emphasize the water saving by hotels and public facilities; train residents' awareness of water conservation; and promote the direct use technology of urban rainwater.

12.5.1.4.1.3   Drainage System Planning   In accordance with the 11th FYP of Wuyishan City and the requirements of Program Planning in 2020, the suggested measures are as follows: build drainage systems for rain and sewage diversion; transform the drainage systems of the old city that have fallen into disrepair and carry out a comprehensive construction of urban sewage pipe network; and construct a sewage treatment system covering the whole city with the daily sewage treatment capacity of 5,000 t. The treatment capacity of artificial wetland sewage treatment plant reaches

10,000 t, and sewage pipeline reaches 30 km; encourage the use of urban rainwater recharge technology.

For rural drainage system, establish and promote ecosan system. The main concept is that the family collects and treats manure, urine, and gray water separately, which eventually will be used for agricultural irrigation, fertilizers, and household biogas fuel. The advantages of the ecosan system are low cost, simple technology, environmental protection, and expense savings for farmers.

12.5.1.4.2 Planning of Land Utilization

12.5.1.4.2.1 Planning of Nonpoint Source Pollution Control Agricultural nonpoint source pollution control should focus on nitrogenous fertilizers. The suggested measures are as follows: reduce the amount of chemical fertilizer (planning to reduce the amount of chemical fertilizer to 150 kg/hm^2 in 2020); encourage manure recycling and increase the amount of organic fertilizer used; rainfall runoff is the direct power of nonpoint source pollution, so it is necessary to adjust the time of fertilization to reduce the loss of agricultural nitrogen and phosphorus; establish a buffer zone such as vegetation zone, land-lake ecotone (including ponds and wetlands) to intercept pollutants.

12.5.1.4.2.2 Planning of Soil and Water Conservation The objective of soil erosion control is that the controlled area of soil erosion accounts for 30%, 50%, and 80% of the total in 2010, 2015, and 2020, respectively. This would result in the total controlled area to reach 218.5 km^2 in 2020. Specific measures include: protection of forest resources; monitoring of the collapsing hill erosion; the control of rainfall runoff; establishment of soil and water conservation demonstration projects; promoting technical methods and measures for erosion control; construction of roads ecological revetment; improvement of slope farmland; and application of rural ecological model.

12.5.1.4.3 Planning of Rivers

12.5.1.4.3.1 Planning of Water Landscape and Riparian Zone The objective of developing the waterfront space in Wuyishan City is to improve the urban river environment, beautify the urban landscape, and provide an open recreational space highlighting the cultural value of Wuyi Mountain for urban residents and tourists in order to promote the development of the whole city. The goal of urban waterfront space development is: clear water, a green shore, leisurely residents, and flourishing customers. The plan strives to make the urban section of Chongyang Creek a leisure corridor and a cultural corridor, and develop the tertiary industry with the unique Minyue culture in Wuyishan City.

The focus of the riparian zone construction is to play its function as a buffer zone, effectively preventing agricultural nonpoint source pollution. In addition, the riparian vegetation can reduce the water temperature and improve the aquatic habitat quality by providing plant debris as food. The riparian zone should be arranged between the farmlands and waters in key protected areas. We should plan to adopt the riparian zone in the ecological agricultural area. The plan would return the grain plots to forestry within 20 m on both sides of the main river in 2015, with no cultivated lands existing in 2020. The rivers in nature reserves and those that serve as drinking water sources need buffer zones of 50 m.

12.5.1.4.3.2 Planning of Aquatic Organism Protection   The river ecosystems in Wuyishan City maintains good health and is suitable for fish, amphibians, and other aquatic organisms. But these animals are still influenced by the construction of hydraulic structures, tourist disturbances, and the fishing industry. Specific protection measures should include: establishing a system of closed fishing areas and closed fishing seasons; establishing a protection system for aquatic biodiversity and endangered species, and preventing invasive species; strengthening ecological protection and management of water area through methods of water pollution and ecological disaster prevention; ecological compensation for dam construction, ecological restoration, and development of ecological farming; repair and rebuild water ecosystems that have been damaged; and the construction of hydraulic structure should be on the premise of the ecological environment protection.

12.5.1.5   Water Ecological Protection Zoning

12.5.1.5.1 Drinking Water Source Protection

First-order protection of drinking water sources is in accordance with the standard for Grade II water quality, and second-order protection is in accordance with the standard for Grade III water quality. Specific measures include protecting forest vegetation resources of drinking water sources and establish riparian vegetation buffer system on both sides of first-order and second-order protection areas, which should not cut down trees and reclaim waste-land without authorization; in first-order protection water and land area, prohibit new sewage drainage exits, move out the existing drainage exits within a definite time, and prohibit any activities that may contaminate water sources. In second-order protection water and land area, do not allow additional sewage drainage exits, and the discharge of existing drainage exits should reach the standards. In first-order protection area, new and extension projects should be shut down within a definite time; for second-order protection areas, any projects that make drinking water sources substandard in and around the protection area should be relocated within the specified period. Further, contingency plans for sudden pollution incidents must be established to avoid possible losses to the habitat.

12.5.1.5.2 Key Areas of Water Ecological Protection

The focus of water ecosystem protection in Wuyishan City is "One line, one reservoir, and three areas." One line indicates the Chongyang Stream; the one reservoir indicates the Dongxi Reservoir; the three areas indicate the urban area, the scenic spots, and the nature reserve. Concrete work should regard the Dongxi Reservoir as the point, Chongyang Stream as the line, and the urban area, scenic spots, and nature reserve as the plane. The plan for protection will combine point, line, and plane to carry out the protection work.

12.5.2 Planning of Sustainable Energy Use

12.5.2.1 Planning Objectives and Indicators

12.5.2.1.1 Planning Objectives

By adjusting industry and transport energy consumption and optimizing the rural energy structure, the objectives of the plan for sustainable energy use are to establish a stable, reliable, secure, and optimized energy system and to achieve a society based on clean energy. The total energy consumption in Wuyishan City is expected to be about 290 thousand tce in 2020, and the hydropower and biomass energy development and utilization will account for 75% and 6% of the total energy consumption, respectively.

12.5.2.1.2 Planning Indicators

The planning indicators of sustainable energy use and atmospheric environment in Wuyishan City are shown in Table 12.18.

12.5.2.2 Current Energy Consumption Situation

12.5.2.2.1 Urban Energy Consumption

The industrial utilization rate of clean energy is low, accounting for only 50% in 2005, indicating the necessity of further introduction of electric power resources.

Wuyishan City is rich in energy resources. The coal resource is dominated by anthracite, with the economically proved reserves of 6.51 million tons, and the resource amount of 8.26 tons; the theoretical potential of hydropower resources is

TABLE 12.18
Planning Indicators of Energy Sustainable Use and Atmospheric Environment

Indicator	Unit	Reference Value	Status Value	2010	2015	2020
Energy consumption per GDP	tce/million yuan	≤120[a]	18	16	16	16
Utilization rate of clean energy[a]	%	50[b]	50	70	85	98
Utilization rate of rural biogas	%	≥3[c]	0.28	20	40	60
Standardized days of urban air quality	Days better than or equal to Grade II standard per year	≥333	365	365	365	365
The exhaust reaching standard rate motor vehicles	%	—	91.60	100	100	100

Notes:
[a] Urban areas.
[b] Quantitative assessment indicators of urban environment comprehensive improvement during 11th five-year period.
[c] Reference to the standard of eco-county.

0.216 million kilowatt-hours, while the installed capacity of hydropower is 0.161 million kilowatts, indicating this has a huge potential for the future development of clean energy.

Motor vehicles in urban areas are dominated by motorcycles and a small number of buses. The exhaust of motor vehicles is a serious pollutant.

In terms of industry, the manufacturing sector is the major energy-consuming industry. An irrational industrial structure and the lack of a leading industry make energy consumption unstable and volatile.

12.5.2.2.2 Rural Energy Consumption

The fuel structure in rural areas is disorganized, and the main fuel for household use and production are firewood and straw; the growth of electricity demands in rural areas is slow, and the promotion of biogas projects is inadequate. The plan supports vigorously promoting rural recycling in tandem with economic development and utilization of biomass energy.

12.5.2.3 Analysis of Energy Demand and the Influence on Atmospheric Environment

This plan put forward the baseline scenario and four energy control scenarios, and the main contents of scenarios are summarized in Table 12.19.

The energy savings under different scenarios are shown in Tables 12.20 and 12.21.

Table 12.20 indicates that the control and adjustment of motor vehicles contributes the greatest to the reduction of energy consumption, followed by the development and utilization of biomass and clean energy substitute. The effect of energy savings of centralized production and adjustment of the industrial structure is not obvious, but these adjustments will significantly improve the nonstandard production and potential safety hazard, and contribute to the reduction of SO_2 emission and other pollutants.

The emission reductions under different scenarios are shown in Table 12.21. Currently, the air quality in Wuyishan City is good, but the potential crisis caused by the increasing number of motor vehicle and industrial development will become a major threat for this atmospheric environment in Wuyishan City. The current unreasonable energy structure is the direct cause of air quality deterioration.

The energy structure of Wuyishan City should be adjusted and optimized according to the results of the different control scenarios. The energy structure during the planning years is shown in Figures 12.11 and 12.12.

12.5.2.4 Planning for Sustainable Energy Use

With the structural optimization as the primary objective, energy planning in Wuyishan City should focus on the development of the energy structure in rural households and industrial clean energy, improving energy efficiency, promoting new energy, and finally establishing a stable, reliable, secure, and optimized energy system.

12.5.2.4.1 Optimize Urban Energy: Improve Clean Energy Alternatives

Take full advantage of the abundant hydropower resources in Wuyishan City to replace parts of the industrial coal and fuel oil with electricity, and introduce electric

TABLE 12.19
Scenario Design of Energy Planning

Scenario		Contents
Baseline scenario		Assume that under the current economic development and energy consumption trend, according to the existing planning projects in Wuyishan City, analyze the possible energy demand during the planning years, including consideration of the basic energy-saving projects.
Control scenarios	Control scenario 1—Centralized production and industrial structure adjustment	Through centralized production and energy saving in industrial technology, reduce energy consumption per unit of products, achieve an optimized distribution of resources, reduce emissions, and improve industrial energy efficiency. Energy consumption per unit of products will be decreased by more than 10% in 2020.
	Control scenario 2—Intensive program of motor vehicle control and adjustment	Reduce transport energy consumption by regulating the volume of motor vehicles and fuel economy, enhance the development and improvement of public transport industry, control the number of motorcycles in urban areas, and effectively improve the urban motor vehicle exhaust pollution.
Control scenarios	Control scenario 3—clean energy substitute	Improve the alternative energy of electricity to substitute industrial coal and fuel oil, promote the dual-fuel LPG transformation of motor vehicles, and adjust the form of household energy consumption. The substitution rate of polluting energy reaches 80%.
	Control scenario 4—development and utilization of biomass energy	Vigorously develop biomass energy in rural areas. Establish demonstration households of rural energy-saving residences, and development solar, wind and other renewable energy. The utilization rate of biomass energy in 2020 will reach more than 5%, and penetration rate of biogas projects will reach 60%.

TABLE 12.20
Energy Savings Under Different Scenarios (tce/a)

Scenario	2010	2015	2020
Baseline scenario	109,000	188,000	336,000
Scenario 1	698.95	863.31	1,234.1
Scenario 2	7,821.2	20,977.7	41,789.5
Scenario 3	3,252.7	4,579.67	9,724.53
Scenario 4	6,162.3	11,384.59	20,428.25
Total	17,935.15	37,805.28	73,176.38

Note: Scenario 1, 2, 3, and 4 indicate the energy savings, while the baseline scenario indicates the total energy consumption.

TABLE 12.21

Emission Situation Under Different Scenarios (t)

	2010			2015			2020		
Scenario	SO_2	NO_2	TSP	SO_2	NO_2	TSP	SO_2	NO_2	TSP
Baseline scenario	159.46	538.3	746.74	248.99	774.55	1,009.84	403.99	1,094.13	1,272.25
Scenario 1	31.69	22.47	9.20	59.22	31.57	8.99	131.91	62.69	14.28
Scenario 2	16.22	54.04	35.12	44.19	196.9	99	83.2	278.8	192.4
Scenario 3	34.27	123.9	32.2	54.47	141.95	51.18	60.85	220	57.18
Scenario 4	9.42	65.37	195.98	14.44	100.14	300.26	17.76	123.15	369.24
Total emission reductions	91.60	265.78	272.50	172.32	470.56	459.43	293.72	684.64	633.10

Note: Scenario 1, 2, 3, and 4 indicate the emission reduction; the baseline scenario indicates the total emission.

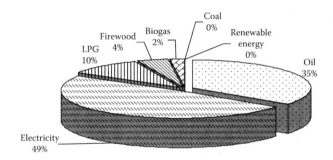

FIGURE 12.11   The energy structure of Wuyishan City in 2010.

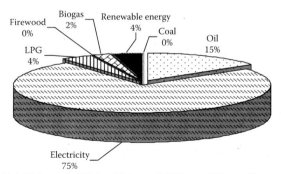

FIGURE 12.12   The energy structure of Wuyishan City in 2020.

boilers; increase the number of buses, control and reduce the number of motorcycles, and promote the conversion of buses and taxis to gas power; improve household energy structures, specifically with a goal of an urban solar energy penetration rate of 5% in 2010, and an urban gas utilization rate of 100%, using liquefied petroleum gas (LPG) pipeline. In addition, vigorously develop solar energy and electric cookers to make full use of renewable energy in Wuyishan City.

12.5.2.4.2 Adjust the Rural Energy Structure: Promote the Development of New Energy

Establish an agricultural recycling economy system as the core source of biogas in rural areas, with the intention that the penetration rate of biogas digesters is expected to reach more than 60% in 2020; develop new and renewable energy sources: on the one hand, develop solar energy, which replaces 2%, 3%, and 5% of the total electricity consumption in 2010, 2015, and 2020, respectively; on the other hand, make full use of the wind resources in Wuyishan City. The government should provide tax, loans, and other economic incentive policies support and guide the development of wind power and other green energy.

12.5.3 PLANNING FOR SOLID WASTE POLLUTION CONTROL

12.5.3.1 Current Assessment of Solid Waste Disposal

The amount of solid waste in 2004 was 1.8 thousand tons, including 0.1 thousand tons of fly-ash, 0.4 thousand tons of slag, 1.2 thousand tons of gangue, and 0.1 thousand tons of other waste. The amount of comprehensive utilization is 1.7 thousand tons, with a utilization rate of 94.4% and an emission of 0.1 thousand tons. Industrial solid wastes mainly come from the nonmetallic mining industry and food and beverage manufacturing.

The total amount of garbage resulting from urban households in 2004 was 30,800 tons, of which 30% was organic and 17.2% were recyclable items. The average growth rate of this residential garbage is 5% in recent years, with a flat tend, which is reflective of the growth rate of population in Wuyishan City.

12.5.3.2 Analysis and Forecast of Solid Waste Production Quantity

To forecast the amount of garbage and industrial solid wastes, two scenarios were set up: a recycling economic scenario and an ordinary scenario. According to the calculation and prediction, the amounts of solid waste generation in the planning target years are shown in Table 12.22.

12.5.3.3 Treatment Plan of Solid Wastes

As a tourist city with a special emphasis on environmental protection of industrial enterprises required for energy generation, there is little dangerous industrial waste discharge at present, and the comprehensive utilization of general industrial solid wastes would therefore have no impact on the environment. However, the current major problems are as follows: (1) Due to the weak environmental awareness of residents, the phenomenon of dumping along the river is still practiced; (2) the location of Laohulong garbage disposal plant is unreasonable, and hence a new disposal plant

TABLE 12.22

Forecast of Solid Wastes in Wuyishan City

Indicator		2005	2010	2015	2020
Total population (10^4)		22.37	33	46	65
Total industrial output value (10^8 yuan)		9.78	21.9	54.5	124.68
Household garbage (10^4 t)	Ordinary scenario	9.8	14.45	21.83	30.84
	Recycling economic scenario		11.56	16.37	21.59
Industrial solid wastes (10^4 t)	Ordinary scenario	0.32	0.72	6.00	13.71
	Recycling economic scenario		0.68	5.45	11.66
			0.014	0.021	0.029
Hazardous wastes (10^4 t)		0.014	0.021	0.029	0.041
Total solid wastes (10^4 t)	Ordinary scenario	10.13	15.19	27.86	44.59
	Recycling economic scenario		12.26	21.85	33.29

TABLE 12.23

Planning Indicators of Solid Wastes Pollution Control in Wuyishan City

Indicator	Unit	Standard Value	Status Value	2010	2015	2020
Treatment rate of harmless urban household garbage[a]	%	95	50	65	80	95
Comprehensive utilization rate of industrial solid waste[a]	%	95	94.4	95	95	95
Solid waste emission per unit industrial added value[b]	t/10^8 yuan	—	500	50	10	1

Notes:

[a] Reference to the standard of eco-city.

[b] Reference to the status value of developed country combined with European standard.

for harmless garbage should be built as soon as possible; (3) inadequate staffing for sanitation services and inadequate sanitation equipment; and (4) the garbage risks caused by the increase of tourists in the scenic spots. The indicators of solid wastes pollution control in Wuyishan City are shown in Table 12.23.

12.5.3.3.1 Treatment Plan of Household Garbage

Based on the separate collection, the treatment method of household garbage in Wuyishan City mainly depends on the aerobic and anaerobic compost, combined with the burning of combustible material, supplemented by landfill. The garbage disposal model is shown in Figure 12.13.

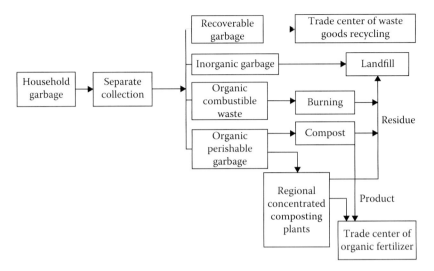

FIGURE 12.13 Household garbage disposal model.

The treatment method for rural households is that garbage will be imported into biogas digesters and fermented together with poultry and animal feces. This will not only reduce environmental pollution, but also promote the development of ecological agriculture as the main energy source in rural areas. The plan should build 16,969 biogas digesters in 2020, covering 50% of households. These digesters combined with an ecological farming model can basically solve the problem of waste disposal in rural life.

12.5.3.3.2 Treatment Plan of the General Industrial Solid Wastes

Comprehensive utilization and treatment of the general industrial solid waste should follow the principle that the "polluter pays," and recycle the solid waste of various industrial enterprises as much as possible. The disposal model is shown in Figure 12.14.

12.5.3.3.3 Treatment Plan of Hazardous Wastes

The major hazardous waste in Wuyishan City is the medical waste. The disposal model is shown in Figure 12.15.

12.5.3.4 Control Measures of Solid Waste Pollution

Solid waste disposal measures in Wuyishan City should be based on the principle of a recycling economy, and should achieve solid waste reduction, recycling, and risk reduction. The specific measures are as follows:

1. Comprehensive utilization and reduction measures of household garbage
   Separating collection and recycling is one of the most important reduction measures. Currently, the number of recycling sites is small, with deficient economic performance at the same time. It is necessary to establish a new waste collection system.

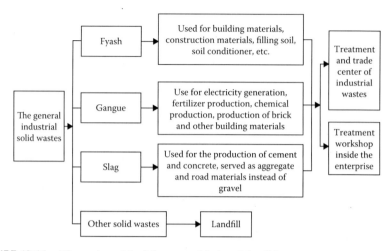

FIGURE 12.14 Disposal model of the general industrial solid wastes.

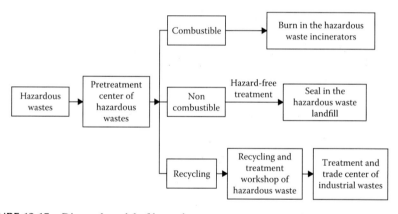

FIGURE 12.15 Disposal model of hazardous wastes.

Utilizing garbage incineration for power generation will make resource utilization more efficient and rational, and the processing costs lower, in order to create the highest economic benefit without secondary pollution. The caloric value of garbage in Wuyishan City is increasing year by year, providing the possibility for using garbage incineration to generate power.

2. Comprehensive utilization and reduction measures for industrial solid wastes

To reduce the amount of industrial solid wastes generally, the following measures are suggested: first, speed up the implementation of cleaner production audit projects and increase the utilization rate of materials from the production process; second, use beneficiated materials to reduce solid waste generation; third, develop material recycling processes. The strategy of industrial solid waste treatment should be transferred from pipe-end treatment to the control of the whole production process.

The current comprehensive utilization rate of mine tailings is relatively low, which should be improved by using this mine filling for tree planting, brick making, paving, and so on.

12.5.3.5 Planning for Solid Waste Treatment Engineering Facilities

12.5.3.5.1 *Refuse Transfer Stations*

The new transfer stations should adopt the compression type; part of the waste can then be carried out by compression truck, and is therefore not totally dependent on the transfer station. Existing underground containerized garbage collection stations should be gradually converted or removed. Plans should include building seven new compression transfer stations.

12.5.3.5.2 *Household Garbage Disposal Dumps*

According to the topography and the construction situation of landfills in Wuyishan City, it was determined that Chengxilong site should be the processing base of urban household garbage, about 4 km from the urban area. The recent scale for daily processing is 120 tons, and the daily generation capacity is 2×10^4 kW·h, while the long-term scale of daily processing is 177 tons, and the daily generation capacity is 3×10^4 kW·h. In addition, old dumps of refuse should be covered after the new dumps are put into use.

12.6 PLANNING FOR AN ECOLOGICAL CULTURE SYSTEM

12.6.1 Planning Objectives and Indicators

12.6.1.1 Planning Objectives

Guided by the principles of feature, connotation, culture, and harmony, a variety of cultural plaques should be constructed: revolutionary history and culture, architectural history, Zhuxi's culture, Liuyong's culture, religious culture, Minyue culture, tea cultivation, habitat, production culture, and so on; build speedy, open, and safe corridors between plaques; and finally brand the city identity with tourism, culture, education, and impression.

12.6.1.2 Planning Indicators

The target value of each indicator is shown in Table 12.24.

12.6.1.3 Current Situation and Gap Analysis

As can be seen from the above-mentioned indicators, three indicators are up to the standard, including environmental compliance rate in tourist district, protection rate of historical and cultural legacy, and public satisfaction on the issue of the environment, indicating that Wuyishan City has a strong ecological foundation in its environment. The indicator of popularization rate of environmental publicity and education is far from the standard. Some indicators such as the popularization rate of secondary education and the proportion of ecological residence lack related statistics, which reflects the hysteresis quality of ecological culture in these fields.

TABLE 12.24

Evaluation Indicators of Ecological Culture Construction

Indicator	Unit	Reference Value Urban	Reference Value Rural	Status Value	2010 Urban	2010 Rural	2015 Urban	2015 Rural	2020 Urban	2020 Rural
Environmental compliance rate in tourist district	%	100		—	100		100		100	
Protection rate of historical and cultural legacy	%	100		100	100		100		100	
The popularization rate of environmental publicity and education	%	>85		28	50		85		95	
Public satisfaction rate over environment	%	>95		99	99		100		100	
The popularization rate of secondary education	%	≥99		—	95	92	100	96	100	99
Hospital beds per million persons	—	—		76	80	78	82	80	85	83
Buses per million persons	—	—		9	12	11	16	14	20	16
The proportion of ecological residence	%	—		—	15	10	25	20	40	35

The main problems are as follows: the overall educational level is low; the development of environmental publicity and education work is not deep enough; there is poor public awareness of the need for environmental protection; eco-residential construction is lagging behind, and ecological concepts are not fully reflected in housing construction; the level of service facilities and transport environment need to be improved; and greater investment and attention should be offered for ecological culture construction in town.

12.6.2 Content of Ecological Culture Construction

Referring to "matrix–plaque–corridor" mode in landscape ecology, the plan will build a framework of the ecological culture construction in Wuyishan City, and carry out the corresponding construction activities for each component, ultimately forming the ecological culture system (Su et al. 2008; Yang and Xu 2007).

12.6.2.1 Matrix Construction

The matrix of ecological culture refers to the backgrounds and basic ideas of different cultures, which play an important role in determining the cultural tone and the basic developmental direction. Therefore, the specific measures are as follows: first, strengthen the environmental protection publicity and education, achieving the popularization rate of 50%, 85%, and 95% in 2010, 2015, and 2020, respectively. Organize diverse publicity and education activities for different subjects so that residents can understand and accept the importance of the eco-city construction and the basic concept of an ecological culture. Second, the ecological legal construction not only strengthens the binding constraint of environmental protection and ecological construction, but also improves government's capacity and efficiency for ecological construction.

12.6.2.2 Construction of Plaques

In ecological culture, the plaque refers to the culture point, and is at the core of the ecological culture construction. In terms of space, the plaque is the gathering of various cultural characteristics, spreading in various parts of the territory of Wuyishan City, including the Da'an revolutionary memorial resort in Yangzhuang Town, the old Caodun neighborhoods in Xingcun Town, the ancient cultural district in Xingtian Town, Xingxian Zhuxi's culture district in Wufu Town, Liuyong's culture district in Shangmei Village, Wuyi's higher education park, the ecological residence district in Tongmu Village, and the Xiandian industrial parks in the new Wuyi area. In terms of cultural connotation, Wuyishan City retains a wide variety of culture including revolutionary history and culture, architecture, Zhuxi's culture, Liuyong's culture, religious culture, Minyue culture, tea culture, habitat, production culture, and so on. For each plaque, focus on the development of a special culture and form a flourishing cultural system.

12.6.2.3 Corridor Construction

The corridors of ecological culture refer to the contact channels between different plaques. In terms of space, they are the linear or strip area, and the network formed by these channels. In terms of content, they include roads, green space, and landscape construction on each channel.

1. *Build a green network:* Ensure adequate green space when constructing a beautiful and pleasant ecological environment. Per capita public green area is expected to reach 12 and 20 m² in 2010 and 2020, respectively. In addition, green spaces should focus on having continuity and being systematic. Consider common greening as the base, road greening as the backbone, and city parks and public green spaces as the focus, which ultimately forms a complete green space system by the combination of point, line, and plane.

2. *Improve the transportation system:* Build an accessible, safe, comfortable, and clean traffic system. Give priority to the development of public transportation, whose proportion in the urban transport structure should be more than 40% in 2020; construct a multilevel, three-dimensional integrated transport system.

3. *Develop cultural joint route:* On the basis of cultivating the cultural connotation of each district, strengthen intercultural relations, conduct cultural combination, and develop joint routes such as folk culture route and a Wuyi celebrity route.

12.7 ECOLOGICAL RISK ANALYSES IN THE PROCESS OF RAPID URBANIZATION

12.7.1 ANALYSIS OF RAPID URBANIZATION PROCESS

According to the data analysis of the Statistical Yearbook of Wuyishan City, the urbanization rate change is shown in Figure 12.16. The rapid urbanization development stage was between 2001 and 2003, with the average growth rate of 26.7%. In 2005, the urbanization rate had reached 44.91%, with an urban population of about 100.5 thousand. The amazing rate of urbanization has caused enormous pressure on the urban system of Wuyishan City.

As a tourist city, the urbanization in Wuyishan City has its particularity and universality. The developmental characteristics are as follows:

1. The urbanization in Wuyishan City is catalyzed by tourism, which artificially raises the level of consumption. The living standards of the local people are not fully elevated, and the quality of urbanization is still to be improved.

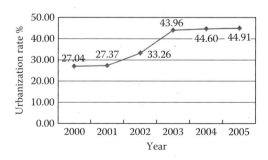

FIGURE 12.16   Development trend of urbanization rate in Wuyishan City.

2. Wuyishan City is dominated by the tertiary industry, which will be influenced easily by tourism. There exist certain risks for the development of this industry.
3. The imperfect urbanization in Wuyishan City mainly lies in the lack of infrastructure and industrial pollution. With rapid urbanization of Wuyishan City, a potential ecological crisis will be inevitably stimulated.
4. Tourism will have impacts on city's structure and function, which pose great potential ecological risks.

12.7.2 ECOLOGICAL RISK ANALYSIS AND FORECAST

The urban natural environment, economic environment, and social environment have a great impact on the ecological risk status and development trends of the urban system. The ecological risk analysis of Wuyishan City will be carried out based on the four impact factors. First, it will build a multilevel evaluation system, as shown in Table 12.25.

The above-mentioned indicators were divided into five evaluation levels (Levels 1, 2, 3, 4, and 5; the larger the value, the higher the degree of risk), and the level of risk was evaluated for every single factor. Then the ecological degree of risk for every factor was calculated on the basis of their different weights and the ecological risk trends were analyzed in the planning years. Results show that the vulnerability of Wuyishan City is strong with its value of 1.47, followed by the degree of regulation, mainly due to the low nonengineering degree of regulation. Both the level of disturbance and the exposure levels are low. On the whole, the current degree of ecological risk for Wuyishan City is small with the value of 3.445 (Table 12.26).

TABLE 12.25
Ecological Risk Hierarchy of Wuyishan City

	Risk Factors	Weight	Impact Factor	Weight
Ecological degree of risk for Wuyishan City A	Disturbance B1	0.25	Natural catastrophe C1	0.3
			Environmental stress C2	0.4
			Urbanization C3	0.3
	Exposure B2	0.2	Population exposure C4	0.4
			Organism exposure C5	0.4
			Industry exposure C6	0.2
	Vulnerability B3	0.35	Economic vulnerability C7	0.3
			Environmental resource vulnerability C8	0.2
			Social service vulnerability C9	0.3
			Cultural vulnerability C10	0.2
	Degree of regulation B4	0.2	Engineering regulation measures C11	0.5
			Non-engineering regulation measures C12	0.5

TABLE 12.26
Ecological Risk Degree of Wuyishan City in 2005

Destination Layer	Value	Criterion Layer	Value	Index Layers	Value
Ecological risk degree	3.445		0.225	Natural catastrophe	0.9
		Disturbance	0.2	Environmental stress	0.8
			0.15	Urbanization	0.6
			0.24	Population exposure	1.2
		Exposure	0.24	Industry exposure	1.2
			0.12	Organism exposure	0.6
			0.525	Economy	1.5
		Vulnerability	0.07	Environmental resource	0.2
			0.525	Social service	1.5
			0.35	Culture	1.0
		Regulation degree	0.3	Engineering regulation	1.5
			0.5	Nonengineering regulation	2.5

TABLE 12.27
Ecological Risk under Forecast Scenario

Item	2005	2010	2020	2050
Disturbance	0.575	0.7675	0.8375	1.00
Exposure	0.6	0.64	0.58	0.64
Vulnerability	1.47	1.26	1.1025	0.84
Regulation degree	0.8	0.7	0.45	0.25
Ecological risk degree	3.445	3.3675	2.97	2.73

According to the trend of urban development in Wuyishan City, a scenario prediction of ecological risk was carried out to determine the developmental trend in the planning years and the long-range perspective. The forecast scenarios took several factors into account, including environmental protection measures, economic development, improvement of living standards, and the implementation of the eco-city planning measures. Urbanization development and the ecological risk situation of Wuyishan City in 2010, 2020, and 2050 are shown in Table 12.27.

Because of the effective ecological planning, the ecological degree of risk in Wuyishan City presents a downward trend under the forecast scenarios from 2005 to 2050. The decline is more obvious, especially after 2010, which is closely related to the effective improvement of vulnerability in Wuyishan City. Meanwhile, the disturbance is still on the rise, which is caused by intensified human activities and increased pressure on the urban environment resulting from the urbanization. Exposure level shows a stable development, with a small fluctuation.

Based on the above-mentioned scenario analysis, the ecological risks and urbanization development in Wuyishan City are shown in Table 12.28.

TABLE 12.28

Ecological Risks and Urbanization Development in Wuyishan City under Forecast Scenario

Year	Short Term	Long Term	Long-Range Perspective
	Rapid development stage (30%–70%)		Mature development stage (>70%)
The development stage of urbanization	Urbanized quantitative stage	Urbanized qualitative stage	Stable urbanized development
Staged problem	Problems in city construction: lack of infrastructure, economic backwardness, etc.	Problems in city development: industrial pollution, resource pressure, etc.	Problems in the city's regional ecology: resource depletion, biological destruction, etc.
The main cause	Extensive urbanized construction	Excessive development of economy and industry	Excessive expand of habitat demands
Main reasons for ecological risks and problems	Problems caused by great vulnerability:	Problems caused by intensive disturbance:	Problems caused by regional environment and resource issues:
	1. Natural disasters such as soil erosion and floods, and their chain reaction	1. Increasing emissions caused by industrial and agricultural modernization	1. The decline of water resources and energy constraints lead to urban resource problems
	2. Backwards infrastructure, supply deficiencies of domestic facilities	2. Exhaust pollution from motor vehicles	2. The problem of agricultural secondary pollution
	3. The lack of sewage facilities, environmental pollution mainly on domestic pollution	3. Serious desertification and pollution at the downstream area of rivers	3. Reduction of biodiversity; extinction of the ecological barrier
		4. Intensified natural disasters caused by human activities	

Through the above-mentioned analysis of ecological risk and urban development in Wuyishan City, it is clear that the recent issues of ecological risk are mainly due to the lack of infrastructure and the economic level lagging behind the city's development speed, showing the problems of great vulnerability and the lack of resistance and resilience from risks; the long-term ecological risk issues are due to industrial and economic development, urban environmental pollution problems caused by increasing tourism, resulting in the problem of intensified disturbance, and intensive pressures on the city.

12.7.3 CONTROL MEASURE OF ECOLOGICAL RISKS

Ecological planning combined with the risk management will effectively reduce the ecological vulnerability of Wuyishan City, achieving an improvement for urban vulnerability in the short time, a decrease in the degree of ecological risk in the long term, and the prediction and prevention of possible risks in the long-range perspective.

We should also strengthen the establishment of emergency response mechanisms and health information networks in Wuyishan City; formulate contingency plans for unexpected incidents and improve the social security system; protect and restore ancient monuments in accordance with relevant laws and regulations, and strengthen cultural tourism development; strictly prohibit hunting of wild animals and implement zone management for natural conservation; develop the construction of eco-tourism demonstration area; implement "bio-engineering," "reforestation projects," and other measures; carry out comprehensive management of small watersheds and implement reinforcement projects for reservoirs to improve security; strengthen the construction and management force of embankment projects and develop the informed construction of flood control and disaster reduction; and improve environmental protection and management of enterprises and promote public supervision ability and the sense of participation, achieving risk popularization education.

12.8 EVALUATION OF AN ECOLOGICAL CONSTRUCTION PLANNING SCHEME

12.8.1 COST ANALYSIS OF ECOSYSTEM

The Eco-City Wuyishan Plan is composed of the following four systems: ecological zoning system, ecological industry, ecological environment system, and ecological cultural system. There are 40 key construction projects in total, with the investment of 1260.8 million yuan. Project classification and investment are shown in Table 12.29.

12.8.2 BENEFIT ANALYSIS OF ECOSYSTEM

Ecosystem service functions are classified into three categories: resource function, ecological environment function, and social and economic function. The main assessment of key construction projects is the influence on the value of ecosystem service function after the project construction has been completed. Benefits of ecosystems for different systems at different stages of planning are shown in Tables 12.30 and 12.31.

TABLE 12.29
Classification of Key Construction Projects and Investment (10⁴ yuan)

Project Classification		Ecological Space	Ecological Industry	Ecological Environment	Ecological Culture
Estimated investment	2010	7,748	14,450	47,677.6	3,045
	2015	650	16,300	19,573.6	4,940
	2020	1,150	2,100	2,451	5,995

TABLE 12.30
Benefits of Ecosystem in Planning Scheme

Planning System	Increasing Value of Ecosystem Service Function (10⁴ yuan)		
	2010	2015	2020
Ecological zoning system	30,803	184,816	338,829
Ecological industry system	59,191	68,155	69,581
Ecological environment system	54,608	78,414	102,220
Ecological culture system	2,925	5,323	8,110
Total	147,527	336,708	518,740

TABLE 12.31
Increased Benefit for Ecosystem (10⁴ yuan)

Benefits of Ecosystem	2010	2015	2020
Resource benefits	23,599	72,555	121,510
Ecological environment benefits	62,403	191,856	321,310
Social and economic benefits	61,525	72,297	75,919

The value of the ecosystem service function of Wuyishan City in 2005 was 21.06 billion yuan. By implementing the plan for the Wuyishan eco-city, the value of the ecosystem service function will increase by 5.19 billion yuan. The benefits of the ecosystem service function will increase by 24.6% in 2020.

12.8.3   Cost–Benefit Analysis of Ecosystem

Table 12.32 shows the costs and benefits of key construction projects at different stages. Through the implementation of eco-city planning, ecosystem service functions are enhanced. Further, the ratio of benefits to costs is increasing gradually.

In summary, through the implementation of eco-city Wuyishan planning, the value of ecosystem service functions will increase by 27%.

TABLE 12.32
Ratio of Benefits to Costs at Different Stages

Year	Cost (10^4 yuan)	Benefits (10^4 yuan)	Benefit/Cost
2010	72,921	147,527	2
2015	41,464	336,708	8
2020	11,696	518,740	44

REFERENCES

Cai J. L.; Yin H.; Huang Y. Ecological Function Regionalization: A Review. *Acta Ecologica Sinica* 2010, 30(11), 3018–3027.

City Introduction of Fujian, China: Wuyishan. Available at: http://news.alibaba.com/article/detail/business-in-china/100114196-1-city-introduction-fujian%252C-china%253A-wuyishan.html

Fu B. J.; Chen L. D.; Liu G. H. The Objectives, Tasks and Characteristics of China Ecological Regionalization. *Acta Ecologica Sinica* 1999, 19(5), 591–595.

Gao S.; Chen B.; Yang Z. F.; Huang G. H. Network Environ Analysis of Spatial Arrangement for Reserves in Wuyishan Nature Reserve, China. *Journal of Environmental Informatics* 2010, 15(2), 74–86.

Liu K. *Ecological Planning: Theory, Method, and Application.* Beijing, China: Chemical Industry Press, 2011.

Liu G. H.; Fu B. J. The Principle and Characteristics of Ecological Regionalization. *Acta Ecologica Sinica* 1998, 6(6), 67–72.

Ouyang Z. Y.; Wang R. S. *Regional Ecological Planning: Theory and Method.* Beijing, China: Chemical Industry Press, 2005.

Su M. R.; Yang Z. F.; Xu L. Y. Building up Urban Ecological Culture Based on Landscape Ecology. *Environmental Science & Technology* 2008, 31(4), 123–126 (in Chinese).

Wu J. G. *Landscape Ecology: Pattern, Process, Scale and Hierarchy.* Beijing, China: Higher Education Press, 2000.

Yang Z. F.; Li S. S.; Zhang Y. Emergy Synthesis for Three Main Industries in Wuyishan City, China. *Journal of Environmental Informatics* 2011, 17(1), 25–35.

Yang Z. F.; Li W.; Xu L. Y. *Environment Planning Theory and Practice of Ecological Urban.* Beijing, China: Chemical Industry Press, 2004.

Yang Z. F.; Xu L. Y. *Urban Ecological Planning.* Beijing, China: Beijing Normal University Press, 2007.

Zhang H. Y. Landscape planning: Concept, origin and development. *Chinese Journal of Applied Ecology* 1999, 10(3), 373–378.

Zhang J. E. *Ecological Planning.* Beijing, China: Chemical Industry Press, 2009.

Zhao X.; Yang Z. F.; Xu L. Y. Study and Application on the Payment for Ecological Services in Drinking Water Source Reserve. *Acta Ecologica Sinica* 2008, 28(7), 3152–3159.

13 Eco-City Wanzhou Plan

Yanwei Zhao, Linyu Xu, Meirong Su,
Gengyuan Liu, and Zhifeng Yang

CONTENTS

13.1   OVERVIEW OF PLANNING AREA

Wanzhou District is located in the junction of the upper and middle reaches of the Yangtze River, the eastern edge of Sichuan Basin, and the northeast edge of Chongqing City. The region stands in the hinterland of the Three Gorges Reservoir region, 283 km away from the Three Gorges Dam. Therefore, it is a typical city in the reservoir region. The construction of the Three Gorges Dam led to the addition of 263,000 immigrants in Wanzhou District, accounting for one-fifth of the total immigrants in the Three Gorges Reservoir region, making Wanzhou a typical immigrant city. The Wanzhou region has good natural endowments and an urban landscape, characterized by the landscape pattern of "Mountains in the city, the city in the mountains, the river in the city, and the city in the river," laying a good foundation for the development of eco-city construction.

13.2   URBAN ECOLOGICAL ASSESSMENT AND PROBLEM IDENTIFICATION

13.2.1   Comprehensive Evaluation of Urban Ecology

The concept of a city as an organism includes productivity, life state, and ecological potential, which respectively represent the development trends of the city's economic, social, and natural subsystems, as well as the degree of vitality for coordination among the subsystems. Using the integrated index method to evaluate the urban vitality of Wanzhou, the results (Tables 13.1 through 13.5) show that: (1) Overall, the vitality index (2005) of Wanzhou City is 0.6214; that is, a relatively weak state. (2) From the perspective of productivity, the index values for economic structure and economic competitiveness are low, resulting in a great restrictive effect. To enhance the production and serving capabilities of the economic subsystem, additional efforts

TABLE 13.1
Urban Vitality Index of Wanzhou

Index Name	Vitality Index	Productivity	Life State	Ecological Potential	Vitality Degree
Index value	0.6214	0.1742	0.1632	0.1646	0.1194

TABLE 13.2
Urban Productivity Index of Wanzhou

Index Name	Productivity	Economic Development Level	Economic Structure	Economic Driving Force	Economic Competitiveness
Index value	0.1742	0.0666	0.0251	0.0537	0.0288

TABI F 13.3
Urban Life State Index of Wanzhou

Index Name	Life State	Social Equity	Scientific and Educational Level	Health of Population	Life Quality
Index value	0.1632	0.0105	0.0701	0.0418	0.0408

TABLE 13.4
Urban Ecological Potential Index of Wanzhou

Index Name	Ecological Potential	Condition and Utilization of Resources	Environmental Quality	Ecological Security
Index value	0.1646	0.0625	0.0874	0.0147

TABLE 13.5
Urban Vitality Degree Index of Wanzhou

Index Name	Vitality Degree	Management and Control Capabilities	System Coordination Degree
Index value	0.1194	0.0380	0.0814

are needed to improve the economic structure, strengthen the economic dynamics and competitiveness, and raise the level of economic development, particularly in economic restructuring and competitiveness, to resolve the issue of industrial hollowness. (3) Concerning Wanzhou's life state, social equity is a major bottleneck, and the residents' health and life quality still must be increased. To enhance the living and serving capabilities of the social subsystem, more attention must be paid to re-employment in the city and social equity. (4) Considering the ecological potential, ecological security is low due to soil erosion and natural disasters, which are a major risk. For the sake of upgrading the ecological serving capacity of the natural subsystem, the control of soil erosion, prevention of geological disasters, and protection of biodiversity should be focused on. (5) Concerning the vitality degree, to improve the comprehensive serving capacity of the urban complex ecosystem, urban management and control should be strengthened, and the coordination among various subsystems must be increased. Specifically, the ecological costs (e.g., water and energy consumption) of social and economic development must be reduced, controlling the socioeconomic development within the carrying capacity of the ecosystem.

13.2.2 Identification of Urban Ecological Problems

We identified the problems in the Wanzhou ecological space, ecological economy, ecological environment, and ecological human settlements according to evaluation results of the urban ecological status based on the vitality index and related investigations (Table 13.6).

TABLE 13.6
Wanzhou Urban Ecological Problem Identification

Contents	Major Problems
Ecological economy	With irrational industrial structure, the eco-industrial system has not been formed yet; the extensive mode of economic growth leads to low economic efficiency, with limited economy aggregation, the industrial hollowness problem is prominent; with many serious problems in agricultural development, farmers cannot maintain a steady income; and with inadequate tourism development, there are no distinctive tourist attractions.
Ecological space	Disordered development and the farmland returning to forest increase the fragmentation level of cultivated land; urban construction and intense human activities disrupt the connectivity of the forest, reducing the integrated ecological benefits of the woodland; an urban green network with a reasonable structure and optimized overall efficiency has not been formed yet; and the management of nature reserves should be further strengthened, and biodiversity conservation should receive more attention.
Ecological environment	Urban atmospheric environment, water environment, and solid waste disposal all face greater pressure, and environmental protection requires urgent improvement; irrational use of energy has increased threats to the atmospheric environment; water pollution has become more prominent, while the fluctuating zone may also aggravate soil erosion and geological disasters, increasing the water ecological security risks; topography, physiognomy, climate, and other conditions lead to land fragmentation and vertical-horizontal ravines, resulting in serious soil erosion; frequent summer storms are likely to cause dangerous rock/landslides and other geological hazards; and irrational development activities exacerbate the destruction of the ecological environment, increasing its vulnerability.
Ecological human settlements	The urban living layout is not quite reasonable, so it should be improved and adjusted with the new immigration construction; the urban green landscape system has not been formed, resulting in a low rate of green space; a high density of urban architecture and less open space result in an unharmonious ecological space; the municipal waste disposal is inadequate, and the sanitation facilities must be improved; and the complex mountain terrain limits the convenience of the traffic, so an efficient transportation system should be built.

13.3 PLANNING OBJECTIVES AND INDICATORS

13.3.1 PLANNING OBJECTIVES

Wanzhou is planned to be built into an ecological city in the Three Gorges Reservoir region with a reasonable spatial layout, complete infrastructure, clean and efficient economy, beautiful ecological environment, harmonious social systems and living ecology, strong potential for sustainable development, and rich southwest landscape characteristics.

13.3.2 PLANNING INDICATORS

Based on the planning objectives and assessment of the status quo in Wanzhou, as well as the index requirements for Chinese eco-city construction, the planning indicator system was established (Table 13.7).

13.4 URBAN ECOLOGICAL SPATIAL PLANNING

13.4.1 ECOLOGICAL FUNCTION ZONING

The fuzzy clustering approach was used for ecological function zoning. The constant indicator and volatility indicator (indicators of low-frequency fluctuations, natural indicators affected by human activities, indicators of high-frequency fluctuations, and socioeconomic indicators) (Table 13.8) were analyzed with the fuzzy clustering approach, and the results were compared and adjusted. In accordance with the main ecological integration principle, when the ecological integration of the crushing plaques is finished, the ecological function zoning (Figures 13.1 and 13.2) is completed.

13.4.2 LANDSCAPE ECOLOGICAL NETWORK CONSTRUCTION

Based on the island biogeography theory, as applied with grid analysis, the Wanzhou landscape pattern was established with core patches, ecological corridors, and key nodes. The construction focus of the core patches is "five mountains of the city" (a city refers to the urban area within the mountains, and the five mountains refer to Shanzi Mountain, Tiefeng Mountain, Fangdou Mountain, Longju Mountain, and Qiyao Mountain). The construction emphasis of ecological corridors is "a river with eight streams" (the river refers to the Yangtze River, and the eight streams are the Puli, Zhuxi, Rangdu, Xintian, Sibu, Modao, Longju, and Suma). The construction stress of the key nodes falls into the "four reservoirs and three scenic spots" (Longan Reservoir, Ganning Reservoir, Xintian Reservoir, and large forest for water supply conservation in the Yubeishan Reservoir; Qinglong Waterfall scenic spot, Longquan scenic spot, and forest vegetation at the Tanzhangxia scenic spot). Considering the plaque and the node as points and the corridors as lines, a butterfly-pattern ecological network of Wanzhou is formed (Figure 13.3).

TABLE 13.7

Planning Indicator System for Wanzhou Eco-City Construction

Categories	No.	Name of Indicator	Unit	Current Value	2010	2015	2020	Reference Value
Ecological economy	A1	GDP per capita	Yuan/person	8,829	13,000	≥20,000	≥25,000	≥25,000
	A2	Net income of farmers per capita	Yuan/person	2,585	3,972	5,800	7,000	≥6,000
	A3	Disposable income of urban residents per capita	Yuan/person	8,540	10,333	13,188	16,000	≥14,000
	A4	Proportion of tertiary industry in GDP	%	45.2	46.1	>46	>46	≥45
	A5	Water consumption per unit of GDP	m³/10,000 yuan	153	145	125	114	≤150
	A6	Energy consumption per unit of GDP	Tons coal/10,000 yuan	0.87	0.80	0.74	0.66	≤1.2
	A7	SO_2 emission intensity	kg/10,000 yuan	5.84	4.4	1.8	1.0	<5.0
	A8	COD emission intensity	kg/10,000 yuan	11.39	6.3	4.0	2.4	<5.0
	A9	Repeated utilization factor of industrial water	%	57	65	75	80	≥50
	A10	Comprehensive utilization rate of industrial solid waste	%	84.2	88	90	>90	≥90
	A11	Utilization rate of clean energy	%	55	64	73	76	≥50

Ecological space	B1	Patch density of arable land	Person/100 hm²	0.215	0.20	0.19	0.18	≤0.18
	B2	Degree of forest connectivity	\	0.50	0.55	0.60	0.65	≥0.65
	B3	Rate of returning farmland to forest	%	47.6	60	80	100	100
	B4	Rare and endangered species protection rate	%	90	95	98	99	100
	B5	Urban public green area per capita	m²	4.06	7.0	8.5	9.6	≥7.5
	B6	Forest cover rate	%	25.1	30	35	40	40
Ecological human settlements	C1	Education coverage of junior middle school	%	98.5	99	100	100	≥99
	C2	Engel coefficient	%	48.89	45	42	39	<40
	C3	Level of urbanization	%	46	50	55	60	≥55
	C4	Urban housing floor space per capita	m²	26.32	28	29	30	≥30
	C5	Main city road area per capita	m²	7.11	8.5	9.5	10	≥10
	C6	Coverage factor of city gas	%	84	90	95	99	≥92
	C7	Coverage factor of rural biogas	%	8.11	16.5	35.3	51.2	≥28.4
	C8	Urban air quality	The number of days reached the second level of China's air quality standard	329	365	365	365	≥330
	C9	Coverage factor of urban standard noise area	%	100	100	100	100	≥95

(Continued)

TABLE 13.7   (Continued)
Planning Indicator System for Wanzhou Eco-City Construction

Categories	No.	Name of Indicator	Unit	Current Value	2010	2015	2020	Reference Value
	C10	Harmless treatment rate of urban living garbage	%	90	95	100	100	100
Ecological environment	D1	Environmental protection investment in GDP	%	0.12	0.4	1	1.5	1.5
	D2	Utilization rate of surface water resources development	%	9.19	17	22	28	<30
	D3	Water qualification rate in urban functional areas	%	70	90	95	100	100
	D4	Water qualification rate of centralized drinking water resource	%	96	98	99	100	100
	D5	Centralized treatment rate of urban sewage	%	57	85	95	100	≥70
	D6	Ratio of protected area in the total land area	%	15.4	16	16.5	17	17
	D7	Treatment rate of soil erosion	%	38.4	45	70	100	100
	D8	Prevention rate of geological disasters	%	24.8	53.1	76.6	100	100
	D9	Environmental qualification rate of tourism region	%	100	100	100	100	100

TABLE 13.8

Indicators of Wanzhou Ecological Function Zoning

Project	Indicator	Unit	Calculation Method
Pure natural indicator (constant indicator)	Terrain elevation	M	–
	Slope	°	–
Natural indicators affected by human activities (indicators of low-frequency fluctuations)	NDVI	–	NDVI = NIR − R/NIR + R = TM4 − TM3/TM4 + TM3
	Soil erosion intensity	$t/km^2 \cdot a$	= Soil erosion area (t/a)/ administrative area (km^2)
	Water surface area rate	%	= Water surface area (km^2)/ territory acreage (km^2)
	Rate of arable land	%	= Arable land (km^2)/ acreage (km^2)
Socioeconomic indicators (indicators of high-frequency fluctuations)	Population density	Person/km^2	= Population (person)/ acreage (km^2)
	Economic density	10,000 yuan/km^2	= GDP/acreage (km^2)
	The level of urbanization	%	= Nonagricultural population (person)/total population (person)
	Annual income of residents	Yuan	–

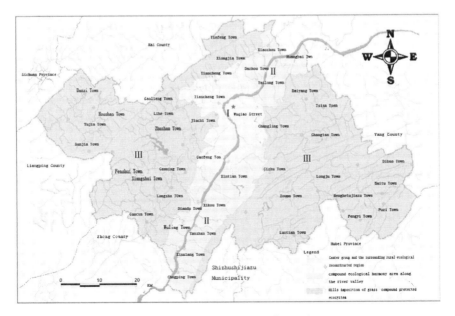

FIGURE 13.1   First-level ecological function zoning of Wanzhou.

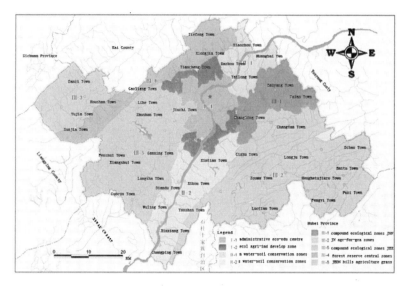

FIGURE 13.2 Second-level ecological function zoning of Wanzhou.

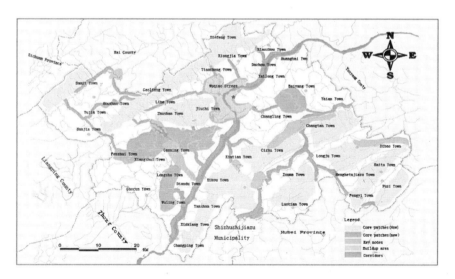

FIGURE 13.3 Wanzhou butterfly-pattern landscape ecological network system.

13.5 URBAN ECOLOGY INDUSTRY PLAN

13.5.1 URBAN INDUSTRIAL DEVELOPMENT CHARACTERISTICS

1. *Economic growth:* In recent years, the economy has maintained a fast and stable growth; it mainly relies on the investment of capital for construction supported by the government.
2. *Economic structure:* The proportion of the primary industry drops continually with an obvious trend of escalation of the secondary and tertiary

industry proportion; the internal structure of industry is unreasonable; the "industry hollowness" puzzles the healthy development of the economy; and looking from the angle of division of income, the urban–rural dual structure is prominent.

3. *Stage of economic development:* From the viewpoint of the industrialization development process, Wanzhou District is in the transitional stage from the initial industrialization stage to the intermediate stage. The industrial development is still insufficient to support the entire economic development, and there exist such issues as small total industry, a low level of industrial structure, a small degree of industrial association, a short industrial chain, weak resource utilization, and the insufficient display of superiority, among others.

4. *Economic development pattern:* The overall industrial ecology level is not high, and the extensive economy development model has not been transformed.

5. *Resource environment pressure:* The land resource supply and demand contradiction is prominent; the ability to develop and use water resources must be promoted, with the potential water savings being huge; the partial mineral resources cannot meet certain needs; and environment pressure from industrial development is revealed day by day.

13.5.2 Planning Scheme for Urban Ecology Industrial Development

13.5.2.1 General Targets

There must be a focus on the development of green foods, the salt gas chemical industry, modern drug manufacturing, the electromechanical light industry, and commercial logistics. Additionally, the pillar industry of the national economy should be cultivated, and the problem of "industry hollowness" must be solved. Based on the main line of economic structure optimization and the core of raising the resources use efficiency, an industrial paragenesis network system is constructed. It relies on three industry zones: Tianzi zone, salt gas chemical zone, and Wuqiao zone. It is based on the ecological agriculture and takes the ecology industry as its leading role, the ecotourism as its characteristic and the venous industry as the link. Therefore, the economic growth transforms from the high-consumption high-emission type to the resource conservation and the environment friendly type, and the economic center status of the Wanzhou Three Gorges Reservoir region is enhanced.

13.5.2.2 Planning Scheme

Guided by the regional economics, the circulation economy, and the industrial ecology theory, four plans for the ecology industrial development were proposed.

1. Cultivate the pillar industry and promote industrial cluster development
   a. *Choice of pillar industry:* One should take factors such as industry foundation, market prospects, resource advantages, environmental protection requirements, national policy for industrial gradient movement, and

solving issues of employment into full consideration to determine the development strategy for a pillar industry concerning green foods, the salt gas chemical industry, modern drug manufacturing, the electromechanical light industry, and commercial logistics in Wanzhou District. Among these, green foods, the salt gas chemical industry, modern drug manufacturing, and the electromechanical light industry are stable industries based on their previous superior development, and commercial logistics is an opportunistic industry with the ability to quickly grow.

b. *Pillar industry cluster development strategy:* Starting from the current state of Wanzhou industry development and based on location, resource, and industrial advantages, the integration of zones, brand effects, path of intensive resource use, industry associations, technological innovation, and the development of human resources should be strengthened. A municipal-featured ecology industrial park (e.g., Tianzi Park, salt gas chemical industrial park, and Wuqiao Park) should be established, and the cluster development of the salt gas chemical industry, electromechanical industry, light textiles, medicine, food, and new building materials should be emphasized.

2. Adjust the industrial structure and optimize the industrial spatial layout

a. *Industrial structure adjustment:* In line with the principle of "optimizing the primary industry, strengthening the second industry, and promoting the third industry," one should set a target and strategy for the harmonious development and overall upgrade of the tertiary industry in Wanzhou (Table 13.9).

TABLE 13.9
Plans for Adjusting the Industrial Structure in Wanzhou

Index Name	Unit	2005	2010	2015	2020
Proportion of the third industry in GDP	%	45.2	46.0	>46	>46
Proportion of industry in secondary industry value added	%	58.2	66.18	69.23	72.54
Proportion of tourism income in GDP	%	3.7	8.5	10	11
Proportion of industrial output value of above-scale industry in the total output value	%	66.3	75	80	85
ISO14000 authentication ratio of scale enterprise	%	—	10	20	50
Proportion of the eco-industrial park value in the total industrial output	%	No ecological industrial park	50	75	90

b. *Optimization of industrial layout:* Establish the general layout of "one center, three belts, and multiple points."

One center—One must strengthen the core status of the urban industry and commerce. Taking the city center as the main body, one should stress the development of commerce, logistics, tourism, finance, information services, scientific research education, and other modern service industries. The five characteristic industries (the salt gas chemical industry, mechanical textiles, green food, modern pharmacy, and new building materials) should be intensely developed, and modern suburban agriculture should be actively cultivated.

Three belts—The ecological agriculture belt should be cultivated along the river, as well as the highly efficient agricultural belt in the shallow hill and the characteristic agricultural belt in the mountain area. The ecological agriculture belt along the river aims to construct the green belts around the reservoir, and the highly efficient agricultural belt aims to develop the processing industry for agricultural products with a leading role being played by small- and medium-sized enterprises. The characteristic agricultural belt in the mountain area focuses on the development of high quality oil, grain, tea, tobacco, out-of-season vegetables, and characteristic fruits.

Multiple points—One should promote the industrial and commercial development of all villages. The implementation of the secondary and tertiary industry development of the eight regional central towns belonging to the project of "develop towns by economy" will be focused upon accelerating the town scale, establishing the raw material base, and actively developing the secondary and tertiary industries and labor-intensive industries in towns. Additionally, the construction of the villages' commercial network will be completed.

3. Develop the circular economy and build the ecological industry system
   a. *Ecological industry:* The venous industry, green food processing industry, modern pharmacy, electrical and mechanical textile industry, and environmentally friendly building materials industry will be vigorously developed, and the salt chemical industry park, Wuqiao ecological industry park, and Tianzi ecological industrial park will be constructed, forming a complete ecological industrial chain.
   b. *Ecological agriculture:* Based on the agricultural productivity layout of "One heart and three belts," one should actively promote the eco-agriculture model, build rural economic pillar industries, and promote the development of ecological agriculture industrialization. At the same time, one must standardize the ecological agriculture development, implement the green agricultural products brand management strategy, and develop ecological sightseeing agriculture, combining it with the three gorges tourism development.

c. *Green services:* Through the development of the commercial logistics industry, one can promote the all-round development of agriculture, manufacturing, construction, transportation, the financial industry, and cultural social undertakings, as well as advance new industrialization and urbanization. The aim is to make Wanzhou become the trade center and material distribution center of the hinterland and become an important circulation city in eastern Chongqing to connect with the eastern portion of Sichuan Province, the south of Shaanxi Province, the northeast region of Guizhou Province, and the west of Hunan and Hubei Provinces.

One must integrate the dominant resources, strengthen the characteristic products, intensively develop tourist resources, protect the environment of tourist areas, and realize the transformation of ecological tourism with the strength of the tourism industry. One should also make efforts to improve the development scale and level of ecological tourism quality, and cultivate the ecological tourism to be the pillar industry of the economy in Wanzhou.

4. *Venous industry:* One should combine the way of concentration and dispersion, actively promote the recycling of industrial solid waste, promote the disposal and recycling of living garbage, and cultivate enterprises for resources recovery based on perfecting the venous industry management.

13.6 ECOLOGICAL ENVIRONMENT PLANNING

13.6.1 ENERGY SECURITY PLAN

13.6.1.1 Energy Security Framework

The main source of energy security lies in energy supply and demand security, energy ecological security, and energy regulation security. These three aspects interact with each other and constitute the energy security system; if any of the three is in a critical condition, it will affect the security of the entire system, as shown in Figure 13.4 for concept.

13.6.1.2 Planning Objectives

Construct a steady energy supply system, a clean energy and environment system, and a thorough energy management system. Realize the regional ecological construction with characteristics of an energy saving society, circular economy, and healthy environment. Reduce industrial, transportation, and life energy consumption and improve energy efficiency. Reduce air pollution and energy ecological costs. Construct a comprehensive energy management and supervision mechanism, and make Wanzhou become a healthy, safe, energetic, and sustainable ecological city with high efficiency of energy recycling and high degree of cleanness.

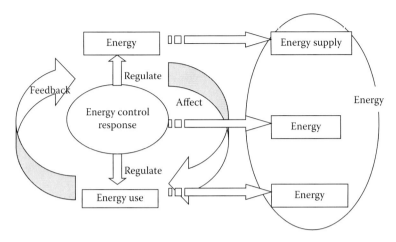

FIGURE 13.4 Energy security framework schemes.

13.6.1.3 Status Quo and Problems

1. The energy structure gives priority to coal and power. In 2005, coal and electricity consumption in Wanzhou accounted for 31% and 39% of the total energy consumption, respectively, followed by oil consumption, accounting for 14% of the total energy consumption. The clean energy use rate is low.
2. Coal consumption relying on high-sulfur coal causes environmental pollution. In 2005, the amount of industrial enterprise coal consumption was 508,000 t (of which, 484,000 t was fuel coal) and was mainly high-sulfur coal containing >2% sulfur, which led to a big proportion of atmospheric pollutants SO_2 emissions and a serious acid rain problem.
3. Wanzhou area is rich in natural gas, which requires further development and utilization. This area is part of the eastern Sichuan oil and gas field, which is one of the three pillars of national natural gas production. Future reserves total nearly 29.4 billion m^3. In 2005, industrial consumption of natural gas was 8.47 million m^3, and the total consumption of natural gas was 70.4 million m^3, accounting for only 7% of total energy consumption. Thus, the development utilization rate is very low, and it must be further strengthened.
4. The development of new energy is insufficient. This area has abundant hydropower resources and contains rich geothermal resources. Additionally, due to the rapid development of the planting and breeding industry in Wanzhou, methane development has many advantages. Regardless, these energy sources must be further developed and utilized.

13.6.1.4 Planning Scheme

1. *The environmental air function division*
   Divide the environmental air function in Wanzhou, and implement division management. See the results in Table 13.10.

TABLE 13.10

Environmental Air Function Partition

Function Zone	Scope	Main Function	Area (km²)	Environmental Air Quality Objectives
A	Line of Daijiayan (Jinlong)-Taibaiyan, Zhanmao Mountain rare botanical garden, and Dayakou forest park Jiuchi Township, Gaofeng Town, and other water reserves areas	Scenic area; cultural and educational residential area; and water reserve area	1215.6	GB3095-1996 level I; execute level I emission standard
B	Other areas except the class A in Wanzhou District	Residential, commercial, industrial mixed districts, and new industrial zone	2241.4	GB3095-1996 level II; execute level II emission standard

2. *Optimize the energy structure*
   a. Reduce industrial coal consumption and decrease the use of high-sulfur coal. Spread the use of clean energy (e.g., electricity, gas, liquefied petroleum gas, and geothermal energy) in Wanzhou for the salt gas chemical industry, green food, textile, and pharmaceutical industry park, and control the total scale of energy consumption. Gradually ban the use of coal or oil for the diet and service industries, and improve the utilization rate of natural gas in the restaurant industry. Until 2015, ≤4 t/h coal boilers and those <10 t/h with more than 8 years of use will be weeded out. For partial boilers, light diesel oil, gas, electricity, and other clean energy sources will be adopted to substitute for coal. By 2020, the clean energy utilization rate will hit 50%.
   b. Improve the gas utilization rate and reduce the cost of natural gas use Introduce advanced technology for natural gas drilling, transport, and use, and effectively improve the efficiency of natural gas drilling and use. Speed up natural gas pipeline work, and improve the gas pipeline utilization rate. Design a reasonable layout, increase the number of natural gas stations, promote the dual-fuel cars, and improve the natural gas utilization rate.
   c. Promote rural methane project construction and introduce ecosystem circulation patterns
      Establish a methane-oriented agricultural circulation economy system; meanwhile, change the homes of livestock, lavatories and kitchens; and

extend the mode of the comprehensive utilization of breeding, biogas, and planting. It will achieve 60,000, 110,000, and 185,000 systems in 2010, 2015, and 2020, respectively. Meanwhile, focus on the development of large- and medium-sized breeding methane tanks, with the aim of 100, 184, and 309 tanks in 2010, 2015, and 2020, respectively. Build wood-coal-saving stoves, renovate kitchens, and improve the digester utilization rate, achieving a normal utilization rate >65% in 2020.

3. *Promote clean energy tanks*
   a. Develop and adopt technology to use clean coal

   Focus on the development of coal gasifiers and promote their industrialization development. Construct a coal gasification demonstration project near the coal mine, and promote coal gasification technology.

   b. Implement and supervise the desulfurization project

   Heavy pollution industries, such as the power and chemical industries, must construct desulfurization facilities or use low sulfur fuel and desulfurizing agents. For enterprises without the ability to concentrate the gas supply, >10 t/h coal boilers must be equipped with desulphurization facilities, requiring 60% desulfurization; 4–10 t/h coal boilers can switch to using circulating fluidized bed boilers; 4–10 t/h oil boilers can consider reinforcement of sulfur agents to decrease the sulfur dioxide emissions. By 2010, all of the ≤4 t/h coal boilers and those <10 t/h with more than 8 years of use will be weeded out, and clean energy (e.g., light diesel oil, gas, electricity) will be adopted or central heating will be used. By 2020, the industrial enterprise desulfurization effect will achieve >80%, and the desulfurization device operation rate in the power plant should be ensured with >95% of the normal power time.

4. *Improve energy efficiency*
   a. Develop the circular economy construction in industrial parks, and reduce the waste of energy consumption

   By reducing energy consumption of per unit product, energy ladder utilization, cogeneration, gas-steam combined power generation, and recycling waste heat, improving industrial energy efficiency will occur. In 2010, 2015, and 2020 the 10,000 yuan industrial products energy consumption will be controlled at 0.58 t of standard coal/10,000 yuan GDP, 0.44 t of standard coal/10,000 yuan GDP, and 0.30 t of standard coal/10,000 yuan GDP, respectively.

   The electric power industry: it is proposed to develop fuel gas-steam combined cycle power generation, using natural gas as the main fuel, with complete combustion. Technology to reduce NO_x emissions will also be adopted, which will meet national discharge standards.

   Salt gasification industry: the natural gas is the main raw material, and clean production process technology is adopted, using centralized gas and heat supply, without separate boiler settings. Through cogeneration and extend lines, peak power load will be alleviated and the efficiency of the use of ash will improve.

b. Control motor vehicle traffic

Central inspect/maintenance systems for vehicles can be adopted, with centralized yearly inspections, separation of maintenance and detection, and remote supervision, and the economic measures should also be adopted to control the speed of the growth of vehicles, especially the motorcycle. Improve the urban planning and transportation system, and give priority to the development of public traffic. Encourage the development, purchasing, and use of alternative fuel cars, and suggest that taxis and buses use clean energy. Strengthen motor vehicle control technology, and ensure vehicle emission control system reliability, achieving 80,000, 120,000, and 160,000 km in 2010, 2015, and 2020, respectively.

13.6.2 WATER SECURITY PLAN

13.6.2.1 Assessment of Water Safety Status Quo

On the basis of a full data collection, use the water safety evaluation system to carry out a comprehensive evaluation, and the evaluation conclusion is as follows:

1. Overall, most indices still have a gap from the standard; in 2005, Wanzhou water security comprehensive index was 0.774 with a level of Class III and was in a general condition (see Table 13.11). Engineering measures are needed to avoid the unsafe factors in water resources, water environment, and water ecology security.
2. In 2005, Wanzhou water safety was in the ideal level, the social economy water requirement achieved the ideal safety level, and the social economy water use safety level is good. In the future, engineering measures are needed to reduce water consumption of 10,000 yuan GDP, and irrigation water utilization rate should be improved.
3. In 2005, Wanzhou water environmental security was in general level; water environmental bearing load and water quality are relatively bad, while water pollution is well controlled. It needs to improve the urban sewage concentration treatment rate through the engineering measures, to make a strict control of pollutants, and to improve the water quality success rate of water functional areas.

TABLE 13.11
Comprehensive Evaluation Result of Water Safety Situation in Wanzhou

| | | | Water Safety Composite Index | | |
Security Level	Class	WSI	Water Safety	Water Environment Safety	Water Ecological Safety
General	III	0.774	0.928	0.703	0.690

4. In 2005, Wanzhou water ecological security was in general level; Wanzhou surface water resources development level is relatively low, the degree of ecological water requirement guarantee achieves the ideal security level completely, but with serious soil and water loss, low level of management, the water ecological security was seriously influenced.

13.6.2.2 Water Safety Trend Analysis Prediction

Set two development modes, namely the conventional development and optimization development, to forecast, analyze, and plan the development trend of water safety of objective years. Take maintaining the status quo as a scene scheme I—conventional development simulation; consider the direct and indirect impact of population development and development planning on water resources, water environment, and water ecological security; make statistics prediction of other relevant variables and comprehensively simulate water safety tendency in the state of optimization development, which is scene scheme II—optimization development simulation.

1. Water safety trend analysis using the conventional development simulation
   In the conventional development simulation, comprehensive index calculation results of planning years for Wanzhou water security is shown in Table 13.12.
   According to the conventional development plan, the comprehensive level of Wanzhou water safety in future is level III, and the comprehensive index drops compared with that of 2005. The main reason for water safety index drop lies in the insufficient removal of massive emissions. It will cause an excessive overweight of water environment bearing load, a poor water environmental quality, a decrease of water environmental safety level, and a decrease of comprehensive level of water safety.
2. Water safety trend analysis by optimization development scheme
   In the optimization development plan, comprehensive index calculation results of planning years for Wanzhou water security is shown in Table 13.13.

TABLE 13.12

Comprehensive Evaluation of Water Safety by Conventional Development Plan

Planning Year	Water Safety Composite Index (WSI)			
	WSI_{Res}	WSI_{Env}	WSI_{Eco}	WSI
2010	0.932	0.550	0.690	0.724
2015	0.932	0.380	0.690	0.667
2020	0.932	0.383	0.690	0.668

TABLE 13.13

Comprehensive Evaluation of Water Safety by Optimization Development Plan

Planning Year	Water Safety Composite Index (WSI)			
	WSI_{Res}	WSI_{Env}	WSI_{Eco}	WSI
2010	0.940	0.952	0.725	0.872
2015	0.980	0.993	0.850	0.941
2020	1	1	1	1

In the optimization development mode, Wanzhou water security in planning year 2010 reached good level and will achieve the ideal level in 2015 and 2020.

13.6.2.3   Water Safety Planning Scheme

1. *Water resources safety guarantee scheme*
   a. *Water supply scheme:* Sixty-six reservoirs with potential safety problems will be renovated to maintain the existing water supply capacity. Most canal systems with serious water conveyance loss will undergo antiseepage renovation to improve their efficiency. Four large, seven medium, and 62 small reservoirs will be constructed. No new inversion projects will be constructed, and only antiseepage renovation to the existing ones will be performed, enabling them to achieve the designated water supply capacity.
   b. *Water saving measures:* For agriculture, such measures as engineering reforms in matching irrigation districts, deficit irrigation schedule improvement, water management enhancement, and water price adjustment policies will be adopted to improve water savings by farmland irrigation. For industry, measures will be employed to increase industrial water efficiency and water saving, including adjustments to industrial structures, renewal of water devices, improvement of production processes, promotion of water saving appliances, and increases in management levels, and among others. Additionally, to increase water use and water saving levels in lives and urban service industries, we can exercise water saving propagation, planned water use management, water price adjustment, and water charge system reform, as well as incentive mechanisms for water saving and promoting water saving technology.
2. *Water environment safety guarantee scheme*
   a. *Concentrated sewage disposal:* In the target planning years, the urban sewage processing capacity will reach 65.32 million, 95.36 million, and 108.96 million t/a, respectively. See Table 13.14 for the planned construction of sewage processing factories.

TABLE 13.14
Planned Construction of Sewage Processing Factories

Sewage Processing Factory Name	Location	Service Scope	Processing Power (10,000 t/d)	Year Built
Mingjingtan	Mingjingtan	Longbao	2.5	Already built
Tuokou	Tuokou	Baianba, Chenjiaba	2.5	Already built
Shenmingba	Shenmingba	Gaosuntang, Zhoujiaba, Shenmingba	6	Already built
Jiangnan new area	Jiangnan new area	Jiangnan new area	4	2010
Xintian Town, Dazhou Town, Rangdu Town	Corresponding town	Xintian Town, Dazhou Town, Rangdu Town	1.1	2005
Gaofeng, Fenshui, Ganning, Longsha Town	Corresponding town	Gaofeng, Fenshui, Ganning, Longsha Town	1.98	2010
Xiongjia, Longju, Changan, Houshan, Baiyang, Tanzishi, Zouma, Changtan Town	Corresponding town	Xiongjia, Longju, Changan, Houshan, Baiyang, Tanzishi, Zouma, Changtan Town	2.67	2010
Lihe Town, Tiefeng Town, Sunjia Town, Changling Town, Yanghexi, Yujia Town, Tiancheng Town, Tailong Town, Xinxiang Town	Corresponding town	Lihe Town, Tiefeng Town, Sunjia Town, Changling Town, Yanghexi, Yujia Town, Tiancheng Town, Tailong Town, Xinxiang Town	2.60	2010
Gaoliang Town, Gaoshen Town	Corresponding town	Gaoliang Town, Gaoshen Town	0.68	2010
Longtanzi	Corresponding town	Longtanzi	1.2	2010
Tuokou (expansion)	Tuokou	Wuqiao, Baianba, Chenjiaba	1	2015
Chenjiaba	Chenjiaba	Chenjiaba	3	2020

b. *Nonpoint source pollution control:* In the short term, a number of demonstration bases for harmless agricultural products will be established surrounding some key water source protection areas (i.e., Mo Daoxi Water Source Protection Area in Wanzhou District, Guandu River Water Source Protection Area, Xin Tian Reservoir Protection Area, Luo Tian River Water Source Protection Area, Gan Ning Reservoir Protection Area, and Pei Wen River Water Source Protection Area) to actively develop organic food and green food.

　　In the long term, the pollution control will include further strengthening the ecological farmland and green farmland sowing area, promoting the use of low-residual chemical pesticides with high efficiency and low toxicity, as well as preventing chemical pollution caused by irrational use of chemical fertilizers, pesticides, agricultural films, and sewage irrigation. In addition, reduction technology for nonpoint source pollution will be vigorously promoted in the vegetative filter strips of some agricultural regions.

c. *Pollution control of key functional districts:* Protection areas such as the Mo Daoxi Water Source Protection Area in Wanzhou District, Guandu River Water Source Protection Area, Xin Tian Reservoir Protection Area, Luo Tian River Water Source Protection Area, Gan Ning Reservoir Protection Area, and Pei Wen River Water Source Protection Area belong to first-order water functional protection areas. Therefore, sewage discharge must be banned, industrial enterprise and livestock poultry breeding industries must be moved or closed, urban sewage must undergo closure, and farmland runoff pollution must be well controlled in these areas to gradually reduce pollutants.

d. *Comprehensive control of the water environment:* The comprehensive control of the water environment mainly contains some key waters' reform and comprehensive project (see Table 13.15).

3. *Water ecology safety guarantee scheme*

a. *Water and soil loss control:* The subregional control method is employed to control water and soil loss in Wanzhou District. The specific control requirements are confirmed (see Tables 13.16 and 13.17).

b. *Ecological restoration of hydro-fluctuation belt:* Artificial vegetation reconstruction should be combined with tourism landscape construction, which is the proper way to restore hydro-fluctuation belt. With reference to the current research findings of green species in hydro-fluctuation belt, vetiver grass, distylium Chinese (Fr.) diels and myricaria: flora are good green species to reconstruct hydro-fluctuation belt, and hydrophilic revetment can be used to restore hydro-fluctuation belt. Artificial revetment will be implemented in the urban hydro-fluctuation belt, while natural repair will be relied upon in rural reservoirs, achieving a new ecological balance.

TABLE 13.15
Comprehensive Control of the Water Environment

Name of the Project	Measures and Environment Benefits
Project of Long Bao River in Wanzhou District	Via the comprehensive project of burying the sewage pipe network, dredging the river way, and feeding livestock and poultry, surface water, soil, and groundwater polluted by domestic sewage will be improved, nonpoint source pollution will be controlled, and the water quality of Long Bao River will be increased. By implementing greening regulation on both banks of the river way, the area of the forest vegetation in this river basin will be increased to 15.99 km^2, and the soil erosion will be reduced to 134,000 t.
Project of Xin Tian River Ecological Environment in Wanzhou District	By completing the 43.5 km^2 soil and water loss control, the river basin's soil and water loss will be kept under control. After the project, 130,000 t of river sediment will be decreased annually. The forest coverage rate will be increased from the current 16.14% to 20.4%.
Project of Xiang Du River Environment	After implementation of the project, 270 km^2 of water environment will be improved and keep urban major drinking water under effective protection.
Project of Wu Qiao River Ecological Environment in Wanzhou District	Once the project is completed, 64 km^2 of soil and water loss area will be controlled, basically controlling the river basin's situation. Through prediction analysis, after the treatment, the annual decrease of soil erosion and sediment will become 422,600 and 250,000 t, respectively. Also, 3317.02 hm^2 forest and grass area will be increased, making the forest and grass coverage rate of 16.64%–43.8%. Increasing forest and grass can effectively prevent land degradation, purify the air around the city, and improve the microclimate of the farmland.
Project of Riverbank of Ru Xi River Segment	By riverbanks renovation and treatment, as well as soil and water conservation and prevention, the water environment of the whole watershed can be improved significantly.
Project of Pu Li River in Wanzhou District	The specific measures are to perfect the sewage collection and handling system and to strengthen the treatment of nonpoint source pollution in agriculture, with the purpose of effectively increasing the water environmental quality. Restoration of forest on banks is employed to reduce the possibility of soil erosion and water loss.
Project of Si Bu River	By implementing a series of engineering, biology, and soil conservation measures, the basin's vegetation coverage rate will increase from the current 38% to 90.38%, and 700,000 t of river sediment will be diminished. Effective treatment of domestic sewage and comprehensive control of nonpoint source sewage will increase the quality of the water environment in the Si Bu River.
Project of Zhu Xi River	By processing the main pollution sources, the basin's water pollution can be completely treated, reaching water standard category III. By dredging partial river segments, protective renovation of banks, and building a flood embankment, the whole basin's soil and water loss (with an area of 153.45 km^2) will be well treated, effectively protecting and rationally exploiting water and soil resources.

TABLE 13.16
Subregion Restoration for Water and Soil Loss in Wanzhou District

Name of the Subregion	Space Distribution	Area (km²)	Area (%)
Key prevention protection region	Water sheds and their two sides of Tie Feng Mountain, Fang Dou Mountain, Qi Yue Mountain, and Shan Zi Mountain	850	24.59
	Small basins of Kui Hua Village, Zhu Shan Village, and Jiu Chi Village		
	Small basins of Long Xi Village, Wan Jia Village, Lu Shan River, and Bai Sheng		
Key supervision region	Wanzhou Inner City, expansion area for immigrants, and industrial park	96.42	2.79
	Key Mining District of massively constructed project area and early warning point for landslide		
Key control district	All the villages and counties in Wanzhou District	2510.58	72.76

Due to the water level's seasonal fluctuation, seasonal management for the hydro-fluctuation belt should also be emphasized in the ecological restoration of Wanzhou while restoring the artificial vegetation.

13.7   ECO-RESIDENTIAL CONSTRUCTION

13.7.1   Existing Problems in Eco-Residential Construction in Wanzhou District

1. *A serious shortage of urban greening:* The amount of public green space is small and unevenly distributed in the inner city of Wanzhou. The old area lacks open green space along streets, and the new area is seriously insufficient in public recreational areas. In addition, the waterfront green

TABLE 13.17
Method to Control Water and Soil Loss in Wanzhou District

Stage	Protected Area (km²)	Distribution	Monitored Area (km²)	Control Area (km²)	Distribution	Total (km²)	Control Rate (%)
2005– 2010	207	Jiuchi, Lushan, Tiancheng Town, Wuqiao Street, Yujia, Qiaoting, etc.	96	1252	Hilly areas of the Yangtze River Valley, MaTou Mountainous low hills, deep area of ShanZishan Mountain	1555	45
2011– 2015	373	Tiefeng, Tiancheng, Zouma, and Xintian		491	Low groove and valley areas of TieFengShan and FangDouShan mountain	864	70
2016– 2020	270	Longju, Dibao, and Puzi Town		768	Low valley and groove areas of LongJushan and QiYueShan Mountain	1038	100

space construction is not sufficient. The current per capita public green land area is 4.06 m², which still represents a large gap compared with the per capita public green space standards of the Chinese health city and ecological city. The green space system for prevention and protection in the main city zone is unsound, which causes difficulty in urban environment maintenance.

2. *The traffic environment must be improved:* The highlighted traffic environment problems are mainly embodied in the road network construction system and the bus system.

The road network construction system has the following problems: road network density is too low, the structure is unreasonable, and the proportion of main road is too high. Restricted by the mountainous construction, the level of urban roads is low, with an imperfect system. The transit traffic, entry, and exit traffic, along with the city traffic, are mixed and interfere with each other. The walking traffic system is inefficient.

Furthermore, spot investigation and related report data demonstrate that public transportation development in Wanzhou has lagged behind and, hence, cannot meet the travel requirements of the residents.

3. *Infrastructure construction is imperfect:* The living garbage disposal has a single pattern with a poor suitability, whose processing power cannot meet the needs of future urban development. By 2015, the Chang Ling garbage disposal field, which is the only one that has currently been put into use, will become fully loaded, and thus a newly constructed one is urgently needed. Moreover, the number of refuse transfer stations and collection stations in the main city zone is too few, which cannot meet the urban garbage collection and transference needs. Domestic wastes have not been classified, the transference is untimely, and the garbage enclosure is adverse, causing secondary pollution over the transportation process.

   The per capita public facilities land is currently 8.1 m², which is significantly lower than that in equivalent cities. From the layout, we can see that the public service facilities are mainly focused in the Gao Xuntang area, whose regional urban function is highly concentrated and repetitive. The public facilities in the new development zone are severely lacking, which causes inconvenience to the public and cannot form a suitable urban living atmosphere. The public facilities layout is not harmonious with the crowded urban space structure. Therefore, public service facilities must be built in new zones, realizing the "people-oriented" concept of residential environment construction.

4. *The old zone is too dense, and the residential environment is poor:* Gao Suntang Zone is the old area of Wanzhou District, with a gross floor area ratio of 1.7 and a gross building density of 29%, and even some building density is as high as 52%. The gross population density in the central area of the old zone is approximately 80,000 people/km², which matches many big cities at home and abroad.

   In this zone, most land functions are incorporated with all properties, including live use land, public facilities land, industrial land, storage land, outbound traffic, road square land, municipal utility facilities, green space, and special land and waters. Moreover, the buildings are dense, sunshine, ventilation, and lighting are poor, the fire prevention span is deficient, and the environment is poor.

13.7.2 RESIDENTIAL ENVIRONMENT PLANNING SCHEME

1. *Construct residential system:* Methods to demolish violation buildings, move the old district industries, and administrative office land conversion are adopted to control appropriate construction. Based on land suitability analysis, the implementation of a scientific residential area layout will help to construct a residential system with a rational layout and a quiet and harmonious district.

2. *Reconstruct an urban green space system:* Strengthening the construction of lump leisure green space and extending the lineal leisure green space in Zhou Jia Dyke, Pi Pa Dyke, Jiang Nan New Zone, and Bai An Dyke will be

employed to enhance the public green space in streets and vertical green-ing. Further, flexible and mobile methods should also be employed to make use of every bit of space. Bridges, corners, and roofs should be utilized to add "point-based" small-scale leisure gardens and greening in the street. By 2020, the per capita public green land area will reach 9.6 m².

3. *Build urban green traffic:* A road network structure suitable for the urban multicentered group space structure should be established. A walk road system should be vigorously promoted, and a three-dimensional walk trans-portation system should be built. Bus lines connecting the new zones' group space should be strengthened, and rapid bus lines should be appropriately developed. The public traffic system is perfectly formed.

4. *Perfect urban sanitation infrastructure construction:* The urban public service facilities system should be reconstructed by strengthening business financial facilities, cultural education facilities, and sports entertainment facilities. At the same time, we should establish refuse transfer and collection stations, perfect facilities system for transfer and collection, realize its gar-bage enclosure processing, gradually establish a classified collection system, and increase the proportion of compost and incineration. By 2015, the rate of urban life garbage that can be harmless treated will be as high as 100%.

5. *Construct a harmonious eco-residential district:* Waterscape and false rocks simulating a natural landscape should be properly constructed to opti-mize the district's landscape design. Under the standard guidance of *Main Points and Technology Guidance Regulations for Green Eco-Residential Area* by the Ministry of Construction, we should commit ourselves to con-structing green eco-residential areas. By 2020, 30% of the residential areas will reach the eco-residential standard requirements. Community libraries, museums, stadium, and comprehensive cultural squares should be built to fulfill modern urban cultural requirements, relying on the reconstruction and management of the Community Management Department for commu-nity culture.

14 Eco-City Jingdezhen Plan

Yan Zhang, Lixiao Zhang, Yanwei Zhao,
Meirong Su, Gengyuan Liu, and Zhifeng Yang

CONTENTS

14.1 DESCRIPTION OF JINGDEZHEN

Jingdezhen is a Chinese city located in the northeast of Jiangxi Province;
N 28°44′~29°56′, E 116°57′~117°42′. The city has a 2000-year history of ceramic
production and is famous internationally as the "capital of porcelain." Although the
ceramic industry was once the primary influence on the economy of Jingdezhen, its
prominence has waned in recent times, being replaced by modern industries such as
automobile manufacturing, aircraft manufacturing, electronics, chemicals, machinery
manufacturing, Limited Company for photovoltaic production (LTD PV), and ecological food industry, which now account for nearly 90% of the total economic activity.

During the transition period from a traditional economy to an industrial one,
Jingdezhen encountered many ecological problems such as the inconvenient location
of ecological functions and other problems relating to ecological spatial planning;
small economic output and extensive economic development resulted in substantial consumption of resources; serious local environmental pollution; delayed construction and poor sustainability of habitat infrastructure; and other environmental
problems (Yang et al. 2007). To solve these problems, Jingdezhen requires an eco-
city plan; however, the basic characteristics of natural resources and economic and
social development must be identified scientifically and rationally before an efficient
strategy can be implemented.

14.1.1 NATURAL RESOURCES

Jingdezhen is located in a typical hilly red soil region of southern China with the
terrain sloping from northeast to southwest. The average elevation of the urban
areas is 320 m, and the peak elevation is 1618 m. The city is subject to a subtropical monsoon climate, characterized by an annual temperature of 17°C, an annual
rainfall of 1763.5 mm, and an annual average of 8 hours of daylight. As a consequence of these climatic conditions, Jingdezhen has a rich forest resource that
covers 60.90% of its land area. The dominant vegetation is subtropical evergreen
broad-leaved trees, of which some are key national protected species such as *Taxus*
and *Ginkgo biloba*. In addition, the forests harbor more than 20 species of state
protected animals such as the Clouded Leopard, Golden Jaguar, Macaque Monkey,
and Golden Pheasant.

The total land area of Jingdezhen is 4,911,662 Chinese acres, and the pattern of
land utilization is shown in Figure 14.1. The main mineral resources of Jingdezhen
include tungsten, coal, but particularly, porcelain stone and kaolin, which have
proven reserves of 0.5 and 5.2 million tons and prospective reserves of 2 and 5.5
million tons, respectively. Jingdezhen has plentiful water resources, with an average annual total water supply of 4.85 billion m^3 and a per capita water resource of
3586 m^3, although the groundwater resources are relatively scarce. The water quality
of Jingdezhen City is good, generally having class II status.

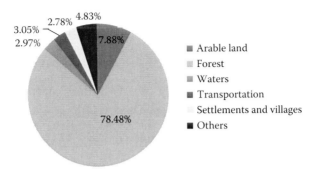

3.05%
2.97%
2.78% 4.83%
7.88%
78.48%

■ Arable land
▨ Forest
▨ Waters
■ Transportation
▨ Settlements and villages
■ Others

FIGURE 14.1   (See color insert.) The percentage of various types of land utilization in Jingdezhen.

14.1.2 ECONOMIC DEVELOPMENT

From the early days of New China, the unique ceramic industry of Jingdezhen dominated its economy. However, after 60 years of construction, Jingdezhen now has a diverse industrial structure, with fully developed machinery, electronics, building material industries, and other light industries. In 2007, the GDP of Jingdezhen was 10.43 billion yuan (RMB), of which primary industry accounted for 1.09 billion yuan, secondary industry 5.43 billion yuan, and tertiary industry 3.90 billion yuan, with the contribution of these three industries to total GDP being 10.48%, 52.08%, and 37.44%, respectively. Jingdezhen has become an important industrial city, and its energy consumption mainly results from its industrial activity. Coal is the main industrial energy resource for Jingdezhen, and coal consumption in 2007 was 3.16 million tons, accounting for more than 60% of the total energy use in Jingdezhen. The main household energy sources are liquefied petroleum gas and pipeline coal gas, which is a consequence of the absence of a natural gas source. In 2007, the values for per capita household energy consumption were 2.09 kg coal, 15.97 kg liquefied petroleum gas, 51.03 m^3 pipeline gas, and 419.78 kWh electricity. The economic development of Jingdezhen is currently still heavily dependent on material and energy inputs, which result in pollutant emissions, and great effort is required to optimize the economic structure and promote the conservation of energy resources.

14.1.3 SOCIAL DEVELOPMENT

The jurisdiction of Jingdezhen covers one county, two districts, and one county-level city. In 2007, the total population of Jingdezhen was 1.56 million with an urban population of 0.62 million accounting for 39.50% of the total. The city transport infrastructure is dominated by public transport vehicles. The Changjiang River is the main channel for water transport and was once the most important mode of transportation for goods; however, it has now become an auxiliary channel, due to the rapid development of railway and road networks.

Jingdezhen was an internationally famous ancient capital of porcelain and played an important role in the history of human civilization. Although many ancient porcelain artifacts have been excavated, with many more remaining unearthed, more importantly, evidence of the large-scale and complex porcelain manufacturing process still remains today. Despite many vicissitudes and transformation, relatively complete ruins of ancient porcelain factories have been preserved in Jingdezhen, which are a rich cultural heritage unique not only to China but also to the rest of the world.

14.2 ASSESSMENT AND PROBLEM DIAGNOSIS OF ECOSYSTEM HEALTH

It is necessary to make an evaluation of ecosystem health and diagnose problems before the planning and design of an eco-city.

14.2.1 ASSESSMENT OF ECOSYSTEM HEALTH

14.2.1.1 Index System for the Evaluation

The index system adopted to evaluate the ecosystem's current status should be consistent with the parameters used to measure the success of the planned eco-city projects, to facilitate a comparative analysis before and after implementation. The index system contains five components, namely, ecological economy, ecological space, ecology, ecological habitat, and ecological culture (Table 14.1).

14.2.1.2 Evaluation Methods

In this investigation, the eco-city was selected as a benchmark for the ecosystem health assessment, where a single-factor distance index was constructed to evaluate the relative distance of the current status from an ideal state, reflecting the health of elements of the urban ecosystem. The distance index can be calculated by standardizing urban ecosystem health assessment indicators (Yang and Sui, 2005). In accordance with the previous research, the maximum distance between the subject and the eco-city was allocated a value of 1 and the minimum distance a value of 0, generating index values between 0 and 1 with a lower index value indicating a healthier urban ecosystem. The grading standards and levels of urban ecosystem health are shown in Tables 14.1 and 14.2.

14.2.1.3 Results

The results of the ecosystem health assessment of Jingdezhen City are listed in Table 14.3.

TABLE 14.1
Grading Standards of Urban Ecosystem Health

Judgment of Health Status	Status of Urban Ecosystem Health				
	Pathological	Unhealthy	Sub-healthy	Healthy	Very Healthy
Distance index value	1.0~0.8	0.8~0.6	0.6~0.5	0.5~0.3	0.3~0

TABLE 14.2
Characteristics of Urban Ecosystem Health Levels

Level	Status	Characteristics (Yang and Sui, 2005; Zhang et al. 2008)
I	Very healthy (ideal state)	A complete ecosystem structure; no significant coercion and interference; capacity for self-sustaining and development; virtuous cycle of energy and material exchange between environment and production for the urban system as well as living components in the system; high degree of harmony between human beings and nature; a comfortable environment provided by a system for human production and healthy living
II	Healthy (second best state)	A relatively complete ecosystem structure; little coercion and interference; capacity for self-sustaining and development; good human services provided by the system; somewhat virtuous cycle of energy and material exchange between human beings and nature; good production and living conditions
III	Sub-healthy	Serious coercion and interference from outside the system; increased rate of deterioration
IV	Unhealthy	Serious damage to ecosystem structure; serious coercion and interference; cannot be self-sustaining; serious degradation of ecosystem services; poor human production and living conditions
V	Pathological (worst state)	Incomplete ecosystem structure; serious coercion and interference; difficult to remediate the ecological status; nearly collapsed ecological services; adverse effects to human beings from serious environmental problems

TABLE 14.3
Results of Ecosystem Health Assessment of Jingdezhen City

Indicators	Ecological Economy	Ecological Space	Ecological Habitat	Ecology	Ecological Culture	Integrated Health Index
Distance index	0.711	0.488	0.741	0.498	0.526	0.593
Health level	Unhealthy	Healthy	Unhealthy	Healthy	Sub-healthy	Sub-healthy

From these results, it can be concluded that

1. The status of the five subsystems was unbalanced and that the overall condition of the ecosystem was sub-healthy. The ecological space and the ecology subsystems were slightly healthier than the ecological culture subsystem, while the indices for the ecological economy and the habitat subsystems were high and categorized as unhealthy.
2. The low index values for ecological space and ecology subsystems probably resulted from the abundance of natural habitats and biological diversity.

3. The external structure of the ecological economy subsystem was good, but the internal structure could be further optimized. Similarly, the capability of economic development to ensure the sustainability of this subsystem should be enhanced. Currently, Jingdezhen has a low economic level compared to other cities in Jiangxi Province, and therefore, policies and support should be directed toward Jingdezhen to achieve a balanced development.

4. The ecological economy subsystem has the lowest level of health of all the five subsystems. This could be improved by injecting investment, optimizing the living components to enhance the service level.

5. The integrated health of Jingdezhen is in a critical state and requires remediation by focusing on the weak links identified by the subsystem indices and reducing the intensity of coercion and interference. The main obstacles to the development of eco-city are the low health levels of the ecological economy, habitat, and culture subsystems. Therefore, great effort should be targeted to develop the economy, improve the urban infrastructure, and direct the ecological civilization of the city.

14.2.2 Problem Diagnosis

Based on the results obtained from the survey and health assessment of the ecosystems of Jingdezhen, the main ecological problems of the five subsystems were identified and analyzed (Table 14.4).

14.3 OBJECTIVES AND INDEX SYSTEM

14.3.1 Objectives

14.3.1.1 Overall Goals and Milestones

In accordance with the concept of harmony between nature and human beings, the carrying capacity of the ecosystem should be strengthened and supported by the establishment of healthy and safe natural spaces, the creation of a clean and efficient ecological economy, the construction of a beautiful and pleasant environment, the building of a green and harmonious ecological habitat, as well as the introduction of training programs targeting a prosperous and advanced ecological culture. This would steadily improve the health level of the ecosystem and develop Jingdezhen as a sustainable eco-landscape garden city.

These objectives can be achieved by completing the goals in three steps. The first step is to build the city in an ecological manner by constructing and managing the city according to the assessment indictors for the "model city," giving special consideration to the key issues, areas, and project during the planning. The second step is to form a stable and safe urban ecological landscape pattern and initiate the construction of an urban ecological economic system. During this stage, the overall quality of the environment must achieve the recommended standards, and the function of natural ecosystem services should be greatly enhanced (Zhou et al. 2012). In addition, the quality of the urban living environment can be further improved, integrating ecological concepts into the production, lifestyle, consumption, and environment of Jingdezhen City. The final step is to achieve the integration and support

TABLE 14.4

Identification of the Main Problems of Ecosystem

Subsystems	Problem Diagnosis
Ecological space	1. Rationally dividing the city into various ecological function blocks is essential to optimizing the development of city and implementing adaptive management. Currently, the natural background and ecological constraints are not fully taken into account in urban development and resource utilization, which leads to confusion regarding the ecological function level of each block and has a detrimental influence on urban construction.
	2. Optimization of the overall urban landscape pattern is urgently required. The abundance of rivers and mountains and the pattern of the landscape are not taken seriously in ecological construction, resulting in uncertainty regarding key nodes, ecological corridors, and localized ecological fragmentation. Therefore, the blocks cannot form an intact ecological network, which is not conducive to the protection of regional ecological security.
Ecological industry	1. The eco-industrial chain network is not perfect. For example, pillar industries such as automobile, machinery, ceramics, and chemical industry are rarely linked, which is not conducive to the formation of industrial clusters or energy saving and the development of a circular economy.
	2. Subject to the construction of industrial chains and the development of a circular economy, resources cannot be used intensively and therefore eco-efficiency is low.
	3. The industrial structure must be further optimized. The economy of Jingdezhen is still in a stage of rapid industrialization, and the tertiary industry contributes a low proportion to total GDP, with a declining trend over recent years.
Ecology	1. Water resources and aquatic environment
	The public consciousness of water conservation is poor due to the abundance of relative rich water sources. The sewage load is deteriorated by the low utilization efficiency as characterized by the low efficiency of irrigation water (0.48), the recycling rate of industrial water (59.41%), water use per 10,000 industrial added value (150 m^3) and domestic water saving equipment penetration rate (65%).
	The capacity of the centralized sewage treatment is low and the centralized treatment rate of domestic sewage is only 41.48%. Part of sewage is directly discharged into Changjing River, leading to eutrophication during the dry season. The sewage treatment plant in the industrial park has not been operational and the industrial wastewater stabilization standard rate is only 95.86%. The wastewater discharged from some enterprises, such as coking and pharmaceutical industries, do not meet the standards required, resulting in excessive concentration of ammonia nitrogen, total nitrogen, and total phosphorus in the downstream of the Changjiang River.
	The control of rural nonpoint source pollution needs to be improved. The safety of rural drinking water is directly threatened by serious nonpoint source pollution caused by direct discharge of domestic sewage and agricultural runoff of chemical fertilizer and pesticide residues as well as from soil erosion and sand dredging of the river channels.

(Continued)

TABLE 14.4 (Continued)
Identification of the Main Problems of Ecosystem

Subsystems	Problem Diagnosis

2. Energy and atmosphere

During the "11th Five-Year Plan," the development of industry was still highly dependent on a coal-dominated energy structure, although the energy efficiency had been improved to some extent. The industrial structure of Jingdezhen is dominated by high-energy consumption enterprises. The energy consumption per unit of industrial added value remains at a high level for many enterprises exceeding the designated allowance.

Coal consumption and the expansion of the cement industry have led to an ongoing increase in industrial emissions and high emission intensity. For example, in 2008, the industrial dust and pollutant emissions increased by 7.4% and 12%, respectively. In 2008, the emission intensity per 10,000 yuan industrial added value was 0.029, higher than the national average, even though the SO_2 emission decreased by 4.55%.

Acid rain has been mitigated in recent years. However, the frequency of acid rain is still greater than 70%, and the average pH value is less than 5. This bad situation is attributed to the natural factors, the surge in the number of cars, as well as the operation of power plant and coke gas plants at full capacity.

Dust and particulate matter are the main air pollutants. The development of urbanization and the rapid expansion of real estate have been accompanied by a large amount of construction waste leading to serious dust and total suspended particles (TSP) pollution. In the 1990s, the government began to remove coal-fired garden kilns and transformed them to coal-fired tunnel kilns. The exhaust from motor vehicles and other pollutants have also been controlled since 2008. Currently, the air quality is better than in the past and the concentration of major pollutants has declined, although the levels of dust are still excessive.

3. Solid waste

A safe and efficient waste disposal system is required to promote the integrated treatment and utilization of household waste. The Songjia Mountain waste disposal plant, which has a daily treatment capacity of 390–420 tons, cannot meet the current demand; however, the Nianyu Mountain landfill currently under construction and which has a daily treatment capacity of 1000 tons will remedy this problem. Household wastes are dealt with by land disposal methods such as landfills. Relying solely on these methods result in a low proportion of recycling and a high risk of secondary pollution to the air and groundwater.

The resource utilization level of solid waste should be improved. Currently, the rate of resource utilization is only 8.5% despite a harmless treatment rate of 92.64%. The comprehensive utilization rate of industrial solid waste is 88.88% and can sometimes even reach 100% for materials such as smelting waste, fly ash, and slag. However, the comprehensive utilization rate of tailings and gangue are only 11.2% and 30.6%, respectively.

The supervision and management of hazardous waste disposal as well as the census of hazardous waste should be improved. There is no information available regarding the total, characteristics, distribution, and processing of hazardous waste, not to mention effective management strategies for tracking and monitoring the situation. Currently, only a small proportion of the hazardous waste is handled onsite by companies themselves and 97.3% of waste must be transferred to other sites.

4. Ecological land

Natural landscapes represent a high proportion of the total land use of Jingdezhen but are unevenly distributed and occur mainly in the north of Fuliang County. The development of the ecological functions is not ideal and the eco-control function is weak. Despite a high proportion of green spaces in urban areas, the ability of antijamming and resilience is low, again due to uneven distribution, poor connectivity, as well as artificial cultivation and monoculture planting.

Nature Reserves and Wetland Reserves require protection. Jingdezhen City is an important area for conservation of biodiversity, but only two areas have been designated as provincial-level nature reserves, the Yaoli nature reserve and the Huangzihao black deer nature reserve, and urban construction is seriously encroaching on wetlands. Sensitive habitats should be identified, especially urban wetlands to delineate appropriate habitat areas and protect and register higher-level nature reserves at the higher level; that is, at the province or nation level.

Ecological habitat

1. Living environment of city center

Housing conditions and the poor layout of residential area should be improved. The land use in the city center is poorly defined because of the complex interaction between industrial, commercial, and residential areas. Currently, the urban housing construction area per capita is 28.96 m^2, which is much lower than that of an ideal eco-city.

The slow development of the municipal infrastructure constrains the improvements to the living environment of the city. Factors such as poor road conditions, disorganization of motor and nonmotor vehicles, and lack of parking lots lead to chaotic traffic conditions. The development of a sustainable urban environment is restricted by the low penetration of gas and a small-scale water supply network.

The development of green communities is lagging behind urban development. Jingdezhen has created only one national green community and five provincial green communities.

2. Living environment in rural areas

There are currently no sewage treatment facilities in rural areas, which leads to the direct discharge of sewage into the environment. Household waste cannot be collected effectively and is often treated by open dumping or dumping in waterways. Much work is required to improve the rural living environment.

The disorganized layout of rural residential areas results in a waste of arable land and presents a barrier to the construction of infrastructure, as exemplified by the condition of poor rural roads and the infrastructure of water supply.

Ecological culture

1. Jingdezhen City has actively implemented publicity campaigns and education regarding environmental protection. For example, more than 85% of primary and middle schools emphasized an awareness of environmental protection during 2008. So far, Jingdezhen has established one national green community, five provincial communities, nine green schools, as well as a number of green hospitals and towns. But continued intervention is still required to develop an ecological culture, especially within the family and among private enterprises and agencies.

2. Jingdezhen has a long history of porcelain manufacture. However, traditional methods should be improved to adapt to the emerging competitive markets, while protecting and utilizing the unique traditions of porcelain and highlighting and promoting the ecological connotation embodied in this industry.

of various functional areas. A harmonious ecological habitat should be created by advocating green consumption and an overall enhancement of the ecological civilization of Jingdezhen. After the completion of these three steps, Jingdezhen will be a sustainable and livable eco-city.

14.3.1.2 Key Components

The eco-city plan includes five key components: ecological economy planning, ecological space planning, ecology planning, ecological habitat planning, and ecological culture planning.

14.3.2 INDEX SYSTEM

An index system for the eco-city plan was created according to the planning objectives, the results of assessment of the current status, as well as the indicators recommended by the national eco-city construction and other related research. The index system includes five parts as detailed in the preceding paragraph, which are subdivided into 37 indicators, as listed in Table 14.5 (Liu et al. 2012). Of these 37 indicators, 22 are derived from the eco-city construction index system established by the Ministry of Environmental Protection and 15 are unique to Jingdezhen City.

14.4 CONTENTS OF ECO-CITY PLAN FOR JINGDEZHEN

14.4.1 ECOLOGICAL SPACE SYSTEM PLANNING

Eco-space planning is a rational and scientific process based on the carrying capacity of natural resources and environment, which is often neglected in traditional planning (Cook, 2002; Yu and Xiao, 2011; He et al. 2011). This process requires a preliminary ecological sensitivity analysis to determine eco-sensitive points and areas before implementation of the eco-city plan.

14.4.1.1 Ecological Sensitivity Analysis

Environmental sensitivity describes the likelihood and extent of changes (deterioration or improvement) to a regional ecosystem in response to human activities. When performing an ecological sensitivity analysis, the possible environmental problems must first be determined and sensitive issues identified. Then, an index system can be established, and finally, the standards of sensitivity levels can be defined and an environmental sensitivity classification map of the studied area constructed, using either direct superposition or weighted superposition methodologies.

1. *Identification of Sensitive Issues and Sites:* The main environmentally sensitive issues are soil erosion, biodiversity conservation, as well as the protection of historical and cultural sites. The frequent problems of soil erosion in Jingdezhen are related to the red sand soil prevalent in the area and the continuous heavy rain in the summer. The forests of Jingdezhen contain many national key protected trees and animals, and biodiversity conservation should be taken into account during the planning of the eco-city.

TABLE 14.5

Index System for the Eco-City Plan of Jingdezhen

Type	Number	Indicator	Status	Unit	2010	2015	2020	Reference Value
Ecological economy	A1	Per capita net income of farmers	4,472	Yuan per person	5,500	6,500	8,000	≥6,000
	A2	Ratio of tertiary industry in GDP	34.5	%	35	37	40	≥40
	A3	Freshwater consumption per unit of industrial added value	46.6	m³/10,000 yuan	40	30	20	≤20
	A4	Effective utilization coefficient of agricultural irrigation water	0.48	—	0.55	0.60	0.65	≥0.55
	A5	Energy consumption per unit of GDP	1.09	tce/10,000 yuan	1.0	0.9	0.8	≤0.9
	A6	Pass rate of enterprises enforced to implement clean production	100	%	100	100	100	100
	A7	SO_2 emission intensity	14.8	kg/10,000 yuan	10	5	4	<5.0
	A8	COD emission intensity	3.7	kg/10,000 yuan	3.5	3.0	2.0	<4.0
	A9	Recycling rate of industrial water	60.6	%	68	80	85	≥80
Ecological space	B1	Forest coverage	60.9	%	61	62	63	≥40
	B2	Urban public green area per capita	9.9	m²	13.5	15	17	≥11
	B3	Ratio of protected area to total land area	11.0	%	14	17	19	≥17
	B4	Ratio of natural land	79.3	%	81	82	83	—
	B5	Ratio of rare and endangered species protection	100	%	100	100	100	100
	B6	Species index	1	—		0.9		≥0.9

(Continued)

TABLE 14.5 (Continued)
Index System for the Eco-City Plan of Jingdezhen

Type	Number	Indicator	Unit	Status	2010	2015	2020	Reference Value
Ecological habitat	C1	*Urbanization rate*	%	52.9	55	60	65	≥55
	C2	*Quality of acoustic environment*	—	Generally eligible	Generally eligible	Eligible	Eligible	Achieving the standard of functional areas
	C3	*Safe treatment of urban living waste*	%	92.8	94	95	98	≥90
	C4	*Urban gas penetration*	%	88.4	90	92	94	≥92
	C5	*Qualified rate of urban water supply*	%	99.7	99.8	100	100	100
	C6	*Penetration of rural sanitary toilets*	%	91	92	95	98	≥95
	C7	*Ratio of clean energy for rural household energy*	%	62.8	65	70	75	≥50
	C8	*Ratio of green community*	%	13.5	20	35	50	≥50
	C9	*Bus ownership per 10,000 people*	—	10.9	11	12	13	11
	C10	*Classified collection rate of urban solid waste*	%	0	20	30	50	≥50
Ecology	D1	*Quality of water environment*	—	Eligible	Eligible	Eligible	Eligible	Achieving the standard of functional areas, no worse than Class V water
	D2	*Quality of air environment*	—	Eligible	Eligible	Eligible	Eligible	Achieving the standard of functional areas
	D3	*Pass rate of centralized drinking water sources*	%	100	100	100	100	100

	Code	Indicator	Unit					Standard
	D4	*Centralized disposal rate for urban sewage*	%	41.5	80	85	90	≥85
	D5	*Utilization of industrial solid waste*	%	92.13	95	96	98	≥90, with no hazardous wastes emissions
	D6	*Ratio of environmental input in GDP*	%	1.9	2.5	3.2	3.5	≥3.5
	D7	Rate of soil erosion control	%	70.9	72	75	80	—
	D8	Water consumption from major rivers	%	13.9	15	18	20	<40
	E1	*Public satisfaction rate for environment*	%	85.63	88	90	95	≥85
	E2	Ratio of enterprises disclosing environmental information	%	82.5	85	90	95	>90
Ecological culture	E3	Enrollment to higher education	%	—				≥30
	E4	Coverage of environmental issues in primary and middle school education	%	100	100	100	100	>95

Note: The indicators in *italics* are those recommended by the Ministry of Environmental Protection.

In addition, Jingdezhen is one of China's most zfamous historical cities and has a large number of cultural heritage sites and scenic spots that should be protected.

2. *Evaluation Indicators:* The index system of the ecological sensitivity assessment should therefore take into account the various impact factors affecting soil erosion, biodiversity protection, and heritage conservation to evaluate the significance of these three indicators (Table 14.6).

3. *Analysis of Evaluation Results:* After single-factor evaluation, these indicators were classified into four levels and the single factor classification map constructed (Figure 14.2). According to the weighting of the impact factors, various layers were superimposed and calculated using ArcView and ArcGIS software, which graded the environmental sensitivity on a scale of 0–4 with higher values indicating higher sensitivity. The resulting values for ecological sensitivity generally ranged from 1 to 3.50, with an average of 2.08, indicating that Jingdezhen is a region of high ecological sensitivity and showing clear spatial differences.

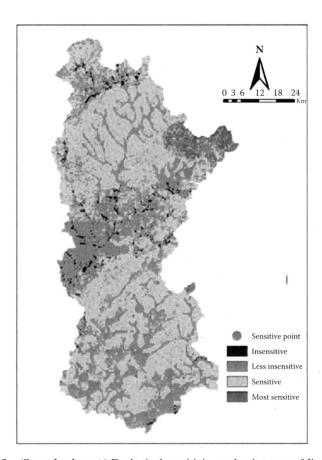

FIGURE 14.2   (See color insert.) Ecological sensitivity evaluation map of Jingdezhen.

TABLE 14.6

Impact Factors of Jingdezhen Ecological Sensitivity Assessment (Hong et al. 2011)

Weight	Sensitive Issue	Weight	Impact Factor	Grading Standards			
				Most Sensitive	Sensitive	Insensitive	Least Sensitive
0.6	Soil erosion	0.2	Precipitation	>2000	1800–2000	1600–1800	<1600
		0.3	Slope	>30	20–30	10–20	<10
		0.3	Soil type	Yellowish brown soil, yellow red soil, yellow soil	Red soil	Limestone soil, brown limestone soil, alluvial soil, lakes, and reservoirs	Paddy soil
		0.2	Land use type	Sloping land, tea plantations, construction land, unused land	Arable land	Woodland, grassland	Water
0.4	Biodiversity conservation	1	Nature reserve	√			
	Sensitive sites			Cultural relics			
Rating assignment of evaluation factors				4	3	2	1
Grading standards of evaluation results				>2.5	2~2.5	1.5~2	<1.5

a. *Most sensitive areas:* The most sensitive areas were mainly located in the eastern and northern regions, accounting for 24.01% of the total area. These regions are characterized by an abundance of natural resources and a complete forest ecosystem, which are an important ecological buffer for Jingdezhen and have significant influences directing and controlling the ecology.

b. *Sensitive areas:* These areas mainly occur in north-central region to the southeast of the urban areas and account for a relative large proportion (38.19%) of the total area. Soil erosion is the most important sensitive issue in this region.

c. *Insensitive areas:* The insensitive areas account for the lowest proportion (only 5.68%) of the total area and are mainly distributed around the urban areas. These regions are rarely threatened by soil erosion because of their flat terrain, strong antierosion properties, and high vegetation coverage.

d. *Least sensitive areas:* The least sensitive areas account for 32.11% of total area and are mainly located in the agricultural plains on either side of the river basins. Having a good water supply and fertile conditions, they also have a relatively high vegetation coverage. However, serious soil erosion occasionally occurs as a result of their topography, soil type, and seasonal rainfall as well as the effects of human activities. Since the ecological sensitivity of these regions is not high, they can be preferentially selected for future development and utilization.

14.4.1.2 Ecological Function Zoning

Ecological function zoning is the process of dividing areas into different functional zones, based on regional eco-environmental elements, environmental sensitivity analysis, and the spatial distribution discipline of ecosystem services.

1. *Ecological Function Zoning Method:* A two-level division of the entire ecosystem service functions was implemented according to the similarities and differences of natural eco-regions as well as features relating to environmental and socioeconomic conditions (Gao, 2006). According to the hierarchical management system advocated by the National Development and Reform Commission during the "11th Five-Year Plan," the regions could be divided into four main ecological function regions, namely, optimized development region, key development region, limited development region, and prohibited development region. Each of these regions was then subdivided into a second level, composed of several regulatory zones, based on the types, structures, and processes of their ecosystems.

2. *Division of Ecological Function Regions:* The ecosystem was divided into four ecological function regions as follows: (1) prohibited development regions, where all types of construction and development activities are strictly forbidden; (2) limited development regions, where the most important ecological function regions are protected, and others, where only limited development is allowed; (3) optimized development regions, where

ecological construction is advocated; and (4) key development regions, where ecological construction is the dominant component.

Roads, rivers, and administrative boundaries were adopted as the basic boundaries for each region.

a. *Prohibited development regions:* The prohibited development regions include nature reserves and scenic spots such as the Yaoli and Huangzihao black deer nature reserves as well as heritage sites such as the Hutian porcelain kiln and kaolinite clay mining sites and areas of protected drinking water sources in Changjing River, East River, and West River.

   These areas must be protected and managed in accordance with relevant laws and regulations that, for example, limit the scale of crop plantations or deforestation and guide the migration of workers and residents in core regions. The natural forests should be protected more rigorously, including attention to species of particular importance and the integrity of the forest ecosystem as a whole. Similarly, the formation of pivot protection units is required for the protection of cultural relics. The adverse effects of tourism on regional ecosystem must also be managed. In addition, there should be no further development activities in the localities of the Changjing River, East River, West River, or other protected water source areas to enhance the prevention and restoration of soil erosion in sensitive and damaged areas.

b. *Limited development regions:* The limited development regions include the mountain areas in the eastern part of Jingdezhen and ecologically fragile areas characterized by steep slope and serious soil erosion. In these regions, the exploitation of mineral resources should be managed strictly by only allocating specific sites for mining. The natural forests, forest ecosystems, and rare species should also be protected and the farmland carefully managed to prevent surface source pollution from soil erosion. In addition, the existing farmland should be conserved and expanded further to achieve the integrated production of food, wood, and tea, side by side with the mining industry (Figure 14.3).

c. *Optimized development regions:* The optimized development regions mainly comprise the urban areas of Jingdezhen, which form not only the core region of construction and administration of eco-city but also the central area of financial and tourist services. The infrastructure therefore requires improvement to enhance the quality of life, and the construction of "software" and "hardware" infrastructure can stimulate the development of tourism services. Outside of urban areas, landscape corridors can be developed to reduce habitat fragmentation and enhance the overall integrity of the environment in the Jingdezhen area. In mining areas, resources should be exploited rationally, and in general, the land available for development should be managed efficiently to strictly control the access of new projects and avoid environmental destruction as well as promote land reclamation and greater intensity of land use.

d. *Key development regions:* These areas include several towns to the southwest of the city. To stimulate economic development in these

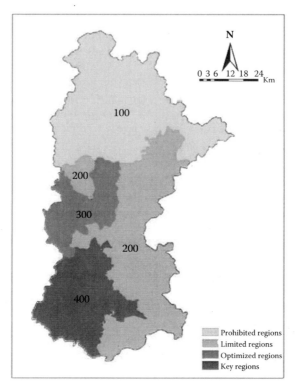

FIGURE 14.3   (See color insert.) Map of ecological functional regions in Jingdezhen.

areas, better access to markets and facilities for vegetable storage and processing should be provided. Action is also required to improve the air quality and beautify the landscape in general, such as regulating mining sites and the emission of "three wastes" and completing pollutant reduction projects ahead of schedule as well as expanding the provision of green spaces.

3. *Ecological Regulatory Zones:* After creating the first level of regional division, the four types of ecological region were subdivided into several regulatory zones and appropriate regulatory guidelines and protection measures formulated, based on the type, structure, and processes of their ecosystems (Table 14.7 and Figure 14.4).

14.4.1.3   Planning of Urban Landscape Pattern

1. *Problems Associated with the Landscape Pattern of Jingdezhen:* Considering the description of basic characteristics and the environmental health evaluation detailed above, it can be concluded that problems still exist in the landscape pattern of Jingdezhen, which can be summarized as follows:

   a. The pattern of the ecosystem is clear, but the development of ecological functions is not ideal. For example, the proportion of natural habitat is high, but it is mainly distributed in the north of Fuliang County, which

TABLE 14.7
Ecological Functional Zones of Jingdezhen

Ecological Functional Region	Ecological Regulatory Zone	Scope	Main Ecological Function	Regulatory Measures
Prohibited development regions (100)	Soil conservation areas and reserves for endangered species in the north of Fuliang County (101)	Forest ecosystem and endangered species reserves in Xihu Township and Jinggongqiao Town, north of Jingdezhen	Water conservation, climate regulation, soil conservation, biodiversity protection	Controlling the scale of crop plantations and deforestation; enhancing the protection of natural forest and important species as well as the overall integrity of forest ecosystems; accelerating the restoration of areas sensitive to soil erosion and damaged areas
	Drinking water source reserves and endangered species reserves in the upper reaches of Changjiang River (102)	Reserves on either side of the Changjiang River	Water regulation and storage, biodiversity protection	Prohibiting the destruction of the aquatic environment and other activities conducive to conservation of water sources, protection of forest, river banks; prevent soil erosion and protect water quality
	Protection of drinking water source areas and soil conservation areas in the upper reaches of the West River (103)	Reserves along the West River	Water regulation and storage, biodiversity protection, soil erosion	Prohibiting the destruction of the aquatic environment and other activities conducive to conservation of water sources, protection of forests and river banks; prevent soil erosion and protect water quality
	Key nature reserves (104)	Yaoli, Huangzihao, and other nature reserves, as well as cultural heritage sites such as the Hutian porcelain kiln, and kaolinite clay sites	Biodiversity conservation, water conservation, climate regulation, and encouraging a spiritual civilization	Enhancing soil erosion remediation, conservation of damaged areas and ecology reconstruction. Guiding the development of regions of restricted use and strictly controlling land utilization to maintain the regional ecosystem balance. Rational development of ecotourism sites.

(Continued)

TABLE 14.7 (Continued)
Ecological Functional Zones of Jingdezhen

Ecological Functional Region	Ecological Regulatory Zone	Scope	Main Ecological Function	Regulatory Measures
Limited development regions (200)	Ecological reserves in the middle and lower reaches of the West River (201)	The middle and lower basin of the West River	Water conservation, soil conservation, and biodiversity protection	Limiting the disturbance of the aquatic environment and other activities conducive to conservation of water sources, protection of forests and river banks preventing soil erosion and protecting water quality. Enhancing the protection of areas prone to soil erosion and remediation of damaged areas as well as the reconstruction of ecosystems
	Central mountainous areas and ecological reserves along the East River (202)	River basin of the East River and the central mountainous areas	Water conservation, soil conservation, biodiversity protection	Limiting the disturbance of the aquatic environment and other activities conducive to conservation of water sources, protection of forests and river banks preventing soil erosion and protecting water quality. Enhancing the protection of areas prone to soil erosion and remediation of damaged areas as well as the reconstruction of ecosystems
	Ecological reserves in the middle and lower reaches of the West River	Sanlong Township	Water conservation, soil conservation, biodiversity protection	Limiting the disturbance of the aquatic environment and other activities conducive to conservation of water sources, protection of forests and river banks preventing soil erosion and protecting water quality. Enhancing the protection of areas prone to soil erosion and remediation of damaged areas as well as the reconstruction of ecosystems
	Ecological restoration of areas affected by mining and tourism (203)	Yongshan Town and Hongyan Town	Soil conservation, promotion of ecotourism, protection of resource supply, water conservation	Rationally locating independent mining sites and strictly regulating mineral exploitation. Limiting the disturbance of the aquatic environment and other activities conducive to conservation of water sources, preventing soil erosion and protecting water quality. Rational development of ecotourism sites and improving the infrastructure of tourism services

Development region	Sub-area	Type	Location	Measures
	Southeastern agro-ecological reserves (204)	Materials provision, soil conservation, hydrological regulation, water source conservation	Low mountains and hilly areas around Lingan Township and Gaojia Town, southeast of Leping City	Transforming traditional agriculture to ecological agriculture by protecting basic farmland and improving agricultural production conditions. Rationally locating independent mining sites and strictly regulating mineral exploitation. Improving the pollution limitation and soil erosion on either side of water catchment areas to avoid agricultural pollution. Manage the production of food, wood, tea, and mining in an integrated fashion
Optimized development regions (300)	Urban ecological restoration areas (301)	Ecological construction of landscapes, buildings, living spaces, and promoting an ecological culture	Urban areas of Jingdezhen City	Regulating the planning of urban construction. Comprehensive treatment of pollution, especially with regard to water pollution and aquatic habitat restoration and construction. Implementing a program of greening urban areas and trunk roads. Achieving an integrated network of natural spaces and improving the quality of the overall landscape
	Agro-ecological restoration arable plain areas (302)	Materials provision, soil conservation, ecotourism, water conservation	Liyang Township and the Nianyu Mountain Town	Transforming traditional agriculture to ecological agriculture by protecting basic farmland and improving agricultural production conditions. Improving infrastructure and the quality of tourism services. Constructing an economic service area integrating tourism, agriculture, and trade
Key development regions (400)	Eco-agricultural areas (401)	Eco-agriculture	Areas surrounding the agricultural technology demonstration park in Leping City	Consolidating the protection of basic farmland according to relevant laws and regulations. Reducing fertilizer and pesticide use, increasing the use of organic fertilizers. Integrated farm production of grain, cotton, and oil, with special emphasis on vegetables
	Eco-industrial parks (402)	Industrial park	The Tashan and Tongjia Industrial Parks	Adhere to the policy of developing fine processing and mining as the main pillar industries. Rational controlling of mining activities and the emission of the "three wastes." Strictly prohibiting the establishment of highly polluting and high-energy consumption enterprises. Improve the overall environmental infrastructure

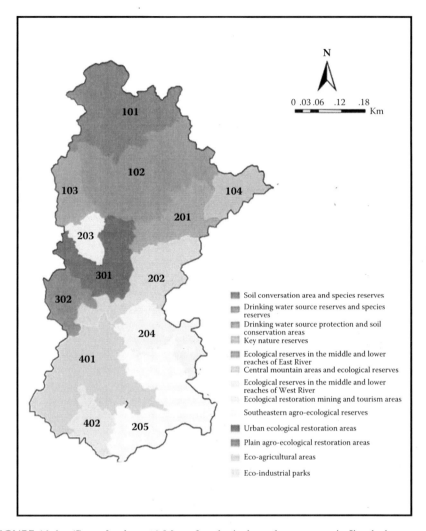

FIGURE 14.4 **(See color insert.)** Map of ecological regulatory zones in Jingdezhen.

limits its effect on the overall environment of Jingdezhen. Furthermore, although there are a high proportion of green spaces in urban areas, their anti-interference ability is poor as a result of their uneven distribution and artificial monoculture cultivation.

b. Although Jingdezhen has a high forest coverage of 60.90%, the forest resource is not abundant, because of its uneven distribution, irregular mix of species, and age structure. In addition, most farmers in mountainous areas use firewood as their main source of fuel, resulting in serious deforestation.

c. Jingdezhen has an abundance of mountains and water sources, features that can be taken advantage of when planning a garden city landscape.

2. *Targets of Landscape Construction:* The goal is to establish a balanced pattern of urban spaces, transportation systems, green spaces, water networks, and a natural pattern to the overall landscape, by the implementation of macro-control for urban ecological districts, while simultaneously achieving rapid, sustainable, and integrated urban-rural development by combining new city planning, land planning, and environmental function zoning (Yang et al. 2010).

   An index system was developed to assist this process (Table 14.8). Three of the factors considered, namely, the proportion of forest coverage, the proportion of protected areas, and the area of public green spaces per urban capita were recommended by the State Environmental Protection Administration while a fourth factor, the ecological land ratio, was also included.

3. *Construction of an Ecological Landscape Network:* According to the Theory of Island Biogeography proposed by MacArthur and Wilson, the increasing fragmentation and isolation of natural habitats into "land islands" results in the extinction of certain flora and fauna. However, an ecological network can be utilized to link various separated ecosystems and to affect the surrounding landscape matrix and plaque community by allowing species to move between different land islands. This approach can be applied to the planning of Jingdezhen according to the concept of ecological network model (see Figure 14.5) to build a healthy and safe ecological network through the construction of ecological nodes (or plaques) and corridors and other beneficial landscape components.

   Considering the specific situation of Jingdezhen City, a network analysis method was used to create the ecological network of landscape features, consisting of core plaques (or core areas), ecological corridors (connecting areas), and key nodes (springboard areas).

   a. *Ecological network of domains:* The ecological network of Jingdezhen consists of core plaque, ecological corridor, and key nodes. When creating the network, full advantage should be taken of the lakes, wetlands, reservoirs, ravines, scenic areas, and other natural components of the landscape.

TABLE 14.8
Index System for Landscape Planning

Indicators	Unit	Standards	Status	2010	2015	2020
Forest coverage rate	%	≥40	60.9	61	62	63
Proportion of protected areas in total	%	≥17	11.0	14	17	19
Urban capita public green area	m²/person	≥11	9.90	13.5	15	17
Ecological land ratio	%	—	79.3	81	82	83

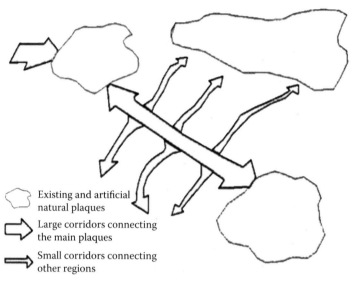

Existing and artificial
natural plaques

Large corridors connecting
the main plaques

Small corridors connecting
other regions

FIGURE 14.5 Ecological network models.

i. *Selection of core plaques (focusing on 10 locations):* Core plaques were selected as areas with good vegetation coverage, specific ecological function, and with a focus on building ecosystems and include (1) perfectly conserved habitat patches such as the Yaoli and Huangzihao nature reserves and (2) new natural plaques such as the nature reserves created at Hongyan Town and Communism reservoir in the southern areas with high vegetation coverage.

ii. *Construction of ecological corridor (focusing on two axis and two bands):* There are long and narrow areas between large natural plaques stretching from the northeast to southwest, which should be connected by corridors to form a connected ecological network. Based on the distribution of roads and rivers, corridors with two axes and two bonds could be constructed as follows:
Two axes—the 206 State highway and the Jing-Wu-Huang highway.
Two bonds—the Changjiang River cultural landscape band and the Lean River landscape controlling band.

iii. *Allocation of key nodes:* The key nodes are points that have an important role in protecting the continuity of ecological process, including (1) the Wuxikou Hydro Project, which can alleviate flooding in the middle stream of the Chiangjiang River and (2) the entrance of Changjiang River and the exit of Lean River, which ensure water quality control and ecological function fulfillment.

iv. *Creation of the ecological network:* Ten core plaques, ten corridors, and three key nodes were identified for the creation of an ecological network. A map of the network was drafted by abstracting nodes and corridors with points and lines, respectively (Figure 14.6a).

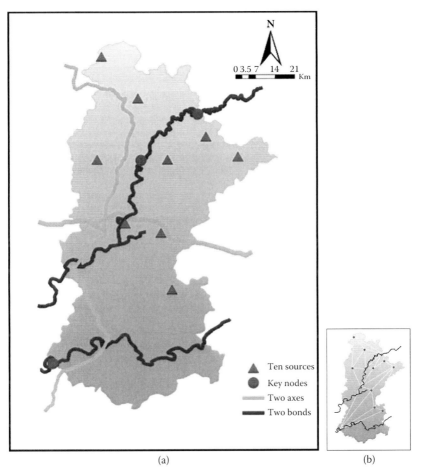

FIGURE 14.6 (See color insert.) Map of ecological network. (a) Main landscape framework construction, and (b) the point-connection network construction.

The map was revised by adjusting the straight lines along the river valleys and adding several corridors to promote the effectiveness of connection (Figure 14.6b).

b. *Ecological networks of city districts:* There is an organic and harmonious symbiosis between Jingdezhen City, the nearby mountains, and the Changjiang River and its tributaries. However, some problems do exist in the ecosystem, such as the destruction of wetlands and other ecological patterns by the development of the City District, the lack of a buffering green belt between the city and various developments, as well as an absence of habitat links between the mountains and the water sources. In the plan for this area, a divergent network composed by two axis, three hearts, four groups, and six belts has been proposed.

The ecological network of Jingdezhen City District is shown in Figure 14.7. The two axes are formed by the riverside green axis along

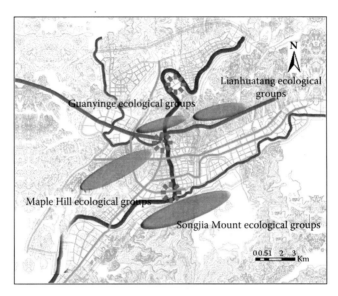

FIGURE 14.7   **(See color insert.)** The ecological network of Jingdezhen City District.

the Changjiang River and east–west development axis (Changnan Avenue–Zhushan middle road–Chaoyang Road–Changhe Road); the three hearts represent the three centers of urban development along the Changjiang River, while the four groups correspond to important natural features of the landscape; that is, the Maple Hill, Guanyinge, Lianhuatang, and Mount Songjia ecological groups. The six belts are formed by the waterfront wildlife corridors along the West and South Rivers and the wildlife corridors along the four trunk roads.

14.4.2 ECOLOGICAL PLANNING OF INDUSTRY

14.4.2.1 Evaluation of Industrial Development Status

1. *General Description of Industrial Development*
   a. *Industrial layout:* The current industrial layout of Jingdezhen City is presented in Figure 14.8.

      The primary industries are concentrated in Fuliang County in the north, which is renowned for forestry and tea production, and in Leping City in the south, which focuses on vegetable production and rearing livestock. The secondary industries, mainly located in the ceramic industrial park, high-tech industrial park, and the Leping Industrial Park, include ceramic production, information technology, and machinery manufacturing as well as chemical and pharmaceutical industries. Tourism is mainly concentrated in urban areas and in Yaoli Town in Fuliang County.
   b. *Industrial scale:* Figure 14.9 shows that the prosperous economy of Jingdezhen has developed rapidly from the year 1998 to 2007. The GDP

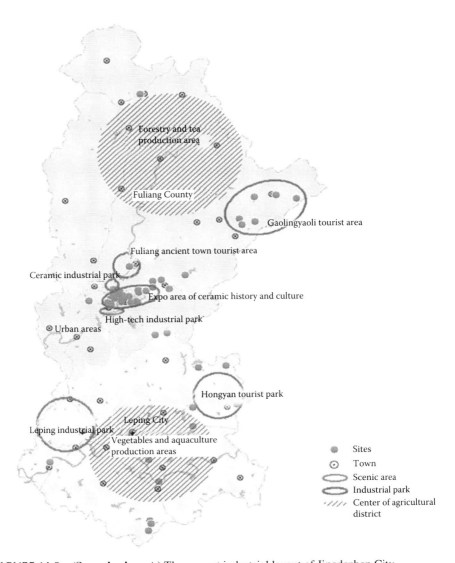

FIGURE 14.8 **(See color insert.)** The current industrial layout of Jingdezhen City.

and per capita GDP recorded for 2007 were 26.19 billion yuan and 16,899 yuan, respectively, and adjusting for inflation, these figures are 3.6 and 3.2 times greater than the figures for 1997. Total GDP increased dramatically in the 1990s and has continued to maintain a high growth rate during "10th Five-Year Plan," even exceeding 20% in 2003. The economy of Jingdezhen is currently in a period of prosperity with a steady growth rate of 16% from the year 2005 to 2007.

c. *Industrial structure:* According to the criteria of the industrialization stages listed in Table 14.9, it can be concluded that Jingdezhen City

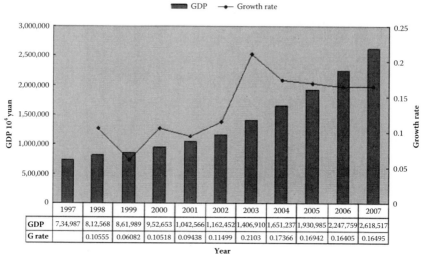

FIGURE 14.9 Total GDP of Jingdezhen City from 1997 to 2007.

TABLE 14.9
Various Stages of Industrialization

Standards	Beginning	Mid-Term	Latter	Of Jingdezhen (2007)
Per capita GDP (dollars)	<600	600–3,000	3,000–10,000	2,414
Agricultural added value divided by industrial added value	<5	5–8	>8	6.04
Percentages of various industries to total GDP	Percentage of primary industry more than 20%; percentage of secondary industry lesser than that of tertiary	Percentage of primary industry more than 20%; percentage of secondary industry higher than that of tertiary	Percentage of primary industry lower than 10%; percentage of secondary industry increases to its highest level	9.3 : 56.2 : 34.5

is still in the mid-industrialization stage. However, the contribution of secondary industries to total GDP is quite large and plays a major role in driving the economy.

The industrial structure of Jingdezhen has maintained a healthy trend from the year 1998 to 2007, driven by rapid industrial development and gradually changing from 28.58:54.58:16.84 in 1980 to 9.3:56.2:34.5 in 2007. In particular, the contribution of primary industries has decreased, being replaced by a growth in secondary industries, which indicates that industrialization has dominated the expansion of the economy. Despite significant improvements during the 1980s, the tertiary industries have remained a low proportion of total GDP, even decreasing in recent years.

2. *Problems in Planning of Industry*
   a. *Industrial layout:* Overall, the industrial layout of Jingdezhen City has become more rational; however, there are still some problems. Firstly, the goal of "moving enterprises into gardens" has not been achieved with the location of some industrial enterprises remaining in built-up areas. For example, it is desirable that the Jingdezhen power company should be relocated as soon as possible as a consequence of its impact on the urban environment. Secondly, there are no ecological buffer belts between the industrial parks and residential areas, tourist sites, or agricultural parks, in particular, between the seriously polluted Leping Industrial Park and the Organic Vegetable Park.
   b. *Industrial scale:* The economic scale of Jingdezhen is still small. For example, in 2007, the per capita GDP of Jingdezhen was only 16,899 yuan, placing it only fifth in the ranking of the 11 cities of Jiangxi Province.
   c. *Industrial structure:* The proportion of tertiary industry has declined in recent years, resulting from both the rapid development of industrial enterprises and the slow development of the tertiary industries themselves. The tourist industry of Jingdezhen City has no obvious advantages that allow it to outcompete its neighboring cities, indicating that the cultural and tourist resources require further development.

14.4.2.2 Planning Targets

1. *Overall Goals*
   The overall goal of the eco-city plan is to comprehensively promote the eco-city transformation of Jingdezhen, to save resources, and to protect the environment during production processes, technology, and consumption process. This should be combined with the industrial characteristics of actively promoting an economic growth program that takes a new eco-industrialization path, improving industrial quality, transforming and upgrading traditional industries, and building eco-industrial chains, orientated around a circular economy and an industrial ecology with high-tech solutions.

2. *Goals for Particular Industries*
   Agriculture: Maintain the current contribution of grain farming and pig breeding to the total economy; conserve its ecological function; create a well-known regional brand of tea; expand pollution-free vegetable production; complete a green recovery process; and generally improve the natural environment.

   Industry: Strengthen resource conservation and environment protection; develop high-tech industries and take a new eco-industrialization path; transform and upgrade traditional industries and create ecological

industrial chains oriented around a circular economy and ecological industry; and transform the five main industries; that is, aerospace engineering,
information technology, chemical and pharmaceutical, ceramic building
materials, and electrical power into regional leaders.

Tourism: Establish Jingdezhen as a tourist city; integrate historical culture and modern civilization; promote the ceramics and tea traditions; and
aspire to be a unique destination and connect with other cities on international travel route of China.

14.4.2.3 Eco-Industrial Development Programs
1. *Industrial Layout Planning*
   a. *Conservation of drinking water sources:* The waste discharge from the
      ceramic industrial park presently released into the upstream of the West
      River should be relocated more centrally to ensure the security of the
      urban domestic water intake. Similarly, the total volume and intensity
      of pollutants emitted from the high-tech industrial park located in the
      downstream need to be controlled to ensure the water quality of the
      section of the river exiting the city.
   b. *Prevention of air pollution:* The prevailing wind near the high-tech
      industrial park, located in southwest corner of the city, already ensures
      that most emissions bypass the main residential and tourist areas, mak-
      ing these areas suitable for further development. However, the direction
      of expansion of the ceramic industrial park must be controlled, with
      new enterprises being built in the northeast of the pack to avoid detri-
      mental effects on the southeastern residential areas during winter.
   c. *Consolidating the centralized distribution of industrial enterprises:*
      The centralized layout of enterprises should be further consolidated by
      moving businesses into appropriate industrial parks. In addition, eco-
      logical buffer zones should be created between industrial parks and
      tourist sites and residential areas.
2. *Planning for Eco-Agriculture*
   a. *Intraindustry restructuring:* To develop an efficient and ecological sys-
      tem of agriculture, the agricultural structure and regional distribution of
      farms must be optimized. In the future, cash crops, livestock, and aqua-
      culture should be developed actively. The goal is to ensure that more
      than 70% of crop output is derived from cash crops and to increase the
      value of aquaculture production to 65% of total agricultural output.

      Particular attention must be given to safe guarding farmland from
      building developments and to increase food production using organic
      fertilizers and creating better agricultural infrastructure. Ten organic
      farm projects without any pollutant emitted to environment will be con-
      structed to facilitate the overall grain production to a stabilized capacity
      of 33 million tons. Farm bases specializing in high-quality vegetable
      and rapeseed; tea production bases should also be established; and ani-
      mal husbandry should be developed further by constructing economic
      zones for the rearing of pigs and cattle and by drawing on the experience

of currently successful businesses. Attention should also be given to developing aquaculture, encouraging both commercial and sport fisheries by adapting convention systems of aquaculture. It is hoped that by 2010 (since this book has been finished in the year 2008, it was hoped to happen in 2010), the area of aquaculture facilities will rise to 78,000 Chinese mu, with the output of aquatic products accounting for 26% of the total agricultural output. In addition, the rational forestry developments are planned including a 50-mu water conservation forest, a 4-mu bamboo plantation, a 100,000-mu forest of fast growing trees, and a 20,000-mu woody herb plantation.

b. *Stimulating the ecological construction of industry:* Firstly, agricultural methods should be improved by creating an extension system of various organizations to promote agricultural technology. An agricultural science–technology park and training farms should be constructed. Focus should be made on the promotion of superior varieties of rice; pest control techniques for rice production and no-tillage agriculture; as well as assistance for vegetable storage and processing, the raising of pedigree pig breeds, agricultural mechanization, and agricultural standardization of agricultural practices, as well as improving eco-friendly methods of animal disease prevention and pest control in agriculture.

Secondly, a highly trained workforce is urgently required for eco-agricultural development. This should focus on four areas, namely, personnel trained in basic agricultural techniques, specialists in the marketing of agricultural products, extension workers, and also versatile personnel specialized in rural resources able to apply all these skills with consideration of local conditions. Finally, the implementation of strict ecological management measures is required to control soil erosion and stimulate the green recovery of mountains, especially the management of mines. More support should also be given for the management and protection of forest resources by further promoting the reform of forest property and state-owned forest farms. In addition, greater emphasis should be given to enforcing and implementing environmental protection laws such as the "Forest Law," the "Wild Animal Protection Law," as well as other forestry laws and regulations.

c. *Implementation of integrated agro-ecosystems:* Eco-agriculture should be developed in Jingdezhen because it is an area prone to natural disasters resulting from soil erosion. Since Jingdezhen is rich in forest resources, it is hoped that a win-win situation can be achieved by combining erosion mitigation with agriculture, forestry, and the rearing of livestock. This system has further environmental benefits, for example, linking the production of animal feed with the livestock industry and the utilization of the resulting animal waste for biogas fermentation, while the residue can be recycled as organic fertilizer into arable production. It is hoped that the biogas produced could help alleviate the energy shortage in rural areas. By the integration of agro ecosystems in this way, a beneficial of state internal recycling can be achieved.

3. *Eco-Industrial Planning*

   a. *Comprehensive ecological transformation of industry:* The investment in environmental projects should be increased, maintaining the current level of investment in environmental protection with the aim of increasing it to 3.5% by the end of planning period. The "Law of the People's Republic of China on Promoting Clean Production" should continue to be implemented, integrating cleaner production into entire process of administration for Jingdezhen. The establishment of new businesses should be strictly controlled by the implementation of "three simultaneous systems" in the management of industrial parks. In addition, outdated and highly polluting companies should be either modified by the introduction of more advanced technology, merged with more efficient companies, or face closure.

   b. *Formation of leading industrial chains:* The characteristics of the industrial systems of Jingdezhen City should be altered to achieve the ecological development of the whole industrial systems through rational development and ecological promotion of the five main industries. The integration and circularity of the whole industrial chain should be enhanced by extending the chain and linking main industries to promote the process of an eco-industrial system.

   The five main industries of Jingdezhen are aerospace engineering, chemical and pharmaceutical, information technology, ceramic building materials, and electrical power. The aviation machinery industry is the most important pillar industry of Jingdezhen. While consolidating its position, diversification into new markets such as parts for helicopters and automobiles should be encouraged to improve competitiveness. With regard to the chemical and pharmaceutical industry, product quality should be improved and adding value should be used to expand its market. As a heavily polluting industry, it should be operating in full compliance with the national environmental policy. The information technology industry is an emerging industry in Jingdezhen. When stimulating its expansion, attention should be given to lengthening its industrial chain and increasing added value, by integrating and supporting the production of both primary and intermediate products as well as finished products. The ceramic building materials industry is unique to the economy of Jingdezhen and holds a crucial position for the further development of the tertiary industry. However, efforts must be made to improve its energy efficiency and reduce pollutant emissions. Meanwhile, the platforms of ceramic detection and information technology services, e-commerce, and ceramic technology innovation of small and medium enterprises (SMEs) should be developed. The electric power industry provides crucial energy security for industrial, economic, and social development, but it is also a highly polluting industry. To reduce emissions, low capacity units should be closed or merged to improve energy efficiency as well as adjustments to the structure of the power generation industry to increase the use of clean energy (Figure 14.10).

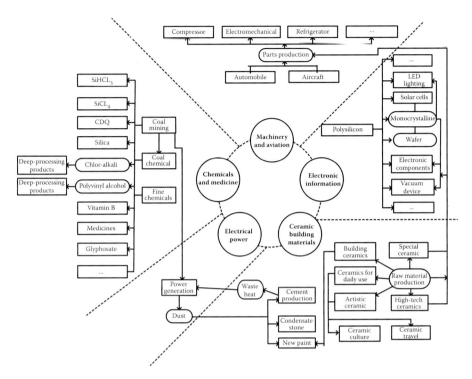

FIGURE 14.10   The framework of industrial system in Jingdezhen City.

c. *Eco-construction of industrial parks:* Three industrial parks will be established when implementing the eco-city plan for Jingdezhen. The first is a high-tech industrial zone, including the three pillar industries: machinery manufacturing, information technology, and electric power production. The second is a ceramic industrial park, dominated by the ceramic industry, the goal of which is to be a provincial level park, providing the combined development of building ceramics, specialized ceramics, high-tech ceramics, artistic ceramics, and ceramics for daily use. Finally, the Leping Industrial Park will be orientated to the pharmaceutical and chemical industry, with the goal of becoming a base for the fine chemical industry. The new enterprises must be migrated into relevant industrial parks according to the principles of scientific planning and optimized layout, which is conductive to form the eco-industrial chain and optimize allocation of resources. Demonstration enterprises should be established for training, which showcase circular economy. In addition, the infrastructure of parks must be improved, including the supply of water, electricity, and heating to promote energy efficiency.

4. *Planning for Ecotourism*

   Jingdezhen is a historical and cultural city, and the core of old city should be protected. To achieve this, land should be developed rationally to avoid detrimental impacts on original and unique architecture. For example, when building residential and commercial areas, attempts should be made to adopt the "Jing School" of architectural style. As a tourist city, the characteristic landscape must also be protected because a beautiful environment is the most important factor that attracts the tourists. The negative impacts of pollution on tourism should be minimized by improving environmental protection facilities and promoting an awareness of the environment to tourists.

The natural beauty of the area and the aesthetic qualities of scenic areas should be conserved by implementing professional management. Attention must also be focused on the environmental impact assessment for scenic spots to achieve optimum utilization and sustainable development. The six elements of tourism, namely, food, shelter, transportation, sightseeing, shopping, and entertainment, must be optimized to improve the quality of tourist services and the environment of tourist attractions. The promotion of cultural tourism requires the conservation of traditional crafts such as handmade pottery, which have been lost in the development of modern porcelain technology. Special attention should be made to preserve the heritage sites and traditional processes as well as the artisans and technicians vital for the training of new artists so that these traditions can be maintained in the future.

14.4.3 PLANNING OF ECO-ENVIRONMENT SYSTEMS

14.4.3.1 Planning of Atmospheric Environment System

1. *Status Assessment of Atmospheric Environment*
   a. *Status analysis:* Jingdezhen can be characterized as having an inland subtropical climate with a mean average temperature of 17.0°C in January, which is the coldest month with an average temperature of 4.8°C, while June is the warmest month with an average temperature of 28.7°C. The annual average wind speed is 2.0 m/s, and speeds less than or equal to 3.0 m/s account for 55.0% of the total.

      There is a close relationship between air pollution and energy consumption. Jingdezhen is rich in mineral and coal resources but lacks oil and natural gas. It has been noted that coal plays a relatively important role in the industrial energy consumption of Jingdezhen as an industrial city; however, the available coal can only satisfy 40% of the industrial demand. In 2007, the consumption of raw coal was 3,159,900 tons, accounting for 63.6% of the total energy budget. Meanwhile, the consumption of refined coal was 1,326,700 tons, accounting for only 33.6%.

      Jingdezhen has three main sources of electricity production: coal, thermoelectricity, and hydropower. However, the latter two provide only a small proportion of the total demand, with hydropower accounting for 4%, while thermoelectricity can only meet the needs of a few high-tech enterprises and does not supply the national grid. Therefore, the electricity supply of Jingdezhen is dominated by coal power with high fuel consumption and heavy pollution. Accompanying the growth of the economy over recent years, a number of major projects and companies have located to Jingdezhen, and it has been difficult for the old power grid to meet the associated increase in electricity demand. Since there is no natural gas pipeline, the main fuels for urban residents are liquefied petroleum gas and pipeline coal gas. In 2007, the average coal consumption per capita of urban families was 2.09 kg, while liquefied petroleum gas consumption was 15.97 kg, and pipeline coal gas 51.03 m^3. Although the overall energy consumption of Jingdezhen is gradually increasing, the energy consumption per unit GDP is declining, with the figure for 2007 being 1.146 tons standard coal equivalents per 10^4 yuan. The energy efficiency of Jingdezhen is a little higher than the national level, but there is still a gap with the average level of Jiangxi Province. However, all things considered, Jingdezhen should be able to meet the target of "11th Five-Year Plan" (2006–2010) (The reference year of this project is 2007 and the target year is 2010).

The air pollution of Jingdezhen results from both automobile exhaust gas and industrial emissions and contains pollutants including smoke, sulfur dioxide (SO_2), NO_2, and so on. The emission mitigation strategy is mainly focused on industrial point sources, even though a proportion of emissions also result from residential sources. In 2007, the industrial exhaust gas emissions totaled 30,339,040,000 standard cubic meters, an increase of 11.97% compared to 2006. With regard to individual pollutants, the total SO_2 emission was 4,840,000 tons, a reduction of 6.7% compared to 2006, and the smoke emission was 8,569 tons, a reduction of 1.75%. A large proportion of the industrial structure of Jingdezhen is composed of industries that have a high energy consumption and emission rates, such as the electricity, coking, and chemical and cement industries. The Jingdezhen Power Generation Limited Liability Company and three other large enterprises alone produced a combined emission of SO_2 and smoke that accounted for 85% and 88.6%, respectively, for the total of these pollutants. Emission reduction measures for air pollution are focused on implementing clean production management and eliminating steel, cement, and papermaking enterprises with high energy consumption and heavy pollution by relocating them to the countryside. In recent years, there has been some progress on the reduction of the emission of some airborne pollutants; for example, the emission of SO_2 fell by 1735 tons in 2006 and 3345 tons in 2007.

b. *Status assessment:* The main pollutants in Jingdezhen are respirable particulate matter and SO_2. However, since the "9th Five-Year Plan," SO_2 emissions have been effectively controlled by a series of preventative measures. As urban construction releases more dust, the main pollutants are now respirable particulate matter and dust. In general, the targets set by the National Secondary Standard for pollutant concentrations have been met with the exception of a few years when there were exacerbating climatic conditions. However, the levels for dust have sometimes exceeded the National Standard as a result of inclement weather; that is, low rainfall and the increasing development of infrastructure and real estate, the construction of which generates large amounts of respirable particular matter. However, with the completion of these projects and improvements to ceramic production technology coupled with the expanding size of urban green areas and the implementation of environment-protecting measures, the air quality has gradually improved. The proportion of days recording a good level of air quality has now reached 100% throughout the whole year (Figure 14.11). However, acid rain, related to increased car ownership and the industrial consumption of coal energy, has been a serious problem in recent years, although in 2008, the trend improved with a rate of 72.73%.

2. *Planning for the Atmospheric Environment*

On the basis of the forecasted demand for energy and pollutant emissions for the near future, the following atmospheric environment plan has been formulated for Jingdezhen.

a. *Energy-saving program:* The energy-saving program can be divided into three levels: firstly, the adjustment of the industrial structure for primary, secondary, and tertiary industries; secondly, the improvement of the energy consumption structure; thirdly, the conservation of energy within highly polluting enterprises.

The tertiary industry of Jingdezhen has declined in recent years. To remedy this, it requires the promotion of tourism to release its untapped potential. Readjustment of the overall industrial structure is also required, which can be achieved by developing the light and high-tech industries and implementing a rationalized structure for both light and

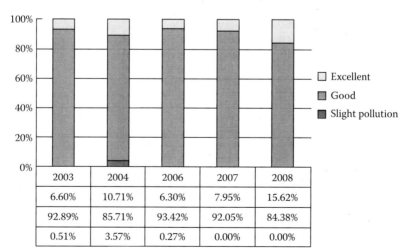

FIGURE 14.11 (**See color insert.**) The proportion of air quality levels in Jingdezhen from 2002 to 2008.

heavy industries. To conserve energy, the outdated facilities for energy production should be eliminated and the renovation of coal-fired industrial boilers and kilns should be promoted. To improve the efficiency of coal use, it is necessary to develop and promote fluidized bed combustion for domestic coal, smokeless combustion, gasification, liquefaction, and other coal cleaning technologies. In addition, the use of renewable energies should be increased and the energy structure optimized. As a result of the "East Gas Transmission" project, it is hoped that natural gas can be delivered to Jingdezhen by 2009 to meet the needs of all urban residents and businesses.

The untapped energy saving potential of vital industries must be developed. The thermal power industry should be developed to a capacity of 600,000 kW and supercritical units, large-scale combined cycle units should replace low-power units. The power generation plants in Jingdezhen should implement "big pressure on the small" project, and two 6,000,000-kW supercritical condensing steam turbines will be constructed in remote areas. Thermoelectric industry should be more widely utilized; for example, for district heating, as well as increasing the role of thermal power units and energy cascades. Cogeneration should also be promoted, for example, heat, electricity, and cooling, or heat electricity and gas triple technologies, to improve the integrated energy utilization efficiency. The second phase in the cogeneration project of Jiangxi Electrochemical Chemical Limited Liability Company plans to construct a 130-t/h circulating fluidized bed boiler and a 12,000-kW generator. The cement industry should transform its existing medium-sized rotary kiln, mills, and drying facilities into energy-saving ones, eliminating the use of shaft kilns, wet process kilns, dry kilns, and other outdated models before 2008. It is also hoped that by 2010 the proportion of new dry cement could surpass 70%. In addition, other energy-saving measures should be implemented, including the rural biogas project, especially the construction of large-scale farm digesters, the adoption of energy-efficient LED street lighting, and other building and motor systems that conserve energy.

b. *Emission reduction program:* The energy-saving planning and management policies of Jingdezhen have been developed to achieve the targets of pollution emission regulations by focusing on reductions in the industrial structure and projects for the management and control of emissions.

Jingdezhen is currently in the process of enforcing energy efficiency for emerging industries and the transformation of the existing ceramic industry, eliminating outdated energy production facilities (Wang et al. 2007; Xie, 2010; Zhan et al. 2010; Zheng and Guo, 2011). Similar updating measures are being implemented for other industries, including electricity, cement, coke, papermaking, brewing, and other industries in accordance with the polices of the Development and Reform Commission and other departments. Clean production and desulfurization projects must be implemented to reduce the emission of SO_2, for example, by accelerating and expanding the construction and operation of existing and newly built desulfurization facilities in coal-fired power plants. In addition to monitoring and updating the existing high-energy consuming and polluting enterprises, the guidelines for newly built enterprises should have more rigorous performance targets and emission permits must be strictly enforced. Enterprises with no treatment facilities and emitting more pollutants as a result of poorly maintained facilities must suspend production until such deficiencies can be resolved. Furthermore, management systems must be perfected and the diligent management of emissions must be promoted.

c. *Pollution prevention programs:* A pollution prevention program has been developed in accordance with the forecasted predictions for the future emissions of SO_2, nitrogen dioxide, and dust to meet the targets air pollution control goals. Besides adjustments to the industrial structure and energy structure that are mentioned in Sections 14.4.2.1 and 14.4.3.1, the management of urban dust, industrial pollution, motor vehicle exhaust, and acid rain must also be considered and an overall strategy developed, which provides pollution prevention engineering programs, pollution prevention technology, and management measures.

The increase of automobile exhaust emissions is the main cause of SO_2 NO_2 pollution and acid rain. It is therefore important to improve traffic management and the organization of the road network to reduce traffic congestion. Meanwhile, adjusting the balance of different modes of transport is also necessary for NO_2 pollution control. Residents should be encouraged to use public transport as the pollutant emission per person is just one-fifth that of private motor vehicles. New traffic regulation should be formulated to restrict the number of private cars, control the number of cars in urban areas, and strictly allocate time for delivery vans and agricultural vehicles to have access to urban roads.

Secondary dust should be managed, and government should pay more attention to the regulations governing construction, road management, and the control of dust transportation. Effort should also be made to expand the number of urban green spaces and increase the green coverage of the city in an attempt to reduce the level of atmospheric particulates. Environmental pollution problems are caused by human activities, and so, adjusting the behavior of residents and encouraging their participation in discussions of environmental policy and improving the degree of environmental measures and policies affecting the public are all effective ways of mediating urban pollution.

14.4.3.2 Sustainable Utilization Program for Water Resources

1. *Environment Status Assessment of Water Resources System*
   a. *Statue of water resources:* The rivers of Jingdezhen can be considered as components of the Raohe River system. The Changjiang River is 81.9 km long and has a drainage area of 3274 km^2, while the Anle River

is 83.2 km long and has a drainage area of 1974 km². Jingdezhen is located in a subtropical monsoon climate zone with an annual average rainfall of 1763.5 mm. As a result, the water resources are plentiful, with an average surface water volume of 5.34 billion m³ and an estimated 0.99 billion m³ of groundwater. The amount of water per capita is 3949 m³, which is higher than the national average. In 2007, the city's total water supply was 0.744 billion m³, of which 93.3% was surface water. The city's total water consumption is 0.744 billion m³/year, of which production water contributes 0.67 billion m³, accounting for 90.59% of the total. The distribution of water between natural water bodies and the three types of industry, primary secondary, and tertiary, are shown in Figure 14.12.

In terms of exogenous agricultural water use, the effective irrigation area of Jingdezhen is 789,200 ha, with a total volume of 0.351 billion m³, providing an average volume of 480 m³/mu of farmland, which is 434 m³ higher than the national average. However, the proportion of water-saving irrigation is only 12.42%, which is just 22% of national average. The effective utilization efficiency of irrigation water is only 0.48, which lags far behind the level of developed countries (0.7–0.8). In terms of industrial water, a higher volume is consumed totaling 150 m³/10,000 yuan. This is 1.3 times higher than the provincial average and 1.15 times the national average. The recycling rate of industrial water is only 60%, which is far lesser than the leading water efficient cities of China. In terms of domestic water, there is an insufficient water supply pipe network, and serious leakage is a common problem. The pipe leakage rate is 29%, which is much higher than the national averages and higher than that in the Jiangxi Province. The domestic use of water-saving appliances is just 65%, and so, there is still room for improvement. In general, the water consumption per unit of GDP in Jingdezhen has declined in recent years, the figure for 2007 being 142.5 tons/10,000 yuan. The overall efficiency of water use has improved, but industries at all levels still have great potential to save more water.

b. *Water resource status:* During the period spanning from 2002 to 2004, vital enterprises were responsible for the highest volume of wastewater in Jingdezhen. However, the growth rate of chemical oxygen demand

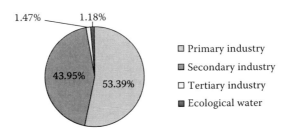

1.47%   1.18%

□ Primary industry

■ Secondary industry

43.95%   53.39%

□ Tertiary industry

■ Ecological water

FIGURE 14.12   **(See color insert.)** Structure of production water in Jingdezhen in 2007.

(COD) associated with industrial wastewater discharge has decreased significantly from 2005 to 2008. The main pollutants of industrial wastewater are ammonia nitrogen, petroleum, cyanide, volatile phenol, and an unfavorable chemical oxygen demand. The main industries responsible are the electricity, pharmaceutical, and paper industries. Domestic water is also A point source pollution. In 2007, the chemical oxygen demand caused by domestic wastewater was 19,700 tons, which accounted for 80% of the total chemical oxygen demand.

In terms of nonpoint source emissions, agricultural runoff is a major source. In 2007, the amount of fertilizer used in rural areas totaled 65,464.1 tons and the amount of pesticide use was 1,175.27 tons. However, the rate of nitrogen and phosphorus utilization in paddy field is just 20%–30% and the surplus can enter water bodies through surface runoff, leaching, and soil erosion. In addition, rural areas discharge 50,890,000 tons of sewage and 2,528,000 tons of animal manure, which are discharged directly into the rivers because of a lack of sewage treatment facilities. Additional nonpoint sources of pollution are the precipitation runoff from the 0.4 million tons of garbage stacked along the river and enterprises around rural areas that lack environmental protection facilities.

c. *Water quality status:* The water quality of the main rivers of Jingdezhen is generally good. The water quality of most rivers is above Class III, including the branches of the Changjiang River. In Changjiang River, 80% branches' water quality are above Class III. In Lean River, 87% branches' water quality are above Class III. The main pollutants in other more heavy polluted rivers are ammonia nitrogen, total phosphorus, volatile phenol, and oil. The Lake Water Factory and three other water treatment plants are tested every 10 days and the water quality is generally good, Class II or higher.

d. *Status of water pollution prevention:* In 2008, the centralized treatment rate for urban sewage in Jingdezhen was only 41.48%, still far behind the national eco-city standards, and the safety of local water cannot be guaranteed. There are currently three sewage treatment plants in operation, including the Watermelon Island Sewage Treatment Plant that has a daily processing capacity as high as 120,000 tons. However, this still cannot meet the demand of domestic waste, and few more sewages are still discharged directly into the rivers. The second phase of these three treatment plants are presently under construction, which will expand the daily capacity by an extra 100,000 tons.

In 2008, the industrial wastewater emission from Jingdezhen was 43.70 million tons, of which 42.86 million tons met the prescribed discharge standards. The wastewater originates mainly from the mining, electrochemical, mechanical, pharmaceutical, and other industries. The compliance rate of industrial wastewater is 98.06%, and the national standard is 100%, and so, it occasionally happens that some of the discharged industrial wastewater exceeds the required standard. As the sewage treatment facilities of some enterprise operate sporadically and since the sewage treatment plant at Leanjiang Industrial Park is still not operational, outgoing water quality sometimes does not reach the standard of sewage treatment plants of other industrial parks.

2. *Water Environment Planning Program*

By analyzing the forecasts for both point source pollution and nonpoint source pollution as well as the water supply and demand balance, and in conjunction with a comprehensive analysis of water environment trends, a water environment planning program was formulated as follows (Yang and Cheng, 2002; Yang et al. 2002; 2004; 2005; 2010).

a. *Sustainable use program for water resources*

   i. *Water resource configuration program:* According to the results of a water supply and demand analysis, it will be a challenge to meet the desired targets within the time frame of the target year. It is therefore necessary to establish an optimal allocation pattern to guarantee the water supply and drinking water quality in Jingdezhen and resolve the conflict between water supply and water demand. However, the Yuxikou Dam Project will be finished before 2012, increasing the dry season river flow and alleviating seasonal water shortages. The Yutian Reservoir Expansion Project should also be completed by 2011 and will safeguard the water supply of Jingdezhen by storing water and guaranteeing the safety of drinking water.

   ii. *Water-saving system construction program:* As a first step, water protection and management regulations must be improved. It is also necessary to establish and improve water-saving measures and regulations and formulate relevant water-saving standards and mandatory water-saving regulations. Departments at all levels should endeavor to save water. Secondly, authorities should alter the patterns of water use and establish efficient water-saving systems. With regard to agriculture, urban ecological agriculture should be developed, the construction of hydraulic infrastructure projects should be increased to construct state-of-the-art water-saving irrigation systems and greatly increase the proportion of water-saving irrigation. The irrigation efficiency targets for the current planning years are 0.55, 0.60, and 0.65, respectively. In terms of industry, water-saving measures should focus on the oil, chemical, papermaking, metallurgy, building materials, and food processing industries. Through the implementation of clean production and the modernization of the original or outdated technologies, the industrial water recycling rate could increase to 68% by the year 2010, with targets as high as 85% for the long term. In terms of domestic use, it is necessary to promote an awareness of water conservation and a change in the social behavior toward water saving. In addition, to meet the challenge of a gradually heavier pollution load, Jingdezhen should redouble efforts to use regenerated water, recycled water, and rainwater sources for urban areas to reduce the consumption of natural sources of freshwater. Wastewater can be utilized for suburban agriculture, urban greening, and the landscaping of rivers and lakes as well as domestic use and the replenishment of groundwater through recycling facilities. Industries should also be encouraged to use recycled water, with a target of 13,000,000 m³ for the year 2010. Jingdezhen has plentiful rainfall, and the construction of rainwater collection pipelines and sewage diversion pipelines should be accelerated to improve rainfall utilization. Further improvements to the environmentally friendly use of water can be gained by studying the effective models of other countries.

b. *Water quality protection and improvement program*
    i. *Centralized sewage treatment program*
        A. *Residential sewage treatment:* The targets for the centralized treatment of urban sewage are 80%, 85%, 90% for each of the planning years 2008, 2009, and 2010, respectively. To help match the forecasted production of urban sewage, the treatment capacity should be increased to 110,000 tons/day, 140,000 tons/day, and 190,000 tons/day, and the investment and construction of the Xiguazhou sewage treatment plant and three others should achieve a capacity of 140,000 tons/day by 2010, meeting most of the sewage treatment needs. Further construction projects at the Leping sewage treatment plant and three others should increase the daily treatment capacity to 270,000 tons/day, sufficient to meet the needs of sewage treatment.
        B. *Industrial wastewater treatment:* Currently, the rate of wastewater achieving the desired standard is only 95.7%. However, this figure must be increased to 100% to achieve the target for the planning period. The discharge management of industrial wastewater must be improved to reduce the levels of COD and other pollutants to guarantee by the efficient operation of pollution mitigating facilities. As this process involves several different sectors, governments should investigate and regulate the separation of priority pollutants and avoid mixing them with concentrated sewage at treatment plants that only employ conventional biological treatment processes. Finally, the treatment rate of industrial wastewater should be improved.
    ii. *Control of domestic and industrial pollution*
        A. *Control of domestic pollution:* Firstly, the process of urbanization should be accelerated to centralize the dispersed rural population within urban areas, concentrating the treatment of

wastewater. Secondly, attempts should be made to increase the awareness of consumers to a greener lifestyle and environment protection. This can guide the public to the automatic selection of green products; for example, detergents without phosphorus. Further measures include economic, legal, and social enticements to reduce the pollution generated by the process of consumption.

B. *Control of industrial pollution:* Firstly, the inefficient industrial production structure should be altered, for example, by prohibiting the development of coking, pharmaceutical, power, papermaking, metallurgy, and other big industries upstream of drinking water sources. Meanwhile, authorities should enforce the closure of small-scale enterprises or those with outdated facilities, modernize existing industrial enterprises, and gradually centralize and relocate enterprises to the corresponding functional areas. In addition, clean production should be promoted and the principle of "4R" implemented. Secondly, managerial strategies should be utilized to strictly enforce environmental laws and regulations, improve the environmental management of highly polluting enterprises, such as the coking, power, pharmaceutical, and papermaking industries. Efforts should also be made to take advantage of the emission reduction potential of individual enterprises and to control the overall industrial production of COD.

iii. *Nonpoint source pollution control program*

Firstly, it is necessary to adjust the structure of fertilizer use by expanding the use of manure, green manure, and other organic fertilizers and promoting the technology of deep placement fertilizer application to control pollution from agricultural runoff. Systems to improve pest prevention and forecasting should be established, and the secure use and management of pesticide should be strictly controlled, Methods of integrated pest management should be improved to increase the efficiency of pest prevention. The sale and use of highly toxic, inefficient, or persistent pesticides should be restrained, and alternatives such as biocontrol and genetically modified crops should be encouraged. The introduction of eco-agricultural methods and modern agriculture technology should be accelerated to ensure greener and pollution-free agricultural production.

Secondly, the domestic pollution generated from farming households and livestock must be controlled. According to the Environmental Integral Management Plan for Fuliang County, the Rural Integral Management Plan for Jingdezhen City, and ecological function zoning requirements, the distribution of livestock should be limited to appropriate locations and special attention should be given to areas sensitive to water pollution near sources of drinking water. The use of manure and recycling of sewage to compost and methane fermentation should be encouraged to bring livestock rearing into a virtuous cycle, a path toward ecological balance. In residential areas, authorities should establish waste management facilities with low costs and a simple management system that performs waste separation and recycling to increase the rate of harmless waste handling to 80% or more.

iv. *Others*

In addition to the program for protection and improvement of water quality, special attention should be given to the protection of potable water sources, the comprehensive remediation of water bodies, and the construction of mechanism for the rejuvenation of watershed ecosystems.

14.4.3.3   Planning of Solid Waste Disposal

1. *Solid Waste Disposal and Recycling Status of Jingdezhen*

a. *Garbage:* According to a 2007 survey, the total quantity of garbage was 151,600 tons and the total disposal of garbage was 150,600 tons, resulting in a garbage treatment rate of 99.3% and a recycling rate of 8.5%. Currently, Jingdezhen has only one waste disposal site although new facilities are in the planning stage. The existing Songjiashan waste disposal site has a capacity of 390–420 tons/day and has a total capacity sufficient for 10 years of operation. However, after nearly 7 years of use, little capacity remains available and so a new site is urgently required. The new project, the Nianyugou waste disposal site, is located about 1 km to the south of the present Songjiashan site and is expected to begin operation in 2012. The projected life span of the new facility is 26 years with a capacity to process 1000 tons/day.

b. *Solid industrial waste:* The solid industrial waste produced by Jingdezhen is mainly composed of tailings, fly ash, coal gangue, slag, metallurgical slag, and other wastes. The amount of solid waste produced during the period of "10th Five-Year Plan" was 5,063,300 tons, of which fly ash and slag accounted for 67.66%. The amount of waste disposal and recycling was 4,537,800 tons, the resource utilization rate was 89.54%, the amount of emission waste was 503,300 tons, and the total amount of stock was 732,500 tons. Jingdezhen generates 1,368,400 tons of industrial solid waste, with a utilization capacity of 1,225,100 tons, and the utilization rate is 88.88%. In 2007, the utilization and disposal rate of industrial solid waste was 92.13%, which was a significant decrease compared with the year 2006 when the rate was 97%. The industrial solid waste of Jingdezhen comes mainly from the electricity, heat production, and supply industry and the chemical fiber manufacturing and is mainly composed of fly ash, slag, coal gangue, tailings, and hazardous wastes. The amount and overall utilization of industrial solid waste produced by industries at all levels is summarized in Table 14.10.

TABLE 14.10

Comprehensive Utilization of Industrial Solid Waste for Jingdezhen in 2007

Industry	Productivity (10,000 tons)	Utilization (tons)	Utilization Rate (%)
Electricity, heat production, and supply industry	84.65	75.42	89.10
Manufacture of chemical fibers	14.00	15.00	100.00
Coal mining and washing	11.56	11.56	100.00
Nonferrous metal mining industries	6.81	2.81	41.26
Petroleum processing, coke, and nuclear fuel	4.30	3.63	84.38
Nonmetallic mineral products	1.78	1.78	100.00
Pharmaceutical manufacturing industry	0.72	0.71	99.58
Transportation equipment manufacturing industry	0.60	0.60	99.83
Chemicals and chemical products manufacturing industry	0.49	0.49	99.75
Papermaking and paper products industry	0.14	0.14	100.00
Special equipment manufacturing industry	0.11	0.11	100.00
Beverage manufacturing industry	0.04	0.04	100.00
General equipment manufacturing industry	0.03	—	—
Electrical machinery and equipment manufacturing industry	0.01	0.01	100.00

   c. *Hazardous waste:* The industrial hazardous waste of Jingdezhen is mainly produced by the Changhe Automobile Factory and other automobile factories and is composed of waste mineral oil generated from the mechanical plant, paint residue generated from spray departments, and industrial sludge with high levels of heavy metals generated by the automotive industry. Hazardous waste is also generated by waste treatment facilities of some environmentally friendly enterprises and the storage of expired drugs. Some industrial hazardous wastes are utilized on site by the producer through incineration or disposed of by landfill with a disposal efficiency of 97.25%. Currently, 80% of medical hazardous waste is disposed of by delivering to specialized disposal companies, and a medical waste disposal site with a capacity of 1000 tons/day is already operational.

 2. *Problems of Solid Waste Disposal*

   a. *Shortfall in the waste disposal capacity:* The new Nianyushan Waste Disposal Site under construction, with a daily capacity of 1000 tons, will be sufficient to meet the needs of future urban waste disposal. However, the current site at Songjiashan only has a daily capacity of 330 tons and is struggling to meet the current demand. This overloading of its capacity creates problems for the control and prevention of pollution.

b. *Improving the level of solid waste utilization:* Although the garbage treatment rate of Jingdezhen has reached 99.3%, the awareness of useful waste resources is poor. A basic garbage collection system has not yet been established and the utilization rate of waste resource is just 8.5%.

c. *Improving supervision and management of hazardous waste disposal:* The availability of surveys and statistical analysis regarding hazardous waste in Jingdezhen is insufficient and basic information regarding the total amount, distribution, processes of generation, and the industries responsible is lacking. Currently, there is no coordinated disposal system, and so little hazardous waste is treated by the producer, and although 97.3% is transferred to specialized disposal companies, the process still needs to be improved.

3. *Planning Objectives and Indicators*

It is necessary to transform Jingdezhen into an eco-city to meet the challenges of development. The ideal of clean production is the motivation while "saving energy and emission reduction" are the methods toward the central objective of improving the efficiency of resource utilization. To accelerate this process, it is necessary to create a system of separated waste collection to facilitate recycling and improve the treatment of hazardous waste and introduce supervision and management regulations to finally realize the goal of "reduce, recycle and reuse" for hazardous waste (Table 14.11). The targets for 2010 include achieving 99.4% urban garbage treatment, greater than 93% utilization of industrial solid waste, and an overall utilization rate of 89%. By 2020, it is hoped that these figures will reach 99.6%, 95%, and 90%, respectively, and it is urgent to ensure that hazardous waste and medical waste are disposed of safely.

4. *Solid Waste Disposal Program*

To meet the targets for domestic garbage and industrial solid waste within the planning period, a solid waste disposal program has been formulated as follows.

a. *Domestic garbage disposal program:* Solid waste can be treated by four methods: landfill, composting, incineration, and recycling.

TABLE 14.11

Solid Waste Planning Indicators for Jingdezhen

Indicator	Status value	Unit	2010	2015	2020	Reference Value
Treatment rate of urban domestic garbage	99.3	%	99.4	99.5	99.6	≥90
Utilization rate of industrial solid waste	92.1	%	93	94	95	≥90 with no hazardous waste

Landfill is currently the preferred treatment option for Jingdezhen. As far as this planning period is concerned, the main focus is the nonpolluting safe disposal of domestic garbage; however, as society develops and people's understanding of composting waste increases, the relevant authorities can consider the construction of waste treatment plants to realize the ideal of recycling and the reduction of waste. In the counties around Jingdezhen, the garbage is simply stacked, which not only uses large areas of land but also has negative effects on the surrounding environment, groundwater, air quality, and soil. Since a high proportion of this area is environmentally important, domestic waste should be disposed of by sanitary landfills and small-scale incinerators.

b. *Disposal program for general industrial solid waste:* In 2007, the overall utilization rate of industrial solid waste was 8.88%, a 1% decrease compared to the year 2006. However, during this period, the amount of industrial solid waste grew by 25% with a corresponding increase in emissions. Strategies must be developed to tackle both solid waste treatment at point sources and the overall utilization rate of solid waste, the specific details of which are detailed in Figure 14.13.

c. *Hazardous waste disposal program:* All urban medical waste should be disposed of at specialized medical waste disposal sites. Enterprises producing hazardous waste should be strictly controlled. Hazardous waste can be delivered to specialized facilities, which have been allocated hazardous waste management licenses, for the correct disposal and utilization. Hazardous waste disposal patterns are summarized in Figure 14.14.

5. *Prevention of Solid Waste Pollution and Control Policies and Measures (Zhou et al. 2011; Su et al. 2012)*

a. *Domestic garbage:* Firstly, it is necessary to increase the overall utilization of domestic garbage. In accordance with the principles of

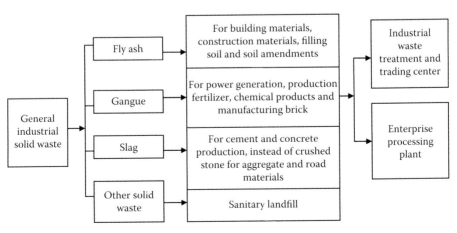

FIGURE 14.13   General treatment pattern of industrial waste.

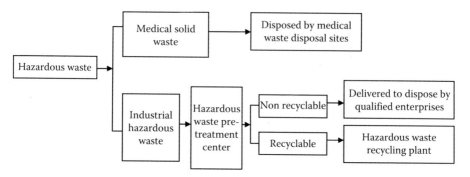

FIGURE 14.14 Hazardous waste disposal pattern.

"harmless, reduction, recycling," authorities should focus on the sources of the waste for appropriate classification and utilization. Garbage collection facilities should be built in both old and newly built residential areas, in office buildings, and in industrial parks. By 2010, the collection rate for domestic garbage in Jingdezhen should reach 25% or more. Domestic garbage contains a lot of recyclable materials such as paper, plastic, fabric, glass, and metal, and so, recycling and reuse are important means of waste reduction. For maximum efficiency, waste should be separated and processed according to type of wastes and processed in an appropriate way by incineration, composting, landfill, or recycling.

Secondly, it is necessary to focus on the safe treatment of garbage. Currently, the waste disposal facilities of Jingdezhen are overloaded, and the existing Songjiashan Waste Disposal Site will soon reach the end of its service life. To ensure the sustainable decontamination and disposal of garbage, the construction and operation of Nianyushan Waste Disposal Site should be accelerated to achieve the garbage treatment targets of 99.4% and 99.6% for 2010 and 2020, respectively. Authorities should strengthen the supervision of landfill sites to ensure the smooth operation of such facilities. Meanwhile, the ecological remediation of abandoned landfill sites should be accomplished by reforestation.

b. *Industrial solid waste:* The industrial planning aims for Jingdezhen include the construction of a high-tech industrial development zone, the Jingdezhen Ceramic Industrial Park and the Leping Industrial Park, all of which should be completed by 2012. A large number of companies will be located in these parks, and as a result, the localized industrial solid waste will reach an unprecedented level. In response, firstly, each enterprise should conduct internal audits and implement cleaner production and increased waste utilization rates appropriate to their production processes as soon as possible. Secondly, companies should be encouraged to use fine materials and improve the purity of raw materials to avoid additional solid waste. Thirdly, material recycling processes must be developed to reduce the amount of solid waste. It is necessary to shift the emphasis from the end treatment of pollution to controlling the whole production process and sparing no effort to realize the goal of maximum recycling and the minimum production of solid waste.

The urban areas of Jingdezhen also produce large amounts of solid waste, including construction waste, coal gangue, and fly ash. The following measures should be taken to improve the overall utilization rate of solid waste in the city.

    i. *Smelting waste:* Smelting waste includes both blast furnace slag and steel slag. Blast furnace slag is characterized by the production of slag granulation water, which can be used as cement materials for both domestic and foreign markets.

    ii. *Fly ash:* Fly ash is also mainly used for the preparation of cement during the production of wall materials such as solid bricks, hollow bricks, hollow blocks, light plate, aerated concrete, gypsum, and other decorative ceramic materials.

    iii. *Gangue:* Transforming soft plastic molded technology and equipment into hard and semihard plastic molded technology and equipment, gangue brick can be on the scale and level.

    iv. *Construction waste:* All unused waste bricks and ceramics can be utilized to cover landfill sites during their restoration.

    v. *Sludge:* Sludge contains large quantities of ash, aluminum, iron, and other components, all of which are indispensable components of building materials. Sludge drying technology for cement is a safe and low investment solution and therefore a useful treatment and disposal method. As such, part of the eco-city plan strategy for Jingdezhen includes the coupling of a large-scale sewage treatment plant with cement production. The construction of a sludge cement kiln is also planned for 2020. The sludge produced by smaller sewage treatment plants can be treated as domestic garbage and can be disposed of centrally by sanitary landfill.

  c. *Hazardous waste management measures:* A Centralized Disposal of Medical Waste Plan for Jingdezhen should be formulated to ensure the operation of the newly built medical disposal center. The authorities should strengthen efforts toward the prevention and control of hazardous waste by updating the charging system for companies producing hazardous waste, establishing a license and registration system for hazardous waste management, and creating a supervision and inspection system to control and monitor the operation of hazardous waste disposal facilities ensuring that they operate in accordance with the law. It is also important to improve the surveying of hazardous waste production for statistical analysis and to compile regular reports regarding basic information such as the total amount, distribution, industries responsible, and processes of generation.

14.4.3.4   Planning for Ecological Protection

1. *Status and Challenges of Ecological Protection*
   a. *Ecological protection status:* Jingdezhen has a subtropical climate
      characterized by high temperatures and abundant rainfall, which pro-
      duces complex geographical environment with dense forest vegeta-
      tion and high biodiversity. According to an environmental survey, it is
      home to 760 kinds of vertebrates, including 88 species of mammals,
      320 birds, 65 reptiles, 38 amphibians, and 157 fish as well as 6000 spe-
      cies of insect. Some of these species are key national protected species.
      However, disturbances from human activities, including uncontrolled
      economic development and indiscriminant exploitation of natural
      resources, are resulting in habitat destruction and the threat of extinc-
      tion for some of these protected species. In recent years, Jingdezhen
      has designated an increasing number of nature reserves to protect bio-
      diversity, using existing nature reserve as a model. In the period span-
      ning from 1997 to 2008, nine nature reserves were created, including
      one provincial reserve, Yaoli, and eight county-level reserves, including
      Denggaoshan, Gongku, and Hongyanzhen. The current area of nature
      reserves in Jingdezhen now totals 53,613.9 ha, accounting for 10.21% of
      the entire land area and representing one of the highest rates within the
      province.
        In an attempt to improve air quality, the government has also
      created many forest parks that now cover 25,269.3 ha and account
      for 4.81% of the total land area. Of these, two are operated at the
      national level—Jingdezhen, Jiangxi Province, National Forest Park,
      and Yaoli National Forest Park—and three at the provincial level—
      Guopufeng Provincial Forest Park, Fuliang Provincial Forest Park,
      and Hongyuan Leping Provincial Forest Park. The forest parks of
      Jingdezhen are in good condition, receive a high level of protection,
      and play a key role in the maintenance of a healthy pattern of the
      ecological landscapes.
   b. *Challenges for environmental protection:* The ecological balance of
      Jingdezhen must be improved. The area around Fuliang County has
      many mountains and rivers, and the landscape of plains and valleys all
      combine to ensure a plentiful supply of freshwater and lush vegetation.
      These natural features mean that this region has a higher proportion of
      nature reserves and ecological conservation land than other areas of
      Jingdezhen. If an overall strategy for conservation is to be achieved,
      future planning should focus on the creation of ecological reserves in
      these other areas to produce a more comprehensive ecological land-
      scape pattern around the city.
        The overall level of protected environments in Jingdezhen is low.
      There are only two national level forest parks—Jingdezhen, Jiangxi
      Province, National Forest Park, and Yaoli National Forest Park—
      and one provincial level nature reserve—Yaoli National Ecological

Protection Areas. There are other county-level nature reserves, but they are small scale, lack funding, receive inadequate attention, and more effort is required to improve their infrastructure.

Human activities can reduce the level of ecological protection, and although ecotourism promotes the creation of nature reserves and forest parks, as well as bringing economic benefits, it can also cause disturbance to natural ecosystems. In addition, land development and the construction of roads also have negative effects.

The management system is not perfect and has many inadequacies. Overall, the management and designation of nature reserves in Jingdezhen is not systematic and needs to be improved. Meanwhile, the local laws and regulations relating to protection of natural ecosystems are lacking, and the enforcement of the existing laws and regulations is poor since it is sometimes difficult to identify the corresponding departments responsible for implementation. These factors all compromise the ecological protection and development of Jingdezhen.

2. *Strategies and Programs for Environmental Protection*
   a. *Planning for the rational implementation of environmental protection areas:* When the protection of important natural environments has been ensured, attention should be switched to the protection of areas of less importance to gradually create the rational distribution and expansion of protected sites, covering various habitats and improving the overall landscape of Jingdezhen according to the related principles of landscaping and ecology. It is hoped that the area of protected land can be increased to the levels of 14%, 17%, and 19% for the three milestone years 2008, 2009, and 2010 of the planning period.
   b. *Reduce the interference in environmental protection areas:* The authorities should take advantage of local natural resources, promoting ecotourism but also directing its management to reduce the risk of ecological disturbance and environmental damage to ecological protection areas. Firstly, the scale of tourism should be controlled according to the environment's capacity. Secondly, the infrastructure protecting ecological protection sites should be improved, and greater attention given to prevent damage to core areas by the encroachment of agriculture and forestry.
   c. *Improving the management of ecological protection areas:* The government should adopt successful strategies from other projects with proven success in environmental protection to establish and improve the management, infrastructure, and laws and regulations governing natural protection areas and forest parks. In addition, monitoring systems and information management systems should be created to organize the assessment of resources, the environment, and society in ecological protection areas.

d. *Creating suitable habitats:* To ensure the structural stability of natural ecosystems and the promotion of healthy ecological functions, it is first necessary to guarantee the stability of ecological protection areas and extend the quality and scale of protected areas gradually. It is also very important to plan the spatial structure of the landscape of ecological protection areas according to the principles and methods of landscape ecology. Corridors, plaques, matrix, as well as other landscape elements should be planned rationally so that the information, material, and energy flow between different natural ecological protection areas can circulate freely to create a suitable living habitat for various organisms.

3. *Key Project for Ecological Protection*
Forest Park Creation Project: There are two existing national forest parks and three provincial forest parks, and so, the ecological creation of forest by the combination of point, line, and surface is required to link them. During the period of planning, there are three goals: the creation of an urban forest park in Jingdezhen, the comprehensive green planning of the city including suburban areas, and the reduction of soil erosion culminating in the creation of a forest eco-city that fully utilizes the conducive climatic and terrestrial conditions. It is hoped that the urban forest coverage rate of 65% can be achieved by 2010, thereby significantly improving the overall environment and beginning a gradual process for the virtuous development of agriculture and natural ecosystems.

Green Circle Construction Project: There are three green circles located in Jingdezhen: the Yaoli Kaolinite Forest Ecosystem, the Fuliang Landscape Leisure Ecosystem, and the Leping Cuiping Lake Diverse Ecosystem. These helped Jingdezhen maintain its ecological balance, optimize its environmental quality, broaden its environmental diversity, improve air quality, and provide a variety of other ecosystem services.

Urban Greening Project: The urban greening project will cover scenic areas, forest parks, main streets, river banks, leisure parks, and the central street gardens of Jingdezhen.

14.4.4 ECOLOGICAL RESIDENTIAL ENVIRONMENT AND PLANNING FOR ECOLOGICAL AND CULTURAL DEVELOPMENT

14.4.4.1 Program for Ecological Residential Environment

1. *Habitat Environment Status Evaluation*
   a. *Evaluation of habitat environment status of city center*
      i. *Evaluation index system*
         The requirements of eco-living can be categorized by the following characteristics: convenient living, comfortable living, environmental beauty, and sustainable development. An evaluation index system can be established considering the impact factors that affect them, including municipal infrastructure, public service facilities, living

conditions, environmental quality, urban landscape, urban appearance, protection and utilization of water resources, solid waste disposal, as well as eco-living to evaluate and analyze the habitat status suitability of the Jingdezhen City center areas (Table 14.12).

ii. *Evaluation criteria and results*

A weighted linear method was used to determine the living standard index L that ranges from 0 to 1. An index value equal to 1 indicates the highest living standard while 0 indicates the worst. For ease of understanding, the living standard index was converted into five grades using a nonequidistant method. Arranged from lowest to highest, the different grades were as follows: bad, poor, acceptable, good, and ideal (Table 14.13).

In 2008, the housing suitability index L was 0.825 in Jingdezhen (Table 14.14), which indicated that the housing provision was acceptable although there was still room for improvement.

By separating the housing suitability index into separate criteria, it becomes apparent that sustainable development is the bottleneck to achieving the eco-living ideal in Jingdezhen. The comfortable living level is already ideal and satisfies the target requirement. However, the level of convenient living and environmental beauty are slightly lower, only achieving the grade good and still leaving room for improvement. Overall, it can be concluded that the path to achieving the ideal living standard is lower consumption and lower emission to achieve greater sustainability (Figure 14.15).

iii. *Identifying main issues*

A. *Convenient living:* In 2008, the convenient living level (L) in Jingdezhen was 0.856. (Table 14.15), which indicates a good level of convient living.

By separating the different components, it becomes apparent that the municipal infrastructure of Jingdezhen is excellent, with the road area per capita already exceeded the requirement and that the freshwater quality and gas coverage rate are both ideal. However, the provision of public service facilities is lagging behind and the medical services need to be improved.

B. *Comfortable living:* In 2008, the comfortable living level (L) for Jingdezhen was 1.024. (Table 14.16), which indicates the ideal level of comfortable living.

By separating the different components, it is apparent that the air quality in Jingdezhen is good and exceeds the national eco-city construction requirements by a large margin. The level of water quality, the level of the sound environment, and people's satisfaction with the environment are all very high. The living space per capita is also close to the requirements designated by the "settlements development technical evaluation index system," and in conclusion, the level of comfortable living in urban areas is high and the living standard level is ideal.

TABLE 14.12

Habitat Environment State Evaluation Index System in Jingdezhen

Target Layer	Criterion Level	Feature Layer	Index Layer	Units	Current Figure	Target Figure	Standard Value Source
Housing suitability index	Convenient living (0.25)	Municipal infrastructure (0.6)	Road area per capita (0.3)	m²	10.65	≥10	Developing and technical assessment indicators of urban settlements (trial)
			Bus number per 10,000 people (0.2)	Standard units	10.39	11	Developing and technical assessment indicators of urban settlements (trial)
			Water supply quality (0.3)	%	99.66	100	Construction elements and technical assessment indicators of green and ecological residential areas
			City gas coverage (0.2)	%	88.39	≥92	Developing and technical assessment indicators of urban settlements (trial)
		Public service facilities (0.4)	TV coverage per capita (0.4)	%	97.8	100	—
			Beds in hospital per 1,000 people (0.6)	Bed	2.73	7	Domestic advanced level
	Comfortable living (0.25)	Living conditions (0.5)	Urban housing area per capita (1)	m²	28.96	≥30	Developing and technical assessment indicators of urban settlements (trial)
		Environmental quality (0.5)	Air quality excellence rate (0.3)	%	100	≥77	EPA ecological city construction standards (2008)
			Drinking water quality rate (0.3)	%	100	100	EPA ecological city construction standards (2008)

Category	Subcategory	Indicator	Unit	Basic standards	Achieving functional areas' standards	Source
Environmental beauty (0.25)	Urban landscape (0.5)	Environmental sound quality (0.2)	%			EPA ecological city construction standards (2008)
		Public environment satisfaction rate (0.2)	%	86.89	≥90	EPA ecological city construction standards (2008)
	Urban appearance (0.5)	Area of urban public green per capita (1)	m²	13.23	≥11	EPA ecological city construction standards (2008)
		Mechanical road sweeping operating rate (1)	%	22.7	40	Status value in Beijing
Sustainable development (0.25)	Protection and utilization of water resources (0.4)	Urban sewage treatment rate (0.6)	%	41.48	85	EPA ecological city construction standards (2008)
		Industrial water recycling rate (0.4)	%	59.41	≥80	Developing and technical assessment indicators of urban settlements (trial)
	Solid waste disposal and recycling (0.4)	Domestic garbage treatment rate (0.65)	%	99.3	100	EPA ecological city construction standards (2008)
		Domestic garbage collection rate (0.35)	%	0	≥50	Construction elements and technical guide of green and ecological residential areas
	Ecological residential construction (0.2)	The proportion of green communities (1)	%	13.5	47	Status value in Beijing

TABLE 14.13
Levels and Grades of Living Standard Index

L	0~0.45	0.45~0.65	0.65~0.85	0.85~0.95	>0.95
Level	I	II	III	IV	V
Livable level	Bad	Poor	Acceptable	Good	Ideal

TABLE 14.14
Results of City Living Standard Evaluation

Living Standard	Level		Housing Suitability Index			
			Convenient Living	Comfortable Living	Environmental Beauty	Sustainable Development
General	III	0.825	0.856	1.024	0.885	0.535

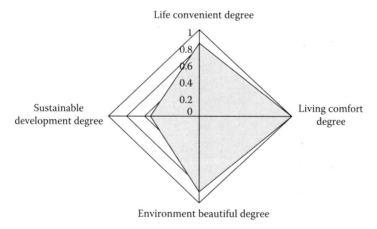

FIGURE 14.15   Living suitability analysis map in Jingdezhen urban area.

TABLE 14.15
Evaluation of Convenient Living Level in Urban Areas of Jingdezhen

Living Standard	Level		Convenient Living	
			Municipal Infrastructure	Public Service Facilities
Good	IV	0.856	1.009	0.625

TABLE 14.16

Evaluation of Comfortable Living Level in Urban Areas of Jingdezhen

Living Standard	Level	Comfortable Living		
			Living Conditions	Environmental Quality
Ideal	V	1.024	0.965	1.083

TABLE 14.17

Evaluation of Environmental Beauty Level in Urban Areas of Jingdezhen

Living Standard	Level	Beautiful Environment		
			Urban Landscape	Urban Appearance
Good	IV	0.885	1.203	0.568

TABLE 14.18

Evaluation of Sustainable Development Level in Urban Areas of Jingdezhen

Living Standard	Level	Sustainable Development			
			Protection and Utilization of Water Resources	Solid Waste Disposal and Recycling	Ecological Residential Construction
Poor	II	0.535	0.590	0.603	0.287

C. *Beautiful environment degree:* In 2008, the environmental beauty level (L) in the urban areas of Jingdezhen was 0.885 (Table 14.17), which indicates a good level for this indicator.

By separating the different components, it can be seen that the area of urban green space per capita already exceeds the requirements of eco-city proposed by Ministry of Environmental Protection of the People's Republic of China (MEP), but that there is still room for improving the urban appearance. More attention should be focused on road cleaning, and the mechanized sweeping operations of the city should be brought into line with other developed cities in China.

D. *Sustainable development:* In 2008, the sustainable development level (L) in urban areas of Jingdezhen was 0.535 (Table 14.18), which indicates a poor condition that requires attention from the relevant authorities.

By considering the individual components, it is apparent that the protection and utilization of water resources, solid waste disposal are not sufficient and barely achieve 50% of the requirements of the ecological habitat resource. The ecological construction of residential areas is also insufficient, and the creation of green communities is still in its infancy achieving only 25% of the required level and lagging behind other developed cities in China.

b. *Rural living environment status assessment*

   i. *Analysis and challenges for comfortable living:* The construction of rural infrastructure should be promoted. The rural road network and water pipelines are insufficient, and the provision of safe drinking water in rural areas is still lacking.

   ii. *Analysis and challenges for convenient living:* The layout of rural residential areas is disorganized and lacks an overall plan. This presents problems for the construction of basic service facilities and also results in the wastage of large areas of land that could be used for farming.

   iii. *Analysis and challenges for environmental beauty:* The sanitation facilities in rural areas lag far behind urban areas. Garbage is not collected on a regular basis and is often discarded in open dumps, which has a negative effect on the rural environment. In addition, the provision of sanitary latrines is still insufficient.

   iv. *Analysis and challenges for sustainable development:* The treatment of environmental pollution in rural areas is insufficient. There are almost no sewage treatment facilities in the rural areas of Jingdezhen, and most of them discharge directly into the environment, although some villages take advantage of the local conditions and use wetland oxidation and oxidation ponds to treat wastewater. The garbage disposal in rural areas is outdated and consists of open dumps and dumping into waterways. The overall utilization rate of livestock manure is low although some is utilized as fertilizer and for the production of anaerobic methane.

2. *Planning Targets and Index System*

   a. *Planning target:* Since the spatial structure of Jingdezhen consists of multicenter urban group, the government should create a residential environmental system with an organized layout, functions appropriate to each group, optimized function, convenience and comfort, as well as environmental beauty. It is also necessary to implement a strategy that includes the joint development of urban and rural areas, improves the residential environment in rural areas significantly, and establishes the sustainable development of residential environmental systems.

   b. *Planning index system* (Table 14.19)

TABLE 14.19
Planning Index System

Target Layer	Criterion Level	Feature Layer	Index Level	Current Figure	Units	2010	2015	2020	Reference Level
Eco-living planning index system	Convenient living	Municipal infrastructure	Road area per capita	10.65	m^2	11	12	13	≥10
			Number of buses per 10,000 people	10.93	Standard units	11	12	13	11
			Water quality rate	99.66	%	99.8	100	100	100
			City gas coverage rate	88.39	%	90	92	94	≥92
		Public service facilities	TV coverage rate	97.8	%	100	100	100	100
			Beds in hospital per 1,000 people	2.73	beds	3.5	5	7	7
	Comfortable living	Living conditions	Urban housing area per capita	28.96	m^2	29	30	32	≥30
		Environmental quality	Sound quality	Basic standards	%	Compliance	Compliance	Compliance	Achieve the function areas' standard
			Public satisfaction rate on the environment	86.89	%	88	90	95	≥90
	Environmental beauty	Urban landscape	Area of urban public green space per capita	13.23	m^2	13.5	14	15	≥11

(Continued)

TABLE 14.19 (*Continued*)
Planning Index System

Target Layer	Criterion Level	Feature Layer	Index Level	Current Figure	Units	2010	2015	2020	Reference Level
		Urban appearance	Mechanical road sweeping operating rate	22.7	%	30	40	60	40
		Rural sanitation	Rural sanitation coverage rate[a]	91	%	92	95	98	≥95
	Sustainable development	Protection and utilization of water resources	Residential water conservation rate	—	%	10	20	30	≥30
			Water recycling facility coverage rate	0	%	10	20	30	≥30
		Solid waste disposal and recycling	Domestic garbage treatment rate	99.3	%	99.4	99.5	99.6	100
			Urban garbage collection rate	0	%	20	30	50	≥50
		Ecological residential construction	Proportion of green communities	13.5	%	20	35	50	47
		Rural energy restructuring	The proportion of clean energy in rural areas	62.8	%	65	70	75	≥50

Note:
[a] Corresponds only to the Fuliang County.

3. *Planning Program*
   a. *Eco-habitat plan for city center*
       i. *Convenient living:* The municipal infrastructure should be improved
           to satisfy the requirements of residents. Specific measures include the
           following. (1) The creation of a green transport system. The authori-
           ties can improve road construction, optimize the public transport sys-
           tem, increase the volume of the public transport services, and increase
           parking lot allocation to a provide convenient travel environment for
           residents. (2) The provision of safe water. The water infrastructure
           should be improved, and the water supply pipeline should be updated.
           The authorities can also optimize water supply facilities and alarm
           systems to protect water safety. (3) The expansion of the gas supply
           pipeline. The gas supply facilities must be improved to meet basic
           demands. In 2008, the urban gas penetration rate was 88.39%, which
           means the comprehensive provision of gas has almost been achieved.
           However, there is still a shortfall to be met and targets of 90%, 92%,
           and 94% coverage have been set for the years 2010, 2015, and 2020,
           respectively.

           It is also necessary to improve public service facilities to enhance
           the quality of life of the city's residence. Specific measures include
           the following. (1) Optimizing the health system. It is necessary to
           improve medical services at the city, county, and community levels
           to ensure that all residents in urban areas have access to a local
           hospital for minor illnesses. The objective is to increase the number
           of beds per 1000 people to 3.5, 5, and 7 by the years 2010, 2015,
           and 2020, respectively. (2) Improving the provision of entertain-
           ment and sports facilities. The authorities should establish a diverse
           range of entertainment and sports facilities at the city, county, and
           community levels and also improve the coverage for radio and tele-
           vision communication. It is hoped that the television coverage will
           be 100% by 2010 and that by 2015 the area of sports facilities per
           capita will be 0.5 m^2 and further increased to 1 m^2 by 2020.
       ii. *Comfortable living:* The government should strictly control the
           strength of buildings and optimize the layout of residential areas.
           The objective is to implement two different categories of layout.
           The first has an optimized layout, is well equipped with municipal
           and public facilities, and set in a good environment. The second
           has a suboptimal layout, lacks a few municipal and public facilities,
           and has an average local environment. These two plans should be
           established according to the system of "eight group functional ori-
           entation" developed by the City Master Plan for Jingdezhen (2008–
           2030). The eight groups functional orientation are as follows: The
           Hongyuan Group, the function of which is orientated to business
           and logistics; Ceramics Science and Technology Park Group, the
           function of which is orientated to industry; The Fuliang Group,
           with a residential and business function; The Chengbei Group, the

function of which is orientated to business, tourism, and entertainment; The High-Tech Zone Group, in which a high-tech industrial park is located; The Xianghu Group, the function of which is orientated to industry, science, education, and housing; The Laocheng Group, the function of which is orientated to historic sites, housing, gold trade, culture, and education; and The Center Group, which is orientated to administration, gold trade, and shopping. One of the main objectives is to gradually increase the area of urban housing per capita with targets of 29 m² by 2010, 30 m² by 2015, and 32 m² by 2020. In addition, noise pollution should be controlled and a quiet environment should be established. Targets have also been set for achieving higher public satisfaction with their environment, which it is hoped will increase to 88% in 2010, 90% by 2015, and 95% by 2020.

   iii. *Environment beauty:* A network of green spaces should be established to ensure that all urban residential areas have access and to increase the overall beauty of the urban landscape. The specific objective is to increase the area of green spaces per capita to 12 m² by 2010, 13 m² by 2015, and 14.5 m² by 2020. Furthermore, vertical construction should be encouraged in residential areas, and the maintenance of green roadside verges should be improved to enhance the appearance of the urban environment as well as other measures to promote the creation of a pleasant urban landscape. The operating rate of mechanical road sweepers needs to be improved. It is hoped that a rate of 30% can be achieved by 2010, more than 40% by 2015, and more than 60% by 2020. The number of garbage collection stations and refuse transfer stations should also be increased and the overall garbage removal system improved. There are currently 18 refuse transfer stations, and the construction of a further 20 is planned. It is hoped that by 2020, the service radius per refuse transfer station will be 500–800 m, the distance between stations being only 0.8–1 km, and each serving a population of less than 20,000 people.

   iv. *Sustainable development degree:* Specific measures include the construction of a water recycling system to reduce the pressure on the aquatic environment, the construction of a waste separation and recycling system to reduce the environmental load of garbage, speeding up the development of green communities, and creating a harmonious living environment.

  b. *Habitat planning program for rural ecological areas*

   i. *Convenient living:* First, an urban and rural road network should be built to facilitate better transit between urban and rural areas. The rural road network must be improved, including the main roads of villages. Secondly, the security and protection of drinking water of rural areas should be established. A safe drinking water project for farmers should be implemented, extending coverage of the water pipe network to all households, building a central water supply for single village or groups of villages to guarantee the provision of a reliable water source. Thirdly, improve the availability of rural communication services. The construction of communication lines and telecommunication stations is required to establish a high-quality urban and rural communication network and the eventual provision of cable services to homes in village locations.

ii. *Comfortable living:* Implementing intensive land use in rural areas. The layouts of some rural settlements in Jingdezhen are disorganized and lack an overall plan. This presents problems for the construction of basic service facilities and also results in the wastage of large areas of land that could be used for farming. It is therefore necessary to protect farmland and optimize land use by the rational planning of small towns and rural settlements and by establishing a unified design and the centralization of buildings. It is also important to construct buildings that are harmonious with the local environment by the appropriate selection of building materials and architectural design.

iii. *Environmental beauty:* The renovation of garbage, sewage, and latrine facilities is required to achieve the orderly disposal of garbage, the safe discharge of sewage, and hygienic latrines. The construction of garbage disposal sites and the implementation of the associated collection and transportation systems for suburban towns and villages should be accelerated. The urban and rural garbage disposal should be integrated following the concept of "collect in the villages, concentrate in the towns, and process in the district (county)." Specific objectives are focused on the rural sanitation coverage rate, which should reach 92% by 2010, 95% by 2015, and 98% by 2020.

iv. *Sustainable development:* The production and consumption ratio of clean energy such as solar, wind, and biomass should be gradually increased. It is hoped that the proportion of clean energy for rural settlements will be 65% by 2010, 70% by 2015, and 75% by 2020. In addition, the construction of biogas digesters in rural areas should be encouraged and the integrated construction of pipelines for urban and rural areas should be accelerated to complete the construction of the county gas pipeline before switching attention to residential areas of towns and the town centers themselves.

14.4.4.2 Planning for Ecological and Cultural Development

1. *Planning Target and Index System*
   a. *Planning target:* The ecological and cultural development of Jingdezhen can be divided into three phases. The first requires the establishment of three concepts: eco-material culture, eco-spiritual

TABLE 14.20

Ecological and Cultural Evaluation Index

Index Name	Units	Standard	Current Figure	2010	2015	2020
Implementation rate of EIA	%	>95	—	95	96	98
Proportion of enterprises providing environment information disclosure	%	>90	82.5	90	92	94
College and university enrollment per 1000 people	Person	—	15.43	20	35	21
Public satisfaction rate on the environment	%	≥85	86.3	88	90	95
Coverage rate of environmental issues in primary and secondary education	%	>95	100	100	100	100

culture, and ecosystem culture. Following this, the concepts of ecological production, ecological consumption habits, ecological ethics, ecological legislation, and public participation can be promoted. Finally, a comprehensive ecological and cultural system can be established by integrating these concepts with the specific ceramic, ecological, and cultural features of Jingdezhen.

b. *Index system:* The milestone targets for each index are presented in Table 14.20.

2. *Current Status and Analysis of Shortfalls*

a. *Compliance status:* According to this index system, the proportion of enterprises providing environmental information disclosure and many of the other indexes do not yet meet the required standards, and there are no data regarding the implementation indicators for the environmental planning assessment. Both these points reflect the current inadequacies regarding the ecological and cultural development of Jingdezhen.

b. *Basis for planning objectives:* The ecological and cultural development of Jingdezhen must be based on the existing situation so that the potential of various advantageous and favorable conditions can be tapped, but also to highlight the weak links that require attention.

   i. Jingdezhen has a long history, over 1700 years, as well as a long tradition of porcelain production. This is a unique historical resource that showcases a great ceramic heritage and historical culture. At the same time, the intangible cultural heritage of Jingdezhen has been combined with modern culture to distil a unique cultural charm.

   ii. There is currently no specific eco-cultural system for the ceramic culture of Jingdezhen, and so more attention should be given to its creation. By improving environmental aspects of the national education, promoting tourist sites that focus on ceramic traditions, and developing eco-agriculture and ecotourism, a comprehensive multifaceted ecological, ceramic, and cultural character can be established for Jingdezhen.

3. *Planning for Ecological and Cultural Development*
   a. *Ecological and cultural development level*
      i. *Eco-spiritual and cultural development:* An eco-spirited and cultural outlook should be developed in Jingdezhen, instilling ecological ethics into the local population.

      Jingdezhen is pursuing a sustainable future and the harmonious combination of social, economic, and environmental development, according to ecological principles. It is hoped that remarkable results can be achieved for not only the urban environment of Jingdezhen itself but also in the outlying towns and villages to establish a shining example of high environmental standards at both the national level and the provincial level and capable of contending for the coveted "National Sanitary City," "China Habitat Environment Award," and "National Green Model City" awards. Jingdezhen has also created "Green Schools" and "Green Community" that raise the awareness of environmental issues in the general public. By investing in these projects, the government of Jingdezhen can strengthen the environmental education of primary and secondary students and rural residents to improve their ecological literacy. This process should be promoted and then the scope of ecological and cultural concepts can be extended.

      By taking advantage of the school, social, and vocational education systems; the participation of the government, private enterprises, and individuals; and by using methods as diverse as classroom education, inspirational experiments, media publicity, demonstration projects, and public participation, the public environmental awareness can be strengthened and the values, morality, and ethics of the population can be refreshed to create a unique, ecological, and modern model.

      ii. *Creation of an ecological and cultural ideology:* The creation of an ecological and cultural ideology for private enterprises and local communities is the first step toward transforming Jingdezhen into an eco-city. Such a mind-set encourages a recycling economy and environmentally friendly production and consumption. A strategy that combines both government and market management should be developed for the protection of the historical and cultural heritage of Jingdezhen. Through publicity and education, a balance can be found between cultural protection and tourism. When considering the ecological landscape of Jingdezhen, attention should be given to the relationship that exists between the ecological and the cultural landscapes (Sun et al. 2009; Zhu 2011; Zhang 2011; Zhan 2012). By protecting both the historical and the cultural heritage of

Jingdezhen and its ecological landscape, a harmony between traditional culture and modern culture can be established, which is reflected by its modern porcelain style.

iii. *Ecosystem and cultural development:* The ecosystem and cultural development of Jingdezhen can be divided into three processes: the establishment of laws and regulations that govern the environment, the establishment of environmental management system, and the development of ecological ethics. In the first case, changes can be made when reviewing and revising existing regulations modifying them to eliminate content that is not orientated toward environmental protection and ecological development of industry. Indeed, regulations should be continuously under review to incorporate an environmental perspective and to formulate corresponding and specific implementations. Secondly, a guide for development projects should be formulated for the main industrial sectors, which encourages the appropriate development of Jingdezhen and limits and prohibits unsuitable projects. This will gradually establish an industrial structure and production methods that are in harmony with the natural environment and improve the management of environmental resources that should be more rigorous to achieve rational use. Finally, the ecological ethics should be promoted, which encourage rational consumption, population control, moderate development, resource protection, and the rejuvenation of the country through science and education.

b. *Key areas of ecological and cultural development*
   i. *Ecological ceramic and cultural systems:* Jingdezhen City is renowned worldwide for its porcelain, and so this historical and cultural heritage should be protected and the local ceramic traditions promoted and conserved for the future. Meanwhile, its ecological culture should be further explored and refined and its tradition of high quality ceramics should be promoted to reflect the deep historical and cultural heritage of Jingdezhen within the context of a modern and advanced life style (Su et al. 2008). On the one hand, full advantage should be taken of traditional ceramic cultural resources, including both physical objects such as artifacts, historical sites, and buildings and immaterial aspects such as the custom and tradition and the expertise of senior ceramics professionals that can be integrated and enhanced by ecological and cultural concepts. On the other hand, key enterprises can be showcased combining elements of production, exposition, exhibition, study, communication, and tourism so that the ecological design concept can be embodied in an ecological and cultural industrial system. Eventually, a unique ecological and cultural system with a comprehensive multifaceted, integrated design, reflecting the many aspects of Jingdezhen, can be created (Figure 14.16).

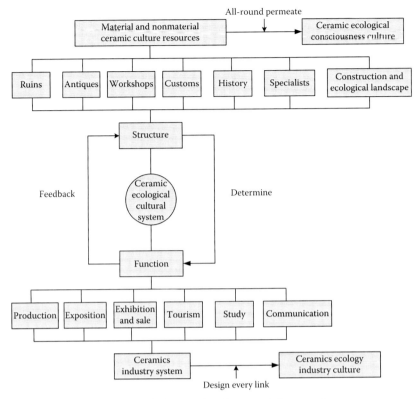

FIGURE 14.16 Ceramic ecological and cultural systems' framework.

ii. *Green industry:* Green industry includes the establishment of ecological agriculture, ecological industry, and eco-industrial parks.

iii. *Green living:* The concept of ecological consumption should be advocated with regard to clothes, food, housing (construction of ecological residential areas), and transport (promotion of public transport).

An ecological consumption protection mechanism should be established, including the establishment of social and material recycling systems, the promotion of material recycling, and the constraint of unhealthy consumption, through both legislation and the implementation of taxes; for example, the creation of a pricing system for tobacco and alcohol that internalizes both environmental and resource costs.

In addition, publicity campaigns and education should be strengthened and the ecological legal system and public participation improved to promote ecological and cultural values.

c. *Measures for ecological and cultural development*
  1. *Improving the environmental awareness of the public:* The cultivation of environmental awareness and ecological ethics in the population is the core of ecological and cultural development and therefore urban development.
  2. *Improving public participation:* The target and implementation of ecological and cultural development must be in accordance with the consent and maximum support and participation of the public. It is therefore necessary to create a public participation system that strengthens the processes that enable people to participate in environmental protection and ecological and cultural development.
  3. *Specific projects:* The planning program should be implemented by the creation of specific projects that provide support for ecological and cultural development.
d. *Results of ecological and cultural development*
  i. *Profound ceramic tradition:* By promoting the material and immaterial aspects of ceramic culture, the appreciation of ceramic traditions can be improved. Combining ceramic ecological awareness with material and nonmaterial ceramic culture, a ceramic ecological and cultural industrial system can be established with a multilink and multiaspect design. Finally, it is necessary to build a unique ceramic ecological and cultural system in Jingdezhen, which showcases the deep cultural connotation of ceramics and the rich heritage, which have given the name Jingdezhen a unique pottery rhyme and porcelain soul.
  ii. *Advancing the green character of Jingdezhen:* "Green culture" is one of the characteristics of today's advanced cultures. The creation of a "green porcelain" concept embodies the innovative spirit of contemporary ideals of progress. Taking a positive attitude toward innovation and adopting the ecological ethics of harmonious development of man and nature can help develop green industries, promote green patterns of consumption, encourage green living, and increase the tendencies of environmental and technological innovation. Green concepts, environmental aesthetics, and ecological ethics and values should be popularized through a series of publicity and educational science activities until they penetrate the various aspects of social life as a reflection of the green porcelain style of Jingdezhen.
  iii. *Unique cultural rhyme:* By protecting the traditional ceramic heritage, the history and culture of the city can be preserved. On the path to promoting the harmonious development of the economy and society, it is necessary to increase the appreciation of the heritage and concepts embodied in the traditional porcelain culture and to integrate this cultural innovation into a modern context that results in a balance between traditional and modern culture

and creates a cultural model with high degree of inclusiveness, innovation, and diversity.

iv. *Harmonious human environment:* A healthy human environment is the heart and soul of an urban civilization and therefore must be in line with the ideas of putting people first and meeting every need of the public. Decisions should be made considering the interest of the majority of people to create a harmonious living environment that will result in a peaceful working and social life as well as social security and stability. The development of such an environmental system requires publicity through government events and an improved public participation system so that a harmonious and peaceful cultural environment can be established in Jingdezhen.

14.4.5 Optimization and Effect Analysis for Eco-City Plan

14.4.5.1 Sensitivity Analysis for Eco-City Plan

Sensitivity analysis can be used to optimize programs to organize their implementation. Evaluating the changes in the health of urban environmental systems in terms of changes in construction investment allows the rational selection and order of priority for project implementation under finite investment and provides a reference for ecological and environmental construction projects (Yang et al. 2002; Su et al. 2008).

The optimization generated by this analysis produces ecological environmental construction investment ranges between $\pm 10\%$, $\pm 20\%$, and $\pm 50\%$, which can help to predict the corresponding changes in urban ecological system health. The specific process is presented in Figures 14.17 through 14.19, where the horizontal axis reflects the changes of investment rate for ecological construction projects (magnitude of change), while the vertical axis indicates the corresponding changes to ecosystem health according to an increase or decrease in investment. The effects of these changes can be subdivided into their constituent components such as ecological and economic investment, eco-habitat construction projects, major solid waste investments, major water environmental projects, major ecological spatial planning projects, and atmospheric environment management investment (Liu et al. 2012). Considering these data, the following conclusions can be made.

From 2008 to 2010, the urban ecological system health of Jingdezhen is most sensitive to changes in ecological and economic investment and eco-habitat construction projects investment, followed by major solid waste investment, major water environmental projects, and major ecological spatial planning construction projects. These corresponding projects and measures should be considered the highest priority.

From 2011 to 2015, the urban ecological system health in Jingdezhen is again most sensitive to ecological and economic investment changes and eco-habitat construction projects investment changes, and because the circumstances effected by major ecological spatial planning construction projects have been improved since

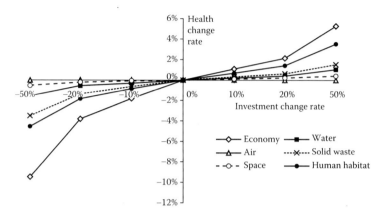

FIGURE 14.17   Investment-effect sensitivity analysis of ecological construction from 2008 to 2010.

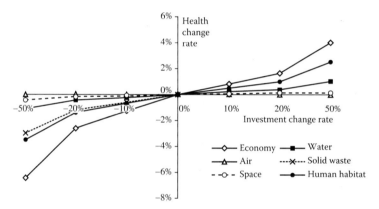

FIGURE 14.18   Investment-effect sensitivity analysis of ecological construction from 2011 to 2015.

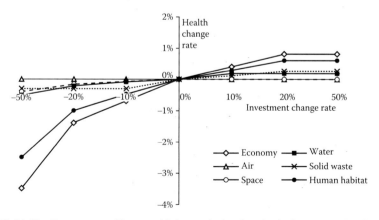

FIGURE 14.19   Investment-effect sensitivity analysis of ecological construction from 2016 to 2020.

preconstruction and management, the sensitivity of the ecological system health status according to investment changes would be relatively low.

From 2016 to 2020, the sensitivity of major solid waste investment, major ecological spatial planning construction projects, and atmospheric environment management investment are not apparent. As the decrease in investment level will reduce the health degree, it is necessary to keep investing and to maintain the health of the city.

According to the sensitivity predicted, the engineering optimization can be prioritized, as illustrated in Table 14.21.

1. *Cost Estimate of Planned Programs:* In accordance with five important areas and two support systems, the key construction projects for the eco-city transformation of Jingdezhen include key ecological and economic construction projects, ecological spatial planning and construction projects, ecological and environmentally important development projects, eco-habitat development projects, and ecological and cultural projects, combined with industrial upgrading, scientific and technological innovation, urban-rural integration, and openness to the outside world. This comprises a total of 31 engineering projects, with a total investment is 14.29 billion yuan, and the related projects involve the cooperation of several government departments including the Jingdezhen Municipal Environmental Protection Bureau, the Municipal Government, and Forestry Department as well as local private enterprises and other organizations.

    The investment scale of the various stages of eco-city development can therefore be estimated according to the type of development projects implemented. For the planning year 2008–2010, the ecological construction investment will be 2.38 billion yuan; for 2011–2015, the value would be higher, 8.07 billion yuan; and for 2016–2020, the value would be less, only 3.84 billion yuan. This accounts for 1.36%, 2.93%, and 1.02%, respectively, of the city's GDP during each phase of the planning period, which are all within the city's expenditure capacity.

2. *Planning Effect Analysis:* After implementing the planning programs, the trend for the healthy development of urban ecosystem in Jingdezhen can be estimated using the ecosystem healthy assessment model to provide a forecast for the short-term (2010), mid-term (2015), and long-term (2020), as presented in Table 14.22.

    According to the results of the program of engineering investment and the ecosystem health assessment model, the short-term (2010), mid-term (2015), and long-term (2020) trends for healthy development of urban ecosystem in Jingdezhen can be estimated as follows (Li et al. 2005; Wang et al. 2007; Liu et al. 2009; Su et al. 2009a,b; Zhang et al. 2009; Zhang and Yang 2009; Su et al. 2010).

3. *Urban Ecosystem Healthy Assessment in Jingdezhen:* According to the results of the urban ecosystem healthy assessment for 2008, Jingdezhen was classified as being in a sub-healthy state. However, from 2010, it is predicted that the situation will gradually improve, becoming healthier and

TABLE 14.21
Budget Planning and Engineering Optimization

Construction Types (Respective System)	Engineering or Compensation/Incentives (Number of Projects)	Total Cost (100,000,000 yuan)	Budget Planning (100,000,000) and Engineering Optimization					
			2008~2010		2011~2015		2016~2020	
			Investment	Preferred degree	Investment	Preferred degree	Investment	Preferred degree
Ecological spatial planning major construction projects	Population source construction and protection projects	3.1	1.3	4	1	4	0.8	4
	Landscape construction shaft ecological corridor projects	1.15	0.65		0.5			
	Landscape cultural construction projects	1.5	0.8		0.7			
	Drinking water protection construction projects	1	0.6		0.4			
construction projects	Soil and water conservation projects	3.9	1.6		1.3		1	
	Forest park construction projects	1.6	1		0.6			
	Landscape restoration projects	2	1		0.7		0.3	
Important water environmental planning projects	Wuxikou Hydro Project	23.3		3	23.3	2		2
	Yutian Reservoir Expansion Project	2.98			2.98			
	10,000–50,000 acres water-efficient irrigation facilities renovation projects in Gongku, Xingfu, Yutian, and other four units	1.2					1.2	
	Water-saving technological transformation projects of industrial enterprises	8					8	
	Life-saving appliances installation projects	0.4					0.4	
	Urban water supply network transformation projects	0.8	0.8					

		Total	Near-term	Mid-term	Long-term
	Industrial wastewater treatment projects	1.2	1.2		
	Sewage treatment plant	4.41	4.41		
	River rehabilitation projects	0.6			0.6
	Automatically monitoring system of surface water and drinking water and online monitoring system of key pollution sources	0.25			0.25
	Drinking water safety in rural areas	0.54	0.54		
Solid waste key planning projects	Urban living garbage collection projects	0.2	0.2		
	Sludge utilization projects	0.1		0.1	
	Legang waste disposal site in Leping	0.3		0.3	
	Sanlong waste disposal site in Fuliang	0.14		0.14	
	Waste materials recycling and trading system	0.25		0.25	
Eco-habitat construction planning projects	Green space construction projects	3	2	1	
	Residential areas green projects	0.9	0.4	0.3	0.2
	Road network construction projects	60		40	20
	Sanitation facilities construction projects	10	5	3	2
	Gas facilities construction projects	1.5	1	0.5	
	Medical facilities construction projects	7	1	3	3
	Ecological residential areas construction projects	0.008	0.002	0.003	0.003
	Rural renewable energy comprehensive development and conservation construction projects	1.5	0.3	0.6	0.6
	Estimating total investment (100 million yuan)	142.828	23.802	80.673	38.353
	Financial investment accounts for the target year of GDP (%)		1.36%	2.93%	1.02%

TABLE 14.22

Health Development Trend for the Urban Ecosystem of Jingdezhen after Program Implementation

Index	Eco-Economic System	Eco-Space System	Eco-Habitat System	Eco-Environment System	Ecological and Cultural System	Integrated Health Index
2008	Unhealthy	Healthier	Unhealthy	Healthier	Sub-healthy	Sub-healthy
2010	Subhealthy	Healthy	Subhealthy	Healthy	Healthier	Sub-healthy
2015	Healthier	Healthy	Healthier	Healthy	Healthy	Healthier
2020	Healthy	Healthy	Healthy	Healthy	Healthy	Healthy

healthier. By 2015, Jingdezhen will reach the healthier level, and by 2020, the pathological and unhealthy trend will have been completely eradicated with the healthy level having been achieved.

In conclusion, the implementation of the eco-city planning program will lift Jingdezhen out of the sub-healthy level by 2015 and into the healthy level by 2020.

REFERENCES

Cook E. A. Landscape Structure Indices for Assessing Urban Ecological Networks. *Landscape and Urban Planning* 2002, 58(2–4), 269–280.

Gao J. C. Some Tentative Ideas to Developing Ecological Tourism of Jingdezhen. *Journal of Jingdezhen Comprehensive College* 2006, 21(3), 106–108 (in Chinese).

He J.; Bao C. K.; Shu T. F.; Yun X. X.; Jiang D. H.; Brwon L. Framework for Integration of Urban Planning, Strategic Environmental Assessment and Ecological Planning for Urban Sustainability within the Context of China. *Environmental Impact Assessment Review* 2011, 31(6), 549–560.

Hong B.; Liu S.; Li S. H. Ecological Landscape Planning and Design of an Urban Landscape Fringe Area: A Case Study of Yang'an District of Jiande City. *Procedia Engineering* 2011, 21, 414–420.

Li F.; Wang R. S.; Paulussen J.; Liu X. S. Comprehensive Concept Planning of Urban Greening Based on Ecological Principles: A Case Study in Beijing, China. *Landscape and Urban Planning* 2005, 72(4), 325–336.

Liu G. Y.; Yang Z. F.; Chen B. Emergy-Based Urban Dynamic Modeling of Long-Run Resource Consumption, Economic Growth and Environmental Impact: Conceptual Considerations and Calibration. *Procedia Environmental Sciences* 2012, 13, 1179–1188.

Liu G. Y.; Yang Z. F.; Chen B.; Ulgiati S. Emergy-Based Urban Health Evaluation and Development Pattern Analysis. *Ecological Modelling* 2009, 220(18), 2291–2301.

Liu G. Y.; Yang Z. F.; Su M. R.; Chen B. The Structure, Evaluation and Sustainability of Urban Socio-economic System. *Ecological Informatics* 2012, 10, 2–9.

Su M. R.; Chen B.; Chen C.; Yang Z. F.; Liang C.; Wang J. Reflection on Upsurge of Low-carbon Cities in China: Status Quo, Problems and Trends. *China Population, Resources and Environment* 2012, 22(3), 48–55 (in Chinese).

Su M. R.; Yang Z. F.; Chen B. Set Pair Analysis for Urban Ecosystem Health Assessment. *Communications In Nonlinear Science and Numerical Simulation* 2009, 14(4), 1773–1780.

Su M. R.; Yang Z. F.; Chen B. Urban Ecosystem Health Assessment Based on Vitality Index and Set Pair Analysis. *China Population, Resources and Environment* 2010, 2, 122–128 (in Chinese).

Su M. R.; Yang Z. F.; Chen B.; Ulgiati S. Urban Ecosystem Health Assessment Based On Emergy And Set Pair Analysis—A Comparative Study Of Typical Chinese Cities. *Ecological Modelling* 2009, 220(18), 2341–2348.

Su M. R.; Yang Z. F.; Chen B.; Zhao Y. W.; Xu L. Y. The Vitality Index Method for Urban Ecosystem Assessment. *Acta Ecologica Sinica* 2008, 28(10), 5141–5148 (in Chinese).

Su M. R.; Yang Z. F.; Xu L. Y. Building up Urban Ecological Culture Based on Landscape Ecology. *Environmental Science & Technology* 2008, 31(4), 123–126 (in Chinese).

Sun X. X.; Hu X. Q.; Feng X. H.; Liu H. W. Transformations in Jingdezhen Ceramic Industrial Kilns and Furnaces and Their Effects on the Ecological Environment. *China Ceramic Industry* 2009, 16(2), 44–46 (in Chinese).

Wang X. B.; Jiang P. H.; Ni Y. X.; Dong R. B. Study on Ecological Compensation Mechanism of Jingdezhen Ceramic Industry. *China Ceramics* 2007, 43(11), 13–16, 19 (in Chinese).

Wang Z. Q.; Long Y. Y.; Feng X. M.; Ding S. T. Economic Evaluation on the Benefit of Ecological Public Welfare Forest in Jingdezhen City. *Jiangxi Forestry Science and Technology* 2007, B10, 28–30 (in Chinese).

Xie Z. M. Thinking of Ecological Planning of Jingdezhen Ceramic Cultural Landscape. *Journal of Jingdezhen College* 2010, 25(1), 116–118 (in Chinese).

Yang Z. F.; Cheng H. G. Models in Simulation System of Urban Industrial Water Pollution Control. *Acta Scientiae Circumstantiae* 2002, 22(2), 213–218 (in Chinese).

Yang Z. F.; Hu T. L.; Su M. L. Urban Ecological Regulation Based on Ecological Carrying Capacity. *Acta Ecologica Sinica* 2007, 27(8), 3224–3231 (in Chinese).

Yang Z. F.; Shen Z. Y.; Xia X. H.; Zeng W. H.; Cui B. S.; Hao F. H. Water Resources Renewable Ability Theory and Its Appliance in Yellow River Basin. *China Basic Science* 2002, 5, 4–7.

Yang Z. F.; Su M. R.; Zhang B.; Zhang Y.; Hu T. L. Limiting Factor Analysis and Regulation for Urban Ecosystems—A Case Study of Ningbo, China. *Communications in Nonlinear Science and Numerical Simulation* 2010, 15(9), 2701–2709.

Yang Z. F.; Sui X. Assessment of the Ecological Carrying Capacity Based on the Ecosystem Health. *Acta Scientiae Circumstantiae* 2005, 25(5), 586–594 (in Chinese).

Yang Z. F.; Xu Q.; He M. C.; Mao X. Q.; Yu J. S. Analysis of City Ecosensitivity. *China Environmental Science* 2002, 22(4), 360–364 (in Chinese).

Yang Z. F.; Yin M.; Cui B. S. Study on Urban Eco-Environmental Water Requirements: Theory and Method. *Acta Ecologica Sinica* 2005, 25(3), 389–396 (in Chinese).

Yang Z. F.; Yu S. W.; Chen H.; She D. X. Model for Defining Environmental Flow Thresholds of Spring Flood Period Using Abrupt Habitat Change Analysis. *Advances in Water Science* 2010, 21(4), 567–573 (in Chinese).

Yang Z. F.; Zhao Y. W.; Cui B. S.; Hu T. L. Ecocity-Oriented Water Resources Supply-Demand Balance Analysis. *China Environmental Science* 2004, 24(5), 636–640 (in Chinese).

Yu B.; Xiao X. The Ecological Protection Research Based on the Cultural Landscape Space Construction in Jingdezhen. *Procedia Environmental Sciences* 2011, 10, 1829–1834.

Zhan J. The Cultural Landscape of Ceramic Workshops in Jingdezhen. *Scientia Geographica Sinica* 2012, 32(1), 55–59 (in Chinese).

Zhan J.; He B. Q.; Cui P.; Jiang L. Jingdezhen Ceramic Firing and Ecological Landscape Change. *Journal of Ceramics* 2010, 31(4), 637–644 (in Chinese).

Zhang L. X.; Yang Z. F.; Chen B.; Liu G. Y.; Liang J. An Analysis on Urban Ecological Competition Capability with Biophysical Accounting Method. *Acta Ecologica Sinica* 2008, 9, 4344–4351 (in Chinese).

Zhang M. Barrier Analysis and Countermeasures for the Ecological Transformation of Jingdezhen Ceramic Industrial Park. *Jiangsu Ceramics* 2011, 5, 5–7 (in Chinese).

Zhang Y.; Yang Z. F. A Method of Analyzing the Interactions in an Urban Metabolism System. *Acta Scientiae Circumstantiae* 2009, 1, 217–224 (in Chinese).

Zhang Y.; Zhao Y. W.; Yang Z. F.; Chen B.; Chen G. Q. Measurement and Evaluation of the Metabolic Capacity of an Urban Ecosystem. *Communications in Nonlinear Science and Numerical Simulation* 2009, 14(4), 1758–1765.

Zheng S. H.; Guo L. The Problem and Strategy about Ecology Development in Jingdezhen Ceramic Industry. *Shandong Ceramics* 2011, 34(5), 33–36 (in Chinese).

Zhou X. C.; Xu L. Y.; Yang Z. F. Municipal Solid Waste Life-cycle Analysis and Its Management Planning Research. *Chinese Journal of Environmental Management* 2011, 2, 33–37 (in Chinese).

Zhou X. C.; Xu L. Y.; Yang Z. F. Optimization of Low-Carbon Municipal Solid Waste Processing Model. *Acta Scientiae Circumstantiae* 2012, 32(13), 498–505 (in Chinese).

Zhu Y. L. The Study on Ecological and Economic City Construction of Jingdezhen. *Territory & Resources Study* 2011, 4, 51–53 (in Chinese).

15 Assessment of Sustainability for a City by Application of a Work Energy Balance and a Carbon Cycling Model

Sven Erik Jørgensen and Michela Marchi

CONTENTS

15.1 INTRODUCTION

This chapter presents how to apply work energy capacity, also denoted exergy or eco-exergy, and a carbon cycling model for the assessment of the sustainability for a city. The theoretical background knowledge and the definitions, which are necessary to understand the analysis and how the analytical results can be interpreted, are given in Section 2.5. A work energy balance for the assessment of sustainability has been applied on the Danish island Samsø, situated in Kattegat between Zealand and Jutland. The area of the island is 114 km², and it is 26 km long and 7 km wide at the maximum width. The island has about 4100 inhabitants with about 7% working in the agriculture sector. The second most important sector is tourism. The experience gained here will be transferred to the application on the sustainability of a city. It cannot be applied directly but indirectly, as there is of course a difference between a city and an island. A carbon model has been applied on Siena Province, Tuscany, central Italy (Marchi et al. 2012). Siena Province has an area of 3821 km² and is predominantly hilly (hills 93%, mountains 7%). Agriculture is an important activity in this regional system, although the significant increase in tertiary sector in the last 50 years has been due to the development of trade, public administration, and tourism. Siena, the core of Siena Province, is a famous touristic medieval town in Italy with about 90,000 inhabitants. It means that the experience gained by the application of a carbon cycling model can be transferred indirectly to towns (cities), which want to use this tool for the assessment of the sustainability. In fact, the carbon cycle model of Siena Province (a fairly large district) can be adapted to smaller systems, such as cities. Guidelines on how to apply the experiences gained by these two projects on a general assessment of sustainability of a city are presented in this chapter.

For the past 15 years, Samsø has introduced the application of alternative energy and at the same time made a significant effort to reduce the energy (or rather exergy) consumption. The energy consumption on Samsø is about 130 MJ/capita and about 50 MJ/ha. The average energy consumption of Denmark is 160 MJ/capita and 200 MJ/ha; Denmark has a low energy consumption compared with the gross national product per capita due to an early and relatively massive introduction of green taxes on energy. The alternative energy sources applied on Samsø today are mainly wind energy for the production of electricity, straw and solar panel for district heating. Figures 15.1 through 15.4 illustrate these three applications of alternative energy. Meanwhile, Samsø Energy Academy has been erected with the objectives to arrange exhibitions, workshops, and courses about renewable energy and publish as widely as possible the gained knowledge about the use of alternative energy and energy saving.

Energy Academy has the address Strandengen, Ballen, Samsø (see Figure 15.4), where an administration building has been erected following ecological principles. Rainwater is used for flushing toilets, and the building has a low energy consumption rate as it is very well insulated. A small panel of solar cells demonstrates the

FIGURE 15.1 **(See color insert.)** Solar panel used for heating. The district heating station is shown in the background.

FIGURE 15.2 **(See color insert.)** Wind mills off-shore, Samsø. Wind mills on land have also been erected on Samsø.

FIGURE 15.3 **(See color insert.)** Storage of straw applied for district heating.

usefulness of this alternative energy technology, partially covering the demand for electricity of the building.

The heating is supplied by the district heating station that is using straw (see Figure 15.3). The electricity consumption is low due to the selection of low energy consuming electrical devices and fittings. The electricity produced by the solar cells is in support to that by wind mills.

Similar initiatives have not been made in Siena due to strong historical and landscape constrains, although there are endeavors to make Siena an eco-city. Siena Province, on the other hand, is focused on the concept of "sustainable energy" through saving, energy efficiency and renewable energy use, and reducing

FIGURE 15.4 **(See color insert.)** The Samsø Energy Academy building, Strandengen 1, Ballen, Samsø.

long-term negative impacts from continuing power generation by the combustion of nonrenewable fossil resources. The following objectives have been set:

- Upgrading and improving the efficiency of geothermal plants
- Exploitation of solar energy (thermal and photovoltaic) and a local production and use of hydrogen from renewable resources to replace traditional fossil fuels for transportation
- Exploitation of wind energy
- Use of biomass (wood, herbaceous and wood crops, manure, and agro-industrial residues) to enable local production of biofuels for the transport sector
- Improving efficiency, saving energy, and using renewable resources for new building and renovations of all types, especially public
- Allocating the landfill a residual role, giving priority to the maximization of recycling and recovery of materials, incinerating the previously selected fuel fractions and generating energy

A carbon model based on a good database, as mentioned previously, has been developed, and it is able to illustrate clearly the advantages that a carbon model offers in our endeavor of transforming a city into an eco-city: it is easy to get a quantitative answer to such question as: How much can the overall carbon emission be reduced by introduction of these, and these specified changes?

Section 15.2 will present the energy and exergy balance for Samsø briefly and give some guidelines on how to apply an eco-exergy, exergy, or work energy balance to assess the sustainability of a city.

Section 15.3 will briefly present the carbon cycling model developed for the Siena Province, while Section 15.4 will discuss how a carbon cycling model can be applied to generally ensuring a more sustainable development of a city. Section 15.5 will make a conclusion on how a work energy balance and a carbon cycling model can be applied to any city to promote a sustainable development. This section will also present a list of recommendations in the form of questions that should be answered and followed to ensure a more sustainable development of a city.

15.2 AN ENERGY BALANCE FOR SAMSØ AND THE PRINCIPLES FOR A GENERAL EXERGY BALANCE

An overview of the use and production of energy on Samsø today can be presented, as shown in Table 15.1. The numbers are from 2005 and have only been slightly improved since.

Table 15.1 shows that almost 90% of the energy consumption is covered by alternative energy. A few tractors are using sunflower oil, but it is a general problem to replace the fossil fuel used by the vehicles.

The exergy (work capacity) balance should include the agricultural production. It is based on solar energy that is considered free. Eco-exergy is, however, used for landfills used for garbage and for the consumer durable goods. Samsø gets, however, due to the agricultural production and the extensive use of alternative energy, a positive exergy balance and it can be concluded that Samsø has a sustainable development. For more details, see Stremke and van den Dobbelsteen (2012).

The principles applied to set up the exergy balance are shown in Figure 15.5. Based on these principles (for further explanation, see Jørgensen 2006), the following quantifications can be developed:

1. *The loss of eco-exergy or exergy (work capacity) due to the consumption of fossil fuel* is calculated by the addition of the chemical free energy (the work capacity) of the fossil fuel and the loss of eco-exergy due to the dispersion of the gases resulting from the chemical processes.

   The exergy loss due to the dispersion of the components of fossil fuel is found by the following calculations: If we consider 1 g of coal that contains 1% of sulfur and 99% of carbon (coal contains also ash, but let us not consider it in our calculations), the exergy loss due to the dispersion can be determined by the following calculations (Equation 15.1):

$$0.01 \times (8.314 \times 300/32) \ln (0.01/50 \times 10^{-9})$$
$$+ 0.99 \times (8.314 \times 300/12) \times \ln (0.99/4 \times 10^{-4}) \qquad (15.1)$$
$$= 1617 \, J \approx 1.6 \, kJ$$

TABLE 15.1
Samsø: Local Energy Use and Production

Type of Energy	Consumption (TJ/year)	Production (TJ/year)
Electricity	286	386
Heating	140	66
Gasoline, including ferries	86	0
Total	512	452

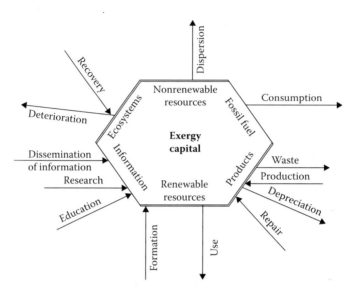

FIGURE 15.5 The principles applied for the development of an eco-exergy (exergy or work capacity) balance.

where 50×10^{-9} and 4×10^{-4} represent concentrations (expressed as ratios, i.e., no units) of sulfur dioxide-S and carbon dioxide-C in a typical town atmosphere. The value for gas constant R is 8.314 J/K, and 300 K is the absolute temperature. The chemical exergy content of 1 g coal is about 32 kJ. The loss of exergy by dispersion is, therefore, only 5% of the direct loss of chemical exergy by burning coal. As all the calculations will have a higher uncertainty than 5% and the quality of coal may vary more than 5%, it seems acceptable not to include the dispersion exergy loss by the use of fossil fuel or as alternative to multiply all exergy losses due to the consumption of fossil fuel by a factor of 1.05 to compensate approximately for the exergy loss due to the dispersion of the formed gases in the atmosphere. It means that the eco-exergy lost by the use of fossil fuel becomes (Equation 15.2):

$$\text{Loss of eco-exergy by fossil fuel} = \text{consumption in kg} \times F \times 1.05 \text{ MJ} \quad (15.2)$$

where F (chemical exergy content) is 32 for coal and 42 for oil or natural gas.

2. *The deterioration of ecosystems:* Eco-exergy can be used as ecosystem health indicators and can be used to express ecosystem services (see Jørgensen 2010). Table 15.2 illustrates this approach: the calculations of ecosystem services based on eco-exergy for various ecosystems. The annual production contributes positively to the eco-exergy balance, while the deterioration of ecosystems contributes negatively to the eco-exergy balance and can be calculated from the following Equation 15.3:

TABLE 15.2
Work Capacity Used to Express the Ecosystem Services for Various Types of Ecosystems

Ecosystems	Biomass (MJ/m²·year)	Information Factor (β-value)	Work Capacity (GJ/ha·year)
Desert	0.9	230	2,070
Open sea	3.5	68	2,380
Coastal zone	7.0	69	4,830
Coral reefs, estuaries	80.0	120	96,000
Lakes, rivers	11.0	85	9,350
Coniferous forests	15.4	350	53,900
Deciduous forests	26.4	380	100,320
Temperate rainforests	39.6	380	150,480
Tropical rainforests	80.0	370	296,000
Tundra	2.6	280	7,280
Cropland	20.0	210	42,000
Grassland	7.2	250	18,000
Wetland	18.0	250	45,000

Note: Work capacity is calculated as biomass (MJ/m²·year) × the information factor (β-value).

Eco-exergy lost by deterioration of ecosystems

$$= \text{ha lost} \times \text{biomass/ha} \times \beta \times 1000 \text{ MJ} \tag{15.3}$$

where β is the value related to the information factor (number of non-nonsense coding genes) stored in the biomass of every taxonomic group (Jørgensen et al. 2005).

Ecosystems give a major increase of eco-exergy, and therefore it is very significant to have green areas in cities and towns. For this reason, we must protect and maintain these green areas in urban systems.

3. *The use of renewable resources:* the formation of renewable resources are calculated separately by multiplication of the annual consumption of the various resources with the exergy content of each renewable resource (Table 15.2).

4. *Loss of exergy due to depreciation of the products* such as cars, houses, refrigerators, TVs, computers, and so on. The exergy content of these products is approximately declining 10%/year, which implies that the exergy content times 0.1 is lost per year. The number of these products per capita is known for most industrialized countries. The exergy requirement for these products is very high due to the structural complexity of the materials used for their production. Table 15.3 gives the exergy requirement according to Hayes (1997) for producing various materials from ore or scrap. Usually, it is easy to get information about the average purchase in a city of consumer durable goods. In Denmark, consumer durable goods are mostly made by steel, which

TABLE 15.3

Exergy Requirements in Producing Various Materials from Ore and Raw Source Materials

Product	From Ore (MJ/kg-product)	From Scrap (MJ/kg-product)
Glass	25	25
Steel	50	26
Plastics	162	112
Aluminum	250	8
Titanium	400	Not yet applied
Copper	60	7
Paper	24	15

Source: Hayes, E.T., EPA, Athens, GA, 1997.

is about 1500–2000 kg per family of four persons. It is, therefore, relatively easy to calculate the depreciation (the previously mentioned 10%) per year.

5. *Loss of exergy due to waste* is highly dependent on the waste handling method. If landfills are used, the loss of eco-exergy corresponds to the loss of agricultural land and it can be calculated by the use of Equation (15.3). If incineration is applied, the corresponding saving of heating can be added as a positive contribution to the eco-exergy balance. The loss of exergy by the use of paper must be considered (high amounts), as well as the low exergy value gained by incineration of paper. It would correspond to about 20 MJ/kg paper (it is approximately the average of the exergy requirements for paper listed in Table 15.3). The loss of exergy due to the consumption of paper is obtained by Equation 15.4:

$$\text{Eco-exergy lost by our use of paper} = \text{kg paper/}$$
$$\text{capita} \times \text{the number of inhabitants} \times 20 \text{ MJ} \qquad (15.4)$$

The loss of exergy by the use of plastic and glass can be calculated similarly, as the capita use times the number of inhabitants times the exergy requirements in Table 15.3.

A comparison of the exergy requirements by the use of ore and scrap clearly shows that reuse and recycling give double benefits: the use of limited resources is reduced and exergy is saved. For aluminum, in particular, the energy requirement by the use of scrap instead of ore is considerable.

The gain of exergy can be calculated by adding the following contributions:

a. *Exergy of products*: When the products are used, they are depreciated. When the products are produced, eco-exergy is gained as production in kilograms multiplied by the exergy requirement; see Table 15.3. It is of course presumed that the various energy forms are covered by point 1 shown previously.

b. *The formation of renewable resources*: It is calculated in the same way as the losses; that is, the annual growth is multiplied by the eco-exergy of the considered renewable resource.

c. *Repair of products*: Which gives the product a higher exergy values, is decreasing in most societies due to the high costs of labor.
d. *Increase of knowledge*: Our information capital.
e. *Recovery of ecosystems.*

A steady annual net loss of exergy, eco-exergy, or work capacity is of course from a long-term view point completely unacceptable because it would mean that we are reducing, year after year, the sum of human and natural capital, which will imply that the life conditions for mankind are steadily deteriorated. If we use eco-exergy/ loss of eco-exergy as sustainability index, it is of course unacceptable that the index is considerably less than 1.00 for a longer period of time.

Our perception of the evolution is that it has increased the order and exergy on earth, which also can be demonstrated quantitatively. The core question is whether mankind by its massive change of the environment, including the change of nature to man-made systems (towns, factories, roads, etc.), has started a decline of order and exergy, which before or after will result in a major crisis for the life on earth. The content of exergy in a natural system is usually many times the content of a technological system (see Table 15.2). It implies that what is named progress for mankind may be a step backward for the exergy content of the earth. The decline of eco-exergy can continue some decades or even some centuries maybe, but it cannot continue on a long-term basis because the basic for human activities will, thereby, be significantly reduced.

The result of a first approximate exergy balance for the Danish island Samsø is given in Table 15.4. The details of the calculations are presented in Stremke and van den Dobbelsteen (2012).

TABLE 15.4

Exergy (Work Capacity) Balance for Samsø

Item	Consumption (TJ)	Production (TJ)
Electricity	286	386
Heating	145	66
Benzin + diesel	91	
Food	27	
Milk		2
Potatoes		150
Vegetables		112
Grain		271
Pigs		106
Pesticides	1	
Waste	84	
Consumer durable goods	24	
Plastic and paper	33	
Glass	5	
Total	696	1093

15.3 CARBON CYCLING MODEL DEVELOPED FOR THE SIENA DISTRICT

The model (see Marchi et al. 2012) has been developed as a management tool to determine how to understand and regulate the overall carbon budget of Siena Province, obtaining quantitative answers to the following questions:

1. What is the current net emission of carbon dioxide? How will this net emission develop in the next 10 years on the basis of current knowledge of expected changes?
2. How much methane is released into the atmosphere by human activities? How does the contribution of methane influence the carbon footprint of the system?
3. What possibilities do we have to reduce emissions? The model is used to examine the possibilities of
   a. Reducing the number of private cars and improving public transport.
   b. Promoting reforestation; for eco-cities, it could be erection of green areas.
   c. Increasing the efficiency of energy use (electricity and heating), as well as the proportion of alternative resources for the production of electricity and heat.
   d. Increasing the productivity of agriculture and grasslands/pastures; for eco-cities, it could be better management of public green areas and private gardens.
   e. Improving the characteristics of landfills for waste management and improving wastewater treatment.
   f. Changing waste disposal to waste incineration over a period of 10 years.
4. Could emissions increase in the next 10 years due to lack of management?
5. How could this be avoided?

The developed model has five submodels:

a. *Forest*
b. *Cropland*, divided into orchards, vineyards, and other crops
c. *Grassland*
d. *Humans*, considering emissions due to combustion of fossil fuels, industrial activities, imported electricity, local electricity production, and waste management (emission of both carbon dioxide, CO_2-C, and methane, CH_4-C)
e. *Livestock*, to calculate the amount of CH_4-C released to the atmosphere from enteric fermentation and manure management

The selection of submodels reflects that the study encompasses not only Siena (City) but also the whole Siena District. Figure 15.6 shows the general conceptual diagram of the carbon cycle model of Siena Province, indicating a summary of

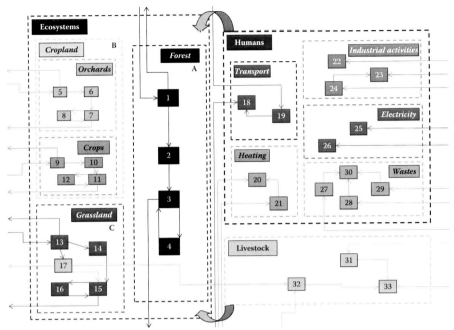

FIGURE 15.6 (**See color insert.**) General conceptual diagram of the carbon cycle model of Siena Province: boxes denote a summary of the state variables (indicated by different numbers), arrows between boxes denote the processes, and arrows entering or leaving the system denote the forcing functions.

the state variables (with a brief description), the forcing functions, and processes involved in the system (Marchi et al. 2012).

It is clear from the conceptual diagram which processes the total model includes, but for those interested in the model details can be referred to Marchi et al. (2012).

Summary of the state variables: (1) carbon in trees; (2) carbon in dead wood; (3) carbon readily released in forest soil; (4) carbon in soil microbial biomass; (5) carbon in fruit trees; (6) carbon in crop residues; (7) carbon readily released in soil; (8) carbon in crops; (9) carbon stocked in grass; (10) carbon in dead plants; (11) carbon readily released in grassland soil; (12) carbon in grazing animals; (13) CO_2-C released to the atmosphere due to combustion of fossil fuels for transport; (14) CH_4-C released to the atmosphere due to combustion of fossil fuels for transport; (15) CO_2-C released to the atmosphere due to combustion of fossil fuels for heating; (16) CH_4-C released to the atmosphere due to combustion of fossil fuels for heating; (17) CO_2-C released to the atmosphere due to combustion of carbonates for the production of ceramics/bricks and glass; (18) CO_2-C released to the atmosphere due to the combustion of fossil fuels for energy production in industry; (19) CH_4-C released to the atmosphere due to combustion of fossil fuels for energy production in industry; (20) CO_2-C released to the atmosphere from electricity produced by burning wastes; (21) CO_2-C released to the atmosphere due to electricity consumption from national grid; (22) CH_4-C released to the atmosphere from decomposition of wastes in landfills; (23) CH_4-C released to the atmosphere from wastewater treatment; (24) CH_4-C released to the atmosphere from compost production; (25) CH_4-C oxidation to CO_2-C (emissions due to waste management); (26) CH_4-C released to atmosphere due to stabled animals; (27) CH_4-C released to atmosphere due to manure management; (28) CH_4-C oxidation to CO_2-C (emissions due to livestock); (A) carbon accumulation in forest ecosystem; (B) carbon accumulation in orchards–vineyards/crops; (C) carbon accumulation in grassland ecosystem.

The model of the Siena Province gave a *negative* carbon dioxide emission, that is, uptake of about 0.25 million tons carbon/year. Particularly, the forest had a huge uptake of carbon, about 0.55 million tons. Without this massive uptake by photosynthesis there would be a carbon dioxide emission of about 0.3 million tons carbon/year. The urban areas have, therefore, a very significant carbon dioxide emission.

The carbon cycling model is an excellent management tool, which Marchi et al. (2012) used to test how much carbon dioxide emission could be reduced by the introduction of various changes. The carbon footprints, expressed in tons of carbon dioxide equivalents (CO_{2eq}), were applied for comparisons. It means that the methane was multiplied by 25 and added to the carbon dioxide because methane has a greenhouse effect that is 25 times the effect of carbon dioxide (IPCC 2007). Methane has actually a very significant effect on the carbon footprint. The oxidization of methane mentioned previously (half life of 7 years) was considered in the calculations as well. The following scenarios were tested, and the emission reduction in percentage over a period of 10 years is indicated (notice that some of the changes are introduced stepwise during the 10 years):

1. Reducing the number of private cars and improving public transportation: 0.64%
2. Promoting reforestation (totally 2.57% stepwise increase during 10 years): 12.36%
3. Increasing the efficiency of energy use (heating and electricity): 2.79%
4. Increasing the productivity of agricultural land: 0.13%
5. Increased use of incineration for waste treatment: 2.62%
6. Improving solid waste and wastewater treatment: 1.72%

Total reduction by the introduction of all six changes: 20.26%

The carbon footprint could be reduced by about 20%, a very significant reduction, in accordance to the results of the carbon cycling model.

For the development of a carbon cycling model of a city, there is a possibility to follow the procedure proposed by Marchi et al. (2012), even if the submodels *Cropland* and *Livestock* are not included. The submodel *Forest* should be replaced by a similar model of public parks and recreational areas (submodel *Parks*). The submodel *Grassland* might consider private gardens, lawns, and ornamental trees in the city and therefore it should be replaced by a submodel *Gardens*. The submodels and the landscape elements considered in a carbon cycle model of a city are shown in Figure 15.7.

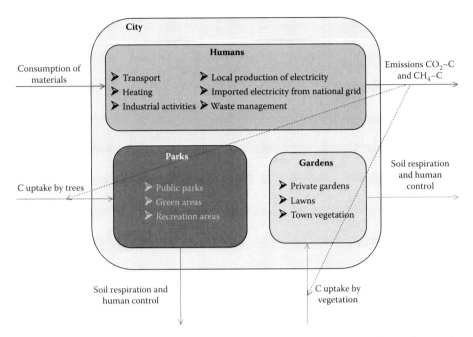

FIGURE 15.7   (See color insert.) Submodels and landscape elements of a carbon cycle model of a city. The arrows indicate the general forcing functions involved in the urban processes.

The three general submodels (*Parks*, *Gardens*, and *Humans*) have the following state variables:

1. *Parks* (public parks, green and recreational areas): carbon in trees and bushes, carbon in dead wood, carbon readily released in soil, carbon in soil microbial biomass, and carbon accumulated in the soil.
2. *Gardens* (private gardens, lawns, and other town vegetations): carbon in grass and all small vegetations, carbon in dead plants, carbon readily released in soil, carbon in soil microbial biomass, carbon in grazing animals, and carbon accumulated in the soil.
3. *Humans* (several different sources of carbon dioxide and methane): the state variables calculated the emission from the different sources that are input to the state variables. Methane is oxidized to carbon dioxide and water with a rate corresponding to a half life of 7 years.

The forcing functions are all the emission sources, the temperature, the precipitation, soil respiration, and the changes of the green areas, the parks, the gardens, and the lawns due to human control and external forces.

A carbon cycle model of a city, as well as that of an administrative district, can be used to provide guideline for the sustainable management of these areas, proposing political considerations and solutions for the greenhouse effect, the biggest environmental problem we face today.

15.4   APPLICATION OF CARBON CYCLING MODELS FOR ASSESSMENT OF SUSTAINABILITY OF ECO-CITIES

The most massive carbon dioxide emission is coming from the use of fossil fuel. A major revision of the carbon footprint is therefore always obtained by changing the use of fossil fuel to the alternative energy, hydropower, wind mills, solar cells, solar panels, biogas, and biomass in different forms. The energy sector is, however, not the only source of carbon emission. Soil can either uptake or emit carbon dioxide dependent on the management of nature, green areas, and agriculture. Furthermore, photosynthesis is an uptake of carbon dioxide by formation of biomass with solar energy as source. If the biomass is not removed or used to replace the fossil fuel as energy source, the uptake of carbon dioxide by photosynthesis is giving a positive contribution to the carbon balance—it means that the carbon foot prints are reduced considerably.

Consequently, it is strongly recommended to develop a carbon cycling model as an important component in the assessment of the sustainability of a well-defined area as, for instance, a city. The assessment of sustainability requires other tools in addition to the carbon cycling model, but the development of this type of model is compulsory because without a carbon cycling model it is impossible to predict the contribution to the greenhouse effect by the city. The expected climate changes are among the most significant factors influencing the sustainable development. Application of an eco-exergy balance has already been mentioned in Sections 15.1 and 15.2, and it may also be considered a compulsory tool for the assessment of the sustainability of a well-defined area. By the use of eco-exergy, it is possible to get a holistic picture of the development, as it will consider the use of various energy sources; the management of nature, solid waste, and water; and the emissions of generally all pollutants. Eco-exergy gives no information about the details of the unbalances or how much a new energy policy could change the unbalance. A carbon cycling model is, however, able to show the unbalances of the carbon cycling caused by the energy policy. Similarly, it is also recommended to include in the sustainability analysis a complete overview of the flows of solid waste and water in the area. The overview should include all sources, all applied treatments, all transportation including reuse and recycling, and the final deposition of the solid waste and the water. The overviews of solid waste and water could easily be developed by a model similar to the carbon cycling model. The three models will probably have some processes in common; for instance, solid waste deposition by landfills may yield a methane emission. Consequently, it could be considered to develop the carbon cycling model and the eco-exergy (work energy) balance on the basis of the information about the solid water treatment and transport and the water cycling. Moreover, it is important to consider solid waste and water, because use of resources is associated with the solid waste handling and water is one of the most important resources for all cities.

The carbon cycling model should under all circumstances as a minimum contains the three submodels presented in Section 15.3 (*Parks*, *Gardens*, and *Humans*), but could be expanded dependent on the pathways for solid waste and water in the area. The most important state variables to include are the main sources of carbon compounds on the one side, but on the other side also the state variables could be changed due to a different management strategy, as it was also shown in Section 15.3 for the Siena Province.

15.5 CONCLUSIONS AND RECOMMENDATIONS

Jørgensen (2006) has developed a global eco-exergy balance, and the results indicate that the loss of sustainability globally is due to the following:

1. Loss of nature (deforestation and natural areas used by mankind for agriculture or construction of roads, building, and towns)
2. Use of fossil fuel
3. Overexploitation of natural renewable resources mainly by fishery and forestry
4. Use of nonrenewable resources, mainly various metals

These four possible losses of sustainability should be considered when a sustainability analysis of a city is performed.

Table 15.4 and Figure 15.5 can be used as basis for setting up the eco-exergy balance, and the three submodels presented in Section 15.3 supplemented by maybe components taken from cycling models for solid waste and water should be the basis for the carbon cycling model. To ensure that the results of the analysis yield a powerful management tool, it is, however, important to consider the scenarios that we would like to develop to obtain a good basis for the right decisions. The scenarios that are desirable to develop are rooted in the recommendations, which we can extract from the sustainability projects that are already carried out. These recommendations can be considered a conclusion of this chapter and can be summarized as 10 questions that the scenarios can quantify. The questions should be included together with the sustainability analyses in all—short-term or long-term—planning of the city:

1. Examination of the allocation of different energy sources, particularly the ratio of fossil fuel to renewable, alternative energy sources is important. Sustainability can be improved considerably by shifting from fossil fuel to hydropower, wind energy, biomass, and solar energy.
2. Which possibilities does the city have for expansion of the green areas and for planting trees and other vegetation in "empty" or "not utilized" areas?
3. Would it be possible to get a more positive contribution to the eco-exergy balance and to the carbon cycling model, if the management of these areas would be improved?
4. To which extent is it possible to improve public traffic and reduce the use of private cars?

5. With which efficiency is the energy use in the various sectors? Could the efficiency be increased? Usually energy efficiency can be increased by a relatively modest investment, which is paid back very rapidly.

6. Which alternatives are available for the present treatment of solid waste and waste water? How would a shift to various alternatives change the eco-exergy balance and the carbon cycling model? Notice that there are many possible alternative methods for the treatment of solid waste and waste water.

7. Would improvements of the infrastructure eventually influence the eco-exergy balance and the carbon cycling model? How?

8. Could reuse and recycling be enhanced? Where and how?

9. How are the renewable resources of the city used? Is it sustainable, meaning that the renewal rate is greater than or equal to the consumption rate? If the answer to this question is no, what can we do to ensure a sustainable use of the renewable resources?

10. Have the long-term plans of the city been tested by the proposed sustainability analyses?

REFERENCES

Hayes, E. T., 1997. *Implication of Material Processing*. EPA, Athens, GA.

IPCC, 2007. *Climate Change: The Scientific Basis*. Report prepared for Intergovernmental Panel on Climate Change by Working Group I. In: Solomon, S., Qin, D., Manning, M., Chen, Z., Marquis, M., Averyt, K. B., Tignor, M., Miller, H. L., (Eds.), Cambridge University Press, Cambridge, UK/New York, NY, p. 996.

Jørgensen, S. E., 2006. *Eco-Exergy as Sustainability*. WIT Press, Southampton, UK, p. 207.

Jørgensen, S. E., 2010. Ecosystem services, sustainability and thermodynamic indicators. *Ecological Complexity*, 7, 311–313.

Jørgensen, S. E., Ladegaard, N., Debeljak, M., Marques, J. C., 2005. Calculations of exergy for organisms. *Ecological Modelling*, 185, 165–175.

Marchi, M., Jørgensen, S. E., Pulselli, F. M., Marchettini, N., Bastiononi, S., 2012. Modelling the carbon cycle of Siena Province (Tuscany, central Italy). *Ecological Modelling*, 225, 40–60.

Stremke, S., van den Dobbelsteen, A., 2012. *Designing Sustainable Energy Landscapes: Theories, Education and Best Practice*. CRC Press, Boca Raton, FL.

Index